RESTRICTED DATA

RESTRICTED DATA

THE HISTORY OF NUCLEAR SECRECY IN THE UNITED STATES

ALEX WELLERSTEIN

THE UNIVERSITY OF CHICAGO PRESS
Chicago and London

The University of Chicago Press, Chicago 60637
The University of Chicago Press, Ltd., London

Published 2021
Paperback edition 2024
Printed in the United States of America

33 32 31 30 29 28 27 26 25 24   1 2 3 4 5

ISBN-13: 978-0-226-02038-9 (cloth)
ISBN-13: 978-0-226-83344-6 (paper)
ISBN-13: 978-0-226-02041-9 (e-book)
DOI: https://doi.org/10.7208/chicago/9780226020419.001.0001

Library of Congress Cataloging-in-Publication Data

Names: Wellerstein, Alex, author.
Title: Restricted data : the history of nuclear secrecy in the United States / Alex Wellerstein.
Description: Chicago : The University of Chicago Press, 2021. | Includes bibliographical references and index.
Identifiers: LCCN 2020033052 | ISBN 9780226020389 (cloth) | ISBN 9780226020419 (ebook)
Subjects: LCSH: Nuclear weapons information, American—Access control. | Defense information, Classified—United States.
Classification: LCC U264.3 .W45 2021 | DDC 623.4/51190973—dc23
LC record available at https://lccn.loc.gov/2020033052

♾ This paper meets the requirements of ANSI/NISO Z39.48-1992 (Permanence of Paper).

CONTENTS

INTRODUCTION: THE TERRIBLE INHIBITION OF THE ATOM 1

PART I. THE BIRTH OF NUCLEAR SECRECY 13

CHAPTER 1: THE ROAD TO SECRECY: CHAIN REACTIONS, 1939-1942 15

1.1 The fears of fission 15
1.2 From self-censorship to government control 26
1.3 Absolute secrecy 38

CHAPTER 2: THE "BEST-KEPT SECRET OF THE WAR": THE MANHATTAN PROJECT, 1942-1945 51

2.1 The heart of security 52
2.2 Leaks, rumors, and spies 64
2.3 Avoiding accountability 77
2.4 The problem of secrecy 82

CHAPTER 3: PREPARING FOR "PUBLICITY DAY": A WARTIME SECRET REVEALED, 1944-1945 97

3.1 The first history of the atomic bomb 98
3.2 Press releases, public relations, and purple prose 105
3.3 Secrecy from publicity 118

PART II. THE COLD WAR NUCLEAR SECRECY REGIME 133

CHAPTER 4: THE STRUGGLE FOR POSTWAR CONTROL, 1944–1947 135

- 4.1 Wartime plans for postwar control 136
- 4.2 “Restricted Data” and the Atomic Energy Act 145
- 4.3 Oppenheimer’s anti-secrecy gambits 158

CHAPTER 5: “INFORMATION CONTROL” AND THE ATOMIC ENERGY COMMISSION, 1947–1950 179

- 5.1 The education of David Lilienthal 180
- 5.2 The “thrashing” of reform 196
- 5.3 Three shocks 209

CHAPTER 6: PEACEFUL ATOMS, DANGEROUS SCIENTISTS: THE PARADOXES OF COLD WAR SECRECY, 1950–1969 233

- 6.1 The H-bomb’s silence and roar 234
- 6.2 Dangerous minds 249
- 6.3 Making atoms peaceful and profitable 270

PART III. CHALLENGES TO NUCLEAR SECRECY 285

CHAPTER 7: UNRESTRICTED DATA: NEW CHALLENGES TO THE COLD WAR SECRECY REGIME, 1964-1978 287

7.1 The centrifuge conundrum 288
7.2 The perils of "peaceful" fusion 300
7.3 Atoms for terror 319

CHAPTER 8: SECRET SEEKING: ANTI-SECRECY AT THE END OF THE COLD WAR, 1978-1991 335

8.1 Drawing the H-bomb 338
8.2 The "dream case": The *Progressive* v. The United States 351
8.3 Open-source intelligence in a suspicious age 368

CHAPTER 9: NUCLEAR SECRECY AND OPENNESS AFTER THE COLD WAR 385

CONCLUSION: THE PAST AND FUTURE OF NUCLEAR SECRECY 397

Acknowledgments 417

Notes 423

Bibliography 507

Archival sources and abbreviations 507
Articles 510
Books and monographs 518

Index 529

INTRODUCTION

THE TERRIBLE INHIBITION OF THE ATOM

I am afraid the scientists have led us into a terrible world.
GENERAL LESLIE R. GROVES, 1948[1]

On the morning of August 6, 1945, the White House issued a press release that would change the world. In an instant, the existence of a vast scientific project was revealed, as well as the fruits of its labor: a "new and revolutionary" weapon, which had destroyed Hiroshima, Japan. "It is an atomic bomb," the statement explained. "It is a harnessing of the basic power of the universe." And prior to that moment of revelation, even the fact that the United States was interested in creating such a weapon, much less had actually created, tested, and now deployed it, had been "Top Secret," the improper release of which could be, in principle, punished by death.[2]

Nuclear weapons have always been surrounded by secrecy, and the American atomic bomb was born secret. From the moment that scientists first conceived of its possibility, through the massive undertaking that was its actual creation, there were efforts to control the spread of nuclear information, including the newly discovered scientific facts that made them possible. This desire for control was born out of fear. For the first scientists working on the American atomic bomb, it was a fear of a dread enemy—Nazi Germany—using said information to build their own weapons. Later, the fears shifted, as officials worried that a premature announcement of the new weapon would lessen its psychological value against the Japanese, and potentially threaten the success of the project itself. Though this secrecy emerged from fears that were originally very specific to the context of World War II, it was easily adapted to the new fears that followed, as new enemies emerged: the Soviet Union, the People's Republic of China, North Korea, even non-state nuclear terrorists. And far more diffuse and varied fears would also promote this desire for control, with consequences ranging from the mundane (diplomatic difficulties) to the apocalyptic (global thermonuclear war).

But from the beginning, the desire for nuclear secrecy contained contradictions and complications. The scientists who had made the bomb, and had become enmeshed in its secrecy, were frequently wary. Some had supported the secrecy entirely, because they too shared the fears that motivated it. But many felt the secrecy, even if it had been necessary, was stifling. And as the war's end grew close, new questions, and new worries, entered into their minds.

The atomic bomb was a product of science and industry, yet the fundamental principles it was based on were well known to scientists prior to the outbreak of war. How could a fact of nature be rendered effectively into a state secret, if any scientist, in any laboratory, in any country, could replicate and rediscover it? Military plans, conceived in the mind of a soldier, can be kept secret indefinitely, but can facts of physics and chemistry?

Many scientists and policymakers further asked whether science should be kept secret at all, and whether attempting to do so could be counterproductive for security. The atomic bomb was not merely the application of science to war, but the result of decades of investment in scientific education, infrastructure, and global collaboration. Secrecy, according to many of the scientists who worked under it, stifled scientific advance. If secrecy were made the norm, would science thrive, or even survive? Which would serve the nation's security more, keeping things secret, or racing forward as fast and as openly as possible?

And the same science that allowed for the creation of nuclear weapons also appeared to offer up the possibility of cheap, abundant, and clean energy generation, among other civilian benefits. Would the fears of military uses of the atom override the hopes of its peaceful applications?

Secrecy had been a defining aspect of the work to create the atomic bomb, but would it be its future? The aforementioned White House press release about the Hiroshima attack, toward the end, addressed these questions, but left them deliberately unanswered. "It has never been the habit of the scientists of this country or the policy of this Government to withhold from the world scientific knowledge," it explained, and noted that under normal circumstances, everything about the work would be released. But the "present circumstances" of the world—one war ending, an uneasy international situation unfolding—meant that the means of producing the atomic bomb had to be kept secret, at least for now. There would be, the statement explained, "further examination" of the question, in order to protect the nation, and indeed the rest of the world, from "the danger of sudden destruction."

The totalizing, scientific secrecy that the atomic bomb appeared to demand was new, unusual, and very nearly unprecedented. It was foreign to both American science and American democracy, and its compatibility with either has always been an area of dispute. But the circumstances of the bomb's creation, and the bomb itself, seemed to mandate the period of secrecy be extended, to avoid an existential risk. And that nuclear secrecy has continued, in evolving but ever-present forms, to our present day. We now find ourselves over seven decades after the end of World War II, and some three decades since the collapse of the Soviet Union, and nuclear weapons, nuclear secrecy, and nuclear fears show every appearance of being a permanent part of our present world, to the degree that for most it is nearly impossible to imagine it otherwise.

This book is a history of nuclear secrecy in the United States, from the first moments that the atomic bomb was seen as a realistic possibility in the late 1930s, through our present moment in the early twenty-first century. It is the story of how a large and varied group of people—scientists, administrators, military officers, politicians, lawyers, judges, journalists, activists, and the broader public—grappled with the question of whether nuclear knowledge should be regarded as something that needed to be controlled, and how many of the fruits of their discussions, policies, and interventions shaped the American national security state that endures to this day. The singular motif that reappears throughout this work is that of tension. The bomb may have been born in secrecy, but that secrecy was always controversial and always contested.[3]

The concerns about the compatibility of science and secrecy were always joined by concerns about the compatibility of secrecy and democracy. The United States has, since its eighteenth-century origins, enshrined Enlightenment ideals of openness and freedom of speech in its core institutions. These ideals have never been treated as absolutes, but they have come with real legal, political, and rhetorical power.[4] In practice, this has meant that while secrecy has flourished in the post–World War II American context, it has never been unlimited in its scope, even with a threat as seemingly expansive and existential as the global development of nuclear weapons.

It has also meant that secrecy reform and nuclear policy have always been in tension with democratic desires. The physicist J. Robert Oppenheimer, who had done much to create both the weapons and their secrecy, referred to the difficulty of public deliberation as the "terrible inhibition of the atom," and it was both a badge and burden to be borne by those with access to the secrets.[5] The secrecy, many like Oppenheimer believed, ultimately contorted American policymaking, and left the American public dangerously ignorant of the evolving national and world situation.

These tensions, between the ideals of science and secrecy on the one hand, and of desires for openness and security on the other, are what make the history of nuclear secrecy in the United States unpredictable, suprising, and, at times, bizarre. In one telling example (discussed at length in chapter 3): while the United States may have been the first country to make an atomic bomb, it was also the first country to release a technical history of the atomic bomb, only days after its first use, and it did so in the interest of both improving democratic discourse and preserving further secrecy. That such a document could be created at all, rendering into plain and unified discussion the work of the Manhattan Project that had been previously enshrouded with code words and a "need to know" compartmentalization, is strange enough by itself, and no other country has done anything similar since. But that the top scientific, military, and political representatives on the project would all agree to its utility, and lobby to the President personally for its release only days after the Nagasaki attack, is a remarkable example of the ways in which secrecy and revelation are not only paired, but can serve many different ideologies and institutional goals.

This book takes as its subject the people and institutions that had as their goal the realization of nuclear secrecy in the world, the means by which they attempted to make the ideal of secrecy "real" (so that some people knew the secret, and others did not), the contexts in which they operated and its influence on their thought and action, and the people who challenged, critiqued, and attempted to reform, undo, or subvert these efforts. It is a history of both the creation of nuclear secrecy as well as the resistances to it, because they have always gone together. And it is also a history of information release as well as containment, for these two actions were, as we will see, frequently two sides of the same coin.

It is not a story of the triumph of nuclear secrecy, nor of the triumph of openness. Rather, it is a messy story, with few clear winners and losers, or heroes or villains. The same people trying to create the new secrecy were also concerned with its ill effects, and with the demands of democracy. Within the institutions that were meant to enforce secrecy, deep debates about the nature, purpose, and means of secrecy were frequently taking place, and reform of the secrecy system has been a goal nearly since it was created. And outside of the national security apparatus were the vast, uncontainable multitudes of the American public, whose willingness to trust that government attempts to control information were done in good faith declined over the course of the twentieth century.

The specificity of the American context matters. The contradictions of secrecy that American scientists and policymakers wrestled with were far less of a concern for totalitarian nations where, unsurprisingly, state security took full precedence. Even the other democratic nations with nuclear weapons seem less deeply conflicted than the Americans: their governmental and social structures seem to accommodate nuclear secrecy far more easily. It is not just that the US has had a hard time finding "balance" between its various ideals, it is that it has not been able to imagine what that balance would look like. Nuclear secrecy may have become deeply embedded in the United States, but it has always been an uncomfortable and often regretted arrangement.

"Nuclear secrecy" does not refer to a single goal, practice, or institution. The English word "secret" is derived from the Latin *sēcrētus*, meaning to cut, to sunder, to separate. Knowledge, however, is ephemeral and immaterial unless it is instantiated in some physical way into the world, whether as something that exists in people's heads, written onto paper, or embedded into some material technology, to pick just a few possibilities. Secrecy is the desire for a cutting out of knowledge from the world, and making that desire into a reality involves very real acts of cutting up society: allowing some people to pass through certain doors rather than others, for example. Sometimes this cutting action is quite literal, such as when a redactor slices out secret lines from a text with a razor, as happened in the past (today they use software for this).

So nuclear secrecy began as a fearful desire, but turning that desire into a reality required the work of thousands of people. Over time, the motivations and justifications for secrecy have changed, as have the various practices and means for enacting that secrecy in the world, as have the institutions and agencies tasked with articulating the motivations and cultivating the practices. While not monolithic, we can regard nuclear secrecy as a "regime," as a bundle of thoughts, activities, and organizations that try to make secrecy "real" in the world, to perform the multitude of acts of epistemological slicing that result in some people knowing things, and other people not.[6]

The American nuclear secrecy "regime" has evolved several times from its emergence in the late 1930s through our present moment in the early twenty-first century. Each chapter of this book explores a key shift in how nuclear secrecy was conceived of, made real in the world, and challenged. Roughly speaking, one can divide the history of American nuclear secrecy into three major parts: the birth of nuclear secrecy, the solidification of the Cold War nuclear secrecy regime, and the challenges to the regime that began in the late Cold War and continue into the present.

Part I (chapters 1–3) narrates the origins of nuclear secrecy in the context of World War II. This was a secrecy initially created as an informal "self-censorship" campaign run by a small band of refugee nuclear physicists who feared that any publicized research into the new phenomena of nuclear fission would spark a weapons program in Nazi Germany. As the possibility of nuclear weapons becoming a reality grew, and official government interest increased, this informal approach was transmuted into something more rigid, but still largely run by scientists: a secrecy of "scientist-administrators" created by Vannevar Bush and James Conant, two powerful wartime scientists, that gradually put in place a wide variety of secrecy practices surrounding the weapons. When the work was put into the hands of the US Army Corps of Engineers, and became the Manhattan Project, these efforts expanded exponentially as the project grew into a virtual empire. And for all of the difficulty of attempting to control a workforce in the hundreds of thousands, the thorniest questions would come when these scientific, military, and civilian administrators tried to contemplate how they would balance the needs for "publicity" with the desires of secrecy as they planned to use their newfound weapon in war.

Part II (chapters 4–6) looks at this wartime secrecy regime as it was transformed from what was largely considered a temporary and expedient program into something more permanent and lasting. Out of late-wartime and postwar debates about the "problem of secrecy," a new system emerged, centered on the newly created Atomic Energy Commission and "Restricted Data," a novel and unusually expansive legal category that applied only to nuclear secrets. This initial approach was characterized by a continued sense that it needed reform and liberalization, but these efforts were dashed by three terrific shocks at the end of the decade: the first Soviet atomic bomb test, the hydrogen bomb debate, and the revelation of Soviet atomic espionage. In the wake of these events, which reinforced the idea of a totemic "secret" of the bomb while at the same time emphasizing a nuclear American vulnerability, a new, bipolar approach to secrecy emerged. This "Cold War regime" simultaneously held that to release an atomic secret inappropriately was to suffer consequences as extreme as death, but that once atomic information had been deemed safe (and perhaps, profitable), it ought to be distributed as widely as possible.

Part III (chapters 7–9) chronicles the troubles that this new Cold War mindset about secrecy encountered from the 1960s through the present. Many of these were problems of its own making: embodying both the extremes of constraint and release, the Cold War approach to nuclear secrecy fundamentally rested on the dubious assertion that the technology it governed could be divided into simple categories of safety and danger, despite its inherently dual-use nature. These inherent conflicts were amplified by the rise of a powerful anti-secrecy politics in the 1970s, which motivated a wide spectrum of people—ranging from nuclear weapons designers to college students and anti-war activists—to attempt to dismantle the system in whole or in part. The end of the Cold War brought only brief respite, as initial efforts to reform the system faltered in the face of partisan politics and new fears from abroad.

Clearly, this is a work of history. I am a professional historian. This means that I traffic primarily in archival sources, citations to which you will find in the endnotes of this book. I have been sometimes asked: How can you write the history of something that is still at least partially secret, much less the secret history of secrecy itself? Wouldn't you need a security clearance to do this correctly? And wouldn't the true history of nuclear secrecy be something that could not be published without endangering national and global security? And even if you can write something about this history, wouldn't the fact that so much is missing make it a paltry offering, and likely to contain falsehoods and omissions?

It is worth noting that while much about the history of US nuclear weapons is still secret, there is an impressive amount that is not. The same forces that created the aforementioned tensions around nuclear secrecy have resulted in a system that, over sometimes very long periods of time, results in a lot of information being eventually released. This is often the case for documents that are not about the weapons, per se, but are about the governance of the weapons. So, as the voluminous endnotes will attest, there is actually a great deal of information available relating to how secrecy was imagined, implemented, and debated internally. And there is more information on even weapons topics declassified than most people are aware of, and this book, in part, describes how that came to be (sometimes officially, sometimes not). Ironically, as the official Atomic Energy Commission historian Richard G. Hewlett pointed out decades ago, the secrecy system actually makes some aspects of this history easier, because it mandates (with severe legal consequences) the preservation of documents that might otherwise have been thrown away, lost, or taken as a souvenir.[7] That doesn't mean that the government will let you look at them, of course.

But there are gaps in our knowledge—and always will be. Archival sources never tell the full story, because not everything is written down, not everything written down is complete, and not everything written down is truthful. People who work on relatively recent history can sometimes supplement their work with discussions with historical participants, though these come with their own problems, such as bad memories, historical grudges, and the living being privileged over the dead. There will always be gaps. This is the case even with history that was not of formerly classified subjects. In the case of once-secret documents, those gaps are sometimes quite literal: a gap will suddenly appear in a sentence, sometimes identified with a "DELETED" stamp, sometimes not. This is the work of the censor, who has "sanitized" the document, removing whatever information they thought would still compromise security, as defined in a guide they have in front of them (the history of these guides is part of the history of secrecy, and emerged in the wake of the Manhattan Project). At times, I have used the Freedom of Information Act (FOIA) to get access to documents that had not yet made their way into archives, but this law compels the censor only to review, not to release: it cannot let one access information the government still determines should be kept secret, and Congress has given the government a lot of latitude in making that determination.[8]

Despite the limitations inherent in trying to write history with an often heavily redacted archival record, I have never sought nor desired an official security clearance.[9] This no doubt leaves many additional gaps in the story, but it also allows me to share what I have found with impunity. This is a trade-off any scholar who works on formerly or currently classified subject matter knows well; even having a clearance does not guarantee that one will see everything one desires, and introduces potentially mammoth difficulties in the publication process, giving government agencies the ability to modify or even veto the text.[10] None of that seemed worth it to me. I have interacted with historians who have had clearances, and for every one who was smug that it gave them a special advantage over those without one (an attitude that I am dubious of, as a clearance can lead to an overestimation of the value of "secret" knowledge), there were others who admitted that it gave them more grief than deep understanding.[11] For me, it ultimately comes down to my aims as a historian: if I can't tell anyone what I know, what's the point in knowing it? I'd rather risk errors (which is easy enough even with a clearance) than be muzzled.

All historians deal with gaps in the historical record, whether caused by water damage, the fires of a war, ill-advised document destruction, or the fact that most of human experience, even in our hyper-documented modern age, is not preserved in a record. What makes secrecy feel different is its intentionality: the information I may want is actually knowable and may even be known, but just not by me, at least right now. Which is frustrating. But there is also a logic to secrecy: the information that is kept secret generally falls into categories of justification (like national security), and so the question becomes, is the information that I care about also information that the censor thinks should be censored? In some cases yes, but in many cases no. The history of secrecy itself is not always still secret; there are places where it intersects with present-day security concerns (for example, discussions about the secrecy of the design of the hydrogen bomb can involve details about said design of the hydrogen bomb), but there are also many places where it does not.

I do not want the reader to take a dim view of "the censor," and I use the term here very much tongue-in-cheek. The censors are people too, often doing their job with great pride and sensitivity; though, as people, they do err. While this book is definitely not a justification or even rationalization of the secrecy regimes that exist, I do mean for it to be a partial resurrection of the censors and their points of view, because their perspectives are frequently obliterated by the same practices of secrecy that they participated in. As a result, their motives and goals are often only inferred by those on the outside, frequently their critics, and so it is the critics' view of the censor that dominates much writing and understanding of secrecy. We can't be neutral toward secrecy, any more than we can be neutral toward the idea of state power in general. But to understand how it works, we must understand it from the perspective of the systems that produce it, as well as those of its inherently more vocal critics. In this book I have attempted to flesh out, and historicize, both perspectives, which I suspect may be frustrating to readers who identify primarily with one or the other.

For all the frustrations involved in working with formerly classified sources, the historian has a major advantage over the people who were living through this history as it unfolded, even those with clearances: time. The secrecy regime in the US was largely set up to erode over time, and even in areas where the erosion was not intended, it occurred. This applies across agencies (though not always equally) and subject matter (though nuclear weapon secrets do not erode automatically, unlike some areas of government activity). The consequence is that I have sometimes had access to a much wider variety of formerly classified information than anyone living in, say, the 1950s, might have had, even those with the top security clearances (because their access was typically compartmentalized).

Thanks to declassification actions and the judicious use of the Freedom of Information Act, we can reconstruct the (partial) archives of multiple agencies and governmental bodies at once, where a historical actor would have likely been limited to one or perhaps two of these files. If something is declassified, I no longer require any "need to know" to know it. I also have access to private journals, correspondence, and sometimes the recorded recollections of my historical sources. So the situation, as tough as the presence of redactions might make it seem, is not really so bad: the historian has a unique vantage point to understand the past, one that the those feeling their way at the time would be envious of, even considering that we are, of course, missing a few things. I have attempted to indicate places where I suspect there is considerable information still missing, and where I have had to make larger interpretive leaps. No history is perfect, and this one is no exception, but I've done my best to tell a coherent story that goes from the earliest days of the Manhattan Project all the way into our present world. No doubt historians of the future (and likely even myself) will learn more as time goes on, but such is how historical knowledge is made, no matter the topic: like any field of advanced study, it stops advancing only when it is no longer of interest to anyone.

Three years after the bombings of Hiroshima and Nagasaki, Leslie Groves, the Army General who had presided over the Manhattan Project, lamented to a secret congressional committee about the impossibility of controlling the dangerous weapons that were steadily emerging: "I am afraid the scientists have led us into a terrible world. I can't figure out how we can keep the knowledge from spreading, except to have a complete iron curtain."[12] Yet even an iron curtain cannot totally keep secrets from spreading, and the US never had an iron curtain. The history of nuclear secrecy in the United States is one about the troublesome quandary raised by fears of dangerous knowledge in a nation where information is anything but easy to control. And it is a history that has not yet concluded.

PART I

THE BIRTH OF NUCLEAR SECRECY

1

THE ROAD TO SECRECY

CHAIN REACTIONS, 1939–1942

The SECRET stamp is the most powerful weapon ever invented.
LEO SZILARD[1]

The origins of nuclear secrecy lay in *fear*: the idea that a dreaded enemy could have a new, enormous source of power at their disposal and that all other nations would be potential victims. The enemy was the Nazis, and the power was, of course, the atomic bomb. This fear guided many decisions during World War II, but one of the very first things it motivated, at a moment when the reality of an atomic bomb was still uncertain enough that many people thought the fear unreasonable, was an attempt at scientific secrecy. In retrospect, what is remarkable about this attempt was that it was initially propagated by scientists who considered secrecy anathema to their interests.

This original secrecy was practiced as *self-censorship*, in which scientists abstained (or didn't, as it turned out) from publishing on topics that they judged "sensitive." But this morphed, surprisingly quickly, into a system of government control over scientific publication, and from there into government control over nearly *all* information relating to atomic research. When the nuclear physicists initiated their call for secrecy, they thought it would be temporary, and controlled by them. They were wrong.

1.1 THE FEARS OF FISSION

Nuclear weapons and reactors are both based on the scientific phenomenon known as nuclear fission: the splitting of heavy atoms (notably uranium) with neutrons. Fission was discovered in December 1938 by the German scientists Otto Hahn and Fritz Strassmann, working in Berlin, and their Austrian collaborators Lise Meitner and Otto Frisch, then living in Sweden. The investigations of Hahn, Meitner, et al., were the latest in a long chain of new discoveries about the nature of matter touched off by Wilhelm Röntgen's discovery of X-rays in 1895, Henri Bequerel's discovery of radioactivity in 1898, the work of Ernest Rutherford on alpha radiation and the structure of the atom, the work of Marie and Pierre Curie on the nature of radioactivity, the revolutions of quantum mechanics led by Niels Bohr, Werner Heisenberg, and others, and, most contemporaneously, the work by Frédéric and Irène Joliot-Curie on artificial radioactivity and the work of Enrico Fermi and his team in Italy on new techniques in using low-energy ("slow") neutrons to create new radioactive compounds.[2]

Hahn, Meitner, and their collaborators were following up on the work of Fermi, who a few years earlier had claimed to have created new chemical elements by exposing uranium to slow neutrons.[3] Hahn, a chemist, had found that the residues of irradiated uranium were not the new, heavy elements that Fermi thought they were; rather, the residues contained a radioactive form of barium, an element roughly half the size of the original uranium. He wrote to Meitner, his physicist collaborator in exile, with his results. She and her nephew Frisch made the physical interpretation of the experiment: the uranium nucleus had not grown from the neutron, as Fermi had thought, but had split into two pieces. They called the phenomena "fission."[4]

This was physically interesting, and scientifically surprising, but not necessarily *scary*. The jump from "fission is possible" to "a nuclear weapon is possible" is a very large one. The amount of energy released from a *single* fission reaction is, from the point of view of an atom, very large. From a human point of view, it is very small: roughly enough energy to move a speck of dust. To turn this into a weapon would require splitting around a trillion trillion such atoms within a millionth of a second. Whether that was possible was uncertain, and even if it *were* possible, it is not clear that it could be accomplished in time for war.[5]

There was one scientist who immediately saw threatening possibilities in the mere discovery of nuclear fission. The Hahn-Meitner results spread rapidly through the global physics community by word of mouth, and finally made it to the ears of Leo Szilard while he was visiting a colleague at Princeton University in January 1939. Szilard, a Hungarian physicist of Jewish background, had been living in Germany when the Nazis came to power. He fled to England shortly after the Reichstag fire, and this experience shaped his worldview. On the day in April 1933 when he decided to flee from Berlin to Vienna, the train he took was essentially empty. One day later, the same train was overcrowded and stopped at the border, and everyone on it was interrogated. Szilard later related the impact this had on his thinking: "This just goes to show that if you want to succeed in this world you don't have to be much cleverer than other people, you just have to be one day earlier than most people. This is all that it takes."[6]

This also summed up Szilard's scientific style: working fast, on the bleeding edge of ideas.[7] The reason he was faster than most in seeing the military implications of fission is that he had been searching for a similar nuclear reaction for half a decade, and had spent more time mulling over the consequences than anyone else. In September 1933, while living in London, Szilard had read in the newspapers of a speech where British physicist Ernest Rutherford had dismissed the idea that atomic energy could be liberated on an industrial scale as "moonshine." Rutherford had merely been repeating what, at that point, was orthodox physics: radioactive transformations could release a lot of energy, but if you couldn't control them, and multiply them on a large scale, then they weren't going to do much. People who spoke of releasing the atom's latent energies in a macroscopic way, Rutherford indicated, were likely talking nonsense. And prior to the discovery of fission five years later, he was right.[8]

But Szilard was a contrarian by inclination, and believed Rutherford was being too conservative. The neutron, a subatomic particle discovered in 1932, held many new possibilities. Because they are electrically neutral, neutrons are much more capable of penetrating the cloud of negatively-charged electrons surrounding the positively-charged atomic nucleus, plunging into the atom's core.[9] Szilard's insight was that if you had a nuclear reaction that was started by a neutron, and then itself produced neutrons that could induce further reactions, you would have the potential for a rapidly growing chain reaction. If one neutron reacted to make two more, and each of those two neutrons reacted to make two more, and so on, you would have an exponential explosion of particles, and energy. It takes only thirty such "doublings" to reach over a billion total neutrons; at eighty doublings, you have a trillion trillion. Find the right reaction and you would have a virtual neutron furnace at your disposal. If you can make the reaction run fast enough, you have a weapon. Szilard became, by his telling, "obsessed" with the idea and its implications, inspired by the far-seeing science fiction of H. G. Wells, who had, decades earlier, written of the possibilities of "atomic bombs" that by their destructive power would not only change the nature of warfare, but the nature of global politics.[10]

But Szilard didn't know of a nuclear reaction that could create such a chain of neutrons, and neither did anyone else in 1933. Szilard did not let that stop his thinking. He instead thought about what he would be able to do if he *did* have such a reaction. By 1934, Szilard had written up a rough outline of how such a process might work, with an early concept of a "critical mass" (the amount of the reacting material you would need for the reaction to become self-sustaining) and the properties of the chain reaction. In order to attract official attention for his work, and also implement some control over it, he filed for a patent with the British, assigned it to the British Admiralty, and *urged that it be kept secret.* This act was arguably the very first instance of nuclear secrecy—even before fission was discovered and atomic bombs were technically possible.

All of this was very audacious on the part of Szilard: he didn't actually have an invention, just an idea that relied on a yet-undiscovered physical process. And his first approach was to make it both proprietary *and* secret, neither of which are compatible with the more idealistic ethos of science. The British physicists whom Szilard wrote to about this idea must have found him eccentric, even fringe. When Szilard tried to sell Rutherford on the idea, he had him thrown out of his office, offended by both Szilard's speculative, dilettantish approach to nuclear physics, as well as his move to try to patent it.[11] The British government was willing to keep Szilard's patent secret, but they didn't show any significant interest in it. It was, as of then, still entirely hypothetical. We can, in retrospect, see the parts of Szilard's schemes that had promise, but there was much that clearly relied on the existence of hitherto unrealized reactions or particles.[12]

Undaunted, Szilard began investigating whether shooting neutrons at various elements would result in more neutrons; it was laborious, tedious, and expensive, and he failed to get any other scientists to take his idea seriously. Given that Szilard was himself an indifferent experimenter, it is not surprising that he did not get useful results. In 1938, in anticipation of World War II, he immigrated to the United States. He lost faith in the idea of his finding the source of a chain reaction—just before he heard about the discovery of nuclear fission.[13]

When Szilard heard about Hahn and Meitner's work, his mind immediately returned to his hypothetical neutron-induced chain reaction. Nuclear fission was initiated by a neutron, but did it create more neutrons as a result, so-called secondary neutrons? The Hahn-Meitner papers did not mention such a possibility. But Szilard was primed to look for the neutrons, not necessarily because he was more clever, but because he was, once again, a day ahead of the crowd. Overnight, his ideas about chain reactions went from science fiction to possibility, if, and only if, secondary neutrons existed.

And in this realization, at this crucial moment, his mind once again turned to secrecy. As he recalled later: "I thought that if neutrons are in fact emitted in fission, this fact should be kept secret from the Germans."[14] Because nothing could be worse to a European, Jewish-descent refugee than the idea of nuclear-armed Nazis, and if in this new discovery of science was indeed a new weapon, then he wanted it to be controlled. It was in these urgent fears, mingled with science fiction and a new physical discovery, that the first collective attempt for nuclear secrecy emerged.

In 1939, the same year that the discovery of fission swept the globe, the prominent British crystallographer and spokesman of science J. D. Bernal put forward the proposition that "the growth of modern science coincided with a definite rejection of the idea of secrecy." To embrace secrecy was to embrace the ways of the Middle Ages—of alchemy and hermetic mysticism.[15] Bernal's views on secrecy and science were colored by his association of scientific secrecy with industry, state control, and military research. And the state control of scientific knowledge ("the far more dangerous form of secrecy") he associated with the Nazis' attempts to dictate the official truths of nature. Secrecy and state control would merge together, he felt, and "the scientist becomes a servant, or more accurately a slave, of the state." Scientific secrecy was not merely an "inefficiency" to Bernal, in other words: it would lead to the total control of science by the state, and even its destruction.

Similarly, in 1942, when the American sociologist Robert K. Merton was attempting to formulate the "norms" of scientific activity, he railed against secrecy. Merton believed a core ideal of the world of science was that no individual held ownership over scientific ideas, and all must be distributed widely and without restriction. Without openness, scientific claims could not be independently critiqued, and the advancement of scientific knowledge would stop. "Secrecy is the antithesis of the norm," Merton declared. "Full and open communication is its enactment."[16] Neither Bernal nor Merton was being at all controversial in these sorts of statements. By the early twentieth century, scientists and especially spokesmen of science tended to see their profession as being defined in part by open, international communication.

But the true relationship between science and secrecy has not actually been so clean cut, as historians and sociologists of science have repeatedly found. Scientists have long practiced secrecy for a variety of reasons, including fear of losing priority, fear of political or religious reprisals, and fear of military misuse. The scientists who did these things were not cranks: among those who have used secrecy to their advantage are such luminaries as Galileo, Newton, and Darwin. In the Industrial Age, scientific knowledge was often regarded as proprietary (even if such a concept impinged on the "purity" of science), and by the time of the First World War, science was associated with possibly dangerous, and thus secret, military knowledge. Merton's and Bernal's pronouncements about science described hypothetical ideals more than literal realities. But even in fields with no commercial, state, or military connections, practitioners of science have long limited how and when they disseminated information for professional reasons, like priority.[17]

But as a barometer for contemporary academic opinion on the practice of scientific secrecy at the time of fission's discovery, Bernal and Merton are excellent guides. Secrecy was viewed as both antithetical to scientific advancement (it would hinder scientific progress) and potentially an existential threat to the scientific enterprise itself. This view is still common today amongst practicing scientists, even when dealing with potentially lethal technologies. The revulsion against secrecy, specifically secrecy proposed and controlled by someone else, was and remains strong.

Physicists of the 1930s who attempted to control their work tended to do so not with secrecy, but with patents. Patents had their own negative associations with industry and profiteering, but academic physicists had found a way around this by assigning them to a neutral, non-profit organization like the Research Corporation, set up for this purpose in 1912. Any commercial royalties would then be channeled into further research, allowing all scientists to benefit. This approach was part idealism, part pragmatism: the idealism argued for a "purity" of academic science, while the pragmatism argued for advancing the careers of the scientists through credit and reinvestment of funds.[18]

It is in this context that we can see that Szilard's practices were in many ways outside the community norms of his scientific colleagues. Szilard's relentless patenting was itself tolerable, but he did not do the work to fully realize his ideas before attempting to put controls on them. That he had pursued *secret* patents was very troubling.[19]

After learning about fission, Szilard returned to Columbia University, where he had been working since he immigrated to the United States. He approached his friend, colleague, and fellow émigré, Enrico Fermi, with his fears. Fermi had been the one who had perfected the means of bombarding materials with neutrons only a few years earlier, and had taken the opportunity of winning the 1938 Nobel Prize in Physics to escape Fascist Italy. No one could better understand the nature of fission, no one could be more interested in keeping nuclear weapons from the Nazis.

Fermi was already planning experiments to find out how many secondary neutrons were produced from fission, if any. If the number of secondary neutrons produced by fission reactions, on average, was more than one, then a powerful chain reaction was possible. If not, then then the whole thing was still just "moonshine." Szilard suggested, in the name of self-preservation, that Fermi agree not to publish his results. Fermi was indignant; Szilard was asking to withhold research on the most cutting-edge work in his field, work directly derived from Fermi's own Nobel Prize–winning research, on the basis that it could potentially be used for ill by the Nazis for a weapon inspired by science fiction. Academic success, then as now, was about "publish or perish," and there are no prizes or awards for being the second person to make a discovery. Fermi believed there was only a one-in-ten chance that a chain reaction was even possible, and the "unknowns" that existed that could get in the way of practical applications were innumerable.[20]

From the perspective of early 1939, Fermi had the facts on his side. Szilard was assuming many things about how the science might work, and about the ability of Nazi Germany to then act on this information, mobilizing the industrial infrastructure necessary to turn this basic scientific research into military applications within a few years. We now know Szilard was right about nature but wrong about the Nazis, but there is no way anyone could have known either of those things at the time.[21]

Fermi's refusal frustrated Szilard, but at least he worked only down the hall, so Szilard would know what he was doing and planning. Who else, other than Fermi and Szilard, might be thinking of chain reactions? The next in line was obvious to Szilard: Frédéric Joliot-Curie, at the Collège de France in Paris. Joliot, as he was known, was ambitious and capable, and worked on the cutting edge of neutron and radiation research. Joliot also had experience with the bitter fruits of missed priority. In 1932, he had barely missed out on the discoveries of both the positron and the neutron, each of which garnered Nobel Prizes for others. In 1934, he and his wife, Irène, discovered artificial radioactivity, finally winning their coveted Nobel Prize. But Joliot knew that the margins for priority in nuclear physics in the 1930s were slim: a few months was all it took for one team of scientists to find what another was looking for. Irène had herself barely missed out on the discovery of fission: the Hahn-Meitner experiment was a duplicate of one that Irène and a collaborator had done earlier in the year but not fully understood.[22]

Joliot's team in Paris had the resources, experience, and imagination needed to test for secondary neutrons, and in February 1939, Szilard received information that Joliot was performing "secret" experiments of some form. Szilard assumed (incorrectly) that only work on fission could be worth that secrecy. He wrote to Joliot and explained (somewhat misleadingly) that scientists at Columbia were considering self-censorship of chain reaction research and suggested that they might request Joliot do the same. Nothing definite was proposed and the letter was in many ways vague. Weeks passed and the French team heard no more from Szilard and considered the matter dropped.[23]

In the meantime, the search for secondary neutrons continued at Columbia. In early March 1939, the experimental setup was complete. Szilard recalled later: "Everything was ready and all we had to do was to turn a switch, lean back, and watch the screen of a television tube. If flashes of light appeared on the screen, that would mean that neutrons were emitted in the fission process of uranium and that this in turn would mean that the large-scale liberation of atomic energy was just around the corner. We turned the switch and we saw the flashes. We watched them for a little while and then switched everything off and went home. That night there was very little doubt in my mind that the world was heading for grief."[24]

It was a high-quality discovery in physics, but one that increased Szilard's fears of a Nazi bomb. As the scientists wrote up the results, Hitler was invading Czechoslovakia. Szilard's argument for self-censorship was taking on more weight. The Columbia physicists met again and a compromise was reached: they would adopt a form of secrecy. Any new papers on fission would be sent to the *Physical Review*, who would register having received it. These registrations could, perhaps, be used to arbitrate later priority disputes. But the papers themselves would remain unpublished until a later date. It was a scheme that, ideally, would satisfy the need for priority without making the work immediately public.[25]

Even though this was only a temporary approach, it was the first proposed *system* of nuclear secrecy, however small-scale and tentative. It was a procedure, but not yet a *regime*: it was still fairly ad hoc, and there were no real consequences for violating it. Any non-adopter could just take their work to a different publication. And even this weak secrecy was controversial among the Columbia physicists. Fermi was still opposed to any form of self-censorship. But Szilard had convinced another émigré physicist, Edward Teller, a fellow Hungarian, of the danger. Outnumbered, Fermi ultimately assented, but he still thought the idea of making an atomic bomb unforeseeable for the near term.[26]

Fermi's conservatism, again, was not due to a lack of vision. So many unknowns remained: they did not know that there were two isotopes of uranium in question, and only one was capable of fission reactions; nor that enrichment was necessary, much less possible; nor that reactors would breed a new fissionable element (plutonium); nor the speed of the reaction; nor the critical mass; nor many other things. It was Szilard who was asking for something extraordinary: a belief that the normal procedures of science should be halted because of a fear that still easily seemed a decade of research away. That the other scientists ended up agreeing with him anyway is a testament to their fear.

The next step would be to tell Joliot about the results and the decision to self-censor them. But just as Szilard was preparing a cable to be sent to France, the Columbia team received notice that Joliot's team had just submitted a research note to the British journal *Nature*, claiming their own detection of secondary neutrons. Fermi was livid. He suggested that they ought to publish their own results immediately. Szilard still thought they should hold back. The French note did not say *how many* neutrons were detected per fission reaction, which was crucial information for anyone thinking about bombs or reactors.[27]

Fermi thought they ought to take the matter to a more senior member of the Columbia department, George Pegram, and have him settle the matter. But Pegram was unsure. Szilard further talked the matter over with other physicists at Columbia. Some agreed that the science was feasible-enough to look worrisome, and the global threat posed by Hitler only loomed larger as time went on. Victor Weisskopf, another émigré physicist, agreed to write to one of Joliot's scientific collaborators, proposing that, like the Columbia scientists, they could use a journal as an intermediary to satisfy both priority and secrecy.[28]

Weisskopf also sent a telegram to the physicist P. M. S. Blackett in England, asking him to persuade the editors of *Nature* and the Royal Society's *Proceedings* to agree to this scheme. Blackett cabled back that he had passed the request on to the journals and that they would "surely cooperate." The Columbia émigré group secured additional agreement from Niels Bohr to make sure that nothing came out of Denmark, though Bohr was dubious about the plan given the public knowledge of fission. Lastly, the Columbia team contacted the heads of American scientific laboratories doing research in related fields to let them know of the new self-censorship scheme. The *Physical Review* agreed to the scheme: not only would they place holds on any fission publications that came across their desks, they would also tell the Columbia physicists who was submitting them.[29]

But the French still remained on the outside, and there were further complications. A group at the Carnegie Institution that was not in on the censorship scheme had detected "delayed neutrons"—neutrons released by the radioactive byproducts of fission, not the fission reaction itself. These would probably not sustain a chain reaction, but a *Science Service* article did not let this get in the way of making exuberant claims about the future of atomic energy. Joliot's team saw this release and concluded it meant that scientists in America were publishing without restraint despite their entreaties to secrecy. They did not know of Szilard's effort to coordinate secrecy among journal editors, and in any case the logic of the secrecy plan remained in doubt. The Germans had their own capable scientists, who were no doubt still actively at work. "QUESTION STUDIED," Joliot cabled Szilard in early April 1939. "MY OPINION IS TO PUBLISH NOW."[30]

On the same day that Joliot cabled Szilard, his team sent a note to *Nature* reporting that they had concluded that the number of neutrons released by the fission of a uranium nucleus was 3.5, well enough to make a fission chain reaction plausible. It didn't necessarily mean that a *bomb* was possible (many uncertainties remained), but at least nuclear reactors, themselves possibly important military technologies, almost certainly were. *Nature* published the announcement soon after.[31]

Once the French team had broken the publication embargo, others felt free to do so themselves. And after reading the short article, scientists in France, Britain, the United States, the Soviet Union, Japan, and Germany started their own research programs and within a year many would petition their governments about the urgent need for state research into fission for military purposes. By the end of 1939, after Nazi tanks had crossed the Polish border, over a hundred scientific papers had been published on nuclear fission, at least a dozen of them relating to the chain reaction and its potentialities. The attempt to use secrecy to control the idea of the nuclear fission chain reaction had failed almost as soon as it had begun.[32] Szilard himself released the paper he had put on hold at the *Physical Review*; regretfully he wrote to Blackett in the UK that "no actions along the lines suggested by Weisskopf will at present be pursued in this country."[33]

How should we regard this early attempt at self-censorship? Typically, the emphasis is on its failure, and not the steep odds of its success. The institutional culture of science in the 1930s did not acknowledge the negative possibilities of science as an argument against publication, and the frantic, overlapping research efforts meant that any work was likely to arrive only weeks ahead of its competitors. A few months after the end of World War II, Ernest Lawrence, the head of the Radiation Laboratory at the University of California, Berkeley, related that work in his laboratory had only narrowly missed the discovery of fission: "If the Germans had not published their discovery, we would have found it within a few weeks. And so, there would have been no gain from the German point of view or from any point of view in not publishing these fundamental discoveries of science. Indeed, on the contrary, science everywhere benefits by wide dissemination of knowledge."[34]

To attempt to create an ad hoc, non-state based, unenforced, international secrecy pact among scientists who viewed one another as competitors was perhaps the wildest of Szilard's many wild ideas. That he managed to get a significant number of scientists and journal editors to agree is a testimonial to his persuasiveness and their own growing fears. His attempt to stifle the publication of information on secondary neutrons failed, but it did not completely die with Joliot's articles. Rather, as we shall see, Szilard's system became the foundation for the nuclear secrecy regime that would follow. By linking scientists across continents, by drawing attention to the possible threats of information, and by setting up a network of journal editors whose attentions had been drawn to the issue of secrecy, Szilard's self-censorship would have a legacy far beyond the question of "secondary neutrons." And while the physicists' suspicion of secrecy would not totally abate, they would quickly become accustomed to working within a secrecy regime.

1.2 FROM SELF-CENSORSHIP TO GOVERNMENT CONTROL

We can distinguish between three phases of Szilard's secrecy attempts. The first was individual self-censorship: an attempt to convince his colleagues to voluntarily hold back their results. The second was his compromise, wherein his colleagues agreed to submit results to journals but secure agreement from the journals not to publish until they told them to. The third was bolder: securing an agreement from the editors of the *Physical Review* to screen all articles on fission prior to publication, whether the submitter of said article was party to the self-censorship pact or not. Each move from phase to phase was slight and subtle, and yet the final result was something quite different from the initial attempt. The locus of control was shifting ever so gradually out of the hands of scientists and into the hands of others.

The French announcement about chain reactions galvanized worldwide physics communities. In the US, Szilard, with the help of Albert Einstein and others, managed to get the attention of President Franklin Roosevelt (after many failed attempts to generate interest at lower levels of government), in part by pointing to Joliot's results, as well as citing apparent German interest in the topic.

In October 1939, Roosevelt authorized the creation of an Advisory Committee on Uranium, headed by Lyman J. Briggs, director of the National Bureau of Standards. This group saw no need for great coordination or urgency and was hampered by either Briggs' own disinterest, conservatism, or desire to keep the matter within a limited scope of discussion. The goal of the Uranium Committee was to investigate whether nuclear technology might be of potential military importance. It was not a *production* effort whatsoever. It was, at most, a feasibility study, and its output would be reports and recommendations—not atomic bombs.[35]

One of the reasons for this lack of enthusiasm was that the technical possibility of making nuclear weapons had started to seem increasingly unlikely. Niels Bohr and John Wheeler had published an authoritative paper in March 1939 on the theory of uranium fission concluding that all the observed fissioning came from just one isotope, uranium-235. Uranium-235 is *fissile*, meaning that it will fission from the same neutrons that are produced by uranium fission, allowing for a chain reaction. But almost all uranium found in nature is composed of another isotope, uranium-238, which is not fissile. Rather, it would absorb most neutrons without fissioning, inhibiting the chain reaction. As uranium-235 and uranium-238 are chemically identical, no easy separation of the two was possible. To separate them physically would rely on the minute difference in mass (three neutrons, a difference of only 1% of their masses), something which had never been contemplated on a large scale. As less than 1% of natural uranium is uranium-235, this seemed to make nuclear bombs less likely to be feasible, though it still allowed for nuclear reactors.[36]

Still, at a meeting of the Uranium Committee in late April 1940, Szilard again raised the topic of restricting publications on fission. Admiral Harold Bowen of the US Navy, an observer at the meeting, suggested that the scientists undertake self-censorship rather than using governmental force. Szilard would continue to self-censor, and he also managed to convince another physicist, Louis Turner of Princeton, to avoid publishing a theoretical paper that concluded that the bombardment of uranium-238 by neutrons would produce a new fissile element, what would later be called "plutonium" by its discoverers. This was a significant thing to try to keep secret, as this would open the possibility that nuclear reactors could be used to make a different kind of fuel for the atomic bomb.

Szilard took this new discovery as an opportunity to again encourage centralized regulation of fission research, but before he could get far, another article appeared in the mid-June 1940 *Physical Review* from Berkeley, announcing the discovery of a new element, neptunium, from uranium bombardment. Neptunium was not itself extremely interesting, other than being new. But it would be easy for most physicists to see that it would undergo radioactive decay into the fissile plutonium.[37] (And, indeed, the Germans did have this insight within a month of publication.[38])

The theoretical possibility of plutonium tipped a physicist at the University of Wisconsin, Gregory Breit, into Szilard's camp. Breit had been affiliated with the Uranium Committee for some time, and having followed the publication of the neptunium paper, he felt that it was time for organized censorship of fission research. Moreover, Breit had recently become a member of the National Academy of Sciences and had been appointed to the Division of the Physical Sciences as part of the National Research Council. He wrote to Szilard to tell him that he had created a committee to review fission publications. Formal controls had been requested "through official channels" but there had been "unavoidable delays."[39]

Szilard and Breit began to consolidate their new secrecy attempt. They again reached out to the British, worried that the fall of France would mean that Joliot would attempt to publish a flurry of articles abroad. Szilard's view of how to encourage further voluntary compliance presaged a later concept: the classified journal.[40] "I feel even more strongly than before that your attempt to prevent publication will break down unless we create a satisfactory substitute in the form of some private publication," he wrote to Breit. "If that is not done there will be a growing tendency toward indulgence and finally practically everything will be published as it has been in the past."[41]

A little over a week later, Breit wrote to Szilard again. It was one thing to prevent publication of new work on fission. But surely some of those trusted scientists working on the problem should see the work, as well? Breit was in favor of "wide circulation" of such work among trusted individuals. But others on the Uranium Committee were not. Who would make such a decision? Who, he asked, would Szilard put on such a list?[42] This is an important point: any secrecy regime with a hope of success must identify who should have access to the secrets and who should not. Were there American scientists who would be indiscreet, or, at worse, treacherous? How would one know?

The system that Breit and Szilard began to assemble was still a very rudimentary secrecy regime. They had started to come up with a system for identifying information they deemed dangerous. They were thinking about which people should have access to that information, for total secrecy would inhibit their own efforts. They had a primitive procedure for dealing with the information they identified as secret. What they still lacked were consequences for violation of secrecy. The Second World War would provide those.

Perhaps surprisingly, from a legal perspective there was very little government secrecy in the United States until the twentieth century: the nation was a late-comer to this particular activity. The founding documents of the country contained no specific authorization for secrecy, and strong protections for openness. All Constitutional support for the vast powers that the American state would later assume derive from rather adventurous assumptions about what may be done by the government in the preservation of "security," and were not transformed into firm legal precedent until well into the Cold War.[43]

Dating the first regulations and laws about secrecy in the United States is tricky. During the US Civil War, for example, there was no formal system of secrecy supported by laws or regulations, though the US Army was given vast latitude over the ability to restrict the movement of journalists and the publication of presses, and to execute suspected military spies. This was a secrecy regime, but it was a largely informal one, capable of being applied capriciously and without pretense that these powers extended into peacetime. One could write "SECRET" on a document, but it gave it no special legal status: it was just an indication, to the reader, that what they were reading should be treated carefully; it was not a reference to specific regulations that applied to a document's handling or use, or laws that dictated what would happen should someone abuse its demand. The US military branches did not adopt formal regulations about access to facilities or information until the late nineteenth century, and it was only in the early years of the twentieth century that careful scrutiny was given to the consistent use of classification categories. The first American secrecy legislation, the Defense Secrets Act, was put into place in 1911, modeled after the British Official Secrets Act of 1911, regulating the taking of photographs and making of sketches of ships and facilities that were "connected with the national defense."[44]

Around the beginning of World War I, US official attention to the formalization of government and military secrecy intensified. Military regulations about the governance of information, and notably technical information, multiplied. The deployment of new technological weapons—airplanes, advanced artillery, gas warfare, and especially submarines—made the threat of loose scientific and technological knowledge especially acute for the first time. Technology, more so than in any previous war the United States had been involved in, was increasingly identified not only as a modern spectacle, but also as a powerful component in deciding the winners and losers of armed conflict. That the

US was woefully unprepared to mobilize technology for its own military purposes at the time played a role in this sudden fear of technical knowledge. At the late time it joined World War I, the US Army lacked gas masks, offensive chemicals, and the expertise needed to train troops to adapt to the new weapon. This was in spite of the Wilson administration's urgent attempts a few years earlier to mobilize the National Academy of Sciences to bring scientists and technical knowledge into better contact with military institutions. Scientists would eventually answer the call, but it was a disordered and not entirely successful collaboration. Several of the young scientists involved in this effort would, in a few decades, play key roles in the organization of government support of science for defense as a new World War loomed, drawing upon the disorganized experiences of their youth as a motivating force.[45]

Shortly after the US entry into World War I, Congress enacted the Espionage Act of 1917, a controversial piece of legislation that is still in force today. The Espionage Act replaced the existing Defense Secrets Act and covered a huge variety of prohibited intelligence-gathering activities, including any that would give information "concerning any vessel, aircraft, work of defense, navy yard, naval station, submarine base, coaling station, fort, battery, torpedo station, dockyard, canal, railroad, arsenal, camp, factory, mine, telegraph, telephone, wireless, or signal station, building, office, or other place connected with the national defense." As this list makes clear, the conception of what was a secret worth protecting was connected almost exclusively with physical locations, as opposed to, say, abstract science. The form of the secret was likewise outlined at length, with prohibitions on the copying, taking, making, etc., of "any sketch, photograph, photographic negative, blue print, plan, map, model, instrument, appliance, document, writing, or note of anything connected with the national defense."[46]

Looked at in this light, with its narrow definition of secrets as being primarily relating to physical locations, it might not on the face of it be obvious that this legislation would be as expansive as the nascent technical secrecy regimes would later demand. Under a very narrow reading of the Espionage Act, any scientific secrets, such as those involved in chemical weapon development, would be secret largely because they were developed in a government facility, not for their inherent danger. However, the vagueness of the "sketch, photograph, [etc.]

of anything connected with national defense" would ultimately prove legally flexible. Coupled with regulations (promulgated largely by Presidential Executive Order) as to what constituted a connection with "national defense," this law would provide the legal backbone for American secrecy.[47] Aside from its later use, the Espionage Act is infamous today for its 1918 addendum, known as the Sedition Act of 1918, which added expansive capabilities for the government censorship of the press.[48]

Also enacted in 1917 was a law to allow for patent secrecy, put forward with the explicit goal of controlling the spread of harmful technology. The way it worked was, by later standards of secrecy, relatively crude. If, during a declared war, an inventor filed an application for a patent that the commissioner of patents considered could be detrimental to the American war effort if it fell into enemy hands, it could be made temporarily secret. Should the inventor disclose the invention elsewhere, they would forfeit their patent claim on it in the United States. At the cessation of the war, or at the discretion of the commissioner of patents, the application would be allowed to continue its normal route through the Patent Office. If it were granted, and had been used during the war by the United States, then the inventor could apply for just compensation (and also prosecute any suits about priority, should any have arisen).[49] The inspiration for this new law was the paralyzing fear of an existential weapon: the submarine, a new technological marvel that was, as a member of the House of Representatives put it during a hearing in 1917, "the most deadly instrument we have got to contend with in this war." The resulting legislation was the first secrecy law to specifically target "dangerous" technical information, but it was, in its initial form, limited only to patent applications, and limited only to wartime.[50]

If World War I marked the birth of the modern American secrecy regime, including the first forays into technical secrecy, World War II was its adolescent growth spurt. It was here that secrecy became routine, and the rules and regulations, habits and cultures that were created during the war, including but not limited to those about nuclear secrecy, tended to persist even after its conclusion into the postwar period and Cold War. After the beginning of European hostilities in 1939, the United States began the slow process of mobilization and preparation even before Pearl Harbor. The importance of background checks and clearances reached epic heights; the Federal Bureau of Investigation (FBI), for one, did background checks on some ten million people. So great were the FBI's pre-computational information management and storage issues that they requisitioned the DC Armory (a mixed-use sports and entertainment auditorium) as an overflow facility to house fingerprint card cabinets.[51]

From a legal standpoint, various statutes in the late 1930s had updated and expanded those from the World War I period. In some cases, such as an updated patent secrecy statute, it was clear that this was done in anticipation of American entry into the war. In others, it may have been just legislative housecleaning. For instance, in March 1940, President Roosevelt issued Executive Order 8381, "Defining Certain Vital Military and Naval Installations and Equipment," which formally adopted the military classification categories of "Secret," "Confidential," and "Restricted" ("Top Secret" would not come until 1944). It was the first of many Presidential Executive Orders that would codify the American classification system, having more legal authority than a simple change in military regulations, and being far easier to modify than congressional legislation. This is the legal framework of the American classification system even today: the Espionage Act provides the punishments and their legal authority, an Executive Order (since World War II, nearly every President has issued an updated order on classification procedures) provides a more formal framework for the broad operation of the system (what the categories of secrecy are, for example, and what they ought to mean; they also can contain guidance on what to do in ambiguous situations, such as whether to favor secrecy or disclosure), and the military and executive agencies use these to produce their own specific regulations governing the minutiae of the maintenance of the regime.[52]

It was in this context that the atomic bomb entered the story: these legal, technical secrecy regimes already existed in the United States, but they were very new and quite untested, and had never been applied to anything as large-scale as the American atomic bomb effort would become.

Prior to the Second World War, the relationship between the United States government and the development of science and technology, even for military applications, was generally ad hoc, unenthusiastically pursued, and not well coordinated. A relationship existed, but it was not a deep one. While the value of technology to American industry was unquestioned, the role of basic science was more uncertain, and the US did not begin to approach the scientific stature of Europe until well into the twentieth century, aided, in part, by the "brain drain" of refugees from Europe in the 1930s. The relationship between the armed services and research scientists was generally poor through World War I, but the importance of scientific and technological innovations in that conflict got the attention of governments worldwide and made it clear, to some anyway, that there was more to the outcome of battles than tactics, morale, and training.[53]

In the early months of World War II, the federal government finally got serious about science. Much of the credit for this shift is typically laid at the feet of Vannevar Bush, the influential scientist-administrator. An electrical engineer by training, Bush had been the vice president of the Massachusetts Institute of Technology until he moved on to head the Carnegie Institution in 1939, and was soon after appointed chairman of the National Advisory Committee for Aeronautics. Along with James B. Conant, president of Harvard University, Karl T. Compton, president of MIT, Richard C. Tolman, dean of the graduate school at the California Institute of Technology, and Frank B. Jewett, president of the National Academy of Sciences, Bush was part of an emerging cadre who believed that organized scientific action could produce palpable results for society both in peace and war. This group formulated a plan, with the help of Roosevelt aides, to create a new governmental organization that would seek to instigate, bankroll, and coordinate research into defense projects by the American scientific community. In June 1940, as the Nazis pressed into France, President Roosevelt created the National Defense Research Committee (NDRC) by Executive Order and named Bush as its head.[54]

It was decided early on that the Committee on Uranium would fall under the organization's responsibility. Briggs maintained his chairmanship, and Bush added other scientific leaders to the Committee's roster. The NDRC Committee on Uranium reported directly to Bush, unlike most of the other NDRC components. It was decided, in deference to Army and Navy standards of security, that only native-born scientists could serve on the Uranium Committee, and the control over scientific publications by Breit's committee at the National Research Council was continued.[55]

Secrecy was not yet the defining component of fission research. It was secret, but not yet a "special" kind of secrecy. Correspondence regarding the program did not use code-names and often did not contain classification markings at all. The fact that Briggs' committee was called the Uranium Committee is itself an example of how unpromising the work still seemed: it advertised its subject of inquiry right in its name. Thus it is somewhat ironic that the main criticism of Briggs' work was that it was *too* secret. As Karl Compton complained to Bush in early 1941:

> As I analyze the situation, Briggs, who is by nature slow, conservative, methodical and accustomed to operate at peace-time government bureau tempo, has been following a policy consistent with these qualities and still further inhibited by the requirement of secrecy. . . . Considered as an element of the present war emergency, speed in attaining the objective is certainly more important than excessive secrecy, as would be abundantly evident if the German scientists should actually get some of the applications into use.[56]

This lack of urgency with respect to secrecy was understandable: whether a bomb could be made to work was still unclear, and those who were most afraid feared the Germans were ahead. In any case, the US was not yet officially a party in the war. Conant summed up a conservative scientific opinion on the matter when he wrote to Bush in April 1941, that "it seems to me that whatever the ultimate outcome of intensive research and development, the inevitable time intervals must be long . . . I should hate to see too many of our limited group of able people committed to the uranium job."[57] To stake a large scientific effort on such an uncertain project would be a poor use of resources and would undermine everything that Bush and Conant had lobbied for when they proposed that they be allowed to organize scientific research for war.[58]

But things began to change quickly over the course of 1941. Separate from the modest fission research work, Bush was quickly outgrowing the NDRC, which had the ability to coordinate research studies but no means of undertaking large-scale development and production operations. Bush once again went to Roosevelt. A new organization was created, again by Executive Order, in late June 1941. With a broader mandate and a larger budget, the Office of Scientific Research and Development (OSRD) was given the power to move its work from the laboratory to the front lines (though, again, the United States was not yet at war), with nobody to report to other than Roosevelt. Bush would once again head the new organization; Conant would head the NDRC, which would persist as a mere advisory group within the larger OSRD. These changes gave Bush wider latitude for all the projects under his mandate, including the nascent uranium work.[59]

At the same time Bush was making his bid for the OSRD, new data from Ernest Lawrence's Radiation Laboratory at Berkeley indicated that a plutonium bomb was probably feasible. Similarly there was increased confidence about the feasibility of enriching uranium on a large scale, with several candidate processes for doing so having been identified. Having two plausible approaches to a weapon elevated Bush's assessment of the importance of fission work. A review of the possibility of success undertaken by the University of Chicago's Arthur Compton for the National Academy of Sciences suggested that making a weapon would be difficult, but not impossible. But more important than either of these was a report from the British equivalent of the Uranium Committee, the MAUD Committee.[60]

Although the MAUD Committee had limited contact with the Uranium Committee and the report was not widely disseminated due to concerns about secrecy, Bush and Conant received a draft of a July report by the MAUD Committee through other channels. The British physicists were confident that the isolation of several kilograms of uranium-235 would indeed allow for a fast-fission nuclear chain reaction—a single bomb with an explosive equivalent to over a thousand tons of TNT—and that existing plans for separating out enough uranium-235 could probably be made to work within two years. It was too big a job for the UK, but probably not for the United States . . . or Germany. The report, it later became clear, was too optimistic about the amount of uranium-235 necessary and the ease and speed with which it could be produced. But for Bush and Conant, the report was stimulating: the British thought a bomb could be made, and they had outlined a program for building one. Bush and Conant, along with Lawrence at Berkeley and Compton at Chicago, began making plans to greatly accelerate the American effort.[61]

In October 1941, Bush finally wrested control of the Uranium Committee from Briggs, and in accelerating the uranium work, he and Conant also began to accelerate and emphasize the secrecy. As Conant recruited scientists to the project, he would now emphasize the highly confidential nature of their work, and urged them to watch what they said to others, even including military personnel.[62] On October 9, Bush went to the White House to meet with President Roosevelt and Vice President Henry Wallace. Bush had told Roosevelt about the optimistic conclusions of the British scientists and gotten approval for a broader research program independent from the rest of the NDRC. It was not yet a bomb-production project, but it was moving into a new phase, where the goal would be to produce proofs-of-concept, pilot plants, and concrete plans for a future, industrial-sized effort to make bombs, should the work prove promising. Roosevelt said that money could be provided at his request "from a special source available for such an unusual purpose," a "black budget" source of discretionary funds not subject to congressional approval. Roosevelt also gave Bush "definite instructions . . . to hold consideration of policy on this matter within a group consisting of those present this morning, plus Secretary [of War Henry] Stimson, General [George] Marshall, and yourself."[63] Bush asked whether the Secretary of the Navy would be included. Roosevelt, Bush later wrote, "looked at me with one of his strange smiles and said, 'No, I guess not, not now.'"[64]

Bush quickly began to reorganize the uranium work in order to swiftly produce a pilot plant that would demonstrate the feasibility of separating uranium-235 from natural uranium and to prove the possibility of breeding plutonium from a nuclear reactor. In early November, Bush met with the members of Roosevelt's small "Top Policy Group" on uranium matters and proposed a large expansion of their effort. This was also the first time that the secretary of war, Henry L. Stimson, became aware of the possibility of a bomb.

Over the course of November 1941, Bush reviewed the entire program and recommended further acceleration. Bush and Conant agreed that it would be moved fully under OSRD auspices. Bush would report directly to Roosevelt on the matter and keep Stimson informed of any progress. The uranium section would be reorganized, with Arthur Compton taking responsibility for basic physics measurements at the University of Chicago, Ernest Lawrence working on electromagnetic isotopic separation at Berkeley, and Harold Urey investigating separation by means of centrifuge and gaseous diffusion at Columbia. Each was a Nobel Prize winner with experience managing large projects. And plans were being made for the Army to begin construction of pilot plants. All of this was finalized the day before the Japanese attack on Pearl Harbor.[65]

1.3 ABSOLUTE SECRECY

In the fall of 1941, both Bush and Conant began to grow increasingly concerned with the question of how to maintain secrecy in a project whose fundamental principles were already well known to many scientists and were even becoming a staple article for science journalists. William L. Laurence, a science journalist at the *New York Times* who had followed the fission story since it broke in the scientific journals, wrote a breathless story about the discovery of fission in the September 1940 issue of the *Saturday Evening Post*, entitled "The Atom Gives Up." He described its discovery as a possible "turning point in human history" and noted that "one pound of pure U-235 would have the explosive power of 15,000 tons of TNT."[66] Stories like Laurence's demonstrated that nuclear fission was easily adapted to the already-existing genre of hyperbolic predictions about latent atomic energy.[67]

These sorts of "wild articles speculating on the possibilities," as Bush would call them, made the goal of secrecy difficult. It meant that in the public domain, there was enough interest in fission that any secrecy would be noticed. In August 1941, John O'Neill, the president of the National Association of Science Writers and science editor of the *New York Herald Tribune*, publicly charged that the government was secretly working on uranium weapons, and imposing censorship on anyone connected to it. He said this amounted to "a totalitarian revolution against the American people," and said that politicians were controlling scientists, pushing them to use their work for war.[68] His remarks were widely reported, and disturbed Bush and Conant enough that they issued a categorical denial that "any development of an atomic bomb is in progress," but admitted that there was work being done on nuclear power.[69] As Bush explained to a British representative that September, the narrow truth was that "there is no program in this country directly aimed at the production of atomic bombs," and that power work had "a relationship to possible work on bombs, but this is inevitable and incidental."[70] Even this was only a half-truth—the work's connection to bombs was certainly not "incidental."

As the work ramped up, the information practices applied to it began to change in ways both subtle and profound. One key change was in the use of code-names. What had once been brazenly referred to as the Uranium Committee became known more cryptically as "Section S-1."[71] It is not clear what "S-1" stood for, if anything; it has been suggested that the "S" was for "Special," which is plausible, and "special" became a catch-all adjective for the later Manhattan Project work.[72] Even today, in the lexicon of American bureaucracy, the term "special" in an organization's name is code for unusual secrecy, even within a system where secrecy has become routine.[73]

Eventually, Bush no longer spoke of "uranium" at all in his letters. When he referred to the work on fission, he became deliberately vague: "the important matter that we have under consideration," to cite just one example.[74] To the British, whose own code name for the project was still "MAUD," he referred to it at one point as "affairs 'concerned with a certain lady.'"[75] In December, Conant suggested that if they called uranium-235 "magnesium" and uranium-238 "aluminum," perhaps they could avoid classifying their letters as "Secret" (then the highest classification) and instead drop down to "Confidential" (which entailed less onerous handling requirements). Bush disagreed with the downgrading, but Conant thought they ought to continue using the substitutions even with the "Secret" classification, "as an added precaution." The secrecy had expanded to the point where everything was either secret or not, with no middle ground.[76]

In addition to the code words, Bush also insisted that all S-1 personnel take an oath binding them to "a pledge of secrecy."[77] Bush put a lot of stock in the oath-taking, though it is not clear if he thought such an oath would be legally binding. When a former project scientist was found to be telling too much of his work to other Americans, Bush consulted the FBI on whether the oath was legally meaningful. When it came to Bush's attention that a former project scientist was discussing more than he should have to other Americans (though it was not espionage), the FBI judged that the oath was "so phrased that prosecution cannot be undertaken for disclosure of secret information by one citizen of the United States to another citizen."[78] Despite this, the use of secrecy pledges continued.

Bush and Conant were increasingly afraid that their secrecy measures were inadequate. In late 1941, Bush received a report that a scientist at Ohio State University with no connection to S-1 "knew the whole set-up in Section S-1 with the names of the Program Chiefs and the general division of work." He had apparently gotten this from one of the S-1 members, Edward Condon. "It begins to look as though the British were right in regard to our inability to hold matters confidential," Bush wrote to Conant.[79] To rectify this, Arthur Compton proposed that they centralize all S-1 work in a single laboratory, perhaps in Berkeley with Lawrence (whose particle accelerators were too large to transport to another lab), where, "both through guards and through inculcation a spirit of secrecy should be possible."[80] Instead, while Berkeley continued to be its own hub of work, Compton's laboratory at Chicago began to absorb other sites' teams and work, like Columbia and Princeton, as part of the newly formed Metallurgical Laboratory.[81]

In early 1942, Conant sent around a letter to all the S-1 leaders on "this all important matter of secrecy" that introduced a new requirement: all participants who would know anything about the ultimate goal of the work would have their backgrounds vetted by military security. He wanted a list of names of everyone who was working on "even a small detail of the problem," who would all eventually be submitted for Army or Navy clearances.[82] In theory this sounded good; in practice, it took considerable work to put such a regime into place. Even the very mundane aspects of secrecy—like using "SECRET" stamps, required organization. C. P. Baker, a physicist at Cornell, after laboriously hand-marking "SECRET" on every page of a lengthy report in the spring of 1942, left a plea on the its final page: "WE NEED A STAMP."[83]

Stamps could be easily provided; the more tricky problems involved people. Many of the scientists working on uranium were, in the eyes of the military, "queer types" (Bush's term).[84] The Navy, which handled many of the clearances for S-1 work early on, balked at issuing a clearance to Arthur Compton, because he had signed his name to a number of petitions at various times in his life, some of which were associated with organizations designated as Communist-fronts. Bush told the Navy representative that it was impossible to exclude Compton from the project; he had "special status." In any case, keeping someone like Compton close made good security sense: it was better to have him inside the project and under its secrecy constraints. Bush mused further that it might be wise to take in the "queer types" and put them under oath, even if they were not given much information about the project:

> There are many individuals in this country having rather complete knowledge as a result of their study of physics, and these are by no means all under control. . . . In fact, I am inclined to believe that should the subject become at all imminent in the sense of promising practical results within a reasonable interval it would be well to take in and put under thorough control practically every physicist in the country having background knowledge of the subject, but the time for this has certainly not arrived.[85]

Making secrecy real, Bush was finding, required entertaining some radical ideas—even the possibility of bringing scientists into the work just to keep them controlled. But merely telling scientists the work was secret was not enough. They needed to be taught how to treat secret documents, and OSRD struggled to do this through circulation of specific rules about what each classification "grade" meant in practical terms.[86] In a number of cases it meant chastising violators until they exhibited a "security mindset." In mid-January 1942, Compton had sent a progress report to Conant outlining the discussion at a conference held at Columbia. Conant sent him a scolding reply because the list of participants included several who had "not been cleared," and "full disclosure was made of the general plans about the whole problem" during the discussion.[87] Secrecy, Bush and Conant were finding, was no easy thing, and it relied as much on trying to control those *inside* the secrecy regime as it did on controlling what kind of information circulated *outside* of it.

One of the novel approaches that Bush and Conant took toward keeping the bomb secret was one that combined secrecy with another, more traditional form of technological control: patenting. Specifically, in order to deal with competing patent applications on nuclear reactors (notably those coming from Joliot's former team, who had since immigrated to the UK), while avoiding a protracted legal fight, Bush began to use a patent secrecy law that allowed him to "put to sleep" any patent applications that had any wartime implications, including relating to atomic energy. Bush and Conant further made sure that every scientist who worked on the uranium matter signed contracts that assigned their intellectual property to the government, and forced the filing of secret patent applications that would be granted only after their subject matter had ceased being classified. This approach would later be taken up with zeal during the full Manhattan Project as well, as a means of technological control that guaranteed government ownership over an entirely new field of scientific and industrial work, and also provided a means of legally squelching inventors who were not part of the project.[88]

In March 1942, Bush sent Roosevelt an enthusiastic report on the status of the uranium work. It was predicted that only five to ten pounds of fissile material (uranium-235 or plutonium-239) might be needed to set off an explosion equivalent to some 2,000 tons of TNT, making them hundreds of times more powerful than the largest conventional weapons of the day, and a variety of promising means were being pursued to extract the uranium-235 from natural uranium. The topic of security and secrecy occupied considerable space in Bush's report. "Preservation of secrecy on this matter is unusually difficult," he explained, because of the amount of information and speculation already in the public domain before it came under government control. To encourage new, secret information staying secret, they had begun to "subdivide" the work, so that "full information is not given to every worker"—a technique that would be better known as compartmentalization, or the "need to know" principle.

Bush concluded by saying that the matter was "under control to a reasonable extent," but was "more vulnerable to espionage than is desirable." The work should, he concluded, "be placed under rigid Army control as soon as actual production is embarked upon."[89] In a typically brief reply, Roosevelt urged Bush to move from the pilot phase to the production phase. On the matter of turning it over to the Army, Roosevelt had only one requirement: "I have no objection to turning over future progress to the War Department on condition that you yourself are certain that the War Department has made all adequate provision for absolute secrecy."[90]

Why Roosevelt would demand "absolute" secrecy is a more interesting historical question than it might first appear. From the beginning, Roosevelt had insisted that uranium work be segregated from other research work, and as it became more feasible, he pushed for greater and greater secrecy. It is tempting to simply say that Roosevelt understood how important the atomic bomb was. But this is unlikely—even the scientists were unsure on this point, and while a single bomb with the equivalent of 2,000 tons of TNT would be impressive, it doesn't necessarily warrant "absolute" secrecy. There were many other secret weapons developed during World War II that didn't require "absolute" secrecy above and beyond "normal" military secrecy. Could he have been worried about the Nazis? Many of the scientists involved were, so it is not implausible. Perhaps, like them, he worried that any inkling of an American program might spur on the Nazi program that was imagined to be occurring in parallel.

The normal explanation for the unusual level of secrecy is the bomb's existential character. That is, the bomb is "special" because it poses problems that threaten the very existence of states and the nature of international order. To some degree this was appreciated at the time, but the believability of this position—before said bombs existed whatsoever, and when they were still at explosive yields comparative to large conventional bombing raids—is itself historically situated in time. Even with the assurances of the scientists, the bomb was a long shot, and whether it would prove to be the existential weapon some of them imagined would have been unclear in early 1942.[91] Either Roosevelt was remarkably prescient, or he had other motivations for his secrecy.

Secrecy came easily for Roosevelt. He was no stranger to diplomatic intrigue and the value of cutting some people, and the public, out of the discussion. The fact that he kept a minimal paper trail for many important decisions, and at times told people what he thought they wanted to hear, has made it difficult for historians to fully assess his internal motivations.[92] Though the documentary record does not reflect this (nor rule it out), it seems highly plausible that Bush may have urged Roosevelt toward this end. Bush was interested in programmatic secrecy: it wasn't one technical fact or another that concerned him, it was the fact that the United States was about to invest a lot of resources into building a bomb. The Axis powers may have been the ultimate motivation for this, but there were clearly more mundane threats as well. Bush had already spent considerable time confronting one of these threats: the United States Congress.

In July 1941, Congress had cut OSRD appropriations by a million dollars, leading Bush to feel that he was "being blocked very decidedly" in his efforts at organizing scientific research.[93] (Congress, of course, was not blocking the bomb, which they did not know about, yet.) In early November 1941, Frank Jewett, the president of the National Academy of Sciences, wrote to Bush that the feasibility of a fission bomb seemed uncertain and that it would take "some millions of dollars" to resolve decidedly. This would be hard to explain to a non-scientist: "Certainly if and when the time comes seriously to contemplate the huge appropriation for a [uranium-235] concentration plant, we will have to be prepared to go before the appropriating body of laymen with a far more convincing story than what I think we now have."[94] Bush agreed with Jewett; the source of this massive amount of funding would require some care, he replied, because "this would be a thing that could hardly be presented to a committee of Congress." Furthermore, if the bomb seemed like it really would be "involved in long-range planning," then it "would have to be handled under the strictest sort of secrecy."[95]

The problem of oversight was not a minor one, nor one that would go away. It would indeed have been very hard to explain something as expensive and speculative as the atomic bomb in late 1941 and early 1942, and became increasingly so as expenditures reached well beyond "some millions." Keeping the bomb secret from people who doubted the wisdom of spending thousands of millions of dollars on physics projects may have been more important to the success of the Manhattan Project than keeping the bomb secret from the Germans. A Nazi atomic effort did not have the power to stop the American work in its tracks—only Congress could do that.

In the spring and summer of 1942, the work on fission quickly expanded. The S-1 project was not yet a production program, but the results looked *very* promising. A workable atomic bomb, available in time for use in the war, was beginning to look like a possibility, perhaps as early as 1944, Bush told Roosevelt. The efforts for implementing Roosevelt's "absolute secrecy" had begun, though they were not without their difficulties.

In May 1942, Gregory Breit, who by now was in charge of the physics of bomb reactions at the Metallurgical Laboratory, wrote a long letter to Lyman Briggs explaining why he, Breit, was quitting. It was not that he thought the work was unimportant. Rather, he had become too frustrated with the pace of it, and Compton's apparent disregard for secrecy:

> I believe that the ultimate importance of the work of Section S-1 is likely to be very much higher than that of most military problems. The tool worked on will exceed ordinary weapons by orders of magnitude in offensive power. So far as I can see, it is very dangerous to have insufficient precautions regarding secrecy. I believe that caution regarding secrecy should be more thorough in S-1 than in any other branch of military research. As a matter of national safety it will be necessary to preserve secrecy not only during the war, but for decades afterwards.[96]

At Chicago, Breit reported, the scientists were openly disdainful of the compartmentalization policy and violated it frequently. In his view, Compton himself was opposed to it and in particular wanted to remove barriers between research on nuclear reactors and on nuclear bombs. Secret colloquia and reports were being distributed to all members of the project, information about sensitive meetings was being leaked. "It has been impossible for me to dissuade him on this matter," Breit complained. "The consequences of following this policy are likely to be disastrous." Bomb work, Breit urged, had to be moved somewhere else where more control could be exercised, and perhaps the Army would need to be called in to enforce real secrecy.

Breit's memo was passed on to Conant, who was already aware that the security situation at Chicago was not ideal. He wrote to Bush, saying that it "disturbed me somewhat but only confirms what I had found." A new organization had to be created to ensure the bomb work was isolated from Compton's indiscreet purview.[97] Bush sent back his own short reply:

> No really serious charges here, but rather a disquiet due to naiveté of [Compton]. There are two steps we might take on making the new plans: 1. Require the set of secrecy rules from each group for approval, together with ways in which *they* will check to see that they are observed. 2. Isolate the bomb portion itself.[98]

Conant was one of the few scientists involved who had previous experience with serious secrecy, having run a secret plant to produce lewisite, an arsenic-based gas meant to be used on Germany during the First World War, in a suburb of Cleveland. The plant was given the nickname "The Mouse-Trap" by those who worked there, because, as was later explained, "men who went in never came out until the war was over."[99] Like many things in the Second World War, despite the weapons being new and "special," many of the ideas for how to control them were old ones, and that Conant was one of those who pushed for further isolation of the scientific work is not a coincidence.

With Breit's departure, a new physicist would be necessary to coordinate the work on the important fast-fission (bomb physics) calculations. Compton already had an idea about who would be a good replacement: the Berkeley theoretician, J. Robert Oppenheimer, who had been brought into the project by Ernest Lawrence only that March.[100] That summer, Oppenheimer hosted a secret summer conference in the physics building of the University of California. Barbed wire was strung up on the balcony outside of the seminar room in LeConte Hall. The theorists at this conference concluded that given their present knowledge, there was no reason an atomic bomb would be impossible with sufficient fuel, though there were still many unknowns.[101]

By mid-June 1942, Bush and Conant felt they had enough evidence to justify shifting the bomb work from mere research into wholesale production. This would involve a substantial change in scope, moving from a science project to an industrial one. They recommended to the

Top Policy Group that construction of plants for isotope separation and reactor development be turned over to the Army Corps of Engineers, with the assistance of civilian scientists. The OSRD would continue to direct the research and development phases. Bush and Conant sent their report to the President on "Atomic Fission Bombs," recommending a full crash project. Secrecy would dominate the effort:

> [We recommend] that the greatest secrecy be exercised in connection with this project, particularly with respect to its purpose, the raw materials used to develop the final product, the final product, and the manufacturing processes involved in producing the final product or products. As soon as actual construction work starts in the field, it will not be possible to conceal that plants are being constructed. It is therefore suggested that these plants be camouflaged under some suitable names, and their purposes be announced in similar camouflaged manner.[102]

Roosevelt approved the recommendations by penning a simple "OK FDR" on its cover sheet. They had entered a "new phase," and the need for secrecy had intensified.[103] The Manhattan Project had begun, and as its military, industrial, and scientific empire expanded across the country, so would also its secrecy.

This "nascent" period of nuclear secrecy, from the discovery of fission through the establishment of the Manhattan Project, is crucial to understanding what would come later. The beginning of secrecy, as noted, was a fear, one that at first seemed improbable to a great many people who were exposed to it, but got increasingly compelling as both science and world affairs marched onward. The scientific aversion to secrecy, or at least to losing scientific priority, was immensely powerful: it took considerable intellectual effort, and the increasingly imaginable threat of world domination by the Nazis, for secrecy to seem like a prudent policy.

The initial secrecy of the scientists was one that they felt they were in charge of. This is in line with general scientific attitudes toward secrecy at the time, and largely even today. Scientists generally agreed secrecy was a bad thing, but they still used it when they thought they were in the driver's seat. It was the idea of *externally imposed* and *systemic* secrecy that inspired the revulsion toward secrecy that many of the scientists felt, and in that respect it is interesting to note how their efforts toward it became a very slippery slope. Szilard's self-censorship effort first put the power in the hands of the scientists, then in the hands of the journal editors (also scientists). Once the government became involved, however, that power began to shift. Under the Uranium Committee, it still technically stayed with (government-employed) scientists, although it was quickly becoming intermingled with the requirements of the military. Vannevar Bush's organizations intermingled them further. And once the Army came into the picture, as we shall see, any pretenses to scientific autonomy became very flimsy indeed.

What is remarkable in retrospect is how few scientists made any kind of principled objections to this secrecy in this period. Those that did reject it (such as Joliot) are, in this telling, typically cast as the villains, an inversion of the more standard view about scientific secrecy. We can see here, though, that secrecy was not yet a way of being or acting that completely spoke to the scientists: they needed to be convinced, they needed the reasoning behind it to be spelled out very explicitly. Its power derived directly from the plausibility of the weapon, which itself connected the idea with the magnitude of the threat. And the Nazis, to be sure, were as dire a threat as any that these physicists had encountered: that so many of these scientific figures contemplating such things were already refugees from Nazism is not at all incidental, and is why European refugees, many of them of Jewish-descent, play such a crucial role in the early period of this story.

Turning this desire for control into action is where the scientists repeatedly stumbled. It is one thing to say that you don't want the Nazis to know something, but how do you make it so? The practice of self-censorship was a weak one, because they could not enforce it effectively: there were no consequences for breaking it. Weak practices, coupled with relatively weak institutions (like scientific journals, which have no strong powers to regulate people), resulted in a non-functional secrecy regime, or at least one that was easily cracked. Only once they got serious about the institutionalization of secrecy did it begin to function, and that involved bringing in the big guns, literally. Government agencies, partnering with the military, and applying mindsets very different than those the scientists were accustomed to working with, would add "teeth" to this secrecy, at the cost of the autonomy of the scientists.

By the time the Army Corps of Engineers entered into the job of making atomic bombs, the American program had already begun to operate under extreme secrecy. Practices were being put into place, one by one: document control ("SECRET" stamps), obfuscation (codenames), misinformation (to the press), compartmentalization ("need to know" policies), personnel discrimination (clearances and background checks), site-isolation (secret laboratories), physical security (barbed wire and guards), publication censorship (initially voluntary, later not), and oaths of secrecy (both psychological and legal in their intent) had all come into place. None of the individual elements were unique to the bomb, though in other OSRD programs, they had been implemented with less fanfare.

But the general character of the secrecy on other OSRD work was still quite different from what would soon come. It is telling that the first item in the general OSRD security directives emphasized not the importance of secrecy, but rather the dangers of too much:

> AVOID OVERCLASSIFICATION. You will save yourself and everybody else unnecessary trouble if you will arrange your correspondence so that it can be sent without classification. Don't classify material indiscriminately. Study the definitions of SECRET and CONFIDENTIAL matter instead of reaching for the nearest rubber stamp. Careful phrasing of communications will permit much correspondence now classified to be sent as open matter by ordinary mail.[104]

The OSRD's interest in personnel clearances was of a different order as well; contractors were trusted to come up with their own security system. The uranium work, from the beginning, had been kept in a different category that sat outside of the normal organizational chart of NDRC and OSRD sections.[105] When the Army took over, the exceptionalism of the nuclear work would continue, but in a way that would encourage some of the most extreme secrecy practices of World War II.

The loss of autonomy over their own work is one of the great "fall from grace" stories that many of the scientists involved would tell in the years after the war, and no one would tell it better than the one who started this chain of events, Leo Szilard, who would later lament, "The SECRET stamp is the most powerful weapon ever invented."[106] But at the heart of the matter, it was a cumulative series of half-steps that led to the secrecy of the Manhattan Project, and practically every policy embraced, and amplified, by the military had already been put in place by civilian scientists before they arrived, inspired by their fears.

2

THE "BEST-KEPT SECRET OF THE WAR"

THE MANHATTAN PROJECT, 1942–1945

Secrecy, like charity, begins at home.
ERNEST O. LAWRENCE, 1943[1]

The project by the US Army Corps of Engineers to produce usable atomic bombs before the end of World War II, code-named "Manhattan," involved a heavy mandate: simultaneously mobilize several hundred thousand people at massive, previously non-existent sites across the nation, toward the end of producing a new and spectacular super weapon without letting any significant news of this work, much less its details and goals, be revealed prematurely. They were trying to keep the secret not merely from the Axis powers, but also from all US allies except the United Kingdom, as well as the national and international presses, the American Congress, and, to various degrees, nearly everyone who was working on the project itself, whose "need to know" extended only as far as necessary to do their direct jobs. Even maintaining such a secret for a relatively short amount of time—under three years—was a gargantuan task.

Though it would be celebrated as "the best-kept secret of war," this mandate was in fact *impossible*: the effort to make the atomic bomb, on the time-scale required, meant that "absolute secrecy" could never be achieved in practice. A close look at the justifiably impressive secrecy practices of the war reveals simultaneously how tenuous and barely-contained the secret actually was, and how in many places, both domestically and abroad, the secrecy efforts failed. Even with its failings, the expansive wartime secrecy infrastructure created by the military and civilian authorities working on the Manhattan Project would signal things to come in the postwar period, and the myths of its success would be foundational to the creation of a new national security state.

2.1 THE HEART OF SECURITY

In June 1942, with President Roosevelt's approval to push the work on fission into a "new phase," Vannevar Bush laid the groundwork for increased cooperation with the US Army Corps of Engineers, who would handle the initial construction work for a uranium enrichment pilot plant near Knoxville, Tennessee. The Corps of Engineers generally worked by establishing "area" engineer districts that would coordinate the local activities in a given jurisdiction. Colonel James C. Marshall, the first head of the project, set up a temporary headquarters in downtown Manhattan, New York City, on Broadway, because of its centrality to the headquarters of many major industrial contractors. A code name was chosen for the project—Laboratory for the Development of Substitute Materials (DSM), but even this vagary was considered too revealing about the nature of the work. Instead, it was given a blander designation: "Manhattan," after its location. The Manhattan Engineer District was formally created as "a new engineer district, without territorial limitations" in early August 1942, its name being only partially misleading.[2]

In September 1942, it was decided that the work, thus far jointly divided between the OSRD and the Army, should be centralized completely under the total authority of the Manhattan Engineer District. At its helm would be a notoriously gruff, undiplomatic, and relentless Army Colonel who had helped Marshall organize the early work: Leslie R. Groves. Of all the figures responsible for the successful wartime development of the atomic bomb, none perhaps can take as much credit as Groves. An engineer by training, he was not enthusiastic about taking on the assignment, as its chances of success were hardly guaranteed, and it was far from the front lines. But once he accepted the job, he was going to push it to completion. For the next three years, he dedicated every ounce of his will toward making the atomic bomb a reality in time for it to play some role in the war. His only precondition was that he be promoted to brigadier general first in order to better enforce his orders, especially with the "many academic scientists involved in the project," who he thought would be reluctant to take orders from a mere colonel.[3] He charted out a program to scale up the project by securing uranium stocks and obtaining a top military priority rating that would allow him to push his project even in the face of opposition from other Army generals as he began to siphon off scarce resources, human and material, into a project whose goals were known to only a select few.[4]

In the fall of 1942, Groves began to search for an isolated site to create a new, secret laboratory that would handle the most sensitive work of the project: the design of an actual atomic bomb. The design of an atomic bomb is not necessarily the most difficult part of making a bomb (though it did prove more difficult than anticipated), but it is perhaps the most difficult to keep secret along the lines Roosevelt demanded, since anyone working on it must know the intended goal of the project. His choice for the head of the new laboratory was the theoretical physicist J. Robert Oppenheimer, a quantum luminary who had joint professorial positions at both the University of California, Berkeley and the California Institute of Technology. Oppenheimer famously proved more successful than anyone would have imagined.[5] One of Oppenheimer's main roles was Groves' liaison to the academic scientists, the "prima donnas" as Groves later described them.[6] Oppenheimer was a scientists' scientist, a theoretician who had never worked on a military project prior to fission, whose erudition was in stark contrast to Groves' crudeness, and whose far-left politics were well known.[7] Yet Oppenheimer was totally loyal to Groves, and pushed every policy Groves told him to push. A dynamic would repeat itself several other times: Groves would mandate a policy, Oppenheimer would champion it, and, if it were something that the other scientists would not accept, he would try to negotiate the compromise.[8] It was Oppenheimer who pushed Groves toward New Mexico for the secret bomb-design laboratory, the scientist having spent time there in his youth, and with the government acquisition of the Los Alamos Ranch School, the Los Alamos laboratory was born.

The bomb work was, in 1942, still seen as a gamble, but the odds were longer than anyone involved realized at the time. The estimates for its budget, staffing needs, and time to completion had been grossly underestimated. The Los Alamos laboratory was initially meant to only have

300 scientists and assistants; by the end of the war, the staff was nearly ten times that. The initial estimation of the cost was around $400 million USD; that was off by a factor of five. The plan was to complete a weapon by early 1944; it was not completed until the summer of 1945. It is of note that the time of completion for the Manhattan Project, a little over three years, is still the fastest nuclear weapons project ever undertaken in the world, despite the fact that the technology was entirely novel and much of the science still unknown.

As the bomb project expanded, so did its security measures. The massive factories to produce the fissile materials that would fuel the weapon were purposefully isolated from surrounding communities, in part for public safety, but largely to ensure security. At Oak Ridge, Tennessee, a massive amount of land was procured in late 1942 for several facilities that would enrich uranium. Dubbed "Site X," Oak Ridge would employ some 80,000 technicians, construction workers, and other laborers with their families over the course of the war. Only a small handful of "residents" knew the true purpose of this gigantic "secret city." At Hanford, Washington, construction of another secret site ("Site W") began in late 1943 to house the first industrial-scale nuclear reactors, along with facilities for extracting plutonium from their spent fuel. Behind multiple layers of armed guards and barbed-wire fence, an atmosphere of strict secrecy was also constructed. "SILENCE MEANS SECURITY," a water tower over Hanford commanded. And at the Los Alamos laboratory, "Site Y," isolated on a remote mesa in New Mexico, thousands of scientists, technicians, and military personnel and their families formed a secret hub for research on the bomb itself.[9]

As a practice of secrecy, isolation had its downsides. Procurements were difficult, especially since items shipped to some places like Los Alamos had to be routed through "front" addresses elsewhere to avoid revealing a secret scientific site out on the mesa. Isolated sites also meant isolated conditions: poor roads, poor infrastructure, and poor access to resources. Convincing tens of thousands of employees and technicians to move to isolated areas to do work they would not understand the purpose of was no trivial task, especially with manpower shortages across the country due to other war needs. James Conant complained in October 1943 that his inability to reveal the purpose of the project made it difficult to recruit top-flight scientists, and that simply appealing to its high priority alone was not enough to convince scientists to leave other important war-related projects.[10] But as more scientists became part of the Manhattan Project, recruiting possibilities improved, as recruitment to New Mexico began to be seen as an exciting and alluring opportunity, a chance to join on to something big and to hob-nob with several Nobel Prize winners.[11]

As the work force grew, Groves became a strong adherent to compartmentalization, the "need to know" principle that each worker should know the minimum amount of information necessary to do their assigned job. In his postwar memoirs, Groves made it clear that to him, not only was compartmentalization key to secrecy, but also it was a tool to control scientific personnel:

> Compartmentalization of knowledge, to me, was the very heart of security. My rule was simple and not capable of misinterpretation—each man should know everything he needed to know to do his job and nothing else. Adherence to this rule not only provided an adequate measure of security, but it greatly improved over-all efficiency by making our people stick to their knitting. And it made quite clear to all concerned that the project existed to produce a specific end product—not to enable individuals to satisfy their curiosity and to increase their scientific knowledge.[12]

Groves didn't invent compartmentalization—the idea was already a standard counterintelligence tactic, and we have already seen Bush and Conant were using "subdividing" in their civilian work. But Groves did take it to new extremes, transforming it into an all-encompassing way of life, and applying it on a scale previously unimagined. At each of the major project sites, compartmentalization operated somewhat differently, because of the different character of the work done at each location. At industrial production sites, like Oak Ridge and Hanford, almost all of the labor force was kept entirely ignorant of the goals of their work. This affected morale, and manpower retention, in a severe way. The civilians who worked at Oak Ridge and Hanford were not doing so under duress; they could leave and do other war work if they wanted

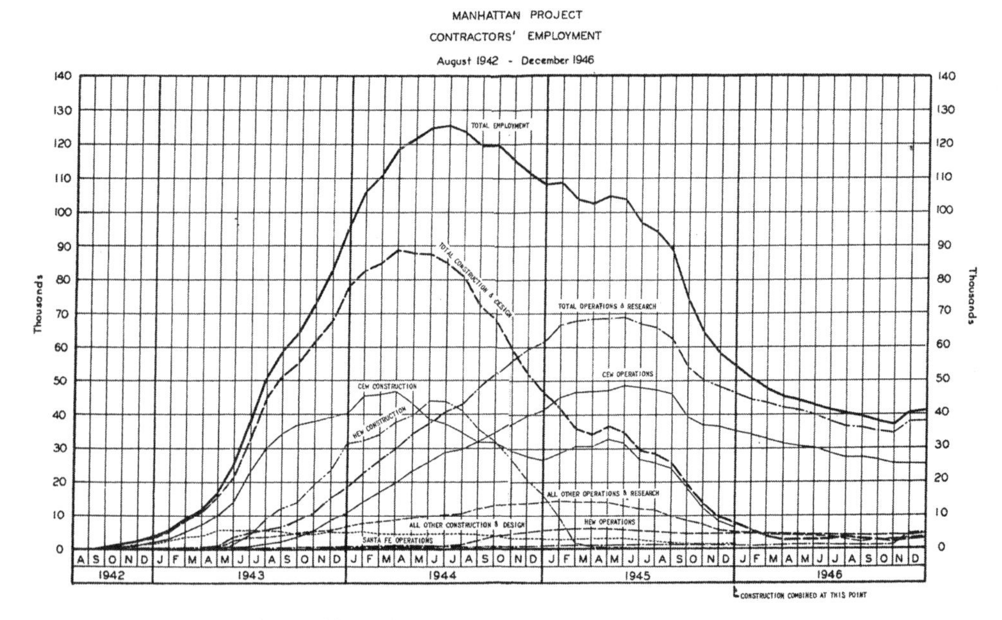

FIGURE 2.1. The monthly employment by contractors on the Manhattan Project, 1942–1946. Source: MDH, Book 1, Volume 8, "Personnel," Appendix A1.

to. The rate of labor turn-over at Hanford, where living conditions were poor, was 20%. At Oak Ridge, it was 17%. Thus, even when the project's overall labor force was growing by tens of thousands of new employees each month, it was also having to make up losses of thousands per month as well. Some 500,000 Americans were at one point or another employed as part of the bomb effort, nearly 1% of the entire civilian labor force during the war, and far greater than the 125,000 people who were employed at the project's peak.[13]

One postwar report at Oak Ridge closely linked the problems of secrecy and the problems of morale:

> The war worker in Oak Ridge, Tennessee has been working under some of the most unique working conditions ever known. Due to the secrecy surrounding the nature of the Project, he never saw the results of this labor. There was nothing in which he could take pride. Thus, one of the common incentives for work was not present. No sense of satisfaction could be realized in a job well done. Naturally, this created quite a problem of morale not commonly experienced.[14]

At Oak Ridge, along with "stay on the job" rallies, other more curious solutions were employed, aimed at finding positive ways for a worker to pass time when they were not working. Allocations were made at Oak Ridge for massive recreational activities, including a badminton tournament and league, a ten-team baseball league, ten leagues (eighty-one teams total) of softball, miniature golf, and table tennis, and twenty-six teams of touch football. Intramural sports, in other words, would be a salve for the psychic wounds created by secrecy.[15]

But there were darker sides to compartmentalization than morale loss. Secrecy could inhibit safety, and the workers were not told about the special dangers associated with radioactive materials. A postwar radio program about Oak Ridge told the story of a woman whose sole job was to wash uniforms and then hold them up to a machine that might click. If the clicking was violent, the uniform would be washed again. Only after the war ended did she learn that the clicking machine was a Geiger counter, and that she had been maintaining safety conditions by measuring radioactive contamination. She reported at the time that she could feel "real proud" of her wartime contribution, but one wonders how many corners were cut out of ignorance.[16] Oak Ridge in particular, with its massive uranium enrichment operations, required huge numbers of people whose war years were spent watching dials, turning knobs, flicking switches, and all the while having no idea what was going on behind the panels, or even what their factories were producing. Such people—frequently women recruited from local cities—were in effect human machines, and used only because automation of these functions had not yet been perfected for the wartime work.[17]

At Los Alamos, compartmentalization took the form of physical security measures and the division of people and spaces.[18] Los Alamos was a gated site that contained an inner, gated Technical Area, which required its own security clearance to enter. A system of colored badges distinguished the different categories of knowledge one might be entitled to. Those with yellow badges, such as security details, could enter the technical areas of the lab but were not to be told any classified information at all. Blue badges were for clerks and warehouse employees who would need to know some classified information, like schedules and rosters and names, but would not be allowed access to technical information. Red badges were for technicians and secretaries whose access to information could be quite deep within the scope of their own job but could never exceed it.

Lastly, the white badge indicated people allowed to know the full scope of what they were doing at Los Alamos, though their access to specific technical information would be decided by their own division and group leaders. For those with white badges, Los Alamos could be relatively free, if they could make a case for their "need to know": scientists had authority to choose what could be talked about, and to whom, once they were within that category.[19] Scientists also signed a different secrecy agreement than other project employees: while all employees signed a "Declaration of Secrecy" acknowledging they would receive sensitive information as defined by the Espionage Act, and thus were under legal obligation to obey security regulations, only "physicists, chemists, and other employees of a similar professional or scientific caliber" were required to state that they "stake their personal and scientific reputation" on such obedience.[20]

Oppenheimer also secured permission from an extremely reluctant Groves to have a laboratory-wide weekly colloquia series where anyone with a white badge could attend to discuss specific scientific problems relating to the project's goals. This sort of activity apparently did not take place at other scientific laboratories associated with the bomb work (like Berkeley or Chicago), and this made Los Alamos a special place, befitting, perhaps, its isolated status. Colloquium topics were largely on practical subjects related directly to weapon design: "Captain Ackerman . . . spoke on preparing shape masses of high explosives for implosion spheres"; "Griesen talked on the X-ray technique of implosion examination"; "Commander Birch spoke on the subject of gun assembly of fissile material with illustrative slides." Topics ranged from the most fundamental (Niels Bohr gave a talk on the physics of a neutron reacting with heavy nuclei) to the most operational (Norman Ramsey talked about how they intended to "deliver" the bomb to its target, and William Penney spoke on "the subject of damage by the blast effect of a gadget").[21]

There appears to be only one case of a scientist quitting the project on account of too much secrecy. Edward Condon, a physicist at Los Alamos, resigned from his work in April 1943 because he felt unduly stifled by the secrecy policies. Writing to Oppenheimer, Condon explained:

Declaration of Secrecy I
Designed for Execution by all Physicists, Chemists,
and other Employees of Similar Professional or
Scientific Caliber

In consideration of the vital interest which the United States of America has in the successful accomplishment of the work being carried on here at ______________________________ : and

In further consideration of my employment by ______________ in connection with such work:

I hereby affirm, without mental reservation, that I bear true faith and allegiance exclusively to the United States of America, that I have secured in the past and will secure in the future to the Government of the United States of America the sole benefit of any developments, experiments, discoveries or inventions here made by me or any information here obtained by me, to the exclusion of any other country, company, party, organization, or person whatever, except as covered by provisions of any patent agreements entered into between myself and. ________________________. Upon the truth of this statement I stake my personal and scientific reputation.

Recognizing the importance to the national welfare of safeguarding all classified information that has not been officially released pertaining to this project or to related work, I hereby agree that I will neither communicate nor transmit, to any person, the performance of whose duties does not require the same, any classified information, documents, notes, memoranda, drawings, photographs, blueprints, plans, maps, models, materials or equipment connected with the project that would convey classified information.

I further affirm that I have not taken or utilized and will not take or utilize any of the classified items listed in the preceding paragraph, other than for official purposes, and then only in accordance with the established rules of the Manhattan District.

I fully understand that a failure on my part willfully or through gross negligence to adhere to the foregoing may involve a violation of the Federal Espionage Act and thereby subject me to punishment thereunder by imprisonment for not more than ten years and, in the discretion of the court, by fine of not more than $10,000.

Witness | Signature Upon Employment

I certify that the importance of safeguarding military information and penalties of the Espionage Act for violation of this declaration were stressed orally to this employee prior to his affixing signature hereto upon separation from employment.

Signature Upon Separation From Employment.

Signature of Official Giving Exit Interview.

FIGURE 2.2. Specimen for the "Declaration of Secrecy" to be signed by "Physicists, Chemists, and other Employees of Similar Professional or Scientific Caliber." Source: MDH, Book 1, Volume 14, "Intelligence and Security—Supplement," Appendix CS-8, ("Security Manual, Manhattan District, 26 November 1945"), Exhibit II.

> I feel so strongly that this policy [compartmentalization] puts you in the position of trying to do an extremely difficult job with three hands tied behind your back that I cannot accept the view that such internal compartmentalization of the larger project is proper. My disturbance was complicated with the feeling that I might sooner or later unintentionally violate such rules through failure to comprehend them fully.[22]

These were not idle concerns: fear that secrecy would decrease morale and hinder progress, especially scientific progress, was taken seriously by Groves, Oppenheimer, and others, though it would not be solved to the degree that many of the scientists desired. It was not that scientists lacked an understanding of practical secrecy—what they chafed at was that they had lost autonomy in the bargain. They were not imposing secrecy, but having it imposed on them.[23]

Compartmentalization would be Groves' most controversial policy amongst the scientists working on the project. Every scientist would later tell stories about how they had to subvert it in order to get their job done. Sometimes "compartmentalization stories" could be offered up as humor, as one scientist related to the *New Yorker*:

> I was directing [two] projects. One was on the separation of isotopes and the other was on chain reaction. People on one project weren't allowed to speak to people on the other. I was in the position of not being allowed to talk to myself.[24]

But many scientists in the postwar period would give bitter testimony to what they saw as the detrimental effects of secrecy on science, and some even credited it with slowing down the work. For many, compartmentalization frustrations represented their chief objection to secrecy, beyond the fences, the tapped phones, the opened mail, and so on. The latter could be laughed away; the compartmentalization was seen as a threat to scientific work itself.[25]

Groves' compartmentalization was applied not only to individual sites, but also between them. Scientists at the University of Chicago, for example, could not receive communications from Los Alamos, with the exception of specific, prearranged categories that did not reveal too much about bomb work.[26] Hanford and Chicago could interact, because Chicago was helping guide the construction of the reactors at Hanford, but Oak Ridge scientists were not supposed to know even the location or purpose of "Site W."[27] When Groves learned that a large number of French émigrés were to be employed at a reactor laboratory in Montreal as part of the British contribution to the Manhattan Project, he set up an almost one-way rule of exchange, so that the scientists in Canada could report their results to the scientists in Chicago but would get very little back in return.[28]

This kind of inter-project secrecy also served Groves' policy aims. By "making our people stick to their knitting," as Groves later put it, he was also removing them from any discussions about the uses of their work. Some, like the scientists at Chicago, would have such discussions anyway, but those reports would not be seen by anyone Groves did not allow them to be seen by. Groves drove not only the creation of the bomb, but nearly every decision about its eventual ends as well. Scientists who challenged this tight control were at risk. None did this more than Leo Szilard, prompting Groves to draw up papers for his internment for the duration of the war; fortunately for Szilard, these were rejected by the secretary of war.[29]

How far would Groves and his counterintelligence agents go to stem threats? Would they have gone further than incarceration? There are rumors, perhaps impossible to confirm, that they might have been willing to take extreme measures in some cases. Jean Tatlock, a former girlfriend of Oppenheimer, died under suspicious circumstances during the war. Tatlock was considered by the security forces to be a Communist or Communist-sympathizer, and in the summer of 1943, Oppenheimer was surveilled as he spent the night at her apartment in San Francisco. In January 1944, Tatlock was found in her apartment by her father, "lying on a pile of pillows at the end of the bathtub, with her head submerged in the partly filled tub." The cause of death was asphyxiation by drowning, and barbiturates and chloral hydrate (the active ingredient of "Mickey Finn" knockout drops) were found in her system. A suicide note in a shaky hand was present at the scene. Did she die by her own hand, or was it something else? It sounds too cloak and dagger to be true, but later revelations about "active measures" taken by American security forces during World War II and the Cold War make it hard to completely dismiss the idea that Tatlock may have been seen as a security risk too dangerous to tolerate.[30]

Groves' intelligence organization had started small, initially as a Protective Security Section whose jobs in 1942 were only plant protection, personnel security, and security education. Within a year, this had been expanded into a full-fledged Intelligence Section, with Counter-Intelligence Corps officers and activities added to the mix, decoupled from normal Army intelligence activity. Although the Intelligence and Security Division had only around 140 officers and 160 enlisted agents working concurrently, over the course of the two and a half years of concentrated Manhattan Project activity, it handled over 1,000 "general subversive" investigations, over 1,500 cases in which "classified project information was transmitted to unauthorized persons," 100 cases of suspected espionage, and 200 cases of suspected sabotage. They also oversaw the protective security of key Manhattan Project figures and the transportation of vital materials, and ran undercover operations to ferret out espionage. In the immediate postwar period, the Division concluded that while "espionage attempts were detected," their actions "had in each case prevented the passing of any substantial amount of Project information."[31]

Groves expanded every practice of secrecy that had already been started by the OSRD and even added a few others. Where the OSRD had routed background investigations through the Navy, Groves now did them in-house as part of his ever-expanding, increasingly autonomous security apparatus. By December 1943, Groves received authorization to assume total control over security at Manhattan Project plants, giving him essentially his own domestic intelligence service; this role later expanded into even a foreign intelligence service with the Alsos project to assess and seize German assets relating to atomic research.[32]

When it came to individual scientists, Groves was more liberal than one might expect, especially when it came to project members who had known associations with Communists. Groves' security services would not approve a security clearance for J. Robert Oppenheimer on the basis of such associations, but Groves overruled them.[33] In the early Cold War, Groves would come under fire, because by the standards of the later 1940s and early 1950s, he had staffed Los Alamos with dozens of scientists with "Communistic" tendencies and backgrounds. In his memoirs, Groves went to great pains to emphasize that for him, the "speed of the accomplishment was paramount":

> [S]ecurity was not the primary object of the Manhattan Project. Our mission was to develop an atomic bomb of such power that it would bring the war to an end at the earliest possible date. Security was an essential element, but not all-controlling. . . . All procedures and decisions on security, including the clearance of personnel, had to be based on what was believed to be the overriding consideration—completion of the bomb.[34]

Such statements are defensive by nature, tinged with the later lived history of McCarthyism. But they do match up with the wartime priorities, and explain some of its lapses. And while the primary wartime focus was on Japan and Germany, Groves spent at least as much time in his counterintelligence reports talking about the Soviets.[35] When Army intelligence indicated that a number of the younger Berkeley scientists, all former students of Oppenheimer, were talking with Soviet spies, Groves found ways to keep them from doing further work. In one case, a scientist's own thesis became classified and unavailable to him, and in another case, a scientist found his draft exemption revoked, and he was inducted into the military.[36] Ironically, Groves' awareness of some Soviet espionage attempts in Berkeley may have blinded him and his intelligence forces to the real espionage going on elsewhere.

Groves' approach to secrecy varied importantly from the OSRD's. The OSRD's approach, focused on coordinating the activities and practices of its scientists, was mostly passive and reactive. It entailed sending out stern warnings, swearing new members to solemn oaths, and farming out issues of personnel security to the armed services. Groves' approach was more active and aggressive: what he wanted was a full counterintelligence effort that would constantly monitor existing security practices for violations and actively search out cases of attempted sabotage or espionage. Eventually, Groves' security plan would even have its own semi-autonomous domestic and foreign intelligence wings that would report only to him, something that distinguished it from any other weapon production project during the war.[37]

Groves saw the bomb project as justifying the greatest security measures, above and beyond any other secret wartime activities. One sign of this was his refusal to even use conventional classification categories in their intended sense. Through March 1944, there were only three official classification categories in governmental use: "Secret," "Confidential," and "Restricted." Each category had an official definition of what it covered, and its own requirements for handling.[38] "Secret" information would be "information the disclosure of which might endanger national security, or cause serious injury to the Nation or any governmental activity thereof," already a broad designation, though not broad enough for Groves. He had already been using a new, in-house category in line with his compartmentalization policy, "Secret-Limited," restricted to only project division leaders.

In March 1944, the Office of War Information issued new regulations, creating a new category of secret information: "Top Secret." It was defined as "information the security aspect of which is paramount and whose unauthorized disclosure would cause exceptionally grave danger to the nation," and meant to be used for secrets the entire war might hinge on, like details of the D-Day invasion. In August 1944, Groves instructed Oppenheimer that all correspondence that indicated the technical nature of the atomic bomb should be reclassified from "Secret" to "Top Secret." Oppenheimer was surprised since, as he wrote to Groves, he had thought the category was reserved for broad war plans, or at least for the schedules of when the bombs might be ready for use. But as with most other matters, Oppenheimer deferred to Groves: "I understand the intent of the broader application and we shall carry it out." If the use of the atomic bomb was crucial to large-scale military strategy, then the very existence of the atomic bomb would be as well. In this way, the core operational secret radiated its importance out to nearly every other aspect of the project—everything would thus be "Top Secret."[39]

2.2 LEAKS, RUMORS, AND SPIES

The Manhattan Project attempted to keep a great many things secret from a great many people. To this end, it was largely successful: when the atomic bombs were used against Japan, it came as a great shock to the Japanese, to the Germans, and to the American people. But in one respect, it was a massive failure: years later it would be revealed that it had been infiltrated by several talented Soviet spies. In 1950, the physicist Klaus Fuchs confessed that while he had been working at Los Alamos on highly sensitive weapon design work, he had also been passing huge amounts of technical information to Soviet handlers. Evidence of more espionage activity would come out in subsequent years: David Greenglass, a machinist at Los Alamos; Theodore Hall, a physicist at Los Alamos; and George Koval, a health physics officer at Oak Ridge, were among the most central of the "atomic spies" in the Manhattan Project, some of whom remained at large for their entire lives.[40] The number of actual spies in the Manhattan Project was quite finite—at most around a dozen, out of a project that had well over 10,000 technical workers and hundreds of thousands of non-technical workers—but important.[41] Almost all of the ones who had access to secrets directly were "moles," scientists or engineers who had volunteered their services to the Soviets, as opposed to trained Soviet agents.

How did so many of these moles slip through the Manhattan Project security system? Groves knew that Soviets were trying to penetrate the project. In one of his final meetings with President Roosevelt, in December 1944, Groves had explained that "there was every evidence the Russians were continuing to spy on our work, particularly at Berkeley." This was not taken as any great scandal: its chief import was as to whether the Soviet Union should be "brought in" on the secret. Groves ultimately rejected this option, believing "that it was essential not to take them into our confidence until we were sure to get a real quid pro quo from our frankness."[42] Secrets, in this framing, were things to be bartered away at a later date. Groves further emphasized this in his notes on the meeting: "I said I had no illusions as to the possibility of keeping permanently such a secret but that I did not think it yet time to share it with Russia." Roosevelt agreed.[43]

Groves' sense that the Soviets had gotten very little information turned out to be dramatically wrong. The problem was not at Berkeley, but at Los Alamos, in the heart of the work. In a hearing in 1950, following the Fuchs revelation, Groves gave a curious answer while testifying before a congressional committee in a secret session. All of the security apparatus at Los Alamos, he explained, was "primarily to guard against indiscretions," not moles.[44] While being self-serving, this explanation does accurately reflect that during the war Groves' secrecy forces spent far more time attempting to prevent leaks than they did in tracking down spies. To track down spies, especially internal moles, one needs to monitor project personnel *very* closely, and to carefully scrutinize who was let into a project. This is a very different approach than trying to discourage rumors and leaks, which circulate much more broadly, or to make sure that outsiders are kept out of project spaces. Groves' chief secrecy problems during the war concerned the latter more than the former, because, as noted, the Manhattan Project was optimized for expediency and needed almost all the scientific manpower that it could get, and Groves feared failure more than spies.

Ironically, Groves' very attempts to limit information leaking out through traditional channels like scientific publishing and news media also made the project's efforts more conspicuous to scientific audiences. As discussed in the previous chapter, much of the science of uranium fission had already been published openly. This was one of Bush's justifications for potentially putting practically everyone with any familiarity with the subject under government contract of some sort, so they would know that the matter was considered secret in official circles.[45] These early attempts to control the scientific literature regarding fission work had continued as the government got involved with the project. Gregory Breit at the National Research Council continued to review scientific papers for any possible fission content, though by July 1942, he was complaining to the OSRD that he hadn't seen any such papers for a long time. Another scientist working for the OSRD suggested that this was probably because "*almost* every man qualified to write such papers are *in* the S-1 project itself."[46]

But outside the scientific sphere there was a much larger and more diversified media environment. Atomic energy had been a big story in the early 1940s; it would not vanish completely just because scientific publications on the matter had stopped. In fact, the lack of scientific publications could itself be suspicious. *Time* magazine noted that scientific meetings in May 1942 were under-attended and exceptionally vague, and that "exploration of the atom—chief interest of physicists—has come to a stop":

> Such facts as these add up to the biggest scientific news of 1942: that there is less and less scientific news. . . . . A year ago one out of four physicists was working on military problems; today, nearly three out of four. And while news from the world's battlefronts is often withheld for days

> or weeks, today's momentous scientific achievements will not be disclosed until the war's end. . . . Pure research is not secret now. In most sciences it no longer exists.[47]

This sudden absence was observed by more than American science journalists. A physicist drafted into the Soviet army, Georgii Flerov, had noticed that Allied countries had seemingly stopped publishing on fission work. From this he deduced that a heavy regime of secrecy had been imposed, which he reasoned must be the result of military interest. In April 1942, after failing to get others to take his concerns seriously, he took the risk of writing to Joseph Stalin himself, arguing that "this silence is not the result of an absence of research. . . . In a word, the seal of silence has been imposed, and this is the best proof of the vigorous work that is going on now abroad." Though Soviet intelligence agencies had heard whispers of the Allied nuclear projects, Flerov's letter finally jarred the bureaucracy into action.[48]

In the United States, it wasn't just scientific periodicals that were beginning to be censored. After Pearl Harbor, Roosevelt put into motion the creation of a program of wartime press censorship. In early 1942, Byron Price, the head of the newly-formed Office of Censorship and a former executive news editor of the Associated Press, issued the first guidelines to govern "wartime practices" of the American press. The news media went along with it, both because of wartime patriotism, and also because the guidelines were entirely voluntary. The first guidelines did not include anything specific to uranium or fission work, only a general prohibition against information about "new or secret military weapons."[49]

The reason the Office of Censorship initially said nothing about the atomic bomb was because they themselves knew nothing about it. The US government's interest in atomic energy was itself the main secret, so alerting all newspaper editors that they should not discuss atomic energy seemed an imprudent move. But by February 1943, as the Manhattan Project's work was beginning to get underway and the big sites were going online, Bush suggested to a military advisor that they consider putting the fission work into the censorship guidelines:

> There would be a certain amount of harm in placing such a proposal before representatives of the press. To me it seems that the good would

> outweigh the harm. Both Dr. Conant and I are afraid that a public discussion of this whole affair might break out.[50]

The final straw in pushing them toward press censorship was the difficulty the Army was having seizing land for its massive sites in Oak Ridge, Tennessee, and near Hanford, Washington. Using the power of eminent domain, the Army had taken ownership of the land and compensated the owners at what the government thought was a fair rate. The landowners, however, did not always agree and demanded through the court system higher payoffs than Groves felt he could allow. This raised a huge problem for secrecy: court records were necessarily public, and press coverage about the seizure of the land, especially at Hanford, threatened to divulge the Army's intense interest in the area, and perhaps even details about the work being done there.

The local Army head of the Hanford project, Col. Franklin T. Matthias, wrote to Groves in early April 1943 explaining that while the local news media were willing to comply with voluntary censorship, they were "very jealous in what they consider their prerogatives in determining what is restricted." The newsmen felt that land use issues were matters of public record and were not specifically prohibited by the voluntary censorship code. Matthias recommended that Groves urge the Office of Censorship to contact all the editors in Washington State and reinforce the importance of censorship regarding Hanford. He also recommended that they consider releasing carefully sanitized stories containing "a minimum of information of a general innocuous nature" to the local press rather than implementing a complete stonewall. His logic was simple, and sound: the journalists found a straight denial to be a challenge, whereas a dull, partially true account would give them the story they needed to get on with their jobs.[51]

The Office of Censorship proposed to send a letter to the West Coast newspaper editors notifying them that the Army interest in the Hanford land fell under the censorship code clause relating to "new and secret weapons" and the processes for producing them. Groves thought this went too far—he wanted a simple blanket order of censorship regarding the Hanford area, not a letter that indicated it was involved in making new weapons.[52] The Office of Censorship replied that the General ought to keep in mind that censorship was entirely voluntary and could not be imposed with impunity.[53]

In late June 1943, the Office of Censorship finally issued a notice to editors at 2,000 daily newspapers, 11,000 weeklies, and all radio stations in the United States that all information about "war experiments" relating to "atom smashing, atomic energy, atomic fission, atomic splitting" or the military uses of heavy water, cyclotrons, and uranium was banned from print or broadcast.[54] And throughout the rest of the war, the Office of Censorship attempted to kill stories that leaked out or were set for print. In several instances, high-profile news sources evaded the ban and published provocative articles referencing government work in the field or the possibility of atomic warfare. In September 1944, Groves had compiled over a hundred press stories with references he considered to be outside the line. But the secrecy desired was so massive that to completely remove all speculations on nuclear physics from the public sphere was impossible, especially after essentially alerting all news organizations in the country that there *was* intense American interest in the subject. Even among major newspapers, stories on uranium and atomic energy continued to be printed up until the use of the bomb against Japan.[55]

Using modern, digitized, searchable newspaper databases, we can actually track the trends of how these "banned terms" were used in many major American newspapers. From the discovery of fission through the end of 1941, dozens of articles were published on "atom smashing" and even "atomic bombs." Toward the end of 1941 and into early 1942, the rate of articles on these subjects dropped dramatically, but there was always a "background rate" of publications on these issues. The Office of Censorship prohibition had no obvious effect on this overall rate of publication. What did seem to have an effect was the acceleration of the American fission research program and creation of the S-1 Committee in late 1941: as more experts working in this field were pulled into wartime research, fewer were publishing articles on the subject, or capable of talking to reporters. This didn't eliminate speculative articles, but it did put a damper on how many new topics emerged.

The main threat of leaks, especially later in the project, came not from scientists inside the project but from people on the periphery of it. In February 1943, for example, the president of the University of California, Robert Sproul, gave a speech claiming that Berkeley scientists were working on a project whose outcome would decide the war. "If we solve this problem first, the United Nations will win," he was quoted

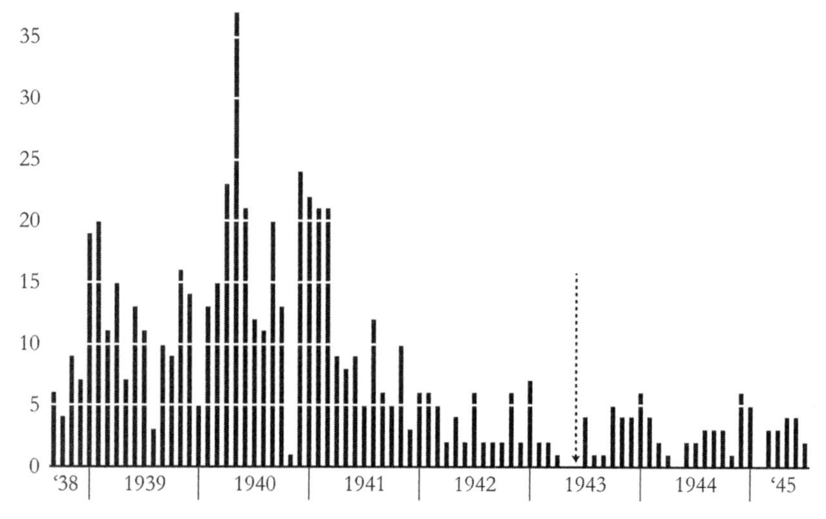

FIGURE 2.3. Articles featuring terms relating to atomic energy matters in eight major US newspapers, September 1938–July 1945. The dotted arrow indicates when the Office of Censorship formally promulgated voluntary censorship regulations on these terms (June 1943). Terms surveyed: "atomic energy," "atomic bomb," "atom bomb," "atom smashing," and "uranium." ProQuest databases surveyed: *Baltimore Sun*, *Boston Globe*, *Chicago Tribune*, *Los Angeles Times*, *New York Times*, *New York Herald Tribune*, *Wall Street Journal*, and *Washington Post*.

in the *Chicago Sun*. "If the Germans—and we know they are working on it—solve the problem first, they will win it."[56] Sproul, however, had almost no first-hand knowledge of what the Berkeley scientists were doing at the Radiation Laboratory. Even though his university was hosting a significant part of the Manhattan Project and was the managing contractor of Los Alamos, he had been told only that the work was important and secret.[57] As a result of Sproul's talk being picked up in newspapers, a Manhattan Engineer District (MED) official wrote Ernest Lawrence an angry letter berating the apparently poor security culture at the Radiation Laboratory:

> It is obvious that such disclosure of information contained in the above-mentioned articles might compromise the entire project. Furthermore, it is not outside the realm of possibility that a statement like Dr. Sproul's could lead to a single disastrous bombing attack upon the University of California. To say the least, it will let enemy agents in this country know where there is a fertile field in which to work.[58]

Lawrence responded by sending out a letter to his entire Radiation Laboratory staff urging increased secrecy. To the MED official, he noted that controlling the press on such a matter would always be difficult, for "popular writers and public speakers, not realizing the harm they do, are always anxious to speculate sensationally in order to catch their audiences' interest." In general, though, Lawrence embraced the call for increased secrecy, and believed it was attainable:

> I heartily agree with you that when the American people know the reasons for secrecy, they can be depended upon to keep silent. . . . Secrecy, like charity, begins at home.[59]

The problem, though, was that the "reasons for secrecy" were themselves a secret. In such a situation, even the military itself could be a source of leaks, since most of its members were also being kept out of the loop.[60] A major incident occurred when the state director for the Selective Service of Tennessee, Brigadier General Thomas A. Frazier, released a statement on new draft regulations in December 1943 which included the assertion that "a major, secret war effort" was being created in Tennessee. "Within the area of the new appeal board is the Clinton Engineering Works"—the formal name of the Oak Ridge project—"in secret war production of a weapon that possibly might be the one to end this war." A version of Frazier's statement was printed in an early edition of the *Nashville Banner* newspaper before it was discovered by the Manhattan Project and stopped from further circulation. When interrogated about the leak, Frazier claimed he didn't understand the fuss: everybody knew there was a giant secret war plant being erected in Tennessee, and there were endless rumors about its purpose. In the end, Frazier got off with a warning, as Groves was not eager to antagonize the Tennessee selective service board.[61]

Frazier wasn't wrong that if you create a project the size of Oak Ridge or Hanford, people will notice. At Hanford, one "local rumor" was that it was a "money-making scheme between Roosevelt and DuPont."[62] In Tennessee, a local American Legion branch claimed the work at Oak Ridge was under much more secrecy than other wartime plants and alleged that it was of "unexampled extent and cost, that great extravagance [was] being practiced in its construction, and that, according to the best informed sources, it will not be completed within the probable duration of the War." This denunciation was circulated to many governmental agencies, to Groves' dismay. Manhattan Project security tracked down its author, an "old man in his dotage" who was "definitely failing," who proceeded to harangue the investigators about the "crackpot or Socialistic scheme" at Oak Ridge.[63] Another rumor floating around Tennessee was that Oak Ridge was being used to print Roosevelt campaign buttons, while another was that the whole thing "was nothing but another of Mrs. Roosevelt's socialistic experiments in housing and community government."[64]

As the Manhattan Project grew in scope and expense, the press control problems got harder. At around the same time that the press around Hanford were creating difficulties, an article in the Swedish newspaper *Svenska Dagbladet* reported on the Allied sabotage of a Norwegian heavy water plant. The sabotage was meant to keep the Nazis from securing enough heavy water to conduct reactor research and development, and the Swedish article reported that heavy water could be used to produce "an explosive of hitherto unheard-of violence."[65] The information was reprinted in a British newspaper, and subsequently republished in a prominent article in the *New York Times*. The *Times* article was, like the Swedish one, oblique in its description of what role heavy water would play in the production of a weapon. "Heavy water or, more correctly, heavy hydrogen water, is believed to provide a means of disintegrating the atom that would thereby release a devastating power," the article explained, deliberately muddling the scientific issue.[66] (Heavy water can be used as a moderator in a nuclear reactor, which can be used to produce plutonium, which can be used in a bomb.) But the idea that the Allies were interested in sabotaging Nazi atomic work could imply that they thought atomic bombs were feasible—and thus that they might be working on them themselves.

Groves worked to find out where the story had come from and how it had slipped by the UK's own press censorship. UK representatives to the Manhattan Project claimed it was a slip-up, though later they would note that the UK press censorship policy did not apply to articles that were first printed outside of their nation, thus providing a "back door" for such material.[67] Vannevar Bush reported to the War Department that he had been "very much startled that such an article could appear, and I have taken steps to see if I can why it was allowed to become published."[68] He was told that General George V. Strong had also been "perturbed" by the *Times* article and had "taken some action in that regard."[69] Specifically, General Strong appears to have written to the Army chief of staff in June 1943 emphasizing the importance of Army-maintained secrecy:

> I deem it essential to suppress all printed speculation concerning research in this problem in order to prevent the enemy from knowing the extent of our efforts or our progress. This cannot be done by force of law but only by securing the cooperation of . . . the various publicity outlets.[70]

The issue prompted the Manhattan Project officials to dabble in a somewhat darker practice: disseminated deliberate *misinformation*. Harold C. Urey, the discoverer of heavy water and an important participant in the US work since the Uranium Committee, sent a letter to multiple inquiring newspaper editors denying that heavy water had any known military application.[71] Urey's denial would itself be a story in the *New York Herald Tribune*, whose editors were no doubt pleased to report that the *New York Times* was repeating unsubstantiated nonsense.[72] Urey's response was calculated to be technically correct—heavy water cannot, by itself, be used as an explosive—but was essentially misleading.[73]

Groves was suspicious of the use of misinformation. It could easily backfire—if an official lie could be spotted as a lie, then it would provoke even more intense interest. In the summer of 1944, the FBI intercepted German agents who were instructed to inquire into the state of American reactor research; the questions they were meant to answer were so basic that it became clear the Germans were not at all aware of the scope of American work at that point. Groves recommended that they be told to report back to their handlers that the US was doing small-scale, academic exploration of fission—and not a denial that any research was taking place, which would be so unlikely as to be provocative.[74] There are a few other instances of deliberate misinformation being used during the Manhattan Project, but compared to other secrecy practices, they are few and far between.

Whenever Manhattan Project security uncovered a rumor that was circulating, no matter how trivial, they chased it down, both to put the fear of the law into whoever was spreading it, and also to see if they truly knew anything. Famously, they even censored Superman: in 1945, Manhattan Project intelligence officers contacted DC Comics about a story about a cyclotron that was unfolding in the newspaper daily strips. Apparently, they did not want too much attention being given to cyclotrons and other nuclear research during the war, and they also didn't want the American public to associate such things with comic strips when the time of revelation finally came. As a result, the plotline was changed.[75] Superman was not the only fictional publication so scrutinized; the March 1944 issue of *Astounding Science Fiction* featured a story that discussed (in a loose sense) using uranium-235 in a bomb, which merited its author and editor similar visits and threats from FBI agents.[76]

The Manhattan Project security officers cataloged and investigated many "typical" examples of leaks or "loose talk." In one such episode, a patent engineer in Chicago decided that his company ought to research the splitting of uranium-235. His supervisor contacted Arthur Compton at the University of Chicago. Compton shared this potential leak with the Manhattan Project security forces, who tracked down the engineer in question. It turned out he had gotten his idea from a pamphlet published by the Moody Bible Institute of Chicago, which had discussed nuclear fission in the context of arguing that "God has given to Christians the gift of the Holy Spirit with energies far more dynamic than those of exploding atoms." This was, the agents later related, a "harmless" case, but indicative of the thoroughness of their efforts.[77]

One of the most curious cases took place in early 1945, as the wartime factories were just starting to produce useful quantities of fissile material. Seven scientists from India were invited to visit scientific colleagues in several major American cities as part of an official program of cultural cooperation. They were warmly received; however, the welcome cooled when Manhattan Project security officers heard that while they were visiting the University of California, one of their party, Professor Meghnad Saha of Calcutta University, began to speak openly about his knowledge of "a large installation near Knoxville" involved with isotopic separation of uranium in order to produce "nuclear bombs." When they were interrogated by security officers, the Indians became, in the words of the officers, "rude and belligerent," insisting that "anyone with the slightest technical knowledge could plainly see that research in this field was going on and that therefore the treatment by the United States Army of this subject as a highly classified one appeared to be a very foolish thing." They resisted any efforts to find out who might have tipped them off, saying that "persistent questioning would be in the nature of an inquisition and they would object to such a procedure." The American agents claimed they had taken every effort to avoid offending the eminent and proud international visitors. The Indians agreed they would not talk further on the issue with anyone else, though they emphasized how silly they felt secrecy was in the face of the obvious size of the project. Groves concluded that the interaction was "most unsatisfactory and certainly indicated no feeling of respect for the wishes of their host, the United States government."[78]

Perhaps the worst of the Manhattan Project leaks, in terms of potential impact, came out in March 1944, and was the chance result of a reporter for the *Cleveland Press*, John Raper, taking a vacation in New Mexico. While in the Land of Enchantment, Raper stumbled upon a massive story: a "mystery city" called Los Alamos, surrounded by barbed wire and armed guards and subject to incredible compartmentalization of labor, doing secret military work led by the famous Dr. J. Robert Oppenheimer that occasionally involved "tremendous explosions." He enjoined his readers to look further: "If you like mysteries and have a keen desire to solve one, here is your opportunity to do a little sleuthing, and if you succeed in learning anything and then making it public you will satisfy the hot curiosity of several hundred thousand New Mexicans."[79]

Raper didn't know what they were working on, exactly, but he knew enough to be dangerous. Much of Raper's article was both specific and correct, and it would have taken little effort for an enemy spy to connect the dots. Groves' response was swift and furious. Aside from stopping syndication of the already-published story and interrogating the reporter, Groves even looked into getting Raper drafted. This failed, as the reporter was in his sixties.[80]

Despite the massive efforts of project security agents, high-profile news stories about atomic bombs still came out. Had the German or

Japanese intelligence services been looking for clues, they would have found them in published sources.[81] In August 1941, the *New York Times* and the *Los Angeles Times* had both reported that the president of the National Association of Science Writers had claimed that the government had "clapped a censorship" on questions relating to the military uses of uranium-235.[82] In December 1943, they would have seen, again in the *New York Times*, that "uranium" had been added to the list of censored commodities.[83] The Japanese might have noticed an article in the *Los Angeles Times* in July 1945, days after the Trinity test, which reported that "correspondents who attempted to describe [an atomic bomb's] effects were told that it would be dangerous to draw the Japs' attention to experiments in the use of such weapons."[84]

The fact that these news stories, along with the sudden drop in fission research publications, were apparently not pieced together by the Axis powers into a coherent understanding that the Manhattan Project existed might be taken as either a success of the Manhattan Project security agents or a failure of Axis intelligence agencies. But it should be noted that the trend here is easier to see in retrospect. Each of these articles also contained significant errors, and the information environment regarding the work on secret weapons in World War II was messy. If one wasn't looking closely, this could all be just part of the background noise of war.

Articles also appeared stating that the Germans had developed atomic bombs, "freeze bombs," beams that could shut off airplane engines, and many other fantastical weapons that turned out to be propagandistic fantasies.[85] Many rumors of Nazi research into atomic bombs also circulated in the American press during the war.[86] In January 1944, Arthur Compton gave a speech to the American Association for the Advancement of Science, later published in *Science*, assuring that it would be impossible for "some new weapon [to] be developed secretly on a small scale which is nevertheless so powerful that those who hold it will have the world at their mercy." Compton's language was careful: he didn't rule out the *weapon*, but he did rule out the *small scale*: "Such a development is difficult to hide. . . . If we are alert we should know of any new military development of this kind before it has become a hazard to nations organized to protect the public safety."[87] In retrospect, it is impossible not to see in Compton's statement a reflection on the difficult security operation he was working within.

When Groves would later claim that their system was meant to deter leaks and rumors more than spies, he might have framed it differently: they spent a disproportionate *effort* at trying to deter leaks and rumors, compared to the spies. The security officers of the Manhattan Project investigated "more than 1,500 'loose talk' or leakage of information cases," which is to say, almost two leaks per day for the entirety of the project.[88]

2.3 AVOIDING ACCOUNTABILITY

In the universe of threats to the Manhattan Project's success, the Germans and Japanese did not actually loom terribly large. Sabotage might have been able to stall aspects of the project, but the amount of duplication and geographic dispersal of the work meant that such efforts would have been unlikely to be very successful. The Pacific and Atlantic oceans protected the United States from long-range bombardment, meaning that its vast facilities were immune from significant enemy attack.[89]

There was, however, a group which Groves and others regarded as truly capable of hindering their work: the United States Congress. As previously noted, Vannevar Bush in particular felt that the fission program could not sustain scrutiny from suspicious congressmen, who would likely regard the entire matter as an indulgence of scientists at best. As the Army scaled up the project, the efforts to keep it outside the purview of external auditors increasingly ran into difficulty.[90]

In December 1942, Bush had requested to Roosevelt that some long-term source of funding be derived for their "special project." Bush pushed for Roosevelt to side-step formal budget requests: "It would be ruinous to the essential secrecy to have to defend before an appropriations committee any request for funds for this project and it is therefore recommended that some time in the spring you request the Congress for the needed funds ($315,000,000); such funds to be expended at your discretion."[91] But these were large sums, with even larger requests expected in the future, and they could not be indefinitely siphoned off without permission from Congress, which in principle controls the government purse strings. Bush met with members of the Congressional Appropriations Committee and gave them a vague outline of the work, on which they did not press him, to help grease the wheels.[92]

But congressional interest would build. In the spring of 1941, Missouri Senator Harry S. Truman had been appointed the chairman of a Senate Committee to Investigate the National Defense Program, soon to be better known as the "Truman Committee." The Truman Committee's goal was to investigate allegations of defense fraud and waste, and made a reputation for itself as impartial, efficient, and thorough.[93] In early June 1943, the Truman Committee heard rumors of a planned new plant in rural Washington State, operated by DuPont. A Truman Committee staffer sent inquiries to the president of the DuPont corporation, asking whether the rumor was true, and asking if he would "kindly state the product produced and the reasons for erecting a new plant for this purpose."[94] The DuPont people passed the request on to the War Department, where it fell to the secretary of war, Henry Stimson, to dissuade Truman from investigating the Hanford plant.[95]

Over the phone, Stimson told Truman that he would "have to trust me implicitly," because he was one of the few people who knew what the plant was for, "and I simply couldn't tell." Truman replied that he only wanted to know that the plant was for a specific purpose. Stimson said that not only was it for a specific purpose, it was for "a unique purpose." Truman agreed to back off.[96] There is evidence, though, that Truman was able to learn a vague idea of the project. In mid-July 1943, Truman wrote to a judge in Spokane about the Army seizure of land around the Hanford site: "I know something about that tremendous real estate deal, and I have been informed that it is for the construction of a plant to make a terrific explosion for a secret weapon that will be a wonder."[97]

Other attempts at audits would come. In September 1943, James F. Byrnes, director of the Office of War Mobilization, wrote to Stimson that he had heard about "the secret Army construction which is comprised under the category 'Manhattan,'" that was costing half a billion dollars. Vast resources appeared to be directed to a project that no one knew the purpose of, which made Byrnes suspicious:

> It would appear probable from these figures that more than one half of Army military construction will fall in the "Manhattan" category. . . . I know that the War Department may have some enterprise so important and so secret that it might be unwilling to divulge the purpose or details even to the Office of War Mobilization. However, you and I should assure ourselves that the projects included under "Manhattan" are of such

> character and that zealous officials do not use the convenience of the high priorities and secrecy attached to "Manhattan" for the purpose of securing material for unrelated projects.[98]

Again, Stimson had to personally intervene so that the project was not put under greater scrutiny. In November 1943, it was the House Military Affairs Committee investigating reports of massive construction at Oak Ridge.[99] That December, the Truman Committee sent another investigator to look into the Hanford project, because of "various rumors" another senator had forwarded to Truman. Two days later, a *Washington Post* columnist reported that the War Department was "doing all it can to throw secrecy barriers in Truman's path," and that the "half-a-billion dollar affair in the State of Washington" under investigation was "one of the largest single projects that's to be built from scratch in the Nation's history," a potent illustration of the dangers of these audits—even without getting information, they could draw attention.[100]

In January 1944, it was Senator Robert Taft who wanted information, worried that the Tennessee project was an attempt to lure key industries away from Ohio. In March 1944, four separate congressional inquiries were launched, as a result of complaints from constituents. The Truman Committee would *again* have to be dissuaded from probing, having sent an "ugly letter" to the War Department (in Stimson's reading of it). Stimson reminded him of his previous agreement to leave it alone. Truman was also told by the undersecretary of war that "it is the most important and most secret project that we have." In his diary, Stimson recorded that Truman "threatened me with dire consequences. I told him I had been directed by the President to do just what I did. Truman is a nuisance and a pretty untrustworthy man. He talks smoothly but he acts meanly."[101]

In February 1945, Congressman Albert J. Engel wrote a scathing letter denouncing the expenditure, and indicated that he had heard rumors that it involved "breaking down the atom" and that the resulting weapon could "'destroy Berlin and keep it burning for a year,' and other statements equally fantastic."[102] Unless he was given more information, Engel threatened to strike their budget requests down in a public session and make the scuttlebutt public. He tried to secure permission to visit the secret plants, but was turned away. Engel thought this a particular affront: "I told him that it seemed rather strange to me that while [the Undersecretary of War] permitted 60,000 workers, male and female, blacks, yellow and white, Mexicans, Chinamen and Negroes, men of every race, creed and color to go these plants daily, I, a Member of Congress and a member of the subcommittee which had the responsibility of handling these tremendous funds, was not permitted to even see what is going on."[103]

James F. Byrnes again attempted an audit in early March 1945, as the project approached the $2 billion mark, "with no definite assurance yet of production." Byrnes offered that while he knew that the project was "supported by eminent scientists," even such scientists "may continue a project rather than concede its failure." Feelings might be hurt, he continued, but two billion dollars "is enough money to risk such hurt."[104] Groves and Stimson immediately nixed the request for an independent audit—it would require a huge group of people to do an audit responsibly, Groves argued, and there were hardly enough nuclear physicists and chemists in the country who were not already connected to the project, making independent review impossible.[105]

In almost all these instances, Stimson had to personally intercede to deter congressional investigators with varying success. As the war went on, resources kept being directed to the mysterious "Manhattan District," whose name appears in dozens of congressional reports as a manpower-hungry, multiple-sited organization with "the highest production and labor priority ratings granted any war-production activity." Those trying to audit the project knew only that it was expensive and secret, and hadn't yet been credited with doing anything in Europe or Japan.[106]

In February 1944, Bush, Stimson, and Groves agreed that they should inform a few key congressmen about the nature of the project so that others could be referred to them whenever attempts to audit were made, and to make budget appropriations smoother.[107] The undersecretary of war had complained to Stimson around the same time that "it has been a constant problem, as you know, to get the necessary funds for this project from Congress and at the same time to safeguard the secrecy of the project. . . . There is a growing restlessness and impatience among Members of Congress on account of the size and cost of the project and also on account of the fact that they can find out nothing officially about the nature of it."[108] The audit attempts were starting to come in at a rate of nearly one a month.

On February 18, Bush, Stimson, and Army Chief of Staff George Marshall went to discuss the project with the senior members of the House of Representatives. They emphasized the importance of the work and its high level of scientific backing, and said that they were "probably in a race with the enemy." Bush indicated the "general magnitude of the destruction that could be caused" by the bomb; the House congressmen indicated their support.[109] That June, Stimson, Bush, and Major General George J. Richards, similarly met with high-ranking members of the Senate. Stimson told the group how the bomb would be made and that it might determine the outcome of the war "in case of deadlock." He further pointed out that they *had* thought they were in a race with Germany but were no longer so sure. The senators agreed that they would "maintain complete secrecy," and that they could guide the budget appropriation requests through the hearings.[110]

In early March 1945, Groves made an account of the total number of congressmen who had been officially informed about the bomb: seven. He felt that perhaps they ought to take a small delegation of senators and representatives to visit the sites at Hanford and Oak Ridge, so they could see that the worst rumors about waste or misspending were unfounded. Los Alamos was completely off limits. This visit would be desirable, he noted to Stimson, "from the security standpoint alone, as well as from the standpoint of minimizing the Congressional investigation with which we will be faced after the war."[111] This fear of postwar investigations animated Groves. In December 1944, he had sent an aide to look at the accounting records of the major project sites to see how they would be viewed after the war. The aide returned to Washington after a number of weeks and reported to the undersecretary of war that: "If the project succeeds, there won't be any investigation. If it doesn't, they won't investigate anything else."[112]

Shielding the bomb project from congressional inquiry, and from inquiry from other parts of the US government, had become part of an overall attempt to keep the bomb secret. Their primary fear was that a group of laymen like those in Congress would have a hard time accepting the expenses and effort being devoted to a speculative project on fission bombs. Though that fear can look well justified, when the War Department did seek out congressmen to let them "in on the secret," they got strong support. Sociologists of secrecy have long noted that the ritualistic aspects of being "let in" on a secret can reinforce loyalty—the oaths, hushed ceremony, and grave warnings—seemingly heighten a sense of tribal belonging and encourage complicity, and this may be part of the story.[113]

The number of congressmen so "let in" was purposefully kept to a minimum, some 1% of the total members of Congress. This was related to a second fear, that congressmen were poor secret-keepers and inveterate leakers (as true now as it was then). Truman's aforementioned letter to the Spokane judge in mid-July 1943, sent without any classification stamps, through the regular mail, and transcribed by a secretary without a security clearance, was exactly the sort of thing they were trying to avoid.[114] Even worse was the fact that the Army would have had difficulty enforcing secrecy regulations on congressmen. Scientists and technicians could be threatened with the Espionage Act, but congressmen were harder to intimidate, and attempts could easily backfire. For Groves, Stimson, Bush, and even Roosevelt, the importance of the bomb trumped the usual demands for transparent democratic institutions. It is this legacy that is perhaps the most chilling aspect of the Manhattan Project. Yet it is this aspect which, in the postwar period, was the least examined: Congress was willing to accept that it had to be left out of the loop, and that security might necessitate a reduction in accountability.[115]

2.4 THE PROBLEM OF SECRECY

The bomb was born under Roosevelt's dictum of "absolute secrecy," but for many of the Manhattan Project scientists and administrators, this was considered a strictly temporary condition, one brought on by the requirements of wartime. The difficulty of knowing what to do about secrecy after the bomb was revealed became frequently referred to as "the problem of secrecy": How could science be free in a world with an atomic bomb? How could the bomb be controlled if so much of its creation was based on science that could be researched by anyone, anywhere? What would the future hold for the bomb, for science, for the ideals of transparency and publication?

At various junctures during the wartime project, schemes for postwar control of the bomb were explored by project participants. Some of these would be implemented to some degree; some would die quiet deaths within the secrecy system they proposed to overthrow. Some look familiar to us today, while some are so out of joint with what actually developed that they can be surprising. Looking at these wartime visions of postwar control, two things stand out sharply: the meaning of the "atomic bomb" was not yet a fully known thing, and the achievement of long-term technological control through the control of knowledge or information—which is to say, secrecy—was considered by most of those involved in the bomb's creation to be a futile enterprise.

One of the most fecund sources of discussion about postwar secrecy was the scientists at the Chicago Metallurgical Laboratory from 1944 through 1945. The Met Lab had been crucial to early work on the Manhattan Project, especially in the construction of the first nuclear reactor (the "Chicago Pile") and helping to design and debug the reactors at the Hanford site. However, concerns about security, and about the political convictions of some of the Chicago scientists (like Leo Szilard), led Groves and others to create a new laboratory (Los Alamos) to do the design work of the weapon itself. As the project went on, many Chicago scientists migrated to other wartime sites, but many were kept at the Met Lab in a sort of scientific holding pen, where they would still be under the constraints of security but their access to weapons work was limited. The result was that many of these scientists, well acquainted with the bomb program but unable to further contribute to it technically, began thinking seriously about what would come next.[116]

The fact that they were being compartmentalized out of actual policy influence may have itself led the scientists to feel especially embittered toward secrecy. In an April 1944 report on the future of the Met Lab's work, the Chicago scientists included a lengthy study of "Peacetime Plans" by the Princeton physicist Henry DeWolf Smyth. Much of Smyth's image of the future was colored by the fact that scientific work had been "internationally and freely published" before the war. The atomic program of the United States, he argued, had begun with European instigation, and would require more than simply American effort to maintain. As a result, any improvement in American nuclear research, including military work, "must ultimately depend [on] the encouragement of fundamental unrestricted research, and the training of men capable of doing such research."[117]

Smyth saw the bomb as coming directly from basic and fundamental research, and thus reasoned that in order to stay ahead, the United States would have to further invest in more basic research. That the Manhattan Project approach could not continue into the postwar period he considered obvious, as its monopolization of university scientists would make the training of new scientists impossible, and no good research scientists would want to join up with such a practically-minded, government-run entity. He further argued that that the "present rules of secrecy would prove stultifying in the long run if they continued to apply to all phases of the work as at present." The "secrecy problem" was "so difficult," he felt, that it required "considerable discussion" apart from the rest of the postwar issues. His proposal for the postwar period included separating the research objectives into separate laboratories, each with differing degrees of classification. But as he admitted, it was still "pretty vague."[118]

Another Met Lab Committee took a more strident take. In July 1944, Arthur H. Compton had appointed a committee chaired by Zay Jeffries to look into the future of atomic energy (which they were at that point calling "nucleonics"). In November 1944, a final report had been prepared by the Jeffries Committee that contemplated a variety of postwar prospects for the field and its organization. They too feared a secret arms race, but their emphasis was on the problems the United States would face if it attempted to stifle its scientists. They found it "both unlikely and undesirable that the whole development of nucleonics should be restricted to those government-sponsored laboratories, under the protection of continued wartime secrecy," and that "full information on most phases of the subject should be released just as soon as possible from the standpoint of national security."[119]

Smyth would himself return to this issue in a March 1945 report on "The Problem of Secrecy and the Future of the DSM [Manhattan] Project." This was written after Smyth had worked for some time on writing up a technical history of the Manhattan Project intended for future public release (discussed in the next chapter), and talking with project scientists about both their work and secrecy. This added some practical depth and nuance to his views. He started from a familiar proposition: too much information was out there to be restricted by secrecy in the postwar period, and government work on the atom would necessarily decrease as the Manhattan Project inevitably disbanded. "The problem, therefore," he reasoned, "is not one of continuing secrecy in an organization like the present one but of what degree of secrecy can and should be maintained in peacetime work in this field."[120]

Smyth divided up the possible positions on secrecy into two major points of view: the "idealistic" view ("dear to scientists") that the work of science required the objective study of nature and dissemination of information without restriction, and the "military or nationalistic" view, which held that future wars were inevitable and that it was the duty of the United States to maintain the strongest military position.

Smyth found the first point of view an easy one from a policy standpoint, since it dictated simply eliminating all secrecy. It would be more tricky to figure out the boundaries of the second point of view, because "at first sight it might appear that the best policy [for military superiority] would be to maintain the present secrecy restrictions, but further examination of the consequences of such a policy shows that it would be neither practicable nor desirable." The problem was similar to the one he had outlined in his first report: university scientists would return to the universities after the war, and new people would not be drawn into a field stifled by secrecy. He dismissed as impossible that "the present cloak of secrecy" should be maintained by swearing all scientists into a secret "guild." America was, and would remain, a free country.

Thus, what was necessary was to demarcate which parts of the current Manhattan Project would return to free circulation after the war, and which parts would stay cloistered. On this point, Smyth considered the many scientific and technical developments that had been achieved and weighed the values of their disclosure or retention. Furthermore, if Manhattan Project papers could be downgraded in classification to "Restricted" or "Confidential" (which have far fewer restrictions than "Secret," much less "Top Secret") it would allow wider circulation during the "transition" period that would inevitably occur after the wartime regime ended.[121] As will be clear in the next chapter, Smyth's shift toward a demarcating approach likely owes much to the fact that he had been recently engaged in trying to make exactly these sorts of judgments for his technical history of the project; it is one thing to imagine eliminating secrecy in the abstract, it is another to actually try to apply these principles to something concrete.

Smyth's formulation of "the problem of secrecy"—as a problem of university scientists being incompatible with military requirements for secrecy, and as a problem of balance between staying ahead and keeping quiet—would eventually become the dominant postwar approach to thinking about secrecy. While it took the safety of technical secrets seriously, it also held that scientific progress could not tolerate such controls, and that ultimately any form of scientific secrecy would prove to be ineffectual, given the fact that science comprised universal truths. That such a mindset could come to being by those within the throes of the "absolute secrecy" wartime regime is not as unexpected as it might sound: it was a combination of pre-war sentiments about science and secrecy with the recognition that, like it or not, a massive corpus of secrets had been created and would, to some extent, likely continue to exist afterward.

Most wartime scientific assessments of secrecy, at least the ones that got written down, doubted the practicality and efficacy of secrecy. But there was at least one curious exception to this. William A. Shurcliff, a physicist working under Vannevar Bush, wrote numerous unsolicited memos on postwar secrecy. Shurcliff had been working as a senior technical aide at the Liaison Office of the OSRD, where he was in charge of disseminating intelligence information about enemy technical developments to the relevant divisions of the OSRD, and was brought into the Manhattan Project in May 1942 in connection with his work as a "censor" of private patents that intersected with project goals.[122] Shurcliff was informed about the bomb, but had an outsider's perspective: his role was administrative, and he was not actually involved in the direct work on the bomb, though he was informed about its scientific aspects and was a physicist himself.[123] He was also a frequent writer of unsolicited memos, which Bush, his boss, seems to have accepted with a mixture of tolerance and gratitude.

Shurcliff wrote internal memoranda on the subject of secrecy multiple times. In an "Informal Memorandum" on postwar atomic policies in March 1944, Shurcliff argued that maintaining secrecy over the long term would probably be called for, though some leakage was inevitable (it would be "almost irresistible" for people "in the know" to reveal it at some point), and that no doubt confusion would be caused when scientists "rediscovered" classified principles:

> "Rediscoverers" located in [the] USA may be expected to tender their "discoveries" to the government and to feel puzzled and even resentful if the government expresses no interest or places the entire matter under secrecy and then expresses no interest. By repeatedly pestering the government (and perhaps press and congressmen also) with ideas already well known to the government and perhaps already in use by it, the rediscoverers will soon be able to make a fair estimate of the secret project.[124]

Bush himself read Shurcliff's above memo and thanked him for it, noting that any information flow between the two "has to be a somewhat one-way affair," but that he would be interested in hearing more of Shurcliff's thoughts on the matter in the future.[125] Bush also forwarded Shurcliff's memo to Conant, noting that Shurcliff's memo had contained "a few interesting thoughts, but I think none of great importance that we have not already been thinking about."[126]

In December 1944, Shurcliff wrote another memo, this one an analysis of the claims that "Maintaining secrecy on details of the present weapon will not insure security" and that "Security will come from 'keeping ahead.'" Shurcliff sent this memo to the physicist Richard Tolman, a technical advisor to Groves and the chair of a Committee on Postwar Policy, for whom Shurcliff also occasionally worked. Shurcliff noted that this point of view had been put forward to the Committee by numerous scientists. But from Shurcliff's perspective, while the statements were "more true than false," it was "apparent that they are seriously inadequate and to an appreciable extent misleading."[127]

Shurcliff argued that the first thesis, that security could not be ensured by keeping the bomb secret, was misleading because you probably *could* extend the time it would take for another nation to develop the bomb by holding the "secrets" close. Whereas the anti-secrecy scientists framed the issue as whether secrecy could be used to *totally deny* another nation the bomb, Shurcliff noted that merely *buying time* could be an end unto itself. Shurcliff admitted that "popular interest" items would leak out, and that there may be espionage, but noted that from his experience with dealing with captured enemy secrets in the Liaison Office, it was quite hard for secret information to be translated into a technological understanding unless "reasonably-intact specimens" were available. This is a sophisticated point: as historians of science and technology have noted for many years, "explicit knowledge" (e.g., blueprints, formulae, and other written information) is by itself often insufficient for technology transfer. Physical specimens, local conditions, and "tacit knowledge" (e.g., know-how, experience, and patterns of thought) are often required for successful replication of complex technical inventions.[128]

In general, Shurcliff felt that any secrecy that would delay enemy countries probably *would* translate into security gained for the United States. He acknowledged that a dedicated nation could simply re-create all the work that the United States had done, but he judged the scale of this endeavor to be sufficiently large that it would be detectable. On the second thesis, that security would come from "keeping ahead," he felt that in the Atomic Age, perhaps this did not matter: if the bomb could destroy entire cities, what difference did it make if it was an older, primitive model or a new, modern model? Total destruction was total destruction. In the end, Shurcliff felt that the conservative position would be to maintain a strong degree of secrecy: "To place one's faith in secrecy may be rash, but appreciably to dispense with secrecy may be even more rash."[129]

Bush considered Shurcliff to have a "pessimistic viewpoint," but one that ought to be considered seriously.[130] Tolman's Committee on Postwar Policy would finish its own lengthy report two weeks later. The intent of its final report was to focus on narrow technical and administrative grounds regarding the future of the Manhattan Project work. Its discussion of the "postwar policy of secrecy" was similar to Shurcliff's analysis:

> [T]he Committee is of the opinion that much information as to scientific and technical results, as to methods of manufacturing active material, as to the nature and mode of use of military weapons, and as to locations and time schedules will still have to be kept as secret as practical in the postwar period. The frequent argument—that the information will leak out anyhow—does not mean that its dissemination cannot be importantly delayed by an appropriate security policy. And the other frequent argument—that military safety depends primarily on keeping ahead in the development of superior nuclear weapons—does not cover the whole story in a situation where even less well developed weapons

may nevertheless be sufficient to produce disastrous results if used just prior to a formal declaration of war.[131]

The Committee, however, did note that too much secrecy in "fundamental research" would be "stultifying." In the absence of an international treaty to limit nuclear weapons production, each proposed piece of information would require "careful consideration . . . to the relative advantages and disadvantages" of its release.[132]

As will be discussed chapter 4, Bush and Conant were both extremely skeptical of the possibility of postwar secrecy and assumed that it would necessarily be avoided. Smyth and the Met Lab scientists were similarly opposed, as were prominent scientists like Niels Bohr. Even the notoriously hawkish Edward Teller would in July 1945 argue that the "only cause" with regard to future nuclear policy he felt "entitled" to work for was in the dissolution of secrecy, arguing that "the accident that we worked out this dreadful thing should not give us the responsibility of having a voice in how it is to be used."[133]

Shurcliff's is one of the few scientific opinions on record that argued in favor of secrecy, but it was essentially his view that Tolman's Committee on Postwar Policy adopted, and the latter's conservatism no doubt was influenced by Groves' expectations. The predominant scientists' view would have required a total commitment and total confidence: if secrets did matter, then releasing them would be disastrous, for they could not be "recalled" once let out. If secrets were of marginal importance to domestic development, the only detriment would be slowed American innovation, but, as Shurcliff pointed out, this was significant only if you believed that a more primitive atomic bomb could be made irrelevant by a more sophisticated one. The positions of the scientists against secrecy were probably purposefully extreme, though, in anticipation that the "conservative" position would likely be the default, and their appeals to the importance of openness as a scientific norm were also blatant appeals for a restoration of their own lost autonomy.[134]

On April 12, 1945, Franklin Roosevelt died, and Harry S. Truman, his vice president since the 1944 election, succeeded him as the new president. Roosevelt had told Truman nothing about the atomic bomb work, something even more striking when compared to Truman's predecessor, Henry Wallace, who had been on the Top Policy Group that had overseen the creation of the Manhattan Project. Roosevelt's motivation for excluding Truman is unknown: it is yet another unexplained, possibly capricious application of secrecy from a president fond of such things.[135]

Just after the first cabinet meeting of the new administration, Stimson pulled Truman aside and cryptically mentioned that there was a new weapon he needed to be told about—at a future date.[136] A full discussion would wait for two weeks, when Stimson set up a meeting with himself, Truman, and General Groves at the White House.[137] Both Stimson and Groves had brought reports for Truman to read.[138] Stimson's was to the point: within a few months, the United States will likely have "the most terrible weapon ever known in human history, one bomb of which could destroy a whole city," but once used, it would only be a matter of time until other nations acquired their own. Long-term thinking would need to be applied to avoid the destruction of civilization.[139]

Groves' memo was considerably longer and more technical. It offered, in twenty-four double-spaced pages, the progress of the work to produce nuclear weapons and a primer on the scientific concepts behind the "weapon of tremendous power" they were building. It gave the schedule of work to be done: one weapon ready to be tested in July, with another ready for use in early August, with more to follow at regular intervals. It described the diplomatic arrangements made with Great Britain and noted that Russia had, since early 1943, mobilized its "diplomatic, information and espionage groups" to "secure particularized information concerning the project." It invoked Roosevelt's personal order for the "extraordinary secrecy and security measures for all phases of the project." Neither Germany nor Japan, it explained, were in any position to use atomic bombs, despite earlier fears about the former. That the bombs would be used was taken for granted; "the target is and always was expected to be Japan."[140] Truman expressed his satisfaction with the work done so far and the direction it was going, as well as his amusement at having spent so much time trying to learn about the project when he was a senator, telling Stimson that he "understood now perfectly" why his inquiries had been repeatedly thwarted.[141]

Over a decade later, in 1958, General Groves attempted to itemize the "major objectives" for his attitudes toward "secrecy in atomic matters" during the war.[142] First on Groves' list was keeping information about the Manhattan Project from the Germans and, "to a lesser degree," the Japanese. There were three intertwined fears here: that the Nazis would intensify their own efforts if they knew the US was working on a bomb; that they might be able to use facts about the American effort to shape their own program; and that if they were aware of the locations of American efforts or the names of principal figures, they could sabotage the work. As it turned out, the German atomic program was quite modest and nowhere close to creating a bomb by the end of the war. (The fact that the Germans were *not* using Manhattan Project–levels of secrecy was interpreted by American analysts as an indication of their lack of seriousness in building a weapon.[143])

The Japanese were an issue for Groves not because he was afraid they would make a bomb, but because of the desire for "military surprise when the bomb was used," a "psychological effect." Groves (and others) believed that a "shock" strategy would be the best means of forcing an unconditional surrender by Japanese leadership and that the bomb's effect would be lessened if its existence were known in advance. The potential of sabotage, especially of West Coast sites, was also part of this consideration.[144]

Next on Groves' list was Russia. Groves inherently distrusted the Soviet Union, though they were an ally at the time, and anticipated that an American monopoly on the means to develop nuclear weapons would be a potential wedge in postwar negotiations. Both Roosevelt and Truman shared this sentiment. He also had indications that the Soviets were interested in American work on fission; at least as early as 1943, Groves was worried about Russian espionage efforts around Berkeley.[145] This also went along with his next "objective": "To keep as much knowledge as possible from all other nations, so that the U.S. position after the war would be as strong as possible." These unspecified "other nations" surely included the United Kingdom, even though they were a partner in the project, and France, which were in Groves' sights during the war as future nuclear nations. In Groves' retrospective accounting, wartime secrecy was explicitly intended, to some degree, to influence the postwar situation.

After specifying the "external" threats, Groves then turned to the "internal." Compartmentalization, he explained to his son, could be used "to keep members of the project working on their own tasks, rather than thinking and worrying about the tasks of others." Again, this was an *organizational* use of secrecy practices, as a form of *labor* practice, reflecting, no doubt, Groves' particular disdain for the unruly academics. If you didn't keep such scientists on track, they would end up treating the project as "an advanced post-doctorate university."

Groves also expressed his desire to keep knowledge "from the hands of those who would interfere directly or indirectly with the progress of the work"—a vague category into which he put any inquiries from scientists outside the project, or even members of Congress. His last "objective" is probably a bit more to the point: he wanted to "keep from having a great political discussion as to how such a weapon could (or should) be used," because "this would have stopped all progress." Lest we attribute this exclusively to Groves' military mind, there is evidence that this sentiment was shared by both Vannevar Bush and President Roosevelt, who constrained any discussion of "policy" to the top levels of the hierarchy. Deliberation on something as important as the atomic bomb would preclude the possibility of its development and use.

In the postwar period, the Manhattan Project would be described in fawning coverage by its employees as the "best-kept secret of the war," largely to give credit to the Office of Censorship for their assistance.[146] Ironically, the Office of Censorship's efforts appear to have played only a minor role in keeping the bomb secret—whatever successes were had were the result of numerous secrecy practices spread out across what was essentially a brand-new industry that spanned the United States. Did Groves succeed at his goals? With some, we can probably give him a passing grade: the Japanese and Germans seemed genuinely shocked by the existence of the bomb, despite all the leaks.[147] Congress never shut the project down. But the Soviet Union was not so effectively kept out of the "secret," and subsequent openings of Russian state archives have revealed that they were able to get their hands on a wide variety of information about weapons design, scientific discoveries, and even fissile material production.[148] So we might clarify our assessment: the Manhattan Project secrecy practices did an acceptable job of maintaining the secret enough for wartime purposes, even if it did not attain all of its goals. While there is no concrete metric of how compartmentalized the project was, my own estimate is that no more than a few thousand of the project's employees (so, on the order of 1%) knew what they were building prior to the attack on Hiroshima.[149]

We should perhaps not be too hard in judging the project's failures to keep all information tightly controlled, however. "Absolute secrecy" was truly an impossible mandate. Never before or since has a project so large, both in terms of resources and employment, tried to keep a secret so basic: that the United States was researching a new weapon. In the Cold War that followed, the United States never attempted such a feat. While weapons details were kept secret, the fact that they were being developed was largely not. The Manhattan Project was an attempt to simultaneously create a new industrial empire from scratch, while also attempting to keep the fact of that creation a totalizing secret from all parties, including the press, an entire branch of the US government, and literally *all* foreign governments besides the United Kingdom. That they succeeded as much as they did is the true wonder, even if was something less than the "best-kept secret of the war."

The sheer novelty of the amount of secrecy involved in the Manhattan Project was something felt both by its participants and by external observers. The newness of it is plainly clear in some of its more mundane traces. Classified "cover sheets" would become a staple of the US national security state, indicating at a glance to anyone who came into contact with them that they were entering into the province of secrets. The "Top Secret" cover sheets used by Groves featured an elaborate hand-lettering more reminiscent of a teenager's binder than national security secrets.

And yet, despite its crudity, it accomplishes the task: aside from being a visual warning, it also serves as a record of document usage, reinforcing a secrecy regime through practices of document accountability and circulation. Such practices would indeed be useful, years later, in tracking down the spies that Groves had no inkling were in his project's midst, allowing the FBI to see which scientists had access to which reports.

Though Groves does deserve much of the credit for the expansive secrecy regime of the Manhattan Project, such an enterprise would not have been feasible were it not for the cooperation of literally thousands of others. A secrecy regime is not merely those who enforce it; it con-

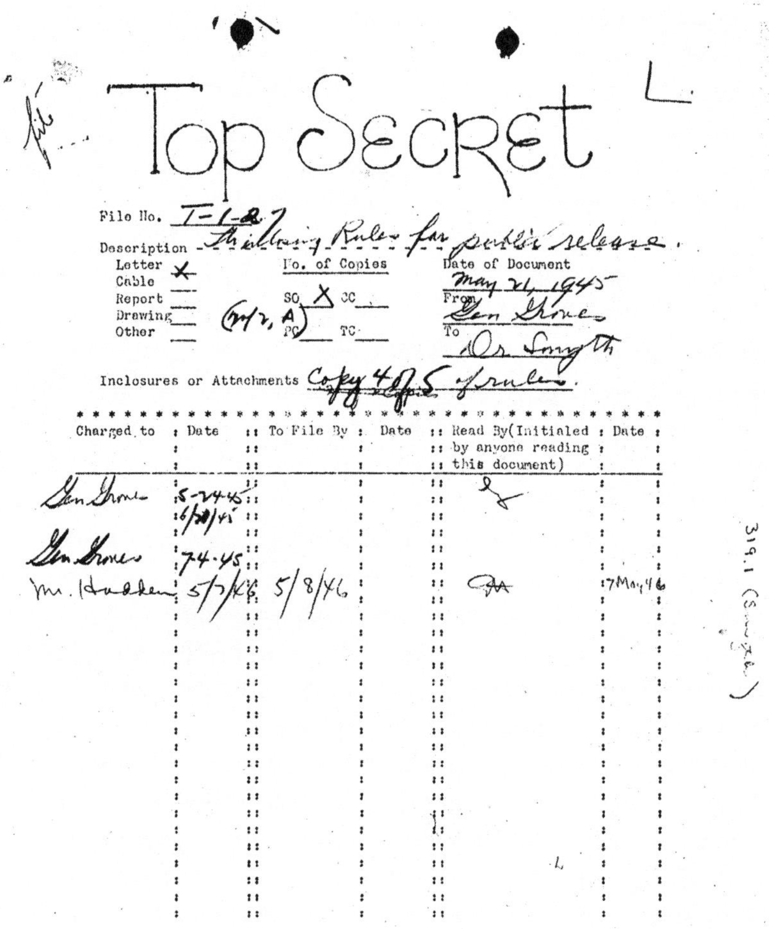

Top Secret

File No. T-1-27

Description Following Rules for public release.

Letter X	No. of Copies	Date of Document May 21, 1945
Cable	SO X CC	From Gen Groves
Report	(1/2, A) PC TC	To Dr. Smyth
Drawing		
Other		

Inclosures or Attachments Copy 4 of 5 of rules.

Charged to	Date	To File By	Date	Read By (Initialed by anyone reading this document)	Date
Gen Groves	5-24-45 6/20/45			[illegible]	
Gen Groves	7-4-45				
Mr. Hadden	5/7/46	5/8/46		[illegible]	7 May 46

L

319.1 ([illegible])

FIGURE 2.4. An example of the "Top Secret" cover sheets present in Groves' files, one of the tools used for the enforcement of the secrecy regime. (The "L" may indicated "Limited.") This particular cover sheet pertains to a letter from Groves to Henry D. Smyth, establishing rules for the Smyth Report. Source: CTS, Roll 2, Target 6, Folder 12: "Intelligence and Security."

tains all of those who live within it, and all of those who agree not to pester it from the outside. Groves was remarkably successful in pushing for the control of information, scientists, and even other parts of the US government and military. In the process, however, he made many enemies, including many of those among the scientists who believed that perhaps such shackles could be tolerated for the expediency of war, but were dead-set on opposing any attempts to continue it into a peacetime period. And the greatest challenge of all to the secrecy regime still awaited: how to manage the fact that in the hours after Hiroshima, the world would not only know that atomic bombs existed, but would enter into a frenzied search for more information.

3

PREPARING FOR "PUBLICITY DAY"

A WARTIME SECRET REVEALED, 1944-1945

It has never been the habit of the scientists of this country or the policy of this Government to withhold from the world scientific knowledge.

"STATEMENT BY THE PRESIDENT ANNOUNCING THE USE OF THE A-BOMB AT HIROSHIMA," AUGUST 6, 1945[1]

The Manhattan Project had been conducted under the mandate of "absolute secrecy." But the use of the atomic bomb on Japan would necessarily result in a huge release of information, first and foremost that the atomic bomb could be, and had been, built. For the wartime scientists, administrators, military officials, and even political leaders, this prospect promised new complications and problems. How much information should be released, and how much should be held back, if any? In what form should releases be orchestrated? Under what conditions could any information be held back, once the wall had been breached? These discussions revolved around a general problem termed "Publicity," and out of these considerations arose fundamental and competing ideas about nuclear secrecy in a world where nuclear bombs were known to exist.

The imminent release of information about the atomic bomb following its use occupied a considerable amount of time of high-level planners, and became tied up into larger policy questions about how the weapons ought to be used, and *if* they ought to be used. Many of the scientists involved in the project, even at high levels, had assumed that the wartime secrecy would be essentially temporary, and evaporate along with the need to keep the Axis powers in the dark. But in trying to outline exactly what a secrecy policy would look like after the first use of the bomb, the Manhattan Project officials would end up taking the first steps toward a permanent, more-flexible secrecy system that could work under peacetime conditions and over a longer duration than the Manhattan Project itself.

3.1 THE FIRST HISTORY OF THE ATOMIC BOMB

No single aspect of the "Publicity" campaign illustrates the dimensions of the issue more than the publication of the Smyth Report, the first technical history of the Manhattan Project. It was a historically unprecedented document: within days of a new, secret technology being revealed and used for the first time, the same government that developed said technology released a book-length treatment of the key phases of its development and an explanation of how it worked. It was also immensely controversial, as later scientists and politicians questioned whether it revealed too much. Despite its rather modest origins, the Smyth Report would become a focal point for the "Publicity" strategy, both in its motivations—an apparently contradictory mix of civic ideals and security concerns—and the difficult problems and questions it raised about how one would divide up nuclear information once one stopped treating it in "absolute" terms.[2]

The genesis of the Smyth Report can be traced to the spring of 1944, a period in which the Manhattan Project was in full swing, though still over a year away from success. That March, Vannevar Bush wrote a memo to James Conant asking about whether they ought to consider getting a "historian" for the Manhattan Project. He was aware that Groves was creating a wholly-secret internal accounting of the work of the project, should he be asked to give an accounting by Congress.[3] But Bush thought it might be worth having this supplemented by "some scientific historian, not that the whole thing could probably be later published, but rather that it would form a good basis for such parts as should be published at the appropriate time."[4]

Conant enthusiastically embraced the thought. He reported having had similar ideas, albeit toward somewhat different ends:

> I feel that such a report should be ready for issuance to the public at the time when the President of the United States is willing to announce that the gadget has been successfully employed, or it has been proved that it can be successfully employed, or at such time as he may be forced to make a revelation, even if neither of the first two objectives have been reached.[5]

Furthermore it would be "of great importance to the security of the essential military secrets and to the furtherance of rational public discussion to have issued at a given time a first-class document." The report could, he argued, serve as a means to rein in leaks of information:

> When such a document were [to be] issued, those in charge of the classified material which has accumulated in this project could decree that no one could reveal or discuss any details of the project which were not covered in the document issued officially through the President. This would tend to put a limit to discussion, both privately and in the Sunday papers, but at the same time provide sufficient material so that the national and international aspects of the project could be debated with a considerable degree of intelligence.[6]

Conant's vision, then, was of a history that would serve both as a basis for public understanding of the bomb as well as an indicator of the boundaries of what could be said in public. It was an idea he felt would appeal to scientists and military men alike: a disclosure, but one that would ultimately retain secrecy. This dual purpose—half democratic idealism, half security statement—sat at the heart of all later "Publicity" efforts. For Conant, what mattered most from a security standpoint was the "knowledge of the design, construction and operation of these plants" for producing fissile material, which he considered "a military secret which is in a totally different class from anything the world has ever seen."[7]

For the historian, Bush initially suggested Karl K. Darrow, a physicist, the secretary of the American Physical Society, and author of several popular books. Conant, however, wanted someone closer at hand: Henry DeWolf Smyth, the chair of the Department of Physics at Princeton University, who had been involved with the Manhattan Project in several different capacities since the summer of 1941 when the work had begun to expand. At Princeton he had researched an abandoned approach to uranium enrichment (the "isotron"), and had since been helping with administrative aspects of the Metallurgical Laboratory in Chicago as its associate director. At Chicago, he had been involved with drafting papers on postwar policy planning, including the thorny question of postwar secrecy and the difficulties it presented given both the "peculiar" nature of scientists, as well as the fact that the atomic bomb was based on scientific facts discovered prior to the war.[8] Smyth later recollected that he may have mentioned a similar idea to Arthur Compton and had discussed it with Conant around the same time Bush had suggested the idea. The idea of a public history of the bomb was "in the air" in the spring of 1944, as the project heads contemplated the trickiness of a post-secrecy future.

Conant convinced Groves of the idea's worth, and Groves asked Smyth to be their historian in April 1944.[9] Smyth accepted and threw himself into the work. He was granted unusual freedom to circumvent the compartmentalization policies that prevented others from gaining a "full view" of the work. He visited the far-flung project sites, looked over their records, and interviewed their participants. The initial report was written in a largely uncensored fashion, for Smyth felt that decisions about the security contents of the report would have to be made in a principled way, and the entire report edited at the end according to these decisions. Only in the area of bomb design itself did Smyth intentionally censor himself from the beginning.[10]

Throughout 1944 and into early 1945, Smyth labored, writing a history of a project that was still very much itself in progress. Drafts of what would become the Smyth Report were circulated to the site leaders, who then provided revisions or suggestions. Oppenheimer, after seeing the chapter on Los Alamos, felt it was "somewhat spotty," given all that had to be omitted about bomb design.[11] The chapter would have even more redactions before Smyth was done, for the original had discussed implosion, which was banned entirely by the time the report was finalized. Oppenheimer, in the end, would be the only project leader to object to publication of his appointed chapter, arguing that it would be better to not release it at all, since it "actually gives a rather misleading impression of the work here."[12] Oppenheimer's objections were overruled, though a short statement was added to emphasize that much was missing from the account.[13] As Smyth wrote to Oppenheimer, "I have not found the writing of this report an easy assignment."[14]

Smyth's job, as Groves saw it, was "writing a description of the entire project including the scientific credits to the numerous scientists who have been responsible for the different phases of the various developments." The work would be written "for public release either in its entirety or in abbreviated form if and when it becomes necessary or desirable to explain to the American people what we have been doing, what we have accomplished and who is responsible for the achievements."[15] This latter focus on *credit* appears to be largely Groves' own obsession and conception of the value of the report. For Groves, giving credit was not just a moral compulsion, but a security need. Groves feared that people seeking credit, especially after the bomb's use, would reveal much in the process.

To this end, it is interesting to note to whom the Smyth Report gives the lion's share of credit for making the atomic bomb: the physicists and physics as a discipline. As the historian Rebecca Press Schwartz has argued, Smyth's focus on physics likely had multiple origins.[16] Smyth was himself a physicist, and clearly did see the physical nature of the atomic bomb as what distinguished it from all other past weapons. Without nuclear fission, there would be no bomb. But this focus also obscured the contributions of chemists, metallurgists, and engineers to the project. Complex scientific-industrial processes such as the atomic bomb involve a multitude of experts of all sorts, and within the Manhattan Project the organizational charts reflected an awareness that physics was hardly king, even though many of the top scientist-administrators were indeed physicists.[17] As a point of fact, even at Los Alamos (much less Hanford or Oak Ridge), engineers and chemists handily outnumbered physicists.[18]

There was another reason for the emphasis on physics: it was, ironically, less secret than other aspects of the bomb. Atomic energy had been discovered before the war, and the basic nuclear physics was not considered a secret. The chemistry and metallurgy both often involved esoteric substances like plutonium that were hard to produce without large-scale facilities, and the engineering is what helped produce said facilities. As a result, the final report by Smyth reads as a dull treatise on administrative decisions about the building of large plants, with small-scale physical explanations of the basic operations. Concentration of attention at the very high level (administrative choices) and the very low level (basic physics) preserved technical secrecy, because the true difficulty of making an atomic bomb involves processes in the middle of these two extremes.[19] Smyth's drawing of a nearly straight line between $E = mc^2$ and the massive production plants deliberately avoided the "know-how" that he and Groves thought mattered the most. As Press Schwartz has argued, the focus on physics made the classification problem easier, though a side effect was that the atomic bomb became exclusively associated with physics, despite the important contributions from other fields.

The question of whether the report would even be released was not decided prior to the bomb's use. It was not a foregone conclusion that releasing such a detailed history, even with its omissions, was desirable for the postwar position of the United States. Bush and Conant supported its wide dissemination, and sought as early as September 1944, nearly a year before the bombing of Hiroshima, to plant the idea of its release into the mind of Secretary of War Stimson. Their push was part of an effort to encourage high-level thinking about the thorny issues of postwar atomic policy and the "problem of secrecy."[20]

As work on the bomb neared completion, the issue of what to release and what not to release became more pressing. Groves, along with Richard C. Tolman, the Caltech physicist who served as his personal technical advisor and chaired the Committee on Postwar Policy, drew up security guidelines for the Smyth Report in May 1945. These were the first attempts by any high-level Manhattan Project participants to draw a line between what was safe to release in the postwar and what was not—the first classification taxonomy of the Atomic Age.[21]

The basic idea behind the "rules for governing the scientific release," was that information would be released only if it seemed like the attempt at secrecy would be pointless. If the information was already public (i.e., published before the war), or could be easily discovered by a small team of scientists, then secrecy wasn't worth it. No information would be released unless it was necessary to understand the project as a whole or was of great value to science. If a given fact or statement met

TABLE 3.1. The Groves-Tolman "Rules Governing the Scientific Release" for the Smyth Report, as formulated in May 1945. Note that to be releasable, a given statement must satisfy at least one of the criteria in both sections "I" and "II." Source: Leslie R. Groves to Henry D. Smyth (21 May 1945), CTS, Roll 2, Target 6, Folder 12, "Intelligence and Security."

The information to be included in a release will exclude all matters connected with the construction of the actual bomb. Any information disclosed must satisfy one of the detailed requirements in each of the two following groups:

I

(a) That it is important to a reasonable understanding of what has been done on the project as a whole

or

(b) That it is of true scientific interest and likely to be truly helpful to scientific workers in this country.

and

II

(a) That it is already known generally by competent scientists

or

(b) That it can be deduced or guessed by competent scientists from what is already known, combined with the knowledge that the project was in the overall successful.

or

(c) That it has no real bearing on the production of atomic bombs.

Or in a limited number of cases (say 5) and these will be reported in a separate memo so that they can be eliminated if desired[:]

(d) That it could not be discovered by a small group (fifteen, of whom not over five would be senior men) of competent scientists working in a well-equipped college laboratory in a year's time or less.

---

the criteria, then it was free to release. If not, then it would stay secret for the time being.

Though these "rules" appear at first glance to be very sensible, they were controversial in the postwar period, and since. There is a narrowness to this approach, one that allows large structural elements of importance to slip through, while smaller, more isolated facts might not. One might say this is a physicist's, and not an administrator's, view of how science works: science as a collection of facts, not an organization of people, or a system of knowledge and technical production.

Critics of the Smyth Report in the postwar period would charge that the real secrets of the Manhattan Project were not in its individual facts, but in the big picture. The Smyth Report, Leo Szilard would later argue, "clearly indicates the road along which any other nations will have to travel," including which methods of uranium enrichment the United States found successful, that the generation of plutonium from reactors is feasible, and a sense of the investment required for each method. "If they do travel along that road," Szilard continued, "they will step by step rediscover what we have discovered, and step by step they will obtain the same results we have obtained."[22]

During the summer of 1945, Smyth and Tolman had gone over the entirety of Smyth's drafts with an eye for final edits and security questions, applying Groves' guidelines as they interpreted them. By the end of July, they had returned it to Groves, with notes about issues that sat on the borderline of the security guidelines, including places where they had been intentionally vague in the name of security.[23] Further editing would continue down to the wire. The physicist William A. Shurcliff worked as copy editor, improving on Smyth's terse, "awkward," style. "He seemed not to have heard of topic sentences," Shurcliff later recalled. A secretary then typed the entire book onto mimeograph forms, and Shurcliff and another Manhattan Project assistant mimeographed fifty copies themselves, unable to delegate the tedious job because no other available clerks had "Secret-Limited" clearance. These copies were circulated to Groves and his staff, and minor changes were made before it was typeset and 1,000 copies were printed under high security—a week before the bombing of Hiroshima.[24] The work was to have a simple, evocative title: "Atomic Bombs." Fears of a leak dictated that the title be left off the cover page until release had been approved. The title would be applied, at the last moment, by a large red stamp.[25]

Aside from its unprecedented nature—never before or since has a history of a secret weapon been written simultaneous to its development—the Smyth Report was remarkable as a catalyst for practical thinking on secrecy and publicity in the Manhattan Project. It imagined a world in which the atomic bomb was no longer an "absolute" secret and challenged both military and scientific members of the project to contemplate what kind of information regime that world would try to maintain. In some ways it was remarkably liberal, describing the infrastructural complex necessary to develop the new weapon and roughly outlining how it worked, while in other respects it is fairly conservative, restricting much of its technical content to what was already known or obvious by global scientists. It was not a comprehensive secrecy regime: it was a one-time release, contemplated for a time when the biggest secret—that atomic bombs existed—was already out. And as we will see later in this chapter, the question of whether it would be released at all was not yet settled.

3.2 PRESS RELEASES, PUBLIC RELATIONS, AND PURPLE PROSE

Though the Smyth Report was the earliest (and ultimately most hotly debated) part of the "Publicity" effort, it was only part of a larger attempt by Groves and others to simultaneously inform the public while keeping much back. Two particular areas consumed considerable effort on this front: the preparation of press releases to be sent out under the name of prominent officials, including the President of the United States, and the creation of ready-made articles to be supplied to the media in the wake of the bomb's first use. Each served similar but slightly different ends: the press releases were meant to aid the push for Japanese surrender and make the reality of the atomic bomb's existence seem unimpeachable, while the newspaper articles were meant to sate the public's desire for information and prevent journalists from straying into classified territory.

The matter of press releases had been raised at the end of 1944 in a series of memos to Secretary of War Stimson. Initial guidelines for a statement to be released by the President "after S-1 is used" had been drawn up by Stimson's assistant, Harvey H. Bundy, in December 1944, as the use of the bomb started to seem like a near-term reality. Bundy's recommendations were general, asking for "a brief statement of the importance of the product and the emphasis that our Government has given to it together with a background of the scientific facts everyone knows."[26] An initial draft of this "possible statement by the President" was prepared in February 1945 (its authorship is not clear; it may have been drafted by Bundy). It was not elegant:

> On—194?, the United States Armed Forces used against the enemy an entirely new weapon of destruction based on a hitherto unproved method of creating the most devastating explosive power. The result has now been reported. It was so great that it becomes apparent that not only

> has this weapon probably changed the very nature of warfare but it also carries with it possibilities of the most vital importance for the future peace of the world. . . . The matter has been surrounded with the greatest secrecy and this secrecy must and will be maintained as to methods of production for some time to come.[27]

The draft was vague (to the point of not specifying what year the bomb might be used!) and poorly written, and it does not appear to have been seriously considered, but it was a start.

In early March 1945, Vannevar Bush and James Conant confronted Bundy and convinced him that planning for the bomb's use needed to take place immediately, including the preparation of public statements. "If the matter is left without adequate planning," Bundy related to Stimson, "there will be confusion and turmoil. . . . There will be almost public hysteria."[28] The target here, it is worth noting, was not an imagined Japanese leadership, but an imagined American public. In response, Stimson created an "Interim Committee," to make recommendations on the "interim" period between the revelation of the bomb and the establishment of a postwar regime, with an ever-larger share of their attention dedicated to "Publicity" issues.

Later that month, Groves wrote to George C. Marshall, the Army chief of staff, to express his concern about the failure of censorship in the period after use of the weapon. Wartime censorship, he argued, had more or less succeeded in holding the line, but there had been "increasing difficulty" in getting journalists to comply. They could not "hope to be successful indefinitely," and "after the initial use" of the bomb, "serious breaks will come . . . these breaks may well develop into situations beyond our control." The antidote was to prepare "carefully written press releases" to be deployed as necessary. Groves added that he was considering employing "a suitable newspaperman prepared to serve, if necessary, as a pool correspondent for all newspapers."[29]

The newspaperman Groves had in mind was William L. Laurence, one of the early founders of science journalism in the United States. Laurence was a Lithuanian-Jewish immigrant who had been smuggled into Germany after the Russian Revolution of 1905, from which he then landed in the United States. After changing his name and becoming a naturalized citizen, Laurence ended up in Cambridge, Massachusetts, where he studied at Harvard. In the 1920s, he was able to break into journalism and by 1930 had become a science reporter with the *New York Times*.[30] Laurence had a messianic view of science and technology and an unabashed "gee-whiz" style of purple prose. He had been one of the early enthusiasts for the possibilities of atomic energy and had published sensational articles about uranium power prior to the censorship order. His breathless article on "The Atom Gives Up," in the *Saturday Evening Post*, from September 1940, had attracted the attention of Bush and Conant for its worrisome speculation, and other reporting on fission by Laurence had made him a thorn in Groves' side. Perhaps this is part of the reason that Groves decided that Laurence might be better off as his employee.[31]

Groves went to Laurence's office at the *Times* in the spring of 1945 and asked if Laurence would be willing to be dispatched for Army press work. Laurence said he would only accept the job if he got "first-hand knowledge" of the work and was given free rein to visit the secret sites. Groves agreed. Laurence was given a few days to settle his affairs in New York and then was officially "indoctrinated" into the Manhattan Project, becoming its only "embedded" reporter.[32]

Laurence got to work helping with press releases. In May 1945, the Interim Committee determined that a "public statement" should be made following the upcoming Trinity test, so that no careful probing was made into the massive explosion that no doubt many people would see or hear some part of from afar. If the test results were meager, a brief notice about an ammo dump exploding would be sufficient cover. But if they were unexpectedly effective or destructive, the president might need to announce the existence of the bomb before any use in the war. They assigned Laurence the job of crafting two statements, with the understanding that they would be reviewed by another experienced public relations agent, Arthur W. Page, a vice president at AT&T.[33]

On the same day as the Interim Committee meeting, someone penned a memo that goes into detail about why the Trinity statement should not be *too* factual. The author, who I suspect was Groves, felt the need to articulate why a successful test could not be followed by an honest press release. "The story can be kept within bounds," the memo's author explained: they didn't *have* to tell the truth, because it was unlikely that the test would *require* them to. "New Mexico is a big place with few people living in it," so the test would not inherently reveal that they had an atomic bomb. To reveal the success of the test would endanger the entire project: "the tremendous value of surprise" would be lost, the enemy might find ways to sabotage or defend against the weapon, and Congress might sabotage the project going forward, among other fears.[34]

Laurence ended up writing four statements for possible release after the Trinity test, labeled A through D, corresponding to a rising level of explosive force. At the lowest level of the result (Form A), the statement, to be released by the commanding officer of the Alamogordo Air Base, was purposefully short and bland:

> Several inquiries have been received concerning a heavy explosion which occurred on the Alamogordo Air Base Reservation this morning. A remotely located ammunition magazine containing a considerable amount of high explosive exploded. There was no loss of life nor injury to anyone and the property damage outside of the explosives magazine itself was negligible.[35]

Form B, by contrast, said that the explosion had been due to "experimentation with high power explosives," with the effects "noticeable some miles" but causing no loss of life or damage. It also claimed that "gas shells" had been exploded by the blast, and as a result there had been a small-scale evacuation of civilians. Form C announced that "some of the scientists engaged in the test" of new explosives had been killed and contained an area to enclose the list of the dead. Form D allowed the Army to release a radius of how many miles of nearby communities were affected by the massive blast of an experiment relating to "improved war weapons against Japan."[36] The Interim Committee was not enthusiastic about Forms C and D, not because they implied catastrophe, but because they connected the test with experimental weapon research.[37]

In the end, the Trinity test, held on July 16, 1945, had required Form B because officials were unsure whether the path of the fallout plume would require local evacuations. No evacuations took place, but soldiers and scientists stationed far from the test tracked both radiation levels and public discussion. The intelligence agents of the project also compiled all local newspaper reporting about the "ammunition dump explosion." In retrospect, the unusual nature of the explosion is apparent in them: many accounts by witnesses comment on the unusual brightness, comparing it to a rising sun on Earth, which is particular to the intense heat of nuclear explosions, and many, prior to hearing the official line about the "ammunition dump," speculated that it was a meteor impact. The Manhattan Project agents reported, ultimately, that the surrounding townspeople accepted the false story.[38]

Laurence had also written, in his typical overcooked style, a long draft statement of a speech to be given by Truman:

> Today marks one of the most important days in the history of our country and of the world. Today, as a result of the greatest scientific and engineering development in the history of mankind, our 20th Air Force has released upon Japan the most destructive weapon ever developed by any nation, a weapon so powerful that one bomb has the equivalent effect of from 5,000 to 20,000 tons or 10,000,000 to 40,000,000 pounds of TNT. . . . This greatest of all weapons, developed by American genius, ingenuity, courage, initiative and farsightedness on a scale never even remotely matched before, will, no doubt, shorten the war by months, or possibly even years.[39]

And so on, praising the release of the "Cosmic Fire" and the dawn of "the greatest age of all—the Age of Atomic Power, or Atomics." James Conant thought it was absurd, "too detailed, too phoney [*sic*] and highly exaggerated in many places." But "there is no danger that it will be used in any such forms," he wrote to Vannevar Bush, because the responsibility for writing a presidential statement had been turned over to the aforementioned Arthur Page, of AT&T.[40] Whereas Laurence was hyperbolic, Page, who would later be known as the "father of corporate public relations," was cautious. He was an old friend of Stimson's, and had been brought in as a consultant to provide a more steady hand.[41] In May 1945, an unknown figure likely associated with the Interim Committee listed the "Objectives" a presidential statement about the bomb should meet:

1. Notify world we have atomic bomb.
   - 1000 to 1.
   - It seems wiser not to mention that bigger ones will follow for we want to have our acts more dreadful than our words to the Japs.

2. Call on the Japs to surrender.
   - Making it clear that if they do not all subsequent slaughter is their guilt.
   - Making it clear that also we do not count on the bomb alone but intend to follow it with the foot soldiers and fleet.

3. Give credit in general to all concerned but leave details to the report from those to whom the details would be more natural.

4. Notify the Russians that they do not get the secret for nothing but might if a proper international organization were effected.

5. Notify the people in the plants that they are not in danger.

6. Get the fact that the British are partners but we have the credit and the plants and have spent the money—before it is brought out by a critic.

7. Hold the second story so it won't compete with the president's message.

8. Choose a military target like a naval base if possible so that wholesale killing of civilians will be on the heads of the Japanese who refused to surrender at our ultimatum.[42]

The last point is noteworthy as this memo appears to have been written not long after the May 10–11 Target Committee met at Los Alamos and made a similar recommendation to target "military objectives."[43] Page's first draft of the presidential statement started with the statement that "two hours ago an American airplane dropped a bomb on the Nagasaki Naval Base and the Naval Base ceased to exist."[44] Further drafts were made into June and through mid-July. It is clear many pairs of eyes were set to it, including other Army public relations officers, and the Interim Committee as a whole.[45]

In a post-Trinity letter to an assistant to Stimson, Page articulated a philosophy behind the importance of a good presidential statement: "I would think that by the 15th of August the destruction in Japan will be sufficient materially and psychologically to justify the President in saying to the Japanese that as natural forces are evidently on our side and as we have no intention of destroying their people or their religion they might well capitulate to the power of the universe."[46] In other words, a presidential statement with especially well-chosen words could play a large part in how the atomic bomb was interpreted, and potentially end the war.

After it became clear he was incapable of writing a press release that sounded like anyone's voice but his own, William Laurence was given a new task better suited for his talents: writing stories about the Manhattan Project and what it had accomplished. In mid-May 1945, he sent Groves a list of almost thirty articles he was considering writing. Most were straightforward: "1. General lead story based on President Truman's first radio address, with further details and background"; "6. An article giving comments of leading scientists and others engaged in the project"; "14. An article on the leading personalities behind the project." Many were discussions of the scientific principles or about specific installations (e.g., the history of the Los Alamos laboratory). Some were speculative articles about scientific progress, classic Laurence material: "4. Another article tracing briefly the various Cultural Ages from prehistoric through historic times, and their principal characteristics in terms of human progress." One suggestion even contained a bit of dark humor about the potentialities of the Trinity test: "23. An eyewitness account (in case eyewitness survives) of the first tests with the bombs." And, of course, Laurence was keen to give a personal, eyewitness account of the first use of the weapons against Japan—the journalistic scoop of a lifetime.[47]

These were not stories Laurence might merely release on his own, exclusive to the *New York Times* and under his byline. The goal was that these stories could be distributed to all newspapers, free of charge and without a requirement for attribution, so that the first week (at least) of the news cycle about the atomic bomb would be both extensive and dominated by a Groves-approved narrative. Groves' "Publicity" strategy was in evidence: selectively release a lot of new, once-secret information, and thereby control what information was available. Groves and others involved with the project believed that lifting censorship on a story of this magnitude would result in an uncorking of pent-up journalistic energy. And this was more than a theory: it was part of the Manhattan Project officials' ongoing experience in trying to enforce a secrecy regime with regard to the press during the war.

By June, Laurence's workload had been reduced to only eight stories. The main two, a "general" story, and the eyewitness account of the use of the bomb, were still big scoops, the latter all the more so because Laurence himself would be indelibly injected into the story.[48] All of the speculative topics had been culled, though Laurence still turned in a story arguing that "in atomic power man at last has a fuel powerful enough to free him from the gravitational bonds of the earth. . . . The Interplanetary Era may be just around the corner." This story was emphatically nixed. "This story is NOT appropriate for War Department use," an attached note explained, and the draft itself was crossed out with "NO" written on it many times, including once by Groves himself.[49] Even Laurence's accepted stories were heavily edited by Groves' people, mainly to tighten them up and to remove Laurence's often florid language—the Hanford project was an "Atomland-on-Mars," which apparently the Army officials found a bit too much to swallow (they replaced it with "the project").[50]

Ultimately, after Hiroshima, Laurence's stories did dominate the early newspaper coverage of the atomic bombs, though they were not totalizing—journalists still did their jobs, even if many of them happily reprinted what was essentially propaganda produced and sanctioned by the US government. Laurence would win a Pulitzer (his second) for his first-hand, ride-along account of the bombing of Nagasaki (he was on one of the observation planes). In recent decades, Laurence's reporting has been used to illustrate the conflict of interests that can be present when reporters embed themselves with the military. Laurence represented himself as an independent journalist, but operated under the

Site W--- 1

Story No. 8

TOP SECRET

(This story on Hanford is for release for morning newspapers R-Day plus 3.)

CLASSIFICATION CANCELLED

For The Atomic Energy Commission

Director, Division of Classification

By William L. Laurence
For the Combined Press

DECLASSIFIED

NND 730039

6-5-74

Richland Village, Wash., — In this Government built, pleasant little town in "Atomland-on-Mars," fifteen miles northwest of Pasco, the inhabitants, largely men and women in their twenties or early thirties, have well-paying jobs and pleasant Government-owned homes on the banks of the swiftly flowing Columbia River. But until a few days ago, when the curtain was lifted on the war's greatest secret, the vast majority of the 17,000 inhabitants did not have the slightest idea of what they were producing in the gigantic Government plants, some thirty miles away.

Until a few days ago no mention was ever made in private conversations of the enormous structures scattered over an area of more than 400,000 acres only a short distance away. Talking to them did not give one the slightest inkling of the raison d'etre of the very town they live in, of the force that brought them all here from Maine to California. They did not know they were dwelling in a pleasant little suburb of "Atomland-on-Mars."

Hanford Engineer Works, as this particular division of "Atomland-on-Mars" is named, is one of the largest and most unique of the Manhattan Engineer District's war construction projects for the production of the raw materials for atomic bombs, the war's most sensational development. It is the center where the greatest miracle of modern alchemy is being performed, where an entirely new element, never known to exist in nature, is being produced in enormous quantities.

TOP SECRET

FIGURE 3.1. Heavy editing of one of William Laurence's articles, probably by William Consodine and other members of the MED Public Relations Organization. This particular article is on Hanford, intended for release on "R-Day plus 3." Laurence's florid "Atomland-on-Mars" was replaced with the more generic "the project," and a sentence extolling Hanford as the "greatest miracle of modern alchemy" was unceremoniously cut. Source: MEDR, Box 31, "Laurence stories."

editorial finger of the Army Corps of Engineers, and used his position to push positions that were favorable to the government. His stories also neglected many aspects of the narrative that made Manhattan Project officials uncomfortable, like the discussion of civilian casualties or radiation injuries, though to be fair, he was probably, like many project participants, unaware of the health effects associated with the bomb. There have even been calls for Laurence's Pulitzer to be retroactively, posthumously stripped.[51]

In any event, Laurence's enthusiasm for the work on the bomb appears to have been genuine, and if anything the Army censors worked to contain it, not censor it: Laurence didn't think the atomic bomb would just end the war, he thought it would irrevocably propel the human race into a glorious utopia in the stars. Laurence was, without a doubt, willingly complicit in the government's propaganda project, but that was part of the reason he was chosen for the job. The troublesome question is not whether Laurence ought to have acted differently—it is hard to imagine him being any other kind of reporter and writer—but whether the Pulitzer should be given to government propagandists of any stripe.

Along with the presidential statement to be released after the use of the bomb, a separate, longer statement was prepared by the office of the Secretary of War. It was to be released some time later to fill in the gaps that would inevitably be left by the pithy presidential release. Drafted by Arthur Page in late June, it covered detailed policy decisions that led to the building of the bomb, including the names of people involved, filling out ten pages of text. This statement was also reviewed by a battery of readers, including members of the British delegation, in order to satisfy the requirements of the Quebec Agreement of 1943, in which the United States and British agreed not to disclose information on the project to third parties without mutual consent.[52] A final version of the presidential statement was sent by Stimson to Truman by a cable on July 31, 1945, noting that the use of the weapon was imminent. Truman immediately cabled back his assent: "Release when ready but not sooner than 2 August." This approval of the press release is the closest thing we have to a positive written order by Truman to use the bomb. It highlights the importance that "Publicity" had within the project that his positive assent was never requested for the use of the bomb, only the press release about it.[53]

But Groves knew that "one-time" releases would not suffice. He anticipated that journalists would still have further questions and want to write unanticipated stories. To prepare for this eventuality, he sought to put into place one more piece of his "Publicity" puzzle: the Manhattan Project Public Relations Organization, whose purpose would be to serve as a liaison with the press for all further inquiries about the bomb. This organization would continue the centralized authority that had existed during the wartime years, at least as an interim solution for the first weeks after the bomb, as Groves was optimistic that a more permanent postwar solution would be passed by Congress in the fall of 1945. As with the rest of the "Publicity" organs, the Public Relations Organization would serve a dual role as a censor and provider of information.

The first plans for a dedicated postwar public relations organization were formulated in late June 1945, later than either the Smyth Report or the plans for press releases and news articles. In a memo to Groves, Lt. Col. William A. Consodine wrote of the plans that he and unspecified others had made for the new "MED Public Relations Program" at a meeting in New York on June 25. Consodine, an Army lawyer, had already served in various roles related to security within the Manhattan Engineer District and had been involved with the "Publicity" effort from early on. The "Program" he outlined was based on a specific timeline of releasing information, assuming a "successful bombing of Japan." First would be Truman's statement, released within hours of the bombing. Local press around Hanford and Oak Ridge would be given a slight lead time to prepare their stories for release, in order to make good on promises made during the war in exchange for censorship. Next would come the Laurence articles, and finally, the Smyth Report, which together would "form the basis of all releasable information on the project."[54]

Beyond the initial release, the organization would seek "coordination" of all public relations efforts, done by informing all relevant agencies to direct all atomic-related inquiries to MED headquarters. "It is imperative," Consodine continued, "that the MED public relations office be the single agency to handle all phases of the project's public relations to eliminate confusion, to insure protection of that part of the project which will not be releasable and to insure proper credit for the work." These were the same justifications for "Publicity" in a nutshell: maintaining order, maintaining secrecy, and giving credit. Regarding credit, Consodine recommended that the Army Air Forces be totally restrained from making any statements on the atomic bomb at all: "While they will be given due credit, they will not be allowed to steal the show." The rest of the memorandum laid out a program including tight accommodations for journalists who might be allowed to visit the project sites after the use of the bomb, including funds for their "entertainment." Consodine further recommended developing contingency plans for two alternative possibilities: the surrender of Japan before the use of the bomb, or an "unpredictable result" of the Trinity test.[55]

Groves revised this plan over the course of July. His plan mirrored Consodine's, but with stronger emphasis on the relationship between "Publicity" and security. The MED Publication Relations Organization, as he had decided to call it, would take precedence over the Security and Intelligence division in order to assure that these two functions (the providing and withholding of information) would not be at bureaucratic cross-purposes. Consodine would head up the overall effort, with other designated officers at the main site tending to local questions. "The entire program," Groves wrote, "is based on the release of the story of the project within specific security limitations." These included not only the boundaries of the Smyth Report, but also questions about "speculation of the post-war usage of present facilities" and "medical speculation."[56]

As Groves prepared the information for release, he made sure the scientists on the project knew that even though the story had been released, they were still muzzled. At the end of July, Groves wrote to Oppenheimer that because "the time is approaching when the announcement of the existence and general purpose of the MED will probably take place," Oppenheimer should prepare to send out a memo written by Groves under his own name to all scientific personnel. Groves' letter emphasized restrictions:

> The official public announcement of the existence and general purpose of the MED and some of the activities of this Site has been made. . . . However, in the interest of national welfare, it is necessary that security be retained over many pertinent phases of the project and especially over the work with which we are interested. The decisions on publication have been made by the highest authorities. However, restrictions established by the decisions will be relaxed progressively as the course of events permits.[57]

The letter then explained that Oppenheimer would forward any specific requests for information releases on to Washington, DC, for approval, and warned that the Espionage Act was still in force. Oppenheimer significantly rephrased the letter to sound less harsh and bureaucratic, and removed the threat of the Espionage Act altogether.[58]

By August 1, 1945, all of Groves' "Publicity" plans were in place, and the first use of the atomic bomb was expected soon. He had his press releases, his scientific history, his newspaper articles, and his plan for postwar information management. This unprecedented effort was a consequence of determining that information management should be continued into the post-Hiroshima period, and of the idea that the introduction of the bomb not only *would* be but *should* be a massive "shock." Indeed, project scientists and officials feared that the Japanese people could misunderstand the significance of the atomic bomb. If they saw it as merely being continuous with existing American tactics, like firebombing, then it might not be a decisive weapon at all, in the same way that gas warfare failed at its promises to shorten the First World War. Even worse would be if the American people and the world as a whole misunderstood the bomb's implications. As Stimson summed up the point of view, the atomic bomb was not a new military weapon, it was "a new relationship of man to the universe."[59]

Many of the scientist-administrators were keen that Stimson, and everyone else, understood this as well. Bush and Conant had tried to make Stimson see that this was a weapon that might fundamentally reconfigure global political power, and there was no hope for the United States to maintain either long-term secrecy or a long-term monopoly. They worried, as did Oppenheimer, that the bombs they had developed were a glimpse at what was to come. Thermonuclear weapons were being discussed behind closed doors, weapons with explosive power measured in the tens of millions of tons of TNT—a thousand-fold increase in power over the Trinity "Gadget."[60] They believed that the only sane answer to the question posed by nuclear weapons was a radical reconsideration of state sovereignty. Such things would not even be imaginable unless people understood the revolutionary nature of nuclear warfare, and so much stood to be gained and lost from the weapon's first impressions.[61]

What loomed ahead was the date when the "Publicity" machine would be set in motion, the actual date of the first atomic bombing. There was no standardized term for what this day would be called. William Laurence called it "R-Day," for "Release Day."[62] James Conant at one point referred to it as "U-Day," presumably for "Uranium Day."[63] Perhaps most direct was one of Groves' security officers, who referred to it—two days before the first bomb was dropped—as "Publicity Day."[64]

3.3 SECRECY FROM PUBLICITY

The strike order to use the atomic bomb, drafted by General Groves and formally approved by the secretary of war, specified that the first "special bomb" would be used "after about 3 August 1945" on one of four targets (Hiroshima, Kokura, Niigata, or Nagasaki). Beyond that, it noted that "additional bombs will be delivered on the above targets as soon as made ready by the project staff."[65] By the end of July, the bomb pieces, including their nuclear cores, had been sent to the island of Tinian for assembly. Weather conditions in the Pacific pushed back the date of the first bombing, against Hiroshima.[66]

Groves received word at his office in Washington, DC, at 11:30 PM on August 5 (local time), that the mission to use the first atomic bomb had been "successful in all respects." Groves spent the rest of the evening writing up a draft report to the Army chief of staff and slept on a cot in his office.[67] The bomb had worked. Hiroshima was destroyed, but it was not yet public: even in Japan, the victims of the bomb were unsure of what had happened, and the Japanese high command knew only that communications to one of their major Army bases had been disrupted.[68]

The next day Groves prepared the final schedule for the "Publicity" blitz, addressed to Chief of Staff General George C. Marshall. As soon as it was determined that "the results of the first bomb have been satisfactory," the statement by President Truman (written by Arthur Page) would be issued. "Within the hour," the longer statement by the secretary of war would be released. "Almost immediately" after it was released, statements would be issued by the United Kingdom and Canada regarding their contributions. "Condensations and descriptive narratives" of the project, drafted by William Laurence, would be released simultaneously for the news media. "Within forty-eight hours thereafter," they would release the Smyth Report, which Groves emphasized was both safe and essential for his information strategy. The Smyth Report was still the one part of this strategy that had not yet been finalized, since Stimson had wanted it cleared by the President himself. Groves again felt the need to underscore the importance of the Smyth Report: "In my opinion, it is essential that this release be made promptly if we are to have any hope of success in our efforts to retain the maximum amount of secrecy on our work." The "Publicity" plan was being set into motion.[69]

The presidential statement was issued from the White House while Truman was still aboard the USS *Augusta*, returning from Potsdam. It was light on the details of the bombing, but emphasized the spectacular nature of the weapon: "It is an atomic bomb. It is a harnessing of the basic power of the universe. . . . We have spent two billion dollars on the greatest scientific gamble in history and won."[70] As one of its last sentiments, it urged the maintenance of technical secrecy:

> It has never been the habit of scientists of this country or policy of this Government to withhold from the world scientific knowledge. Normally, therefore, everything about the work in atomic energy would be made public.
>
> But under present circumstances it is not intended to divulge the

> technical processes of production or all the military applications, pending further examination of possible methods of protecting us and the rest of the world from the danger of sudden destruction.[71]

The bomb was special not only because of its city-destroying powers. It was special because it required continuing scientific secrecy, at least temporarily, something unusual in the contexts of both scientific openness and American political transparency.

The secretary of war release that followed was longer and more concerned with details of the creation of the bomb and reflections on its future implications. Early on, however, it made clear that the "requirements of security" meant that only a "broad" picture of the plan could be revealed. An entire section of Stimson's brief history of the Manhattan Project was devoted to security, noting that the "extraordinary secrecy" had been ordered personally by Roosevelt. But the report concluded, per Bush and Conant, that secrecy could not be an effective postwar strategy: "Because of the widespread knowledge and interest in this subject even before the war, there is no possibility of avoiding the risks inherent in this knowledge by any long-term policy of secrecy."[72]

Then, British and Canadian statements were released. Just as the British had been given reviewing rights to American statements under the Quebec Agreement, Groves' "Publicity" staff reviewed the foreign statements as well. Like the Truman statement, the Canadian statement, by the minister of munitions and supply, C. D. Howe, also emphasized continued secrecy: "It has been necessary to take extraordinary security precautions and while we are anxious to give the people all possible information it is obvious that until some appropriate methods are devised to control this new source of energy that has been developed it will not be possible to divulge the technical processes of production or of military application."[73] In the United Kingdom, former Prime Minister Winston Churchill's statement was similar to the previous ones, albeit with emphasis on the British contributions. It ended with a somber invocation of the "secrets" that had been discovered: "This revelation of the secrets of nature, long mercifully withheld from man, should arouse the most solemn reflections in the mind and conscience of every human being capable of comprehension."[74]

Such was the seemingly paradoxical nature of these "publicity" statements: they advertised their revelations along with the fact that they were holding back vast, powerful "secrets." The explicit purpose of this sort of statement was in warning newsmen and project participants to tread carefully. More implicit was a reinforcement of a new mystique—part of the "special" aura of the bomb.

And all of this was also about extending Groves' control of information. One passage cut from an early draft of the secretary of war's statement made this clear: "The President has designated the War Department as the sole releasing agency of all information concerning this project. The continued cooperation in secrecy of all individuals and organizations associated in any way with the project and of all informational media is required in the interests of national security."[75] In practice, control of information would be a much messier affair.

The press reaction to the announcement of the bombing of Hiroshima was frenzied. The front page of the *New York Times* for August 7, the first news day after the bombing, carried no less than six long stories devoted to the bomb. "FIRST ATOMIC BOMB DROPPED ON JAPAN; MISSILE IS EQUAL TO 20,000 TONS OF TNT; TRUMAN WARNS FOE OF A 'RAIN OF RUIN,'" the top inches of the newspaper blared. "Steel Tower 'Vaporized' in Trial of Mighty Bomb," a story on the Trinity test announced. Other, smaller headlines hit the main themes: "NEW AGE USHERED," "ATOM BOMBS MADE IN 3 HIDDEN 'CITIES,'" "TRAINS CANCELED IN STRICKEN AREA." The *Washington Post* carried eight stories on the front page, though three were verbatim reprints of the Truman, Stimson, and Churchill statements. Other national newspapers gave similar coverage, almost all repeating the same few stories: the raid on Hiroshima, an account of the Trinity test (a Laurence story), and discussions of the immense plants that had been constructed in secrecy. As the days went on, the bomb stayed in the headlines. The Nagasaki attack, of course, got coverage, though by that time the excitement had begun to fade, and it had to share inches with the news of the Soviet invasion of Manchuria.

The Manhattan Project officials attempted to manage further press coverage through the auspices of "Publicity." Several newsmen were allowed to tour Oak Ridge, for example, but were given a stern warning from Col. Kenneth D. Nichols that they needed to stick to the official information so that "the interests of the United States would be fully protected."[76] American newsmen by and large seemed willing to do so.

But not all interactions with the press were as positive. The biggest news crisis came when Harold Jacobson, a physicist in New York with a minor connection to the Manhattan Project, wrote a story on August 8th that was syndicated by the International News Service to several newspapers. Jacobson's story claimed the radioactive effects of the bombing would render Hiroshima uninhabitable for at least 70 years, that all nearby aquatic life would be killed when rains had washed the "lethal rays" out to sea, and that the bombed city would resemble "our conception of the moon." Though most of Jacobson's article was devoted to basic atomic clichés (the promise of atomic energy, etc.), its radiation-themed intro was what got Groves' attention. Jacobson claimed that investigators who entered the city to examine it were "committing suicide."[77]

This kind of story veered sharply away from the tidy narrative of accomplishment Groves had constructed, replacing it with a science-fiction hellscape. It amplified the uncanny aspects of the weapon, and would, Groves feared, encourage sympathy for the Japanese. The story hit headlines alongside the first casualty estimates from the attacks. "200,000 Believed Dead in Inferno That Vaporized City of Hiroshima," the *Boston Globe* blared.[78] Their numbers were inflated—a result of an imprecise methodology—but the sudden confrontation with the human costs of the weapon complicated things for Groves and others in the government. The non-military nature of Hiroshima (it was a city with a military base, but 90% of the casualties were civilian) was enough of a propaganda problem that unidentified "high authorities" in the War Department urged the Office of War Information to stress that the targets possessed "sufficient military character to justify attack under the rules of civilized warfare."[79]

Groves might not have been able to control every narrative, but in Jacobson's article he had an opportunity: Jacobson was decisively *wrong* about the contamination issue at Hiroshima. By detonating their bombs at high altitudes, the Hiroshima and Nagasaki strikes had avoided significant amounts of radioactive fallout on the ground, and what induced radioactivity there would be was predicted to be short-lived.[80] Groves' "Publicity" machine quickly printed a denial in most newspapers on the same day that Jacobson's story came out. In fact, very few papers carried Jacobson's article, while its refutation was reproduced widely.[81]

The denial quoted J. Robert Oppenheimer as saying that based on their experience at the Trinity test, "there is every reason to believe that there was no appreciable radioactivity on the ground at Hiroshima and what little there was decayed very rapidly."[82] Not all project scientists agreed. Robert S. Stone, a physician who headed the Health Division of the Met Lab, wrote in a critical letter to a Manhattan Project military representative, "I could hardly believe my eyes when I saw a news release said to be quoting Oppenheimer, and giving the impression that there is no radioactive hazard. Apparently all things are relative."[83] But Oppenheimer was Groves' man, and willing to put his official stamp on a statement that was at the very least premature, and at worst, as misleading in reassurance as Jacobson had been in alarmism.[84]

Jacobson's connection to the Manhattan Project lent his statements greater apparent authority in the news media, but also exposed him to legal danger. Military officials were dispatched to Jacobson's office in New York, threatening that he had violated the Espionage Act and project secrecy agreements. Jacobson collapsed from the strain. After he recovered, he was interrogated by FBI agents; Jacobson claimed he was relying on his own speculation, not officially-obtained information. He also said he was pleased to hear he was wrong. Jacobson officially retracted his claims a few days later.[85]

Groves turned the Jacobson incident into a short-term gain: his "Publicity" apparatus had managed to take an unsanctioned narrative and turn it into a sanctioned one. But the question of radioactivity would not so easily go away as Japanese reports of radiation sickness, similarly denied by Groves, were later found to be true.[86]

By August 9th, the Smyth Report had been written, screened for security, and edited. One thousand copies had been printed in secret by the

US Government Printing Office, held under lock and key. But would it be released?

The Smyth Report was controversial even while it was being developed, and its release was not actually entirely subject to Truman's own approval alone. The Quebec Agreement of 1943, which outlined the collaboration between the United States and the United Kingdom on the bomb, specified that neither party would "communicate any information about Tube Alloys to third parties except by mutual consent." So at the same July 1945 meeting where the British consented to the use of the atomic bomb against Japan, they were also asked to approve the release of the Smyth Report.

The head of the British scientific delegation, the physicist James Chadwick, voiced extreme unease about the release of so much information in one source. The minutes reflect that Vannevar Bush strongly defended on the grounds that it was "impossible" to keep the basic information secret after its use, and that "the balance of advance lay in giving the maximum amount of scientific information possible without actually disclosing technical data which would be of practical assistance to other Governments." Groves further assured the committee that solid security principles had been developed to govern the amount of material released.

The British ambassador suggested that the Americans draw up a statement for the British government that would summarize their views on the attainability and desirability of secrecy in the short and long term, and speculated that revealing the secrets prematurely might decrease the likelihood of the Soviets agreeing to international controls. Bush argued that the release would tell the Russians nothing they couldn't easily find out on their own, and that "he thought it was better to give the British and American public the scientific information which Soviet Russia already would have or could get without difficulty." The group agreed that Groves would send the British a copy of the security rules applied to the Smyth Report, and that Chadwick would certify assent on behalf of the British government if he was satisfied.[87]

A few weeks later, the British indicated in a memo that they considered giving away information politically troublesome, and again offered that it could hurt international control negotiations.[88] In the meantime, Groves had forwarded a memo by Tolman and Smyth which discussed some of the more borderline secrecy cases in the report. Another meeting took place in early August, only days before Hiroshima. Stimson told the participants that he was interested in a small debate on the "pros and cons" of publication.[89]

Conant defended release based on the "uniqueness of the subject matter" and argued that the thousands of people involved made it impossible to keep the project under wraps once the bomb had been used. As it was, he argued, "rumors were building to a crisis and the situation was explosive," and without some sort of official release that could guide further security practices, "a serious situation might develop." Chadwick, by contrast, remained ambivalent. He offered that it "was difficult to understand our problems in this country." In England, he explained, even with such a large project, "we would not do it," and then asked the key question of secrecy: "Where should we draw the line?" He worried, ultimately, that the Smyth Report was like a detective story, with little hints riddled through it that would give away quite a lot. But Conant argued that Russia would figure out everything in the report within three months anyway, and that if they didn't release it, even more information would end up leaking out.

Stimson voiced his growing concern about the Soviets since his contact with them at the Potsdam Conference. He was now doubtful that international control could be achieved with such a closed, secretive, suspicious nation, and he was now by his own account "much more conservative than General Groves." He further expressed his doubt in the ability of the Soviets to piece together the information effectively, since "the Russians are necessarily slower in their way of life," due to their lack of freedom. Chadwick eventually agreed that there was nothing in the report the Soviets could not get in a short amount of time, though he was still ambivalent about release. Groves and Conant both noted that if newspapers decided to investigate the atomic bomb, a thousand articles pieced together from various sources would emerge. Groves finally concluded that if the report was not published, it would "start a scientific battle which would end up in Congress." Stimson concluded they would have to ask Truman when he returned from Europe.[90]

The next day, Stimson held another conference within the War Department where at least one official strongly opposed publication, but Stimson had by that point decided to recommend to Truman that it be published. Chadwick sent one more memo, noting that their difficulty hinged on how "to judge how far one must go in meeting the thirst of the general public for information and the itch of those with knowledge to give it away." He felt that the report really did give away important information to "competitors," but also that "such assistance to possible competitors is not as much as one might think at first sight." He concluded it would save a nuclear aspirant only about three months' work, out of a three to four year development program. Ultimately, however, the British agreed that if the US wanted to publish it, they wouldn't stand in their way.[91]

The final decision would wait for Truman. On the morning of August 9th, not long before news of the bombing of Nagasaki would arrive, a group of Manhattan Project associates filed into the Oval Office. Stimson, Groves, Bush, Conant, Secretary of State James F. Byrnes, and Stimson's assistant George L. Harrison all intended to lobby for the release of the report. After all the long debates, the final decision was anticlimactic. In Bush's recollection, Truman listened to the arguments for and against publication, sat back in his chair, and stared at the ceiling. "I regret that I have to make decisions such as this," Truman said. Then, after a pause: "You will release the report; the meeting is adjourned."[92]

On the evening of August 11, two days after the bombing of Nagasaki, a thousand mimeographed copies of Smyth's technical report were removed from a Pentagon safe and distributed to members of Congress, the press, and Manhattan Project site leaders. Its title was ungainly: "A general account of the development of methods of using atomic energy for military purposes under the auspices of the United States government, 1940–1945." In the rush to get it out, nobody remembered to stamp its intended title, "Atomic Bombs," onto the first page, and so "the Smyth Report" became its de facto title.[93]

News organizations were informed by a press release from the War Department that "nothing in this report discloses necessary military secrets as to the manufacture or production of the weapon," and that "the best interests of the United States require the utmost cooperation by all concerned in keeping secret now and for all time in the future all scientific and technical information not given in this report or other official releases of information by the War Department."[94]

The initial thousand copies were taken almost immediately; an addi-

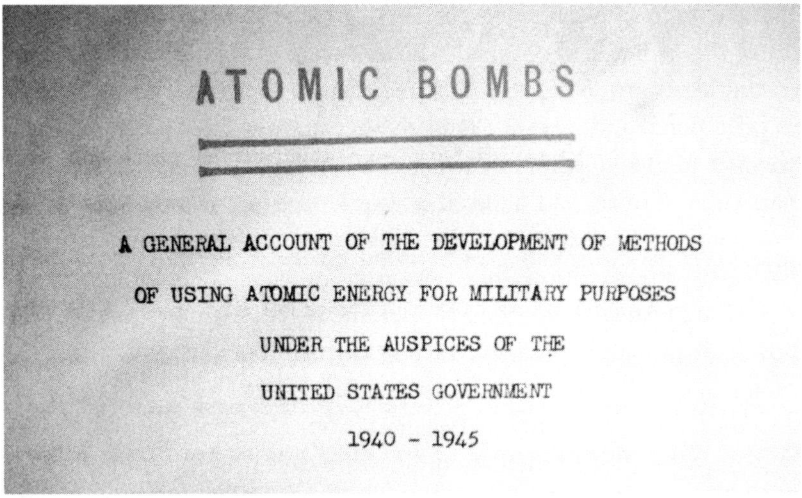

ATOMIC BOMBS

A GENERAL ACCOUNT OF THE DEVELOPMENT OF METHODS
OF USING ATOMIC ENERGY FOR MILITARY PURPOSES
UNDER THE AUSPICES OF THE
UNITED STATES GOVERNMENT
1940 - 1945

FIGURE 3.2. The only version of the Smyth Report that received the stamp of its actual title—"Atomic Bombs." Source: UF767.S5, Library of Congress, Washington, DC.

tional 2,000 were quickly printed. In September, a slightly edited edition would be released by Princeton University Press. The Princeton edition would sell over 125,000 copies in two bindings, over 103,000 by the end of 1946. The book was released with no copyright restraints and was immediately translated into a variety of languages, including German and Russian. Around 12,000 copies were sold at Oak Ridge, Los Alamos, and Hanford, which required special arrangements due to their security and remoteness.[95] The physicist Herbert York, then a young Berkeley transplant to Oak Ridge assisting in Calutron operation, later recalled that he and his fellow scientists devoured their single mimeographed copy, hungry for the "big picture" knowledge they were denied due to compartmentalization: "you could take it apart by chapters, and there were so many of us that wanted to see it, that's what we did, we took the whole thing apart, then just passed it around, just passed it back and forth, reading in a totally random sequence! The Smyth Report . . . it's all there, plutonium, and Berkeley, and everything's there."[96]

The Soviets also read it with interest. Several years later, an American analyst went over the Russian Smyth Report carefully, praising it for its care and detail: "The editorship is excellent. There is every indication that the American text was screened in great detail by the technical editor of this volume. . . . I think it is significant in that we have evidence that at least one Soviet technical man has screened the Smyth Report in great detail and it is very unlikely that some of the references which we have hoped 'maybe they won't notice' have not been noticed."[97] In particular, Groves had allowed a "borderline" item to remain in the first mimeographed version that was later removed in the Princeton University Press edition: "pile poisoning," a technical issue that caused significant problems in the early operation of the Hanford reactors. The Soviets included the line, even while claiming their version was based off the Princeton edition, indicating, in the analysts' eyes, that they were aware of the discrepancy between editions. Indeed, the discrepancy may have even called their attention more closely to the issue.

The Smyth Report aided public understanding of the bomb even within the USSR: the Russian Smyth Report was for Soviet citizens one of the few sources of information about the atomic bomb during the Stalin years, when official Soviet sources were generally mute on the subject. Aleksandr Solzhenitsyn read a copy while in transit to the Gulag, and gave a "scientific report" of its contents to an informal meeting of prisoners interested in science at the Butyrskaya prison.[98]

On the whole, the "Publicity" strategy seemed to have worked the way Groves had hoped it would. There were, of course, some leaks, some speculation, and some problems. On August 10, Richard Tolman sent Groves an analysis of articles about the bomb in both *Time* and *Life* magazines. The *Life* spread contained speculations on the bomb's design, drawing a simple, "gun-type" weapon involving plutonium. Tolman noted that "this is not very much like the actual gun assembly used, and is probably only a good guess." On the other hand, an artist's rendering of the Trinity tower, with a spherical "gadget" on top of it, was, in Tolman's appraisal, "perhaps too near the real thing to be merely a good guess." Not ideal, but not fatal, especially since there was no indication that the information was "authenticated" in any way.[99] Similarly, bits and pieces of information appeared in other publications, but always fragmentary, never confirmed or denied.[100]

The one area where, in Groves' mind, propaganda and pernicious speculation needed to be countered directly were the bomb's radiation effects. The press release accompanying the Smyth Report noted that the sole addition to the Report was a statement arguing that the detonation over Hiroshima had been set deliberately high so that "all of the radioactive products are carried upward in the ascending column of hot air and dispersed harmlessly over a wide area."[101] But by the end of August, reports of radiation sickness and "radioactivity burns" amongst survivors in Hiroshima and Nagasaki had been published in American newspapers. On the morning of August 25, Groves conferred over the phone with Lt. Col. Charles E. Rea, a surgeon at Oak Ridge. Rea concluded that the reports were "good propaganda" and that the reported deaths were due to "delayed thermal burns" from the bomb. In Groves' mind, the Japanese were deliberately spinning the story to "create sympathy." Rea told Groves he'd better "get the anti-propagandists out," though Groves felt that since the reports were coming from American sources, and potentially even American scientists, there was nothing that could be done but to "sit tight."[102]

Groves had been reassured numerous times by Los Alamos scientists that radioactivity would not be a problem if the bomb was detonated at a great height.[103] He arranged to send scientists into the bombed cities immediately after the war had ended to get reliable data on the radiation effects. The results would eventually show that the scientists were both right and wrong: the cities would not be "uninhabitable" over the long term, but the radiation effects were more complicated than their earlier models had accounted for, and the Japanese symptom reports were vindicated.[104]

During the American occupation of Japan (1945–1949), US forces under General MacArthur exerted heavy censorship over both materials published within Japan and materials that could leave Japan. The atomic bombings were considered a particularly sensitive topic, as American officials believed that dwelling on them could incite Japanese enmity against the US. Real discussion about the legacy of Hiroshima and Nagasaki, and the question of long-term effects, were stifled within the country after the United States returned control to Japan, and would especially flourish after the 1954 "Bravo" accident invigorated Japan's sense as a "nuclear victim."[105]

Ultimately, even the Manhattan Project's vaunted reputation for secrecy was a result of the "Publicity" campaign. Not only was the fact that much could not be said reinforced in every press release, but the very idea of the Manhattan Project as "the best-kept secret of the war" was a story circulated by the Manhattan Project Public Relations Organization. Soon after the attack on Hiroshima, Byron Price, the director of the wartime Office of Censorship, repeated the "best-kept" secret claim in a press release, as it had not leaked (significantly) into newspapers prior to the attack. A War Department press release a few days later further hyped the "extraordinary" security measures employed, and furthered the line that the bomb was the war's "best-kept secret."[106] Popular news stories about the success of keeping the atomic bomb secret appeared in a wide variety of periodicals.[107]

All of this emphasis on "the secret," in part made as an overture toward thanking the press for their compliance with voluntary censorship regulations, would later have profound effects on postwar discussions about what constituted effective control of the atomic bomb. Ironically, in November 1945, Groves received a reprimand from Army G-2 about all the articles devoted to keeping "the secret," as longstanding War Department policy was to not discuss military intelligence or counterintelligence methods.[108]

The "Publicity" campaign's success while shaping the narrative was considerable, though not hegemonic. While much of the official narrative of the atomic bombs mirrored Groves', there were also, over the years, competing narratives. These include Jacobson's account of an uninhabitable Hiroshima, or the journalist John Hersey's later gripping account from the perspective of Hiroshima victims in the August 1946 edition of the *New Yorker*, or the various officials, both military and civilian, who expressed doubt that the atomic bomb had "ended the war." Much effort would be later made to "manage" this history, a sure sign that Groves' control over the narrative was not so air-tight. Groves would even go so far as to assist with, and promote, a myth-filled, Army-sanctioned dramatic film about the Manhattan Project from MGM Studios that opened in 1947.[109]

On the other hand, there were few journalistic interrogations that, in the narrow technical sense that Groves fretted about primarily, divulged further "secrets." The press releases achieved their goal of impressing upon the world the "specialness" of the atomic bomb, while the Smyth Report did give—for better or worse—a far more comprehensive view of what an atomic bomb program looked like than one might otherwise expect a country to reveal.

What is most curious about the "Publicity" campaign from today's perspective is the way it embodied an odd mix of idealism and practical effect. The "Foreword" and "Preface" to the Smyth Report laid out the contradictory nature of the effort effectively. The "Foreword," signed by Groves, warned that "all pertinent scientific information which can be released to the public at this time without violating the needs of national security is contained in this volume," and people who disclosed or received additional information would come under the full penalties of the Espionage Act. On the next page, Smyth's "Preface" opens with a testimony to the value of transparency in a liberal democracy: "The ultimate responsibility for our nation's policy rests on its citizens and they can discharge such responsibilities wisely only if they are informed." It appears paradoxical, at first glance, that two opposed concepts—restriction of information for security purposes, and distribution of information for democratic deliberation—could be accomplished in one volume.

And yet, this forging of the two seemingly opposed ideas into one is what would make secrecy persistent into the postwar period and Cold War. If secrecy had been *only* about restriction, without sufficient nod to the importance of dissemination of information for American ideals and policy, either secrecy or freedom of research and expression would have given way over the long term. But merging them served both the discursive goals of making the secrecy more compatible with a more permanent American context as well as allowing the secrecy apparatus to accept the possibility that disclosure could be a form of control as well. The practices of secrecy, and the practices of openness, were thus made into two sides of the same coin, united in an appeal to both security *and* democracy. It was a potent brew, and the new institutions that would emerge to manage these forces in the wake of the Manhattan Project would struggle with its contradictions.

PART II

THE COLD WAR NUCLEAR SECRECY REGIME

4

THE STRUGGLE FOR POSTWAR CONTROL, 1944–1947

The measures which I have suggested may seem drastic and far-reaching but the discovery with which we are dealing involves forces of nature too dangerous to fit into any of our usual concepts.

HARRY TRUMAN, STATEMENT TO CONGRESS, OCTOBER 3, 1945[1]

The dropping of the atomic bombs on Japan, and the subsequent release of information, did not resolve the "problem of secrecy" that gripped the scientists, military administrators, and policymakers toward the end of World War II. If anything, it complicated it. What would be the relationship between secrecy, security, and science? Nearly everyone working on the Manhattan Project seems to have thought that *some* secrecy would necessarily persist, but that *most* of the wartime secrecy had been a temporary expedient. But once the existence of the bomb became public knowledge and its terrible power began to be understood, the stakes multiplied dramatically. Discussions that had begun during the wartime secrecy regime were suddenly happening in the halls of Congress and the opinion pages of newspapers. All agreed that the resolution to the question of postwar control of the atomic bomb was a matter of national survival. But there were several competing visions of the future based on different conceptions of atomic policy, and the answers to these questions were thus unclear.

The creation of a new, postwar nuclear regime brought up fundamental questions of to what degree secrecy regimes were or were not compatible with long-term American national policy. These discussions had started during the war, as part of postwar planning facilitated by the scientist-administrators and military leadership of the Manhattan Project. Congress, however, would want to reason through this on their own terms, creating a new and in some ways bizarre approach to secrecy and the atom. At the same time, powerful actors—notably the scientist J. Robert Oppenheimer—would wage their own, behind-the-scenes battles to dislodge the growing secrecy mindset, but with only very limited success.

4.1 WARTIME PLANS FOR POSTWAR CONTROL

Though there were many discussions by scientists during the wartime project about the future of secrecy, the views of only a few scientists were capable of being presented to the highest levels of policy and politics, as most were constrained by the compartmentalization and strict chain of command that Groves imposed on the project. One was that of Niels Bohr, the internationally famous quantum theorist who had escaped from Nazi-occupied Denmark in 1943 and ended up at Los Alamos as part of the British delegation under the code name "Nicholas Baker." Bohr did make some technical contributions to the bomb while at Los Alamos, but he was most influential in exhorting the scientists, particularly Oppenheimer, to think about the long-term implications of their work.[2]

Bohr was worried about the Soviets. What would they think when they learned they were cut out of a secret as large as the atomic bomb? Would they not simply embark on their own secret nuclear arms race? How could this be avoided? In the summer of 1944, Bohr put some of his ideas down as a memorandum on the need for postwar international agreements that would prevent such an outcome, an early proposal for what would later be known as "international control of atomic energy."

Bohr's memo expressed amazement at the wartime Manhattan Project: "What until a few years ago might have been considered a fantastic dream is at the moment being realized in great laboratories erected for secrecy in some of the most solitary regions of the States." But there were no secrets being developed there that could be kept from others, he argued. He had received correspondence from a Russian colleague that implied, to him, that the Soviets were paying attention to fission.

For Bohr, the way forward was openness between the Allied powers, both during the war and in the postwar period: "The prevention of a competition prepared in secrecy will, therefore, demand such concessions regarding exchange of information and openness about industrial efforts including military preparations as would hardly be conceivable unless at the same time all partners were assured of a compensating guarantee of common security against dangers of unprecedented acuteness."[3]

Bohr's fame gave him better political connections than most project scientists. Through these connections, he had tried in vain in March 1944 to convince Winston Churchill of the need for openness between Allied powers. Later, his old friend, Supreme Court Justice Felix Frankfurter, who knew something of the bomb effort, arranged a meeting between himself, Bohr, and President Roosevelt. Bohr felt this meeting went well and he had made his viewpoint heard, but it turned out that Roosevelt's main worry the entire time was about how Frankfurter learned anything about the project.[4]

Bohr was probably the worst ambassador for science one could imagine: he had a reputation for having no understanding of practical affairs, his speech was notoriously hard to understand, and Bohr saw language less as a means of practical persuasion than as an elaborate philosophical exercise.[5] It's unsurprising that neither Churchill nor Roosevelt seemed to understand his intent. The meeting with Roosevelt did, however, spur a conversation on postwar planning between Roosevelt and Vannevar Bush that September, which may have been the most Bohr could have hoped for.[6]

Bohr continued to develop his views, both in discussions and on paper. His plans for the future focused primarily on the unrestricted scientific and technical exchange between nations ("free access to information"), which could prevent a future arms race. It would not simply protect against the bomb, but serve as a bridge between nations, he wrote in the spring of 1945.[7] It was perhaps a naively idealistic goal, but his emphasis on the importance of total openness would influence many later efforts toward international control. Oppenheimer in particular would be persuaded by the notion, and carried forward the idea into several postwar endeavors, as we shall see. Bohr's emphasis on openness imagined the bomb as an excuse to impose an internationalist, unfettered, idealized vision of the scientific community as a model for world affairs, in direct opposition to the world of secrecy and mistrust that characterized the bomb's actual creation.[8]

Following the conversations about international control that Bohr sparked between Roosevelt and Bush, Bush began a long discussion of these issues with James Conant. In September 1944, Bush and Conant wrote a long memorandum to Secretary of War Stimson urging him to take official action regarding the status of the bomb after its use.[9] They worried that the question of postwar control had not been adequately considered at the highest levels: namely, they worried about the dissemination of information after use ("Publicity"), domestic atomic legislation to be introduced immediately after the war, and the question of international control. Secrecy was core to their argument, as it represented a status quo that was acceptable in wartime but both "quite impossible" and futile in peacetime: "The progress of this art and science is bound to be so rapid in the next five years in some countries that it would be extremely dangerous for this government to assume that by holding secret its present knowledge we should be secure."[10] A little over a week later, they wrote another memo amplifying the theme: the American nuclear advantage was only temporary, maintaining complete secrecy was impossible, and partial secrecy would lead to an international armaments race. "Basic knowledge of the matter is widespread," they summarized, "and it would be foolhardy to attempt to maintain our security by preserving secrecy."[11]

Bush and Conant included in their memo to Stimson a lengthy "supplementary memorandum" that discussed the future military potentialities for nuclear weapons and what they thought a postwar "international exchange of information" would look like. They painted an intentionally "lurid picture": atomic bombs threatened the existence of any nation, and the future possibility of a multi-megaton "Super-Super Bomb" (hydrogen bomb) meant that "unless one proposed to put all one's cities and industrial factories underground, or one believes that the anti-aircraft defenses could guarantee literally that no enemy plane or flying bomb could be over a vulnerable area, every center of the population in the world in the future is at the mercy of the enemy that strikes first."

Any sufficiently industrialized nation could make such bombs, they continued, and it would surely be easier now that the US had demonstrated which methods were feasible. "The danger is that we would never know, if secrecy prevails as between countries, whether this were indeed the case," and thus the US might proceed under the dangerous delusion of its enemies' inability to retaliate. The only way out was free exchange of information, a complete renunciation of secrecy, very similar to the idea of "openness" that Bohr had championed.[12] The United Nations would have to coordinate this activity, and all nations would pledge that "all their scientists would make their results freely available" to one another. Free travel to all technical installations in all member countries would be allowed to the UN representatives. Bush and Conant recognized that this would be "violently opposed" not only by Russia, but also by industrial representatives in the United States as well, who would recognize that this openness would apply to *all* secrecy, including trade secrecy. Still, Bush and Conant felt these problems could be overcome, and that all would bow down before the "terrific potentialities of the new weapons which now lie just over the horizon."[13]

Bush and Conant did pique Stimson's interest in postwar questions, and Stimson became personally quite convinced of the need for some kind of consideration of international control. But getting results from that, with everything else going on in the war, was slow. It was not until May 1945, after Roosevelt's death, that Stimson finally created the Interim Committee to study such issues.[14] Though Stimson would convey the outlines of international control to Truman, and this would be one of the reasons that Truman would obliquely reference the project to Stalin at the Potsdam Conference (not knowing that Stalin's spies had informed him about the work long before Truman himself knew about it), nothing too concrete was begun while the war was still on.[15]

It is worth reemphasizing how radical the positions of Bohr, Bush, and Conant were regarding international control, and how impressive it is that they managed to get any traction at all in high levels of government. The argument was that a new technology—the atomic bomb—would require a remaking of the world. International politics and industrial practices would both have to be forever changed. The Soviet Union would have to open up to inspection, or else. Their imaginations included wars that could end civilization as it was known, and they worried that secrecy might turn these fears into realities. But making practical policy on this matter would prove difficult. In the postwar period, Bush would write to Truman directly on the matter, emphasizing the starkness of their options. There were only "two paths": "international collaboration" or a "secret arms race on atomic energy," the latter of which could lead to "a very unhappy world." He concluded: "Both paths are thorny, but we live in a new world and have to choose."[16]

While Bush and Conant were lobbying Stimson about international control, they were also considering the problem of "domestic control" of the atom: how the fruits of the Manhattan Project would be handled in peacetime. The wartime development of a new atomic industry was run by the military out of a black budget, with no congressional oversight, and was entirely concerned with the short-term problem of making and using the first bombs. What sort of peacetime organization should take over the work, and on what basis? And what would occur in the interim between the establishment of a peacetime organization and the end of the war? Any delay in the handoff of authority would lead to uncertainty and possibly danger, they believed. They didn't want the Manhattan Engineer District to persist, but finding an alternative proved tricky.

Nearly a year before the attack on Hiroshima, Conant wrote a first draft of a postwar bill that would establish a twelve-man "Commission on Atomic Energy." Working with Irvin Stewart, secretary of the OSRD, Conant's bill imagined it as half health-regulation agency and half sponsor of nuclear research and development. This unwieldy body—composed of five scientists or engineers, three "other civilians," two Army officers, and two Navy officers—would have the power to regulate experiments relating to atomic fission, although it was instructed to "exercise as little interference with normal scientific research as it may judge consistent with the national welfare." The draft didn't mention secrecy, but instead saw an extension of patent controls as the primary means of government regulation of atomic energy. In a memo to Bush, Stewart wrote, "on the patent side the draft goes to the limit. It not only attempts to centralize in the Commission all outstanding rights, but would make it impossible for any private right in this field to accrue at any future date."[17]

Further discussions between Bush, Conant, and Stewart focused on the possible public health hazards of private researchers experimenting with radioactive or fissionable substances without oversight. Private scientists creating their own laboratory reactors ran the risk of producing huge numbers of neutrons, or even a meltdown, and the threat of the actions of "overenthusiastic and not well-trained physicists in universities" would require "drastic regulation" to rein in. Moreover, they were beginning to see this as a unique regulatory problem that went beyond simple funding of scientific work, as nuclear fission was a "new art and the potentialities are almost beyond reckoning," Conant recorded. They sent these suggestions to Stimson in September 1944.[18] There the matter rested until Stimson authorized the creation of the Interim Committee in the spring of 1945.

In June 1945, the Interim Committee appointed Brigadier General Kenneth C. Royall and William L. Marbury, both experienced lawyers working with the War Department, to draft a new version of the bill for a "Post-War Control Commission," that, at Vannevar Bush's suggestion, would not be an "operating agency" but rather a "policy and control body which would farm out operations under contract." In other words, it would be an agency that would resemble the funding model of Bush's OSRD, and would be explicitly compatible with the postwar National Research Foundation that Bush would be proposing in his July 1945 report, *Science: The Endless Frontier*.[19] It was *not* an extension of the Manhattan Engineer District, with its totalizing control and extensive secrecy.

The first draft of the Royall-Marbury bill was finalized on July 18, 1945, two days after the Trinity test. It called for the creation of a nine-member Atomic Energy Commission similar to Conant's initial suggestion, but with vastly expanded powers. The commission was granted total supervision and direction "over all matters connected with atomic research, the production of atomic fission, and the release of nuclear energy," limited only by the jurisdiction of the United States. It would claim custody of all ores of fissionable material and all means by which fissional material could be produced and allowed the President to appropriate any property he deemed "necessary and proper for the use of the Commission" through eminent domain and condemnation, including patents. The commission would be allowed to fund related research and grant licenses for research by nongovernmental entities (contractors), and the act would make it "unlawful for any person to conduct atomic research or experimentation concerning the release of nuclear energy, without the consent, and subject to the direction and supervision" of the commission. The commission was also authorized to develop and administer security regulations "governing the collection, dissemination, publication, transmission, and communication of all information, data, documents, equipment and material of any kind relating to or connected with atomic research and the release of nuclear energy." Any violation of any section of the act could incur penalties up to $10,000 in fines, up to 10 years of imprisonment, or both.[20]

This was not what Bush or Conant had in mind. It was more orderly than the Manhattan Engineer District, but it still assumed vast powers over science and research. Bush worried about the commission "asking for more than necessary as a result."[21] In Bush's mind, the commission should be like a combination of the Food and Drug Administration and the OSRD, not a military hegemon. Congress, he felt, would not grant such "sweeping and unrestricted powers." As for the broad power the bill granted the commission to control all information on "atomic research," Bush considered that it "undoubtedly goes too far at the present time and would be unenforceable." The commission should "certainly" control information created at its own plants and in work that it has contracted for, and over information "relating directly to the embodiment of results in military devices," but beyond that, Bush was unsure. The long-term question of secrecy in nuclear science still vexed him, and he was "quite sure that any attempt to control as generally as this section would indicate would be quite futile."[22]

Bush expressed to the Interim Committee his belief that the "censorship and security provisions of the bill were too broad." He suggested that the regulation of information should be limited to things that would "endanger national security," and that the commission should be required to draw up detailed rules to implement such a principle. The Interim Committee as a whole "generally agreed" that the United States could lose its advantage in the field "if publication were too narrowly restricted."[23]

Further drafts were forthcoming, but Royall, now in charge of the editing, was not interested in changing the "basic approach of the document."[24] By the third draft, though, the broad control of the commission "to restrict the freedom of speech and publication" had been limited to "cases in which persons have obtained the classified knowledge through official duties or in the course of employment." Royall himself raised the question to the Committee about "whether a person who independently develops methods or processes relevant to project work, and whose rights are acquired by requisition or purchase, could be forbidden to disclose such information to others."[25] The draft bill's secrecy restrictions by this point had been limited to information that had come into the possession of "any such person by reason of his official duties or in the course of employment by the Committee, its subcontractors, or any other government or private employer."[26] This was still too much for Conant, who railed against the restrictions:

> To my mind the maximum we should aim at is securing through this drastic provision . . . control over *all* information obtained under the auspices of the Commission and all information concerning construction of the bomb or construction or operation of our present plants. The Commission should not have jurisdiction over any other scientific or technical information obtained prior to its establishment. For the security of this information the U.S. Government will have to invoke the sanctions used through the rest of the secret war research.[27]

The restrictions, as they stood, would have trouble in Congress, the courts, and "with scientific opinion." But, interestingly, Conant also thought secrecy restrictions would be creating unnecessary problems, "as it is proposed to release this information anyway."[28]

In a message sent on the day of the Hiroshima bombing, two of Groves' legal assistants further judged that any attempt to extend restrictions or penalties beyond people employed by the commission or one of its contractors "would be impossible of passage, difficult of enforcement, and would simply arouse needless Congressional animosity."[29] Royall was unwilling to weaken the provision, to the frustration of Groves' lawyers, who insisted that the security sections involved a "very serious defect in the statute," but that it would probably not be "appropriate or productive" for them to submit changes to Royall directly anymore.[30] Another major "defect" in the bill was that punishments for passing on information illicitly could be meted out only if it could be proven that it was done "with the intent to jeopardize the interests of the United States." Groves' lawyers recommended adding "or to further the advantage of any foreign nation," as an additional check that mirrored the Espionage Act.[31]

When Bush's General Counsel at the OSRD was finally given a chance to read the legislation, he judged it to be "very well drafted" but wondered about the need for such extreme powers. "Is it necessary, for example, to brush aside all of the checks and balances of our form of Government?" he asked Bush. "Is it not enough that those who control this new source of energy will have tremendous power by the very nature of their function without constituting them as virtually a supra state?"[32] But this was the day after Hiroshima, and these discussions could no longer be a private affair.

Even Groves' lawyers had blanched at Royall's expansive measures, and the scientists were horrified by them: they read them, correctly, as an extension of wartime authority into the postwar period. Bush and Conant believed the bomb might necessitate a changing of the world order, but not the abandonment of checks and balances. But such was the danger in investing so much rhetorical power in the revolutionary nature of the atomic bomb: once endowed with the power of life and death of an entire nation, it could justify a wide variety of far-reaching and radical policies.[33]

While these discussions of postwar secrecy were being conducted, Groves was pursuing a very different approach to an American atomic monopoly. For Groves, the key was not controlling information—that could leak out or be stolen by spies. Instead, he wanted to control the uranium. Without access to large stores of uranium, there would be no enrichment, and no reactors. Over the course of the war, Groves encouraged a secret and extensive effort to identify all known uranium (and thorium) reserves in the world, and through secret contracts and agreements to secure control over them by the Anglo-American Combined Development Trust. By the end of the war, he felt certain that the United States had extended its monopoly forward by decades. The Soviet Union, as far as was known at the time, had very poor uranium reserves, and Groves had made sure that the best supplies were part of his system. This may explain why someone who saw secrecy as so core to his effort was relatively open to the idea that through either domestic or international control, secrecy might be significantly loosened.[34]

4.2 "RESTRICTED DATA" AND THE ATOMIC ENERGY ACT

By September 1945, the final draft of the Royall-Marbury Atomic Energy Act was prepared, with only minor changes having been made from the previous drafts. Bush and Conant retained their reservations but felt that a flawed law could be amended, whereas having no law at all could lead to disaster. And some of their hesitations had indeed altered the final draft: the security section now clearly applied *only* to people who were employed by the commission and not to every scientist who dared to speculate about atomic energy, and the penalties for security disclosures had been made less draconian (disclosures made without intent to harm or without gross negligence were now punishable by only a $500 fine, and/or 30 days imprisonment; those who gave away information with an intent to harm the US could by contrast be fined up to $10,000 and imprisoned for 30 years). It was still a sweeping act, giving the commission the power to physically and legally control all atomic energy research, though it explicitly directed the commission to interfere only minimally with small-scale work done at "research laboratories of non-profit institutions."[35]

At that point, the bill left the hands of the scientists and military lawyers. It was deliberately put on hold for the month of September, as Undersecretary of State Dean Acheson worried that it would foul attempts to negotiate international control of atomic energy. Various congressmen moved to seize the opportunity, introducing their own forms of legislation. All were squelched by the new secretary of war, Robert Patterson (Stimson had retired). The nascent Scientists' Movement, comprising Manhattan Project veterans whose release from secrecy spurred their sense of political obligation, got wind of the more unpleasant aspects of the Royall-Marbury bill and began to organize a powerful lobbying effort against it. And the first tenet of their Ad Council–designed public campaign was the emphasis that was "no secret."[36]

On October 3, 1945, ahead of discussions between Truman and the British and Canadians on international control, the White House released a statement. Aside from praising the bomb's role in ending the war, Truman suggested that the weapon might prove more revolutionary than the invention of the wheel. He urged Congress to pass legislation along the lines of the Royall-Marbury bill, creating an Atomic Energy Commission with strong regulatory and control powers: "the discovery with which we are dealing involves forces of nature too dangerous to fit into any of our usual concepts."[37]

Paired with Truman's statement was the introduction of a slightly modified Royall-Marbury bill into the House of Representatives by Andrew Jackson May, chairman of the House Military Affairs Committee, and an identical bill in the Senate by Senator Edwin C. Johnson. From this point on, it was referred to as the May-Johnson bill, and May scheduled hearings on the bill to begin within a week of its introduction. Despite the exhortations for haste, it would not be a fast process.[38]

The May-Johnson bill gave its Atomic Energy Commission broad latitude to establish security regulations "governing the collection, classification, dissemination, publication, transmission, handling, and communication by any person of information, data, documents, equipment, and material of any kind" related to nuclear fission or atomic energy, so long as that information had been in some way connected with official government research or had been connected to large-scale nuclear experimentation. Any security violation could be punished, without benefit of hearing or criminal prosecution, by dismissal from the government's employ, at the very minimum, and if the violation was "willful" or due to gross negligence, with fines ranging from $500 to $10,000, and/or 5 years imprisonment. Anyone who transmitted secret information, however obtained, to an unauthorized person, with "intent to jeopardize the interests of the United States," or with reason to believe that such jeopardy would occur as a result of his or her actions, would face a maximum penalty of $300,000 and/or a 30-year imprisonment.[39]

The hearings proved contentious and difficult. Numerous people connected to the Manhattan Project testified to the value of the bill. The general argument in its favor was that it had been created by people with knowledge of the work and with attention to the ways of scientists. Groves further stated that if Congress tarried, "irreparable damage" to the American position would result: "we are flirting with national suicide if this thing gets out of control." And all emphasized that atomic energy mandated extreme controls. As Conant put it: "This is an extraordinary bill, drawn for extraordinary circumstances. . . . We are dealing here with something that is so new, so extraordinary and so powerful, gentlemen, that I, for one, feel that we are justified in setting up a commission with equally extraordinary powers."[40]

Secrecy preoccupied the congressional testimony. The emphasis on the importance of wartime secrecy had given them the idea that the wartime work had been about acquiring "the secret," as opposed to a massive scientific, military, and industrial collaboration.[41] Several representatives openly lamented that the "valuable secret" of the bomb might be released by this legislation. Groves tried to disabuse the congressmen of the notion that "the secret" could be kept: "The big secret was really something that we could not keep quiet, and that was the fact that the thing went off." The fact that bombs could be made "told more to the world and to the physicists and the scientists of the world than any other thing that could be told to them."[42]

To the Military Affairs Committee, Groves broke the "secrets" of the project into three categories. First were established scientific facts, "which were not secret at all." These were what the Smyth Report contained. Second, there were scientific developments that went beyond this information, most of which fell into the realm of "applied" rather than "basic" science. These were replicable by other nations but would require time and expense. Third, and most important to Groves, were the industrial techniques, a combination of practical applications of scientific knowledge, managerial practices, and the solutions to all the various challenges ("know-how"). The US was ahead, Groves emphasized, and might remain so for some time, but could not rest on its achievements. Secrecy alone could not guarantee security, and the American military policy couldn't rely on maintaining the present atomic monopoly, as Groves emphasized: "This is a secret that cannot be held; it is just a question of time."[43] It is worth highlighting how relatively nuanced this take on secrecy was, compared to both those who held "the secret" to be something totemic, and those who dismissed secrecy altogether, and it is in contrast with Groves' usual reputation.

In the meantime, the nascent Scientists' Movement had intensified. Its members read the bill in light of their resentment for wartime compartmentalization, and they saw it as an attempt by Groves to extend military power and style into the postwar period. Groves had overseen a project that had led to the destruction of two cities without significant democratic oversight, and had, in the eyes of many of the scientists involved, inhibited normal scientific research while excluding nearly every scientist from policy decisions. To extend this indefinitely was unallowable. The scientists began to organize, publicize, and lobby. They wanted new hearings that allowed people not involved in the actual dropping of the bombs to testify; they wanted a bill that preserved scientific freedom for research and did not allow the military to monopolize atomic energy for its violent ends.[44]

The initial hearing on the May-Johnson bill was scheduled for only one day, but congressional concerns and scientific lobbying led to another day, this time with more contrary opinions represented. Even Leo Szilard was given an opportunity to provide a counter-narrative. Though somewhat rambling, he got his point across: secrecy was pointless, and compartmentalization was counterproductive. Half the secret was that the bomb could be made, and half of what was left was contained in the Smyth Report. Even Arthur Compton seemed to somewhat agree: American preeminence would be best achieved by staying ahead, not by secrecy.

J. Robert Oppenheimer gave an opposing view, arguing in favor of the May-Johnson bill, warts and all, because, he argued, it established a framework for policymaking rather than trying to make final policy itself—the commission it would create would ultimately be the arbiter of policy. He observed that the scientists would oppose any bill that sought to regulate them: "Scientists are not used to being controlled; they are not used to regimentation, and there are good reasons why they should be averse to it, because it is in the nature of science that the individual is to be given a certain amount of freedom to invent, to think, and to carry on the best he knows how."[45]

But the criticisms of the bill were growing. Congressmen resented that the bill had been drafted without any congressional input, and that it gave Congress no input into the nation's atomic program other than providing funding. The witnesses could not simultaneously tell them that atomic energy was the most important thing in the world and also not give them any power or oversight over its operation. Congress had been excluded from the Manhattan Project, but would not accept being excluded going forward.

The question of domestic control simmered in the last months of 1945. The advocates of the May-Johnson bill struggled to keep it afloat, but impetus for atomic legislation had shifted into the Senate. The initial hope to get atomic policy sorted out quickly had failed. Truman himself withdrew support for the bill once he realized that the commission would be largely autonomous of the presidency. Though he had signed off on a statement about the need to disregard "usual concepts," the idea that the president would cede such power to a federal agency was a step too far. The May-Johnson bill was close to death.[46]

In November 1945, Senator Brien McMahon of Connecticut was appointed to head a Special Senate Committee that would take up where the May-Johnson bill had left off, reconsidering the problem of atomic energy. McMahon was no expert on atomic matters; he had angled for the position because it offered power. He drew upon the expertise of James R. Newman, a mathematician and lawyer in the Office of War Mobilization and Reconversion, who had a broad understanding of both public administration and scientific and technical matters. Newman had played a key role in undermining the May-Johnson bill and had ideas about what a replacement bill would look like. Newman's first suggestion was to table immediate legislation in favor of "self-education." He wanted to be sure the members of McMahon's Special Senate Committee, especially McMahon himself, understood the issues under debate. They appointed the physicist Edward U. Condon, the sole scientist to leave Los Alamos because of its excessive secrecy during the war, and who had recently been made director of the National Bureau of Standards, to lead this education effort. Condon's inclusion would have been a major red flag for Groves: Condon, like Szilard, was someone Groves considered a malcontent. During the war, Groves had gone so far to have Condon's passport revoked "in the interests of security," because Condon had wanted to attend a scientific conference in Russia.[47] To have him as the advisor to the Senate Committee guaranteed it would be hostile to Groves, and Groves would be hostile to it.

The members of the Senate Committee were guided through a reading of the Smyth Report and toured the Manhattan Project's Tennessee facilities. In order to get a full understanding of the bomb program, the Committee requested some of Groves' most secret information pertaining to fissile material stockpiles and production rates.[48] "No committee can make reports to the Congress or weigh legislation without being in possession of the facts," Newman and Condon wrote to McMahon, urging him to resist Groves' protestations about the highly classified nature of the information.[49] Groves considered this information militarily important: if the Soviet Union, for example, knew how small the American nuclear arsenal actually was, it could drastically affect the evolving situation in postwar Europe. (They likely did know, but Groves was unaware of that.) Groves had only the assurances of the Senate Committee that such information would be treated confidentially, and he already had reasons to dislike and distrust Condon, Szilard, and other scientists who had associated themselves with attacking the May-Johnson bill.[50]

Groves went so far as to recommend that Secretary of War Patterson discuss the matter with Truman and Byrnes "with a view to persuading the [Senate] Committee that it should accept the limitations expressed herein and furthermore that there should be a line drawn between information given to the Senators alone and that given to the Committee attachés as well."[51] Patterson passed Groves' thoughts on to Byrnes, noting that he agreed that the information requested "would be extremely dangerous to the safety of the United States." He argued that Groves was still operating under an order signed by Roosevelt "to reveal nothing of military importance" and recommended that Truman keep the order in effect.[52]

Newman and Condon, on the other hand, believed that Groves had overstepped his authority and that the War Department had no statutory ability to deny a Senate committee "Top Secret" information.[53] The matter was sent to Truman as an acrimonious dispute between the authority of the Senate and that of the secretary of war. Truman upheld the secrecy, though he cautioned that if the Senate Committee was intent on getting certain facts, he might have difficulty holding them back.[54]

The McMahon Committee's hearings in the final weeks of 1945 were leisurely compared to the hustle of the May-Johnson period. McMahon was working with Newman and others to try to develop a bill that would be a strong contrast to the May-Johnson bill, both in its appeal to the concerns of American scientists and in its contrasts with the military mindset. On December 20, McMahon introduced his bill, drafted largely by Newman and two other lawyers.[55]

The McMahon Act was, in certain respects, similar to the May-Johnson bill, though couched in a language of hope and progress rather than restriction and control. Its creators framed it in the language of civilian control rather than the military control of the May-Johnson Bill. But the Atomic Energy Commission imagined by the McMahon Act was still a powerful entity. It conceived of atomic energy as a state industry, "an island of socialism in the midst of a free enterprise economy," as Newman would later put it.[56] It had an extreme ability to control and own fissionable materials, but its powers were deliberately circumscribed by the opinion of the President. For all its expansiveness, in the area of secrecy, it was initially very liberal, its language emphasizing the importance of scientific research and free dissemination of knowledge, with security and control being secondary considerations.

The first version read to Congress contained only one section that touched on secrecy, entitled the "Dissemination of Information." It specified that "basic scientific information" could be freely disseminated, clarifying that this specific term referred only to results "capable of accomplishment" rather than "techniques of accomplishing them." It was, then, a legal distinction between basic and applied scientific work, although in not quite those terms.[57] The commission would also "provide for the dissemination of related technical information with the utmost liberality as freely as may be consistent with the foreign and domestic policies established by the President," and would have the power to designate information as free to distribute, so long as it did not constitute a threat to the nation under the Espionage Act. Between the overtures to free distribution of research and an unwillingness to declare nuclear information different from any other information under the scope of the Espionage Act, this was a very liberal approach to secrecy. Indeed, the act's "Declaration of Policy" said nothing about secrecy whatsoever and instead spoke only of fostering scientific research and development, "cementing world peace," and dissemination (not restriction) of scientific information.[58] Again, the contrast between this bill and the May-Johnson bill was deliberate, and extreme.

Hearings on the McMahon Act were conducted through the spring of 1946, covering several hundreds of pages of open Senate testimony and yet-unknown volumes in closed executive sessions (they have never been released).[59] Groves continued to worry about "secret" information slipping out during the testimony. Lengthy hearings, especially those featuring secrecy-hating witnesses, seemed like a recipe for leaks. Groves worried that his greatest fears about the dissipation of his Manhattan Project empire were being realized. Worse yet, if any security problems occurred on his watch, they would be seen as further reasons why the military should be left out of future domestic control. One of Groves' deputies reported in February 1946:

> Each day this office receives additional information indicating that there is almost a complete lack of realization on the part of non-Project military and naval personnel, that certain information relating to the atomic bomb remains classified. This attitude, unless quickly corrected, will eventually lead to a serious violation of security. It will also be a reflection upon the efforts of the Manhattan District to protect properly information relating to the bomb.[60]

Groves needed a strong play. On February 16, news suddenly broke that twenty-two people had been arrested in Canada for passing atomic "secrets" to the Soviet Union. This "Canadian spy ring" story was the first of its kind, emphasizing that there were "secrets" and that they had been compromised. The attitude in Congress shifted rapidly: the McMahon committee returned to the old trope of preserving "the secret" in the face of evidence that the Soviets were aggressively pursuing it. There is reason to believe that the news about the Soviet spy-ring had been leaked to the press by Groves himself. This would be ironic if one truly believed Groves thought secrecy was paramount, but we have already seen how much he understood the value of a well-timed release of information.[61]

Two days later, Groves argued in a letter to the secretary of war that there had been a "loss of security" as a result of the McMahon committee's efforts. "Some of the scientists participating in the study of safeguards and controls are also closely allied with the groups now urging relaxation of security rules," he alleged. "These men have urged that their committee work can best be done if permitted to discuss at least the technical problems involved with associates. . . . Briefly, Top Secrets of the Manhattan Engineer District are now known to a large number of people over whom we have little, if any, control or jurisdiction."[62] The "lower echelon academic employees" who made up the Scientists' Movement, as Groves dubbed them, intended to undermine security in order to further their own anti-secrecy ideologies: "Their plan is that the Army cannot put everyone in jail if everyone starts talking about technical knowledge in their possession."[63] The solution, Groves urged, was to reinforce existing restrictions, resulting in a memorandum from Truman (written by Groves) urging the maintenance of secrecy and of War Department supremacy in enforcing controls over information, pending the creation of a true postwar agency.[64]

Groves appeared twice before the McMahon Committee at the end of February, once in private, once in public. In the private session, Groves emphasized the loss of secrets from the Canadian ring and apparently made a strong impression on the Committee's conservative members.[65] At the public hearing, Groves positioned himself as tough on security but liberal on scientific release. He expressed a lack of confidence in the McMahon Act's security provisions. But he also noted that he was already establishing a sound declassification policy that had been drawn up by top project scientists (discussed later in this chapter), and was moving toward having research at the new Argonne laboratory unclassified. McMahon did not hold back his own antagonisms, belittling Groves' record of career advancement in the Army. But McMahon came off looking desperate, while Groves appeared to have the moderate position: security and freedom, sundered from ideological hubris.[66]

In closed committee sessions, the McMahon Act began to change in key ways. Many aspects that had appealed to the organized scientists, like its exclusion of military influence, were stripped away through a series of amendments. Most striking was how the provisions relating to scientific information were drastically rewritten as the Senate Committee struggled to accommodate to a new emphasis on security.

That March and April, significant changes were made to the section on “Dissemination of Information.” The most obvious change was the renaming of the section title to “Control of Information,” signaling a large change in approach. The new bill would allow the military, with approval of the Commission, to create regulations governing the dissemination of information relating to “atomic bombs and other military applications of atomic energy.” This was essentially the approach taken by the May-Johnson bill, except that it required the President to assent to any such regulations. Penalties had also been increased, punishable by up to a $20,000 fine and/or a 20-year prison sentence to anyone who did so with intent to injure the United States or to secure advantage to a foreign nation. Committee notes on the bill indicated repeatedly that their original idea, that the “basic” and “applied” distinction for deciding what should be secret, had been disavowed by Manhattan Project scientists: the real world of nuclear technology was far more dual-use and overlapping.[67]

In early April, Committee members raised concerns about how the Atomic Energy Commission could control the secrets of the bomb without running afoul of the free-speech provisions of the First Amendment.[68] By April 11, to address this question, they had contrived a new draft in which the section on “Control of Information” had been radically altered. The bill’s authors had employed a legislative trick favored by James Newman in earlier drafts: create a novel legal term, then indicate the scope of the term, then indicate that the commission would have powers over that term. In this case, they created a new category of information entitled “restricted data.” “Restricted data” would be defined as “all data concerning the manufacture or utilization of atomic weapons, the production of fissionable material, or the use of fissionable material in the production of power.”[69]

This seemingly straightforward definition is in fact extraordinarily broad: the “secrets” in this case were defined *by their nature* and not by an act of regulation, which make them different from every other category of secrecy in the United States. When an authorized classifier determines that a document is “Top Secret,” for example, the classifier is making a determination about the potential harm of the information and issuing judgment in the form of the secrecy stamp. In the case of “restricted data,” information either is within the definition of “restricted data,” or it is not. Harm has nothing to do with it. And, because "restricted data" was defined as "all data," not "all data created by the Commission" or some other limiting factor, the statute was open to the interpretation that nuclear weapons information is "born secret," no matter who or where the new information comes from. No other statute on secrecy has ever been defined so broadly in the United States.[70]

The Atomic Energy Commission's main information responsibilities were then redefined to "control the dissemination of restricted data in such a manner as to assure the common defense and security," with grave penalties associated with mishandling of it. They could also *remove* information from the "restricted data" category upon concluding that it could be published "without adversely affecting the common defense and security."[71] So the commission could not legally create secrets, it could only release them—a very unusual construction, and again, very different from the way secrecy had worked up until this point in the United States.

What logic lay behind the unusual construction? McMahon later reported to the president that the Senate committee added these specific security provisions because they interpreted the Espionage Act to apply only to *documents*, not the transmittal of *information*.[72] This was deemed too feeble to protect something as dangerous as "top-secret information concerning the atomic bomb," which could come in many forms. Amending the Espionage Act would be a legislative battle in and of itself, so they had taken it upon themselves to craft a replacement that would work for their special subject. The McMahon Committee had repudiated the early attempt to draw a line between "basic scientific information" and "related technical information" and felt that to give the commission the power to "issue regulations" about security would be to face charges of allowing the commission to "act arbitrarily and capriciously." The "restricted data" solution, in McMahon's view, was a worthy compromise, since by setting *all* data as classified by default, the commission's only act could be to *withdraw* classification status. As he put it: "The Commission's withdrawal power can only reduce—it cannot enlarge—the scope of the crime." Thus, he concluded, the needs of "both military security and scientific progress" were satisfied.[73]

The McMahon Committee feared an Atomic Energy Commission that could classify things at will. To avoid this, they had Congress preemptively classify everything potentially related to nuclear weapons or nuclear power, and then gave the commission power to withdraw information from that category. In the decades that would follow, it would create a series of intractable problems for the commission, but despite various challenges from within and without, the "restricted data" clause would survive the entire Cold War and persist into the present, and "Restricted Data" would stand as one of the most distinctive aspects of nuclear secrecy in the United States.[74]

When the final version of the McMahon bill was presented to the Senate in mid-April 1946, it was accompanied by a general report on atomic energy and an analysis of the new bill's components. Although the bill was strict about security, it took as a matter of policy that the "dissemination of scientific and technical information relating to atomic energy should be permitted and encouraged so as to provide that free interchange of ideas and criticisms which is essential to scientific progress." The report acknowledged that this was diametrically opposed to information restriction, but said that juxtaposing these two "considerations of opposite tendency" would "frame a program that will reconcile their apparent divergence."[75] The circumstances created conditions that, as we shall see, the newly created Atomic Energy Commission found difficult to understand and enforce. McMahon's idea that writing deliberately contradictory legislation would lead to enlightened governance is perhaps as much of an indicator of his inexperience as anything else. Unsurprisingly, the Scientists' Movement participants were aghast at these new provisions, but they did not raise a large outcry. They appear to have considered the McMahon bill better than the May-Johnson bill, and had probably concluded that the fight against security provisions was a losing proposition given the present atmosphere. The War Department found the current version of the bill sufficient.[76] Secretary of War Patterson judged "Restricted Data" to be "a different way of expressing the same idea" of secrecy by regulation.[77]

The Senate voted on and passed the bill on June 1, and it was sent to the House of Representatives, where further debate continued. Secrecy was a major point of contention. Some argued that even with its new "Restricted Data" concept, the bill would "give away the secrets." Rep.

John E. Rankin of Mississippi pledged himself to its defeat: "You are not going to wreck my country if I can prevent it; you are not going to take the only weapon we now have to protect ourselves and give it to our enemies. God forbid."[78] The bill narrowly avoided being sent back to committee. The House passed the bill only after adding the requirement that all commission employees desiring access to "Restricted Data" would need a full FBI investigation into their "character, associations, and loyalty," and raising the penalty for deliberate espionage to life imprisonment or even death. In committee, the Senate accepted all security changes made by the House, and the McMahon Act, now the Atomic Energy Act of 1946, was signed by President Truman on August 1, 1946, almost a year to the day after the bombing of Hiroshima.[79]

The staffers who did most of the drafting of the McMahon bill were dismayed at how much the "compromise secrecy section," as one put it, undermined their original intent. Byron Miller, a young lawyer who had worked with James Newman from the beginning to draft the bill, complained that in the end there was a kind of "schizophrenic performance" in Congress, "with a definite swing to military emphasis despite the victory for 'civilian control,'" verging on "war hysteria."[80] Newman, the author of the free research clauses in the original McMahon bill, lamented that "Congress, nevertheless, decided that the dangers of free speech in nuclear science and related technologies could not be risked." The law, he argued, had a "draconic sweep" that revealed "Congress' obsession with the safeguarding of secrets."[81] He concluded:

> Preoccupation with the "secret," instead of the thing itself, will stifle the scientific research from which our real strength is derived, will strengthen the pernicious misconception that we have a monopoly of knowledge in the science of atomic energy, and will beguile us into embracing a fatal fallacy that we can achieve security for ourselves by keeping our knowledge from others."[82]

Nearly a year earlier, Robert Patterson, the secretary of war, had warned that the scientists who sought to derail the May-Johnson bill "[do] not realize that by delaying action and raising all sorts of objections to the present bill, they may very well end up with a much more stringent measure than is now before the Committee."[83] He turned out to be correct. The Atomic Energy Commission that the Atomic Energy Act of 1946 created was every bit as imperial as the one proposed by the May-Johnson Act, except it would face greater congressional oversight in the form of the new Joint Committee on Atomic Energy. Given Congress's conservative tilt toward the preservation of secrecy, this could be little solace to those who had advocated for a more liberal law. Nuclear secrecy had been made legally permanent and expansive through the unique and novel legal concept of "restricted data."

4.3 OPPENHEIMER'S ANTI-SECRECY GAMBITS

There were two other postwar attempts to rein in secrecy in the period between the end of World War II and the real onset of the Cold War that are worth looking at. Both efforts had strong ties to the beliefs and actions of J. Robert Oppenheimer, the wartime head of the Los Alamos laboratory. Durng the early postwar period, he held considerable sway as a government advisor and expert, and one of the ways he attempted to use this influence was to reshape how secrecy was conceptualized. While these efforts, like most attempts at secrecy reform, were unsuccessful, their failures are instructive.

The first of these efforts was in establishing the internal processes of postwar declassification procedure—the means by which secrets could systematically stop being secret. The Smyth Report had been the first attempt at formulating a rationale for determining what information should be released and what should not. In that case, the focus was on information that had already been public prior to the establishment of the Manhattan Project or that could be easily discovered with modest effort. But the Smyth Report was not a system, it was a one-time release, tailored to a very specific context. Manhattan Project officials had not seriously contemplated what a postwar declassification system might look like, largely because they considered this something that would be solved by whatever domestic control legislation was put into place.

But by late October 1945, it was becoming clear to Groves and others that Congress would not quickly enact a new sytem, and that a one-time release was not going to be adequate. This necessitated the creation of a system of "declassification," a term so linguistically awkward and culturally novel that it frequently was placed in scare-quotes in the mid-1940s, and competed with other terms for the same meaning (e.g., "reclassification").[84] Unlike the "Publicity" approach to weapons information, this new system would be a continual, changing process, one that could adapt to its times.

Today, classification and declassification are frequently linked, but prior to World War II no system existed for determining how to remove secrecy in anything other than wholesale measures (e.g., declaring all wartime secrecy orders outdated, which is what largely happened after World War I, with a few exceptions), mainly because the war was the first occasion in which military classification had extended to such a great degree over scientific and technological data, which is harder to declassify than, say, outdated military operations. Toward the war's end, scientists and engineers, as well as industrial organizations eager to use their wartime work to augment their peacetime operations, had begun to call for the release of scientific and technical data.[85] In 1944, the Office of War Information issued regulations that created a requirement for agencies in possession of classified information to "downgrade" secret material as conditions in the war changed, but it still wasn't a system.[86]

The Manhattan Project was not the only wartime project to develop classified scientific and technical information. The OSRD had begun to deal with postwar declassification as early as the summer of 1944. Their major concerns were to avoid flooding the market with scientific papers, and to convince their scientists to put the papers into a publishable form once the war had ended. They also had security issues and patent questions to work out. Vannevar Bush eventually, through work with the Bureau of Budget, developed declassification procedures that were codified in Executive Order 9568 ("Providing for the release of scientific information"), issued by President Truman in June 1945.[87]

Executive Order 9568 required the War and Navy Departments to arrange to release wartime scientific and technical data that had been previously kept secret, "to the end that such information may be of maximum benefit to the public." It allowed for information that still had military significance, as determined by the armed services, to be retained under secrecy. It further authorized the use of government funds in publishing said materials. It was an authorization of declassification in general and a mandate of responsibility for implementing it, but not a detailed how-to. Within the OSRD, a Committee on

Publications was created to coordinate this activity. The goal was to release the majority of useful OSRD information before the OSRD itself was disbanded. Determination of the security status of information was left to the Army and Navy, who would convey their judgments to the OSRD for implementation. But the Army and Navy did not use consistent methodology: the Army would declassify entire projects and fields of research, whereas the Navy declassified information on a report-by-report basis.[88]

Groves' Public Relations Organization would serve, in the short term, as a de facto declassification organization. But just days after the bombing of Nagasaki, the Organization was being deluged with requests from Manhattan Project scientists to clarify or change the information release policy. The requests came from scientists whose work had only peripheral relevance to weapons development, but many peacetime applications.[89] Contractors who had accepted secrecy during wartime were also asking to release information, and private individuals and companies uninvolved in the project were asking about the use of new materials, new processes, and new ideas for their own work.[90]

In early October 1945, the War Department requested that Groves start to "make available all possible by-product information gained as a result of the work conducted by the Manhattan District," supported by Executive Order 9568. Groves dismissed the request, on the basis that atomic energy would remain a "closed" field until Congress had created a postwar organization.[91] But within a few weeks, as the promise of a rapid postwar organization faded, Groves reached out to Oppenheimer for a "sketch of a declassification policy." While admitting that the "problem of declassification" was "among the most difficult," and that there was "no really satisfactory solution" that could be reached, Oppenheimer presented a basic schema for dividing Manhattan Project information into three distinct categories, based on the possible relevance of the knowledge to a potential adversary: "What I have attempted to do is to list items in three categories: those which in my opinion should be declassified[;] those I think that should be declassified, but where controversy may seem to be inevitable[;] and those which at the present time I should feel it impossible to declassify."[92]

Into the category of "Group I. Declassify," Oppenheimer had all basic physics, chemistry, and metallurgy related to the bomb project, including basic research on separating isotopes and "all known cross sections" of all isotopes used. Into "Group II. Declassify (controversial)," Oppenheimer included issues related to neutron diffusion and pile operation, "the basic principle of implosion," and known critical masses. "Group III. Do not declassify," included production rates, exact plant designs, the exact bomb designs, industrial techniques, and the work on the "Super" (hydrogen) bomb.

It was only a "sketch" of a general taxonomy of existing information, with no recommendations for how such a system might be implemented or overarching philosophy for deciding which information went into each category, or how to administer it on a larger-scale. For these aspects, Groves decided to set up a committee headed by Richard C. Tolman, the aging Caltech physicist who had been a close technical advisor of Groves during the war and who had helped develop and implement the security restrictions for the Smyth Report and was chair of the Committee on Postwar Policy. Tolman was considered conservative on matters of secrecy. He was a scientist, but one Groves could trust to create the kind of system that Groves himself would find acceptable.[93] In a telegram, Groves authorized Tolman to form and chair a committee that would make recommendations as to how they should "carry out in an orderly manner the declassification and release of information obtained as a result of the work in the Manhattan Project, if the security of such information is no longer important to the welfare of the United States."[94]

The Tolman Committee would have as its members an all-star group of American scientists who had worked on the Manhattan Project, who were not homogenous in their temperament, specialties, or political sensibilities: Robert F. Bacher (a physicist who had helped design the "Gadget"), Arthur H. Compton, Ernest O. Lawrence, Frank Spedding (a chemist who had developed the means of making uranium metal), Harold C. Urey, and, of course, J. Robert Oppenheimer. Aside from having three Nobel Prizes between them, the Committee possessed experience on nearly every aspect of the industrial and scientific processes behind the Manhattan Project.[95]

Politically, Tolman and Lawrence represented a conservative stance more in line with Groves', while the others ran the gamut from the ambiguously liberal Oppenheimer and Bacher to the very liberal Compton and Urey. But their sensibilities about declassification did not fall along political lines. All shared some agreement on the importance of retaining information, at the very least as a "hedge" to encourage international control, and all were eager not to impose onerous restrictions on the development of science outside the laboratory. Tolman had recorded his views during the war in a letter to Warren Weaver of the Rockefeller Foundation, associating heavy control of science with Nazism, and noting that, after the war, they would have "to set science free again."[96] Compton believed that nuclear science needed to be granted "free rein" if it was to flourish in the postwar period.[97] Compton also worried that if the field of reactor engineering were not opened up to some degree, it would be extremely unattractive to new students.[98] Even a political conservative like Lawrence thought the attempt to "keep fundamental science secret for the national security" would be pointless.[99] At the same time, the committee was to develop a system for controlling information that assumed there would be no international control regulations: they were to make a plan for information release for the "worst case scenario" of an arms race.[100]

In Tolman's office in Pasadena, the group convened for the first time in mid-November 1945. They dubbed themselves the Committee on Declassification, a name evocative of their desire to determine how information was released, even though comprehensive means of determining why it should be retained was explicitly part of their agenda. They began by looking at specific requests for declassification sent to the Manhattan District from industrial representatives, at requests and advice from scientists on the project, at plans for declassification written by scientists at the Metallurgical Laboratory, and at thoughts on the declassification of information relating to the gaseous diffusion plant at Oak Ridge. Based on their discussions, Tolman drew up a draft "Report of Committee on Declassification" and circulated it for committee comment a few days later, before sending off a final report to Groves in early December. Ironically, the report itself would be classified "Top Secret" because, as Tolman explained, "it presents an overall view of the Project," and because knowing even the existence of some of the items on the do-not-declassify list would compromise security.[101]

The "General Philosophy" of the Committee's report expressed skepticism "that the concealment of scientific information can in any long term contribute to the national security of the United States," though members "recognized that at the present time it may be inevitable that the policy of the Government will be to conceal certain information in the interest of national security." They felt that unless war was likely in the next five to ten years, most information not related to the actual production of atomic weapons could be declassified; however, even in that case, the attempt to hold on to that information would "disastrously" weaken the US advance in atomic energy matters in the long run.

The Committee divided scientific information into categories similar to those Oppenheimer had used in his "sketch of a declassification policy" the month before. The first, Class I, was recommended for "immediate declassification" and consisted of scientific data with no direct relevance to military matters or that was easy to reproduce in a small lab. The third category, Class III, was for information that had immediate military applications, either relating directly to the work of making bombs or hinting at the stockpile size or production rates of the United States. Between these two extremes was Class II, "information whose declassification would conduce to the national welfare or to long term national security," but that also "has direct bearing on production or military utilization." Thus, declassification for things in this category "should depend on estimates as to the probability and imminence of war," but would still be expected to be declassified in five to ten years. Oppenheimer would later describe this middle category as the "tough problems," and the categories themselves as "what should be made public, what should by all means not be made public, and what should be worried about."[102]

The Committee then divided the scientific and technical developments of the Manhattan Project among the three categories. In weighing their determinations, the Committee considered eight "positive" criteria that would favor declassification and three "negative" ones that would favor continued classification. Some of these are as one would expect: "advancement of general science" was a positive consideration for declassification, and "would jeopardize U.S. military security" was a negative one. But some are more surprising: "advancement of military aspects of nuclear technology" was a positive reason to declassify, and "would jeopardize patent position" was a reason not to.

TABLE 4.1. The "positive" and "negative" criteria established by the Committee on Declassification to be weighed in evaluating the release of any particular piece of information. Quoted verbatim from "Report of Committee on Declassification."

Criteria Affecting the Declassification and Transmission of Information

In considering the declassification and the transmission of information, the Committee concluded that it was appropriate to apply the following positive and negative criteria:

Positive Criteria

1. Advancement of general science.
2. Advancement of non-military aspects of nuclear science.
3. Advancement of military aspects of nuclear science.
4. Advancement of general technology.
5. Advancement of non-military aspects of nuclear technology.
6. Advancement of military aspects of nuclear technology.
7. Information already substantially known outside project.
8. Information readily obtainable by theory or minor experimentation.

Negative Criteria

1. Disclosure would jeopardize U.S. military security.
2. Disclosure would weaken U.S. position in international discussions.
3. Disclosure would jeopardize patent position.

The Committee's "Classification of Topics" comprises seven pages of rulings, parsing the nuclear field into discrete pieces. In general, most basic science fits under Class I, and all information directly related to military stockpiles, production rates, or bomb use or design falls under Class III. The middle category, Class II, represents a view of what these scientists thought was the most contested information on the boundary between safe and dangerous. These include experimental and practical work on uranium enrichment methods the Manhattan Project found potentially-workable but unsatisfactory, and large amounts of information relating to nuclear reactors, the latter presumably because of their potential for peacetime applications. Almost everything relating to the electromagnetic method of enrichment was placed in Class I, as was all medical information, so long as it revealed nothing about the physical properties of isotopes that were otherwise classified. The Committee singled out the medical effects of the bombs at Hiroshima and Nagasaki for release, "in order that exaggerated statements as to the lasting radiation effects may be discredited on the basis of the true facts."

In addition to general science and technology, the Committee also

TABLE 4.2. Selected examples of recommended categorizations by the Committee on Declassification. Paraphrased and selected by the author from "Report of Committee on Declassification."

Class I. ("Declassify")

- Physical instrumentation (counters, ionization chambers, cyclotrons).
- Methods of applied mathematics and computation, if illustrated on declassified subjects.
- Metallurgical techniques.
- Design and operating characteristics of small experimental piles in which enriched material or heavy water is used, provided the pile generates power at a level under 100 KW. The chemistry of decontamination is not included.
- All nuclear properties and chemistry of non-classified substances.
- Basic studies of chemical effects of radiation.
- General theory of centrifuge.
- Most information relating to experimental and theoretical work on electromagnetic enrichment, except details on enrichment levels, production rates, etc. (which are Class III).
- Basic theoretical work on gaseous diffusion cascade design and kinetic chemistry, although specific applications to the project cascade and to conditioning respectively should not be classified.
- List of non-classified isotopes that are produced in uranium piles, so long as production capabilities are not revealed by showing abundance of isotopes or their rate of production.
- Details of fission product chemistry, omitting reference to separation processes.
- All reports on medical research and health studies, omitting such items as might disclose information beyond other Class I information.
- Medical information as to the effects of the bomb on Hiroshima and Nagasaki.

Class II. ("Worry about")

- Experimental work on centrifuge method of isotope separation and detailed mechanical design.
- Nuclear characteristics, including capture, fission, and scattering cross-sections for all energies of neutrons; number of neutrons produced per fission; spontaneous fission rates, etc.; for all isotopes of plutonium, uranium, protactinium, and thorium.
- Thermal diffusion method as applied to uranium hexafluoride.
- Plutonium extraction and decontamination chemistry, "without reference to larger scale problems."
- Pile theory for production units, omitting reference to actual installations.
- Experimental and theoretical work on converter, breeder, and power piles.
- Critical masses, without reference to weapon design.
- The theory of implosion, without reference to military applications.
- General theories of efficiencies, without reference to specific weapons.

Class III. ("Do not declassify")

- Production plants, overall details, flow sheets, rates of production, operating procedures, and policy. Stocks and reserves of uranium and other classified substances.
- "All specifically military matters."
- All practical aspects of constructing gaseous diffusion plants.
- Use of fission products as chemical warfare poisons.

TABLE 4.2. Continued

Class III. ("Do not declassify") continued

- Specific military and naval uses of atomic power.
- Use of atomic energy for jet propulsion.
- Detailed design of weapons, including fuses, firing systems, detonators, neutron initiators, explosive lenses.
- Rate of production, reserves, and storage of bombs.
- Destructive effects of actual stockpile weapons.
- Use of weapon under water.
- The "super" [hydrogen bomb] as a weapon.

considered "classified substances," the elements and isotopes that had direct relevance to nuclear weapons, and their chemistry, metallurgy, basic physics, nuclear physics, and technological applications. No chemistry, metallurgy, basic physics, or nuclear physics was rated higher than Class II, except for a blanket category of "Special Ordnance Materials developed at Los Alamos," which was entirely considered Class III. All technological applications were rated as Class II or Class III.

Along with recommendations for the immediate future, the Committee also suggested a proposed "mechanism" for declassifications. First, the authors recommended that their initial determination of Class I, II, and III categorizations could serve with minor modifications as a "Declassification Guide" for individual requests. Specifically, they recommended that several "Declassification Guides" be created. Rather than one master guide, dissected sub-guides would cover specific sections of the project: the wartime principle of compartmentalization applied to the guides themselves because they recognized that the full guide "gives an overall picture of the whole project and makes mention in certain instances of extremely secret matters." Secondly, they recommended the creation of a "Declassification Organization," part of the Manhattan Engineer District, that would consist of the directors of various project sites and a set of "Responsible Reviewers," local scientific experts on particular topics approved as declassification experts, who would apply the Declassification Guide and their own expertise to individual declassification requests.[103]

The end goal of these two recommendations—to create a distributable set of Guides and to create the distributed positions of Responsible

Reviewers to implement the Guides—would allow a decentralization of the declassification effort, a necessary action if it were to be implemented on any sort of scale. Until decentralization was possible, the only people who would have the breadth of project experience to give informed judgment on such topics were those in the upper echelons of the project, people too busy to add declassification of documents to their daily routine. Thus, decentralization was necessary to de-skill and scale up the scope of the declassification process.

By the end of 1945, several members of the Committee on Declassification had met directly with Groves and several prominent industrialists to get their perspectives on their proposed system.[104] Tolman reported to those not in attendance that Groves had been "bothered by the fact that he has been criticized for releasing the Smyth Report" and that, consequently, he was bewildered about what to do in the ensuing public discussion of postwar secrecy, where the options ranged "all the way from recommending releasing everything about the Atomic Bomb to releasing nothing."[105] Tolman suggested that his committee could aid Groves by writing up a statement for McMahon's Senate Committee, "setting forth their philosophy of declassification and incidentally expressing their approval of the release of the Smyth Report," and then make a "public announcement of the feelings of the Committee on the desirability of a liberal and forward looking declassification program."[106] They issued a public statement by early February 1946, but it got scant attention in newspapers. People were more interested in stories of spies.[107]

The Committee prepared three more reports for Groves, outlining the bureaucratic procedures for declassification and tackling specific problems Groves had requested they look into (a detailed justification of topics for inclusion in the various categories of the original Committee report and an itemized list of information already known in the public literature about "non-classified" substances on which the report had bearing).[108] Groves had begun to organize a Manhattan District Declassification Office based on the Committee's recommendations, and the proposed Declassification Guides had been created and distributed to a group of Responsible Reviewers recommended by members of the Committee.[109] In February 1946, Groves had informed Secretary of War Patterson on his declassification activities and had Patterson seek approval from President Truman. He also encouraged Patterson to make clear who had instigated it in the first place, if he had the opportunity, and to emphasize that it was not a case of them reacting to external pressures, but an effort originating from within the War Department.[110]

As Groves put the Committee's recommendations into effect, its direct work was completed and it went into what Tolman described as "hibernation." Further revisions to the Declassification Guides could now be made by the Senior Responsible Reviewers. In April, Groves belatedly fulfilled the US obligations under the Quebec Agreement by informing the Combined Policy Committee of his plans to release information, and received preliminary approval.[111] In July 1946, one of Groves' assistants reported to Bacher that "all of the essential elements of the declassification organization are now in actual existence and are functioning as such."[112]

But there were still unanswered questions. It was one thing to construct a "sensible" or "enlightened" system, it was another to make it function in practice. And in July 1946, John H. Manley, the Senior Responsible Reviewer for weapons questions, threatened to resign because he felt that the Declassification Guides were not "guides" at all—they were rulebooks:

> It was my impression that the Tolman Report and the Declassification Guide were to form the basis of a policy which would *guide* Responsible Reviewers in making recommendations for declassification. I am now informed that . . . nothing is to be declassified unless the Guide specifically states it to be declassifiable. . . . I believe that the legalistic interpretation of the Guide as illustrated above is contrary to the interests of the country and to the recommendations of the Tolman Committee.[113]

By August 1946, the Senior Responsible Reviewers were writing to Groves that they felt being asked to refine the definitions used in the Declassification Guide was "acting in the formulation of national policy," and as such not "a proper obligation" of their group.[114]

Even while the first report was being drafted, Spedding had misgivings, noting that "while it does not give the freedom which I had hoped for and which I feel is essential over any long period of time, I believe it is the best we could hope for at the present moment."[115] Almost a year later to the day, he would write to an Atomic Energy Commission (AEC) official that while he was "still in entire agreement with the philosophy put forth at that time by that Committee," "there has, unfortunately, crept into our present procedure for classifying information a second consideration that 'anything the government pays for is classified until it has been written into a document and declassified.'"[116]

Spedding's latter comment is actually something that the Tolman Committee's report recommended explicitly. It was "document-based" declassification, not "field-based." That is, not all information in a given field of research was to be released, but rather individual reports would be approved or disapproved based on the manuals. The reason for this distinction was never articulated, but it seems likely that the intent was to make sure that all pertinent information was properly reviewed by the many interested parties within the MED. This procedure was made explicit in the new "Manual for the Declassification of Scientific and Technical Matters" disseminated throughout the Manhattan Project sites in May 1946, which specified the labyrinthine system through which these documents would wend their way before release.[117]

The declassification system forged by the Committee on Declassification was not immune to criticism, but it was meant to be a significant reorientation of thinking about secrecy from the closed environment of the Manhattan Project. Many things did not come to pass: much of the information that was to be released barring imminent war stayed under wraps because of the long Cold War, though research on topics relating to reactor theory, medical isotopes, nuclear fusion, and the like was eventually declassified in the mid-1950s as part of Eisenhower's "Atoms for Peace" program. In the end, what the Tolman Committee had set up was more a framework for policy than a policy itself.

For the Committee on Declassification, the chief question was how to evaluate *technical* data. Almost nothing the committee categorized could be described as political or administrative, other than in the broad sense that all nuclear technical information in the postwar period had long-reaching political implications. The Committee's criteria for and against revelation were all related to the advancement of science or the preservation of military security; none were concerned with many of the other issues that had been associated with secrecy, such as democratic deliberation or accountability, or the public's own "need to know."

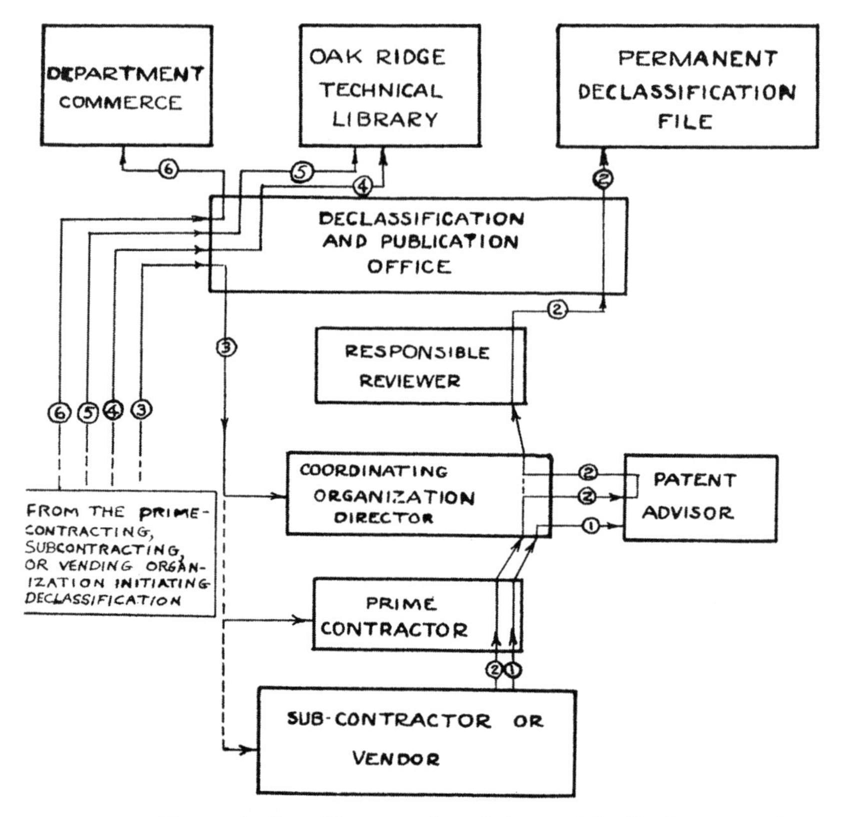

FIGURE 4.1. The complex flow of documents through the new declassification system of the Manhattan Engineer District. Each number is a copy of the document being evaluated for declassification. Source: "Manual for the Declassification of Scientific and Technical Matters" (1 May 1946), copy in NTA, document NV0713951.

The only acknowledgment of a broader political context is a negative one: information should not be released that would "weaken [the] U.S. position in international discussions." Specific elaboration of what it meant to "jeopardize U.S. military security" was lacking, as was what constituted "advancement of general science." It is this self-consciously technical nature of their balancing, based largely on the question of what was already known and what was easy to know, that very likely accounted for Groves' willingness to implement the system as described. It is the sort of system that a group of physical scientists might come up with.[118]

Oppenheimer's gambit, in this case, was in his effort to turn secrecy from a black-and-white question into shades of gray. The tripartite system of classification, which seems to have come directly from him, was a compromise between the "all" that the more activist scientists wanted released and the "nothing" that the military preferred. In practice it did not work as liberally as he might have wanted: there was still plenty of opportunity for conservative judgments about scientific releases, and in the end, the practices of secrecy were broader than the practices of classification and declassification alone. In retrospect, however, the Committee on Declassification's system was immensely influential: their system of guides, reviewers, and the process of declassification were adopted by the Atomic Energy Commission when it took over the work of the Manhattan Project and would, during the Cold War, spread to every other agency of the US federal government, becoming the way in which declassification was handled in the United States, warts and all.[119]

At the same time that the Manhattan Project declassification program was being put into motion, it was being invoked in a different context. Over the course of the spring of 1946, J. Robert Oppenheimer was serving as the primary technical advisor to a committee, chaired by David Lilienthal, that was developing recommendations for Undersecretary of State Dean Acheson on the future international control plan to be advocated by the United States. This was the outgrowth of many conversations, started during the Manhattan Project, about how to avoid a postwar arms race. The final scheme proposed was largely rooted in Oppenheimer's belief that international control had to be based on something other than secrecy and Lilienthal's belief, derived from his experience as the head of the Tennessee Valley Authority, in the power of large, forward-looking, technocratic organizations to create positive social transformation. This would be Oppenheimer's second major gambit for decreasing the stranglehold of secrecy.[120]

The final product, finished that March, was titled "A Report on the International Control of Atomic Energy," but was commonly referred to as the Acheson-Lilienthal Report.[121] The Acheson-Lilienthal Report argued for the creation of an Atomic Development Authority, through the United Nations, which would be granted control over worldwide uranium stocks. Uranium, the report argued, was the single essential link in the large chain of all applications for atomic energy, including military applications. Whether enriched in its U-235 content or used to breed plutonium (or U-233, from thorium), natural uranium was a prerequisite for producing fissile material that could be used in weapons purposes. Thus, if the Authority could control uranium reserves, it could guarantee that no nations under its auspices were engaging in military use of the atom.

The Authority would also authorize and oversee the different types of "safe" and "dangerous" atomic activities of its member states, and would need to put itself in a position through which it could observe possible diversionary activities and keep abreast of changing scientific and technological developments that would affect its classification of "safe" and "dangerous" activities. It was this "positive" approach that invigorated Lilienthal and Oppenheimer in particular: the Authority would not be simply a police body, but an effort to coordinate the global use of peaceful nuclear technology as well.

At its core, the control regime proposed in the Acheson-Lilienthal Report was radically different than one governed by secrets, as it argued that the vector for control of nuclear technology was *material*, not *epistemic*. Without sources of raw uranium, no amount of knowledge could possibly make an atomic bomb. The final report made this argument more implicitly than explicitly, speaking of the need for control and the political problems of secrecy but without contrasting its own conclusions with a regime of secrets. The closest it came was noting that any international control regime necessarily would have to abandon the idea of secrecy and monopoly: under international control, "knowledge will become general, and facilities will neither in their legal possession nor in their geographical distribution markedly favor any one nation."[122]

Furthermore, the final report did concede that theoretical knowledge was at least a temporary hedge, and that some secrets should be maintained. For example, it recommended that the US "monopoly on knowledge cannot be, and should not be, lost at once."[123] It further went on to state that while some information could be ready to be disclosed to the United Nations for their work in formulating a policy of international control, full disclosure would have to wait for positive political developments.

But this final, published version, based on the version of the report adopted on March 17, 1946, is actually significantly different in this respect than the draft of the report dated a day earlier.[124] In particular, the final section of the report, which itself had been a late addition, had been changed to be more equivocal about the question of secrecy in the overall control scheme. The original version of March 16 was an unabashed assault on the idea of secrecy, and made the strongest contrast between control-by-materials and control-by-knowledge:

> The security which we believe ultimately feasible is essentially a security growing out of material things; it does not rest on keeping nations or individuals ignorant. We believe that this is the only firm basis for security, and that there can be no thought of international control and international cooperation which does not presuppose an international community of information and fact. Correspondingly, we believe it less relevant to our actual security with regard to atomic weapons to maintain private our knowledge than to keep, during the early years when the international authority is undergoing development and tests, our advantage in facilities and material.[125]

Additional cut pages discussed why secrecy regimes must fail. There were "grave difficulties in setting any rigid limitations in advance on the extent to which the field of knowledge—especially theoretical knowledge—can be monopolized by any nation." Theoretical knowledge would be "one of the first things which any other nation would be able to obtain," because it rested "on a genuinely international body of science." For this reason, "attempts to maintain an American monopoly by secrecy are the greatest threat to any healthy development of our own science and technology." This line was tempered with the acknowledgment that to go from theoretical knowledge to industrial practice was quite a leap, and knowledge could "shorten somewhat the time needed by a new group to solve the practical problems of making atomic weapons, and may perhaps eliminate certain unworkable alternatives and thereby reduce the effort needed." It concluded that the amount of time saved would be small, and therefore the committee had "adopted it as our view that the maintenance of a monopoly on theoretical knowledge should not be allowed to interfere with the establishment of international control."[126]

The differences between this draft and the published version are striking when put next to each other. In the original draft of March 16, the work of the Tolman Committee on Declassification was held up to prove that most of the knowledge that the United Nations would require to have an adequate understanding of international control was already going to be declassified, not because it would benefit international negotiations, but for the interests of the United States' own nuclear development. This conclusion was considered valuable enough that it was underlined: "All of the basic theoretical knowledge which is likely to be required for the discussions of the United Nations Committee and the general planning of the Atomic Development Authority has been recommended for declassification in the interests of our own long-term national security."[127] By contrast, the final (one day later) report says that only some of the needed facts would be published, and then immediately points out the dangers of publishing information prematurely: "We wish to emphasize that the initial disclosures will place in the hands of a nation (should it be acting in bad faith) information which could lead to an acceleration of an atomic armament program."[128]

It is not clear why the attack on secrecy was toned down. The last editing session included several consultants, including Vannevar Bush, General Groves, and Secretary of State Byrnes, who each made suggestions for alterations.[129] Perhaps the philosophy was altered, or maybe they wanted it to seem less extreme to skeptical eyes. I suspect more the latter than the former: the anti-secrecy approach is there in effect, even if the rhetoric had been softened. And for all of his gusto for secrecy, Groves did clearly believe that material controls ultimately mattered more, which was why he had gone to such efforts to secure global uranium and thorium ores.[130]

The Acheson-Lilienthal Report, and even its modified version as the Baruch Plan, was essentially, as the penultimate draft had argued, a scheme of control that focused on the material nature of the atomic bomb and not on keeping any nation ignorant of information. And in some places, even the final version kept some of the original anti-secrecy intent: "When the plan is in full operation there will no longer be secrets about atomic energy."[131]

The Acheson-Lilienthal Report can read as radical (to the point of naiveté) even today. But it should be noted that all later nonproliferation schemes have relied on the same principle: information is hard to regulate, control, and verify, but physical and material things are not. The scheme is radical only if you believe that nuclear technology is primarily made up of "secrets," and those who have tried to imagine practical and technically-informed schemes for their regulation have tended not to conclude that. Facts and plans both transmit easily and are concealed easily; thousands of tons of uranium ore, and the installations necessary to process and use them, do not.[132]

The Acheson-Lilienthal Report was leaked to the press in the spring of 1946, and this resulted in the Truman administration feeling pressured to present it as the official American plan for international control to the United Nations Atomic Energy Commission (not to be confused with the United States Atomic Energy Commission). The Baruch Plan, as the version formally presented became known, after the head of the American delegation, financier Bernard Baruch, failed to attract Soviet interest. There are many debates as to who or what was to blame for its failure, but international control was always a long-shot idea: a total remaking of the international order in response to a new technology, one requiring immense international trust and cooperation, and it would require both of the superpowers to give up a new weapon that promised new levels of security, even as it offered up new risks. The failure of international control was not surprising; that it was briefly taken as seriously as it was is the more remarkable thing, and reflects just how unsettling the atomic bomb was to the existing order.[133]

How should we judge these behind-the-scene pushes for reframing secrecy made by Oppenheimer? His efforts at "declassification" did indeed make systemic changes that continue through to this day, but may have had the inadvertent effect of solidifying permanent secrecy rather than eliminating it. The "rationalized" system that he helped build was good at making incremental releases of information, but poor at dealing with fundamental questions about the purpose and ills of secrecy. In his effort at reframing international control around "materials," rather than "secrets," Oppenheimer's influence is perhaps more subtle in the long term, but important. While international control attempts died a quiet death in the late 1940s, as we will see in later chapters, there have been recurrent overtures to the materiality of the bomb: the secrecy vs. safeguards debate that arose in the 1960s would embody similar concerns, and indeed the later regimes for enforcing the Nuclear Non-Proliferation Treaty would focus almost exclusively on materiality as its vector of regulation. It is interesting to contemplate what might have occurred had Oppenheimer been more successful in the 1940s with such an injunction, and if it thus did not need to wait until the late 1960s and beyond to take more effect.

The "problem of secrecy" and the question of postwar control were posed by the scientists, but answered by the politicians, in the end. The systems put in place, whether the "restricted data" clause of the Atomic Energy Act or the declassification procedures that spread beyond the Manhattan Project, did not totally embrace the most liberal views of the Scientists' Movement or even those of the wartime administrators. Secrecy would gain strength in the late 1940s and beyond, much to the consternation of those involved with regulating nuclear matters. But these newly created systems did not quite embrace the most fearful articulations of control, either. The US Atomic Energy Commission, while perhaps an "island of socialism," would not be a closed state, and was not a military organization, either. Restricted Data* was confining but could be reduced; declassification procedures were ponderous but would grind along; international control would fail in the first instance but make a return in the guise of nonproliferation.

The postwar system attempted to have everything both ways. Science needed to be open, but the bomb needed to be contained—a statutory formulation that was acknowledged as contradictory by its congressional author. In this, we will see the nascent beginnings of what will evolve into the bipolar "Cold War secrecy regime," which attempted to

* From this point on in this book, all instances of the term Restricted Data, whether capitalized or not, or in quotations or not, refer to the legal category defined by the Atomic Energy Act.

merge two extremes into a new, coherent synthesis. But such a synthesis built on explicit contradictions was bound to lead to tensions, which, as we shall see, is exactly what occurred.

Typically, when we look at how ideas and policies evolve, it is easy to focus on the "winning" arguments to the exclusion of those that were unsuccessful. Oppenheimer attempted to provide two alternatives to that of military secrecy. One, "declassification," was initially posed as a sliding scale of secrecy, one that would, over time, lead toward progressively fewer secrets. But once embodied in practices and institutions, it proved to be only partially capable of this task, and indeed much frustration that is expressed with secrecy in the present period comes from those who have to interact with the baroque and conservative declassification system. Still, as a transition from the "absolute secrecy" of the Manhattan Project, and the "burst" approach of the Smyth Report, it did put secrecy on a more systematic foundation, although one that appears to have made its permanent existence seem tenable.

Oppenheimer's attempt with international control was a more sustained destabilization. Regulating the materiality of the bomb—its factories, materials, laboratories—would radically re-center the postwar regime from the wartime one, and the added "tangibility" of such regulation would reign in the near infinite spread that the idea of secrecy-as-security enables. In this gambit, Oppenheimer was far less successful, both because international control failed (which was not really his fault), and because his broader discursive measures had far less penetration into the popular or policy spheres.

All of these postwar systems were the creation of imperfect men for imperfect times, a muddled synthesis of the many ideas and proposals that had been circulating within classified circles for several years. The US Atomic Energy Commission was given a deliberately ambiguous mandate: be a force for technocratic good in the world through simultaneous application of dissemination and restriction of scientific and technical information, with dangerous consequences for failure at every level. Rather than being a boon, this deliberate ambiguity would become a curse.

5

"INFORMATION CONTROL" AND THE ATOMIC ENERGY COMMISSION, 1947–1950

And so we are in a circle—chasing our tail.
DAVID LILIENTHAL, 1945[1]

On the day that the US Atomic Energy Commission (AEC) officially took control of the American nuclear weapons complex in January 1947, their first act was to discuss secrecy policy. Meeting with Richard Tolman about the work of the Manhattan Project Committee on Declassification, the AEC commissioners agreed that its procedures seemed sound and readily adopted them. The AEC, under the influence of its first chairman, David E. Lilienthal, would officially embrace the idea that as an organization it would aim to serve as a positive, technocratic force in the world, and deliberately distance itself from the military obsessions of the Manhattan Project. This was an organization, its leaders believed, that would not be swayed by emotional, totemic fears of "the secret."

But despite its liberal leanings, the question of secrecy would preoccupy the organization and its chairman. Attempts to reframe secrecy as "information control," and attempts to reform the fundamental mindset surrounding all things atomic, would come up against the fact that the AEC was politically vulnerable and the world was becoming very complicated very quickly. Rather than creating the progressive approach to secrecy its chairman desired, the Lilienthal AEC morphed into one of the most secretive bureaucracies in US history, and questions of security breaches and laxity plagued it from the beginning. Despite this, reform efforts quietly pushed on, behind the scenes, out of public view, until they were shattered by three "shocks" in late 1949/early 1950: the detection of the first Soviet atomic bomb, the revelation of Soviet penetration of the Manhattan Project, and the acrimonious H-bomb debate.

5.1 THE EDUCATION OF DAVID LILIENTHAL

The choice of David E. Lilienthal as the first AEC chairman set the tone of its early years. Lilienthal was a Harvard Law graduate who had been a Roosevelt appointee and had run the Tennessee Valley Authority (TVA), first as a director and then as chairman, from 1933 until 1946. As true-believer New Deal liberal technocrat, Lilienthal trusted that government resources, combined with technical expertise and industrial know-how, could be used to benefit the American populace, uproot poverty, further democratic participation, and promote public service. At TVA, a New Deal megaproject centered around hydroelectric power, Lilienthal's unapologetic liberalism had led him to acquire both deep admirers and powerful enemies. As a public administrator, his manner and voluminous diaries showed that he consistently worked over ideas, searching for "radical" solutions that would move beyond positions of impasse, tired talking points, and conventional conclusions.[2]

Lilienthal had no involvement with atomic energy prior to the bombing of Hiroshima, though he knew that energy generated by TVA was being used to support a "Mystery Plant" in eastern Tennessee, one that he noted in his diary must have been of "great portent" due to the quality of the scientific personnel consulted about the site.[3] In late September 1945, he attended a conference at the University of Chicago on the future of atomic energy policy. The "problem of secrecy" was a major topic of discussion and of Lilienthal's own comments both at the conference and later in his diaries. He expounded on the dangers involved in the connection of science with weapons of mass destruction: "We are, I rather assume, going to have a whole series of crises as a result of increasing scientific knowledge that is adaptable to blowing the hell out of the world."

Lilienthal found the physicist James Franck's formulation of secrecy persuasive: whatever "secrets" there were in the manufacture of the atomic bomb were, at best, "trade secrets" that would allow an American monopoly only for five years or so at the most. He framed his own feeling on secrecy in the terms of democratic deliberation:

> You must realize we are caught in a circle. . . . First: Unless the people are properly informed of the facts, the resulting public policy regarding the atomic bomb will be neither sound nor enduring. Second: The people cannot be informed so long as the scientists are prevented from speaking out. Third: *Whether* the scientists may speak out is a matter of public policy. And so we are in a circle—chasing our tail.[4]

Lilienthal's identification with Scientists' Movement talking points would grow stronger as he was drawn into Undersecretary of State Dean Acheson's work on the committee that would formulate the Acheson-Lilienthal plan for international control. He was primed to agree with Oppenheimer that there were no significant secrets of the atomic bomb, and that secrecy was a sin associated with the military.[5]

But once Lilienthal was "indoctrinated" into the world of classified information, his confidence in this view wavered. As he wrote in his diary:

> There were things that have never even been hinted at that are accomplished, or virtually accomplished, facts, that change the whole thesis of our inquiry, and of the course of the world in this generation. None of this can be written down. These are the very top of the top secret of our country; some of them are likely to remain secrets for some time to come. . . . This is a soul-stirring experience.[6]

Lilienthal was sympathetic to the scientists' view that secrecy was hampering the progress of technical inquiry and development. But having seen the "secrets" up close, he felt that he could no longer dismiss the problem as one manufactured by the Army in order to assert political power. He remained suspicious of the Army, however, and Groves in particular, and perceived that the Canadian spy scandal in February 1946 seriously endangered the possibility of "rational dealing with the problem."[7]

The Acheson-Lilienthal Report made Lilienthal a major public figure on atomic energy matters, and in September 1946, Truman asked him to be nominated as the first chairman of the new Atomic Energy Commission. Lilienthal accepted, and by October the full roster of commissioners had been lined up. Robert F. Bacher, a physicist who had overseen the assembly and engineering of the bomb at Los Alamos, and served as a member of the Committee on Declassification, was the sole scientist. William W. Waymack was a newspaper editor; Sumner T. Pike had been a businessman and member of the Securities and Exchange Commission. Rear Admiral Lewis L. Strauss, who was the token Republican member of the commission, and would be Lilienthal's main antagonist, had been Herbert Hoover's personal secretary during World War I, had worked for the Navy Bureau of Ordnance in World War II, and had been associated very early on with medical applications of nuclear physics.[8]

In early 1947, Lilienthal endured acrimonious confirmation hearings. Much of the acrimony originated with a senator who had a longstanding grudge against Lilienthal relating to his TVA work, but some of it was also caused by renewed interrogation of some of the principles of the McMahon Act and of the relationship of the AEC to Congress. Lilienthal emphasized that it was the AEC that would have to clean up the military's messes. When asked by Senator McMahon whether the Smyth Report was the "biggest giving-out of information" in the history of the bomb project, Lilienthal jumped on this as an example of how the military, not a civilian agency, had released "secrets" into the world.[9] This testimony distressed James Conant, who felt that Lilienthal had walked into a trap set by McMahon and the Szilard contingent of the Scientists' Movement to criticize Conant, Bush, and Groves. Lilienthal would claim he'd been unaware of the backstory and internal controversy between the Manhattan Project administrators and the Chicago scientists, though he recorded in his diary that he still felt his work to "blow the top off of the 'security' myth"—that a civilian commission could not be trusted with secrets—was necessary.[10]

The confirmation hearings forced Lilienthal to take a hardline approach on secrecy and security. He stressed his toughness on the matter, in contrast with the scandals that took place under the Manhattan Engineer District's control. Throughout the hearings, he attempted to avoid traps of "the constant threat of 'Red' scares, witch-hunting, spy charges, alarms about leaks, and charges that we had, deliberately or carelessly, lost 'secrets' that never really existed in reality."[11] It is this attempt at balance that makes Lilienthal the most important and problematic figure in the early Atomic Energy Commission. The AEC under Lilienthal was trying to satisfy two different and at times incompatible ideals simultaneously, all while under the increasingly hostile eye of the Joint Committee on Atomic Energy and the ambivalent eye of the Presidency.

Lilienthal was neither a pure idealist nor a pure compromiser, but someone who tried to be both at the same time. His overriding, ideological, and heartfelt commitment to civilian control of atomic energy, and his feeling that any major failures would give the military increasing and permanent influence in these early years of his agency, meant that he would make hard choices that, by the end of his tenure, would tranform him into exactly the kind of creature of secrecy that he once abhorred.

The first meeting of the AEC commissioners was held at Oak Ridge in November 1946, two months prior to their taking over atomic control. The meeting covered basic administrative matters and policies. One of these was the "Classification and Handling of 'Restricted Data,'" which set out the basic principles of handling secret documents, compiled from War Department practices. They made no special allowances for Restricted Data, other than a requirement that it be marked as such. It was made clear that this policy, which followed the procedures established in the Manhattan Engineer District, was "an interim system" for classification guidance, one that would be replaced "pending a comprehensive review" of the responsibilities and authority of the AEC along these lines. The procedures were duly approved and circulated.[12]

A few weeks later, the first pushes for new policy began. At the fourteenth meeting of the AEC, Lilienthal suggested that "some thought be given to a coordinated security and public information program" and recommended that Carroll L. Wilson, the AEC general manager, draw up suggestions for review.[13] In early January 1947, a staff paper on plans for an "Information Program" suggested that the AEC should turn to an outside panel of newsmen and journalists for advice.[14] In the spring and early summer of 1947, a series of staff papers were developed on the organization of classification authority within the AEC. They advocated a liberal attitude toward the problem of secrecy, interpreting the Atomic Energy Act of 1946 as advocating an extreme *minimum* of secrecy. The staff documents recognized that the Atomic Energy Act also called for restricting information. But even here they saw a way out: "It is quite clear to us, however, that insuring the common defense and security requires far more than the mere withholding of technical data which might be misused by those with other than peaceful intentions."[15] If the requirements of the Atomic Energy Act were to preserve "security," they reasoned, then that might best be done in many instances by discouraging secrecy, not promoting it.

In early June 1947, the AEC appointed an "informal panel" to discuss the organization of their information services and to recommend a possible "Director of Information."[16] The panel would be headed by Milton Eisenhower (president of Kansas State College, and the younger brother of the future US president), and included George Gallup (head of the American Institute of Public Opinion), Warren Johnson (a physicist and administrator at the University of Chicago), Eric Hodgins (former vice president of *Time* magazine), Raymond P. Brandt (head of the Washington bureau of the *St. Louis Post-Dispatch*), and John Dickey (president of Dartmouth). Their instructions were to meet with Lilienthal and Waymack to discuss "our problems in the area of information control and public education."[17] The meeting resulted in a short set of recommendations. The Eisenhower Panel said that the AEC should see its role as being made up of three "Tasks": the Positive Task (putting out information to the press and public), the Service Task (responding to queries for information), and the Security Task (as it functioned differently for the press and public, as opposed to strictly scientific declassification). They suggested William Laurence, the *New York Times* reporter, and Gordon Dean, a lawyer who had once been partner with Senator Brien McMahon, as potential candidates for the top job. Their main suggestion regarding these tasks was that it should *not* be in the bailiwick of the Classification Branch of the AEC, that declassification and public relations should be kept administratively separate.[18] The reason for this, Dickey elaborated separately to Lilienthal, was a fear that unifying these functions would cause them to either veer toward too much conservativism in secrecy, or too much release. Separating the functions might produce the balance desired.[19]

Later AEC staff papers on this issue, while agreeing with the Eisenhower Panel that the AEC should make dissemination of public information a key part of its regular business, argued instead that a unified office of declassification and dissemination, where the functions of information release and constraint were run by the same people, was necessary. One paper argued that this is how outsiders, especially the news media, would perceive as the way classification worked when they sought information from the commission. Canvassing opinions within the AEC, they found that many research scientists were afraid of declassification being housed within the Division of Security because they would be biased toward restraint, and the security officers objected to its being housed within the Division of Research because they would be biased toward release. Housing declassification in the general manager's office would increase that office's workload substantially. Thus, they concluded, a new division should be formed, which they proposed be called the Office of Information Control.[20]

"Information Control" would be the buzzword of choice regarding this idea, which seemed to them to be a clever bureaucratic fix to a thorny philosophical and organizational problem. AEC staffers recognized from the outset the difficulty of this balancing act but felt the agency was already making progress. The goal, they concluded in late summer of 1947, was to create an organization that would "avoid the disadvantage of making censorship the sole principle of operation." With a competent staff, they would "strike a wise balance which will win the confidence of the fearful and at the same time give useful public and technical information service outside the Commission group and a useful technical information service within the Commission group."[21]

But they agonized over the name. Normal government bureaus had Offices of Information. But the AEC's situation was one of both control and dissemination. So perhaps Office of Information Control, as previously suggested? Except that this emphasized the "control," which felt like secrecy. Office of Information Control and Dissemination, while unambiguous, was also too cumbersome. In the end they judged the likeliest candidates to be the "Office of Technical and Public Information" and "Office of Information Services."[22] The AEC commissioners approved the plan in September 1947, though they too deferred on the name. Eventually they went with the wordy "Office of Technical and Public Information," though the "Information Control" mindset, with its emphasis on both restraint and dissemination, ran through it.[23]

The first AEC director of technical and public information was appointed in the fall of 1947. Morse Salisbury had been the director of information for the Department of Agriculture, had run the public information program of the UN Relief and Rehabilitation Administration. He was not, in other words, a "security" man. His job, the AEC announced to the public, was to coordinate the AEC's responsibility "for insuring that all information issued to the public regarding the Commission activities has been declassified and properly cleared with respect to restricted data as defined by the Atomic Energy Act."[24]

The new Office of Technical and Public Information would oversee three previously separated functions of the AEC: a small Declassification Branch, which would screen materials for security information as well as make recommendations for security changes; the Technical Information Branch, which would prepare information for public distribution; and the Public Information Branch, which would "provide general information service and security guidance to all media of public communication." Declassification would rely on around 100 part-time Responsible Reviewers around the country, but the overall Office was expected to be large, at a projected cost of around $900,000 per year in personnel costs alone in 1948.[25]

The "Information Control" approach to defusing secrecy was not as successful as Lilienthal had initially hoped. Attempts to administratively separate the functions of "secrecy" and "security" had been a recurrent frustration, both practically and conceptually. "Security" was to be put in the hands of men with military or intelligence experience, who would make sure that plants could not be broken into, be sabotaged, employ disreputable people, or be prone to theft of material or documents. "Secrecy," on the other hand, was to be a subtler art, involving determinations of what the security was meant to protect.[26] The civilians and scientists would determine the rules, and the G-men would enforce them. In practice, the lines would get considerably more crossed. When the final public report for the Operation Crossroads nuclear tests was being authorized in late 1947, for example, the AEC director of military application vetoed much of the information that related to bomb effects, under the argument that it clearly fell under the "utilization of atomic weapons" definition of Restricted Data.[27]

The Office of Technical and Public Information undertook a wide variety of activities, including contact with journalists and textbook editors and overseeing the disclosure of public information. An account of an office staff meeting from late 1948 gives some of the flavor of that work: they agreed to an outside request to review a paper on integrating atomic energy information in a "Negro educational program" in Baltimore; they reported on progress on a special report on nuclear waste disposal; they discussed changes to the "Weapons Classification Guide"; they began work on a "Weapons Effects Handbook"; and finally, they agreed to work with the Social Science Research Council on a monograph series for college and university teachers.[28] One minute they would be discussing grade-school education, the next they would be revising top-secret classification guides. Such was the unified and broad scope of "Information Control."

In several instances, Lilienthal had major conflicts over screening the contents of a report with the Military Liaison Committee, which was meant to coordinate AEC and Department of Defense activities. In the summer of 1949, the military had requested that the AEC omit all information relating to progress in reactor design and radioactive waste processing. In a draft letter, not sent, Lilienthal railed that "the Commission feels security of the more basic and valuable information is served by publication of sufficient information to permit intelligent public discussion," and that "it would be completely impractical and probably prejudicial to security to attempt to carry out such a program without making available to the public general data as to the nature of the reactor development effort or without delineating the areas of publishable information." In the version actually sent, he agreed to make almost all of the changes, and all protests to the contrary were watered down, one of the many compromises made during this period.[29] In another case, the military and the Joint Committee on Atomic Energy (JCAE) both marveled at why the AEC would release photographs of its facilities, even ones that had been long-released before, noting that in World War II such photographs of enemy industrial facilities were crucial to the planning of strategic bombing and sabotage.[30]

And yet, the AEC increasingly found its primary job to be suppressing information, including information they agreed with. In late 1948, Rear Admiral William S. Parsons, a major figure in the Manhattan Project, proposed publishing a "dragon-slaying" article meant to discourage the American public from putting too much emphasis on the power of the atomic bomb to provide security against the Soviet Union. The AEC discussed this article during one of its regular meetings. Lilienthal felt that the intention was "unquestionably to induce in the minds of the American people a sounder attitude toward the dangers of atomic warfare and toward the possibilities of effective civilian defense," something he had no quarrel with, but that its unsettling effect would be felt not just by the American people, but also by their allies in Western Europe, and thereby potentially embolden the USSR. Commissioners Strauss and Bacher shared his worries. Bacher felt while it "seemed to him to present a picture that was by no means inaccurate," it would be "unfortunate" if the position was advocated by someone as well connected as Parsons. The article, though accurate, was ultimately censored.[31]

Another episode involved experimental work with radioactive substances on human subjects who had not given "informed consent." Some of the most ethically problematic of these involved the injection of plutonium into terminally ill patients between 1945 and 1947 without the patients' knowledge, not because the injections would provide any therapeutic benefit, but because the scientists desired information on the rate at which plutonium was excreted by the human body in order to set occupational exposure limits in Manhattan Project and AEC facilities.[32] In mid-1947, Robert S. Stone, director of the Health Division of the Metallurgical Laboratory during the war, inquired about the publication of several Manhattan Project Technical Series volumes relating to health and radiation. AEC General Manager Carroll L. Wilson replied that any experiments that had been undertaken where human beings were "unwitting subjects" would probably not be released, since to do so "might have an adverse effect upon the position of the Commission."[33] Eventually, in November 1947, the AEC's Advisory Committee on Biology and Medicine sent a judgment to Wilson, noting that they were against "the atmosphere of secrecy and suppression" in medical work and recommended that no studies should be undertaken in which substances would be injected into patients for non-therapeutic purposes or without consent. Wilson forwarded these recommendations to Stone, noting that since the studies did not meet these guidelines, they would remain classified.[34]

Similarly, AEC attempts to distribute public information about radiation and nuclear waste were watered down to avoid presenting the matter in an unfavorable light. Salisbury felt that the goal was to "inform citizens generally on a subject about which public information is needed in order to dispel misconceptions and allay possible latent hysteria." These "dual and in some ways conflicting goals" of the early AEC, as the official Nuclear Regulatory Commission historian J. Samuel Walker has written, prompted reports that were "not entirely candid." Walker further noted that the AEC's concern about "hysteria" was out of proportion to public attitudes at the time, and that "its abridged candor undermined public confidence over the long run" at least on the waste issue.[35] This was the challenge of "Information Control": though it sought to change the AEC's information policy from heavy restraint to heavy distribution, it had to reckon with the fact that not all atomic energy information was pleasant to hear.

The AEC also found that its legal requirements to enforce the Atomic Energy Act and its conflicting goals and ideals could lead to complicated questions about private speech as well. The Restricted Data clause of the Atomic Energy Act of 1946 applies to "all data" relating to nuclear weapons irrespective of its origins. This "born secret" interpretation means that the AEC is charged with regulating the output not only of its own scientists but also of *anyone* in the country who might speak about information deemed restricted. This was not merely an academic issue: very early on, the question arose of restrictions on scientists and journalists not under formal AEC control or obligation.

The first example of this kind of problem went back to the postwar months when the Manhattan Engineer District was still in control. In the winter of 1945, faculty at the University of Pennsylvania Department of Physics, who were not in any way connected with the Manhattan Project, began a series of seminars on the physics of nuclear fission as revealed by pre-war literature and the Smyth Report. These were compiled into a volume entitled *Nuclear Fission and Atomic Energy* that was meant to enable physicists "to obtain a semi-quantitative understanding of the phenomena of nuclear fission," as the foreword explained. None of the physicists involved had any access to classified data. Their ignorance, the volume emphasized, enabled them to speak where others, for legal reasons, had to be silent. Furthermore, their work would show that much of the postwar fuss over secrets was an act of futility: "In a sense the fact that this book could be written by physicists having access to no material not freely available to scientists the world over makes it clear that Nature is the only possible guardian of her own secrets. . . . Nature will not be a party to man's attempt at discrimination between nations, races, or individuals."[36]

The book contained a chapter on the physics of "Fast Neutron Chain Reaction" that went into bomb design, including a discussion of the implosion method, which had not yet been declassified. Even proponents of openness thought specifics about nuclear weapon designs should be kept close. As Henry DeWolf Smyth told Congress in late 1945, even those against secrecy were "not here recommending that we publish the technical details of the manufacture of atomic explosives or tell how the atomic bomb is finally put together," for "this is the only 'secret' of the atomic bomb that we should keep to ourselves for the moment."[37]

The Pennsylvania physicists submitted the manuscript to the War Department for review, apparently voluntarily. It was sent to reviewers at Los Alamos, who were instructed to examine it "as though it had been written [as part of] the Manhattan Project."[38] Col. Kenneth D. Nichols, in passing the matter on to the AEC in early 1947, considered the matter a key test "in formulating policy for safeguarding and declassification of secret information in accordance with the Atomic Energy Act," and AEC Commissioner Waymack remarked in notes on the volume a few months later that "it underlines the desirability of approaching this whole 'Security Problem' in the broadest way, and getting mobilized behind a sane general policy a moral force greater than the mere caution, whim, or indecision of the [AEC]."[39]

Ultimately, the Los Alamos reviewers recommended that the AEC do nothing. To censor the manuscript would highlight the information the AEC considered "dangerous" and thus confirm that it did contain secrets, a revelation that itself would be illegal. To give approval would verify its contents, thus allowing it to "assume the significance of the Smyth Report." As the reviewer concluded: "I would urge as forcefully as possible that since this document does not specifically involve any work done by the project, the War Department should say so, and make no further comment on the manuscript."[40]

The "no comment" approach would become standard AEC policy for private speech, with an important caveat. If the AEC felt that it had a high chance of compliance, it would find ways to indicate what sections it had difficulty with. If it had low trust that the violator would bend to its will, it would say very little at all. In the case of the Pennsylvania scientists, the AEC did request that the bomb design chapter be cut. In the final publication, the "Fast Neutron" chapter abruptly ends after suggesting the problem is difficult, with no mention of implosion whatsoever.[41]

At times, the AEC's goals of providing security advice to the press would, when combined with the "no comment" policy, result in contradictory exchanges. When a chemistry professor at Oberlin College attempted to publish an article with the International News Service on "The Secrets of the Atomic Bomb," which contained fairly accurate estimates of the amount of fissile material used in a bomb, the heights at which the bombs detonated over Japan, and the current stockpile of US bombs, Salisbury replied to the publisher that if the "secrets" given were accurate, they were considered classified under the Atomic Energy Act, and if they were not accurate, then the publisher would be misleading his readers. International News Service declined to publish the article.[42]

The author of the article submitted an even longer article to *Harper's Magazine* on "How to Make an Atom Bomb," which again made its way to the AEC in draft form. In this instance, Salibury's associate director replied that the AEC policy remained that they could only say "no comment" on the specific speculations, since to do otherwise would indicate what was sensitive information. But the reply continued: "The areas in which we can make no comment are as follows . . . ," and then listed the precise topics of concern in the article, with references to the paragraphs that contained the sensitive information. *Harper's* made the changes, and the article was duly declared to be "unclassified" by the AEC Declassification Branch.[43]

The AEC could, and did, investigate whether articles that had been published constituted "legitimate" security violations; that is, whether the information in the article had come from an official source who was not authorized to divulge the information, though it is not clear that they ever prosecuted anyone in this category. In the case of high-ranking military officials, whom the AEC saw as the source of the most important leaks, the AEC had little leverage. When a Major General told the United States Conference of Mayors that they could expect atomic bombs to have "40,000 tons high explosive potency," *Time* magazine proclaimed the AEC must have increased the efficiency of fission weapons two-fold over what they had been in World War II. The AEC considered this a major leak, but hesitated to even bring it up informally with the secretary of defense and felt that it would be "futile and really harmful" to try to file a formal complaint with the military, even though they felt that if "any of our [AEC] boys" leaked in this manner, they would be severely punished.[44]

In early March 1947, President Truman announced a new "loyalty-security" program that was designed to guarantee that Communists had not infiltrated the US federal government. This was a general anti-Communism program of massive scope and scale: all government employees would require an investigation into their "loyalty" by the Civil Service Commission, and the careers of all federal employees would be imperiled if it was found that they had inappropriate political associations, such as membership in, or even "sympathetic association" with, any organization that the US Attorney General deemed "totalitarian, fascist, communist or subversive." The very vastness of this approach, and the wide net that it cast, disturbed many, even though the most flagrant abuses would await further anti-Communist fervor.[45]

Even prior to this, though, the most controversial provisions of the AEC's attempts to control information were in the procedures for authorizing people to have access to Restricted Data. The Atomic Energy Act stipulated that nobody could be an AEC employee unless the FBI had investigated his or her "character, associations, and loyalty." This was one of the harsher additions to the McMahon Act when it was sent to the House of Representatives in the summer of 1946, while the Canadian spy scandal was on congressmen's minds. The FBI, however, did not "rule" on whether people got security clearances. They simply "investigated," and the AEC would then make the determination about the clearance status of the individual. In theory this division between investigation and administration would give the AEC considerable latitude in personnel matters. In practice, it would involve idealistic staffers and commissioners in the sordid rumors trafficked by the FBI.[46]

The policy regarding personnel clearances was hastily assembled in late 1946 but was not operational until February 1947. Col. Charles H. Banks, an intelligence officer for General Groves, made the initial suggestions for the AEC's postwar system, proposing that all personnel be required to fill out a Personnel Security Questionnaire (PSQ). When the AEC took over, it assigned these duties to Thomas O. Jones, who had been a security officer at Los Alamos. Jones drafted regulations that would establish three types of clearances based on exposure to restricted data. Contractor employees with no access to restricted data would be given a "P" clearance and would only later be given an FBI investigation. Business visitors to AEC installations who had no access to restricted data would be given an "S" clearance. Finally, all AEC employees, whatever their access to restricted data, would require a full FBI investigation before they received their "Q" clearance. Though the "Q clearance," which is still used to designate those personnel who have been cleared to access restricted data, is a name with a mystical and mysterious sound to it, it is merely the remainder lopped-off from the bureaucratic initialism "PSQ."[47]

Lilienthal felt the ethical weight of his new role as a judge of "character." In June 1947, he griped to his diary that having to work on clearance issues made him wonder "why in the name of hell and good sense I am willing to have anything to do with so ugly and insane an enterprise, much less accept chief responsibility for it." Most cases were handled at lower levels, but borderline cases and causes célèbres were bumped up to the commissioners. Lilienthal felt that they forced him "to play God and decide on ex parte evidence of FBI detectives whether Mr. A.'s or Mrs. B.'s loyalty, character, or associations are such as to justify permitting them to access . . . 'restricted data.'" He felt the work was in opposition to the Constitutional guarantees of fair trial, cross-examination, and open examination of evidence. He found the FBI files themselves to be nothing but records of gossip and "hearsay, most of it opinions." Liliethal lamented that a career could be cut short because "ten years ago a scientist contributed to the defense of the Scottsboro boys, or believes in collective bargaining or the international control of atomic energy." In short, he concluded, "this process makes me sick at the stomach—a lot more when I find myself part of it than when it is operated against me."[48]

It has been estimated that the AEC commissioners devoted a third of their meetings to personnel security in the first two years.[49] With the exception of secrecy hardliner Lewis Strauss, most of the commissioners believed in a liberal approach to personnel clearances, but were hemmed in by the demands of the far more conservative members of the Joint Committee on Atomic Energy and fear of scandal. This was a time of public fears about the threat posed by "subversive" and Communist "infiltrators," the loss of the "atomic secret" by means of espionage, and the political naiveté of "scientists." Though only a small number of scientists would be denied security clearances in this time, the publicity accorded to a few examples, with their apparent infringements of civil liberties, humiliations, and potentially career-ruining outcomes, led the scientists both inside and outside of AEC installations to characterize the period as one of witch-hunts and arbitrary hysteria.[50]

Aside from the ethical issues, this need for clearances posed practical difficulties. The standard military classification system had "grades" of classification indicating the seriousness of the information in question. Information that was only "Restricted" (not to be confused with Restricted Data) or "Confidential" could be circulated within the military establishment far more freely than information marked "Secret" or "Top Secret." In contrast, Restricted Data, the term applied to all nuclear information the AEC deemed "unpublishable," had no grades; it was not a classification category in the same way that "Secret" and "Confidential" were. It could, on top of being Restricted Data, also be "Secret" or "Confidential."[51] Yet, access to a document that contained Restricted Data, even if the document was only considered "Confidential" from a security point of view, required a single-scope FBI investigation for access, the same as "Top Secret" clearance, even if the person in question already had been investigated by the national military establishment.

This meant that anyone in the Air Force who worked with physical atomic bombs would need a Q Clearance, because as the exact shape, size, weights, center of balance, and explosive yields of the weapons were Restricted Data. Lilienthal mused on this in his diary in the spring of 1947 when the AEC was asked to clear the entire Eight Air Force:

> Wow! Complete information security with all the whole Air Force "clear"! What a change from the day when Groves wouldn't even inform the top military forces.[52]

This clear impracticality led over the course of the late 1940s to special military clearances, sub-categories of Restricted Data that were approved for operational use, and articulated agreements over the relationship between the makers and users of bombs.[53] Many of these issues would not be resolved until well into the 1950s.

Additionally, the number of people who required access to Restricted Data was overwhelming to the FBI, who were not consulted prior to having this responsibility thrust upon them. The AEC employed hundreds of thousands of people, from research scientists to construction workers, all of whom might need access to some secret information. The number of AEC requests grew from 2,000 a month in February 1947 to 8,000 a month in July 1947, the most the FBI could handle at one time. From January 1947 through the end of April 1949, the AEC had the FBI investigate over 140,000 individuals for clearances; approximately one out of every eight hundred adult Americans.[54] The FBI would later lobby to have the Atomic Energy Act changed—not because they opposed secrecy, but because in their zeal for rigor, Congress had set up an impossible system.

Lilienthal's desire for a fair and realistic personnel security policy was hampered by the political context and the fact that the primary forum he had for explaining himself was the often-hostile Joint Committee on Atomic Energy. In late 1947, the AEC assembled a Personnel Security Board, headed by former Supreme Court Justice Owen J. Roberts. Difficult issues were on the table: Should a rejected employee get to appeal the decision? Was being rejected on an AEC security check a more substantial blight than being denied a job in private industry? What were the boundaries of "character" and "associations"?[55] At the same time, the AEC worked to decentralize its security operations, with the hope that this would streamline clearance reviews.[56]

At the heart of the personnel crises that haunted the Lilienthal Commission was one of the trickiest questions of secrecy: once information was divided into what could be known and what could not, how did you determine *who* could know it? The Lilienthal AEC was vulnerable on this point. The JCAE was always on the prowl for signs of mismanagement and weakness, and the AEC had few political allies. The fear of scandal pushed the Lilienthal AEC toward conservative approaches to secrecy because when secrecy is working well, nobody knows; when it fails, or even appears to fail, then the scandal really begins.

5.2 THE "THRASHING" OF REFORM

In the AEC's first year, the agency was consumed with creating new policies and procedures while trying to put the sagging nuclear infrastructure of the Manhattan Project onto firm peacetime footing. By the next year, though, the criticisms were already coming in fast and hard. One somewhat unusual source of these was the General Advisory Committee (GAC) of the AEC, headed by Lilienthal's friend J. Robert Oppenheimer, which issued a scathing report in June 1948. Among other complaints, the scientists reserved special enmity for the AEC's handling of secrecy:

> The GAC has not understood the basic policy at the root of the rules at present in force with regard to security and secrecy of information. We most strongly recommend that the Commission, if necessary through the work of an ad hoc panel, make a fundamental study of the issues involved, particularly with regard to the use of secrecy as an instrument of maintaining security.[57]

This was added late in the session in which the GAC had written up its grievances, at the instigation of its members Glenn Seaborg (the chemist behind the discovery of plutonium), Oppenheimer, and Enrico Fermi, who desired to remove "the fetish for security with its ridiculous aspects," according to Seaborg's diary.[58] Ironically, the memo in which the general manager informed the other AEC staff members of the GAC's complaints was classified as "Secret," a relatively high grade of classification, though it was later downgraded to "Official Use Only."[59]

Lilienthal and his staff saw the GAC's criticism as an opportunity to push for the reform they already desired. Lilienthal did not agree that the AEC did not have a clear "policy," but did feel the agency suffered from a lack of courage to "stand up against the fear and fear-begotten emotions that have swept the country, and that are being inflamed by almost every event, and by reactionary forces." The solution was to reverse the equation so that the burden was on keeping things secret, not on releasing them. One front in this battle would be on personnel clearances, as described above. The other would be a new statement on the AEC's secrecy policy that would unburden the organization and clarify its new stance.[60]

The first result of the AEC staff's reevaluation of secrecy was a brief report on the "General Discussion of Security Problems in the AEC," authored by the Executive Office of the AEC Program Council.[61] The report considered two criticisms of the secrecy program the AEC had inherited from the Manhattan Engineer District: 1. "We may be fooling ourselves by our secrecy measures," and 2. "We are biting off our nose to spite our face." It argued that secrecy had hurt public deliberation on atomic energy issues, had created an "unhealthy" climate for AEC operations, and had led to "general disaffection" amongst the AEC scientists. The staff paper further noted that existing AEC policies regarding secrecy were a "mosaic of practices proceeding mostly on the basis of precedent," with a "lack of clear assumptions at the most fundamental level." In a survey of staff opinion, everyone recognized that there should be a "proper balance between advancing our own progress and retarding that of a competitor," but beyond that, "there is an obvious dearth both of guidance from above and agreement on fundamentals below."[62]

In some respects, the paper was very cautious; it never used the term "enemy" or "Russia" or "USSR" to describe the opposition power. Rather, it used the generic term "competitors." The paper advocated categorizing information based on its value to the US program *and* its value to the competitor's program. Information that had a high value to the US program and would serve to generally improve US scientific standing but had a low value to the competitor should be disseminated (though not necessarily published openly). Information that the competitors already knew should also probably be disseminated. Information that was of high value to competitors and low value for dissemination should be kept secret. Of course, in order to make many of the above determinations, the AEC would have to know what the competitor already knew or would find useful. This would require information that was not readily available or that was hard to interpret: foreign intelligence or assumptions based on past US efforts. It also considered conducting experiments to see how easy it was for scientists outside the AEC system to derive classified concepts.[63]

The report concluded that the AEC should approve in principle a "Statement on Security" that could be disseminated within the organization and begin a study of the fundamental problems of secrecy, first by canvassing the positions of people "of long acquaintance" with atomic energy matters. The "Statement on Security" argued that, in terms of AEC priorities, "the Commission considers that the greatest assurance of the common defense and security of this country lies, first, in the most rapid and widespread advance of our own technical position and, second, in delaying the advance of competitors."[64]

The commissioners met and read the proposal, and while their specific responses were not recorded, they did not yet approve the "Statement." Instead, they suggested that the AEC staff take up the matter with the GAC and the Senior Responsible Reviewers, who oversaw declassification policy. Lilienthal sought comments from others, noting that "the whole question of secrecy is a subject uppermost in our minds," and that what they were after was a "thrashing out of the many issues involved among as many thoughtful people as possible." Copies of the staff report were sent to Owen J. Roberts, the chairman of the AEC's Personnel Security Review Board, Alan Gregg of the Rockefeller Foundation, and the members of the GAC.[65] To Oppenheimer, whom he considered a friend and an ally, Lilienthal added a separate note: "I recognize that the philosophical borders of this problem are so broad and undefinable that we can wander into months of discussion, and reams of speculations. What I would like to do is to get far enough along to take a few definite steps in the right direction and do this by not later than mid-August."[66]

In the meantime, Lilienthal, Strauss, Morse Salisbury, Declassification Branch head Harold Fidler, and a few other AEC staff met with the Senior Responsible Reviewers. Lilienthal told the Reviewers that the commission was in the process of "reconsidering the basic philosophy of its secrecy and security system," and that their experience would be vital. The Reviewers were preparing for their second joint US-UK-Canada Declassification Conference, being held in Harwell, England, that September. These international conferences were meant to standardize declassification procedures amongst the Manhattan Project allies. Lilienthal urged that while they should not commit the AEC to anything, perhaps the Reviewers should float the idea that the AEC was reconsidering its security policies and see what the British and Canadian representatives had to say. The Reviewers were eager for change; they had long been at the front lines of applying classification judgments and felt that things had stagnated after the initial exuberance of the Committee on Declassification's report. Lilienthal told them he would appreciate if they would draw up evidence of how secrecy had impeded US progress, for "while it is almost axiomatic that secrecy does retard scientific progress," Lilienthal "had encountered difficulty in obtaining supporting examples from technical people." With those in hand, he could argue for a "more liberal secrecy policy" in the AEC.[67]

Lilienthal's archival files do not suggest he got the evidence he wanted. Showing harm caused by secrecy is actually quite difficult: the demonstration of harm relies on counterfactual speculation about how things might have proceeded had the secrecy not been in place. The only Senior Responsible Reviewer who responded to Lilienthal's request was physical chemist Willard F. Libby, who argued that secrecy was far more harmful during peacetime than it had been during World War II because most of the excellent scientists had moved outside of the range of government laboratories. The only tangible example relating to the effects of secrecy that he could come up with was the fluorine chemical industry, which he said had been sped up by twenty years thanks to declassification efforts. In general, Libby felt that secrecy did little to actively hurt the AEC's own research agendas, except for the fact that nobody outside the AEC was allowed to work on secret AEC problems.[68]

Oppenheimer wrote a long letter to Lilienthal in August 1948 regarding the ideas that the AEC had developed. He highlighted three points to be considered in future drafts. First, he felt that there was too much emphasis on keeping secrets from "rivals" when the US had no idea what was going on in the Soviet Union or how the Russians would go about making a bomb in the first place. A Soviet bomb project, he cautioned, might look very different from the wartime Manhattan Project.[69] Second, he felt that by dissecting classification into whether individual items should be declassified, they often missed the greater argument that "the mere bulk of secret material is a positive evil." He felt that any balancing criteria should consider whether classifying the item in question would erode the principle of minimizing secrecy.

Finally, Oppenheimer noted that many of the most highly-classified facts were not technical in nature at all, but administrative: the size and nature of the atomic stockpile, sources of raw materials, programmatic expectations and direction, and questions of military-civilian "custody" (which related to how fast the weapons could be mobilized for use and who had authority over their use). On a strictly "technical" basis, there was no reason to release this information, since it had nothing to do with advancing technical research programs. Still, Oppenheimer argued, there might be good reasons for this kind of information to be released or exchanged, and so building a classification around technical benefits or costs would have limitations: "I fear that some of the biggest and most valuable fish will slip through the net if it is woven entirely of criteria bearing on technical things." In general, he noted that he felt that relying on arguments about the importance of "technical preeminence" was being "overburdened a little in this matter of secrecy," for while "it is an important argument for not keeping secret more than has to be kept," "it was "not the only argument" for doing so.[70]

In September 1948, the efforts at formulating a concrete policy continued. David B. Langmuir, a physicist on the AEC staff who had been involved with the secrecy statement work, prepared two memos that Lilienthal enthusiastically forwarded to the other commissioners: a detailed analysis of the "Objectives and Methods of Secrecy," and a "Proposal for Study of Secrecy Problems."[71]

Langmuir's "Objectives and Methods" essay attempted to derive "certain questions of fact and policy upon which the intelligent use of secrecy seems to depend." He considered the ultimate goals: advancing the US position and retarding the "progress in 'dangerous' aspects of atomic energy on the part of certain foreign nations as much as possible." He considered these purposes to be interrelated and illustrated his general argument by means of a diagram. Langmuir painted the

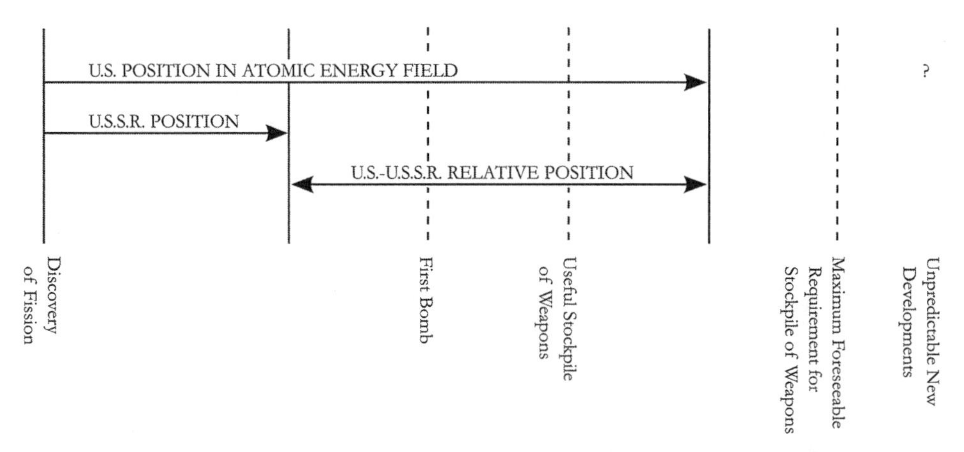

FIGURE 5.1. David B. Langmuir's illustration of the interrelatedness between advancing the US position and retarding the USSR position in the field of atomic energy. Redrawn by author from typewritten original. Source: David B. Langmuir, "Objects and Methods of Secrecy" (15 September 1948), OSAEC46, Box 41, "Basic Security Policy, Control of Info, Vol. 1."

arms race as a linear timeline branching out from the discovery of fission. The US had already achieved its first bomb and a number of other useful weapons and was racing toward maximum stockpile requirements. The USSR did not yet have the bomb, and the gap between the US and USSR's relative positions was highlighted in the chart as the area under dispute. For Langmuir, the chart did not just visualize the presumed arms race, but illustrated three particular policies that could be pursued: trying to advance the US position without worry about the Soviet (the "dynamic" position—no secrecy), trying to hinder the USSR position without worry about advancing US (the "conservative" position—all secrecy), and a mix between the two approaches (the "compromise" position—partial secrecy).[72]

The presentation makes obvious which position Langmuir advocated—who could argue against the "compromise" solution? Langmuir outlined three different ways a "compromise" position might be articulated, again with a diagram illustrating the "flow of information" under different secrecy regimes. Langmuir first considered the case of "extreme secrecy," where no information was passed "through the secrecy wall around AEC projects." Foreign advantage would be limited to a trickle of espionage and their own scientific investigations, but the US position would be entirely reliant on work done within the AEC's own

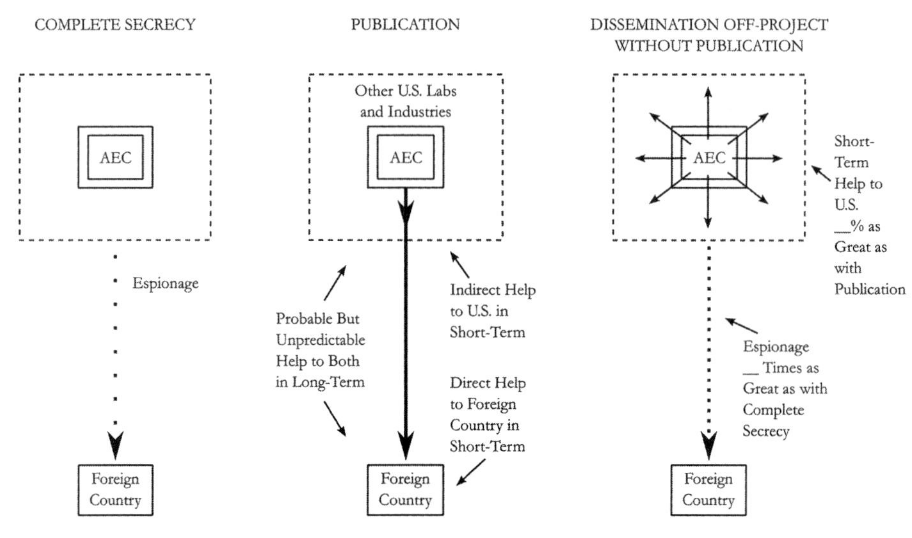

FIGURE 5.2. David B. Langmuir's illustration of three methods of controlling information, along with their probable effects for both US and "foreign" progress. Redrawn by author from typewritten original. Source: David B. Langmuir, "Objects and Methods of Secrecy" (15 September 1948), OSAEC46, Box 41, "Basic Security Policy, Control of Info, Vol. 1."

laboratories. Then he looked at the opposite extreme, "publication," in which nuclear information was widely disseminated. This would advance both the US and rivals in an "absolute" sense but would damage the "relative" position of the US. Finally, he considered an in-between position in which information was distributed to US labs and industries but not otherwise published. Langmuir believed this approach would advance both the absolute and relative position of the US. He noted that present AEC policy was to "publish some material and keep the rest secret" and that no effort had been made to create graduated zones of access to information.[73]

Langmuir's figures reveal the AEC's mindset regarding security: secrecy was a question of how information would move from secured zones to enemy hands. Langmuir felt the pictures raised more questions than they answered. How much would publication help the absolute US position in the short and long terms? How much would the AEC program be helped by a policy of dissemination-but-not-publication? How serious was the problem of espionage? Was secrecy really affecting *anyone's* progress? How should the value of information to a foreign competitor or a domestic publication be assessed? How could the US know what the USSR already knows or does not know?

These were tougher questions than Langmuir alone could answer. He proposed that they survey those with atomic energy expertise about the value of certain pieces of technical information to foreign nations and the advantage of publication to the US. The survey contained no data itself, just hypothetical statements: "The critical mass of an isolated sphere of pure U-235 is __ kg"; "The total number of stages at [the gaseous diffusion plant] K-25 is __"; "The spacing between uranium slugs in Hanford piles is __ inches"; "On 1 June 1948, the US had in stockpile __ atomic bombs." Langmuir composed a list of sixty-four questions, and a scatter plot grid that the answers could be plotted on, measuring "value of knowledge to foreigners" on one axis, and "value of publication to ourselves" on the others.[74]

Lilienthal forwarded Langmuir's two memos to Oppenheimer and the rest of the GAC in October 1948.[75] Langmuir also wrote up a "Proposed System for Secrecy" that clarified his previous views. He declared that his system was a "rational" way of dealing with secrecy and information flow issues. He proposed that any system of secrecy would have to first identify and restrict the "bottlenecks," information that would actively retard Soviet progress if not known. He also argued that to avoid the detrimental domestic effects of secrecy, the AEC would need to empower individuals within the organization to make their own security judgments. A "good secrecy system" would have clear rules enforced by people with competence and authority and would be "evolutionary in nature so that adaption [*sic*] to new conditions will occur organically and spontaneously."[76]

Langmuir's proposal relied, again, on diagrammatic thinking: they should develop complicated flow charts of how information "mattered" in the development of an atomic weapons program. These "obstacle charts" would then be "the map upon which the strategy of secrecy is to be planned." Langmuir's approach "stands or falls upon the possibility of preparing some sort of chart of the type illustrated, and upon the va-

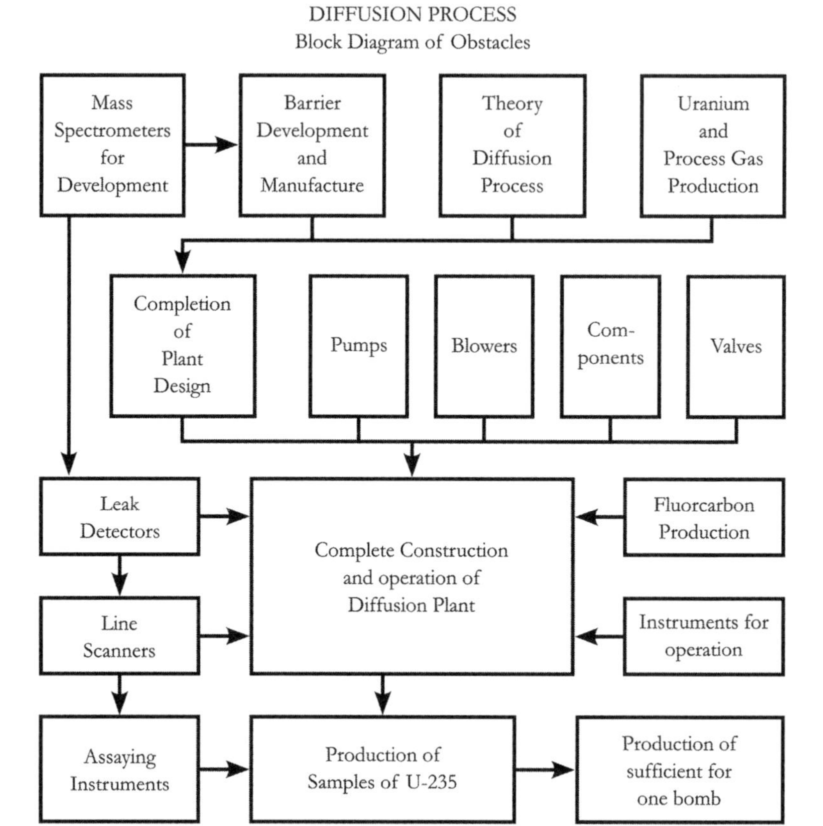

FIGURE 5.3. An example of one of David B. Langmuir's "obstacle charts," showing the necessary scientific and technical requirements for generating a bomb's worth of enriched uranium using a diffusion plant. He drew additional charts for the electromagnetic enrichment process and the plutonium process. Redrawn by author from typewritten original. Spelling of "Fluorcarbon" is in the original. Source: David B. Langmuir, "A Proposed System for Secrecy," AEC 111/2 (14 October 1948), OSAEC46, Box 41, "Basic Security Policy, Control of Info, Vol. 1."

lidity of its use as a framework for secrecy policy." The charts outlined the steps in a program to enrich uranium or breed plutonium, with the idea being that one could estimate the value of each step in potentially helping the Soviets reach a bomb.

Langmuir suggested that individuals seeking declassification would send their work to reviewers, who, armed with these charts and grids, would assess the value of the information to the enemy and to the US on a case-by-case basis. Classification questions would become largely "local," based on individual expert opinions at AEC sites, and the cen-

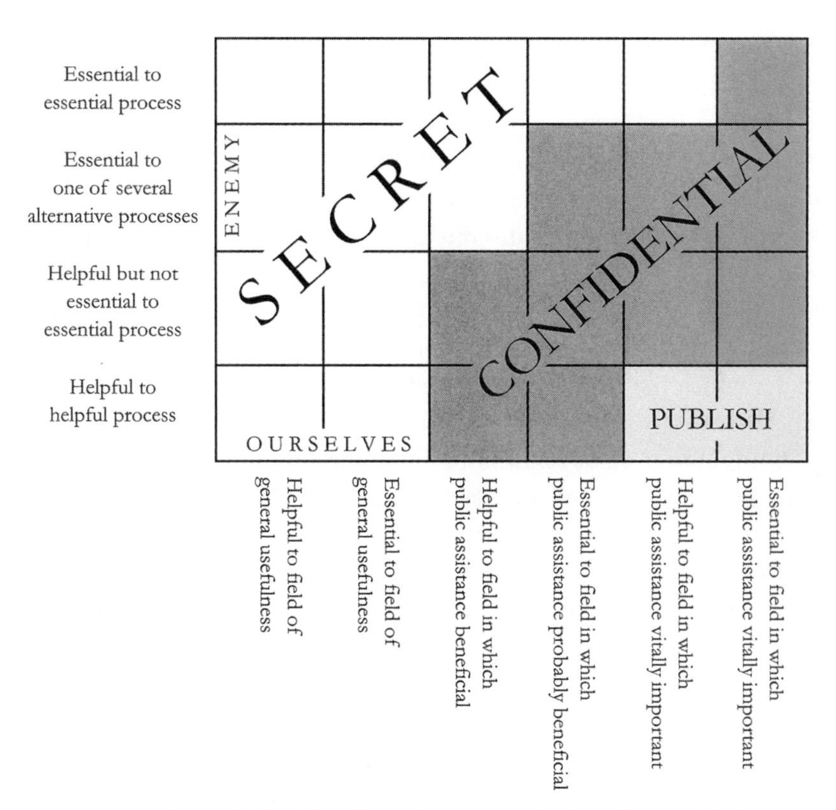

FIGURE 5.4. Langmuir's final chart illustrating the principle of classification of information based on its publication value to the United States (horizontal axis) and its potential value to the USSR (vertical axis). Redrawn by author from typewritten and hand-drawn original. Source: David B. Langmuir, "A Proposed System for Secrecy," AEC 111/2 (14 October 1948), OSAEC46, Box 41, "Basic Security Policy, Control of Info, Vol. 1."

tral declassification office would simply do record-keeping and occasional arbitration on borderline cases. There might be some inadvertent releases, but there would be errors in any system. In the end, Langmuir concluded that a "completely rational and analytical system of secrecy might well turn out to differ only slightly from the status quo, and probably can be shown to be neither very bad nor very good." Langmuir had made a new, complicated system for secrecy but was not himself certain it was worth implementing or that it really resolved the tough questions facing the AEC. A final diagram graphing the total "map" of information sensitivity was further discouraging: almost everything would be "Secret," a few things "Confidential," almost nothing "publish[ed]."[77]

Langmuir presented his system to the AEC in mid-October 1948.

Despite all of the work and effort, it does not appear to have led to significant changes.[78] While his analysis seemed sound, it acknowledged that improving on the status quo was difficult and entailed risks. His approach was not significantly better than that of the Tolman Committee, and it was similarly tilted toward a physicists' view that information could be divided into discrete categories separate from the people, sites, and practices that created it, and flowed in neat, orderly lines. The failures of these kinds of abstract, physicist-centric approaches to classification reform are telling in their own way about the assumptions under which the AEC staff operated; at no point does anyone seem to have thought that consulting sociologists, anthropologists, or psychologists might be of value in thinking about secrecy, or having deeper connections with people involved in the use of technical espionage from abroad.[79]

From then on, the AEC staff focused on creating a statement on the "Control of Information," an evolution of the "Statement on Security" discussed in the summer of 1948. This amounted to refining a position paper without committing to any particular change in policy. Another draft was finished by the end of 1948, authored by Langmuir and Harold Fidler, the AEC's director of declassification. Much of the content was the same: it acknowledged the need for a unified statement on secrecy and proposed a series of balancing factors biased in favor of the release of information. It went far beyond the "technical pre-eminence" issue Oppenheimer was so concerned about. Now reasons for releasing information included stimulating industry, increasing effectiveness of the armed forces, educating "the public in essential subjects," improving administrative efficiency, and stimulating and encouraging scientific research anywhere in the nation.[80]

There were also significant changes in the other direction. In the summer of 1948, the proposed policy statement had made the success of the US atomic program the primary goal and hindering the Soviets a secondary goal. In the winter of 1948, these priorities had switched, and this switch was made explicit:

> The prime objective of secrecy should be to delay the date when a potential enemy possesses an atomic bomb, and therefore to minimize the number and effectiveness of bombs in its stockpile. The production of

> fissionable material, and the techniques of bomb development, production and use are equally relevant to this prime objective. Retarding a potential enemy in these respects is to be regarded as more important than advancing our own progress. Benefits to ourselves of disseminating information are to be given reduced weight, as compared to the value of information to the enemy, and the status of enemy knowledge should be estimated most conservatively.[81]

The end of 1948 marked the beginning of a shift away from Lilienthal's idealism and toward the re-embrace of the idea of a tangible "secret" that could be given away. It is unclear what caused this dramatic shift, but it represented a more explicitly Cold War sensibility at a time when the Cold War was becoming a far more crystallized reality: the Berlin Airlift in the summer of 1948 marked for many the beginning of an openly hostile relationship between the US and USSR.[82]

The AEC staff's ambivalence was revealed in a pair of options appended to the end of the proposal. One option was for the AEC to endorse the statement that the US could make better use of technical knowledge than the Soviets and thus information should be generally released. The other advocated caution so that no technical information was unnecessarily released. When the AEC reviewed the proposal in January 1949, Robert Bacher argued that was a false dichotomy: these were not the only two possibilities. Sumner Pike worried that such a policy statement would give the false impression of classification being an exact science rather than a judgment call. Lilienthal agreed that any "illusion of certainty" was "dangerous," but felt that a policy based on "gut reaction" was no policy at all. He further suggested that questions of classification not be limited only to technical information. Lewis Strauss noted that while the policy statement was primarily concerned with giving information to the "enemy," giving information to "even friendly nations" was dangerous as well. Bacher suggested that the revised statement be modified to discourage overclassification explicitly, because giving information higher classification than it needs both "makes the system of classification ridiculous" and creates administrative difficulties.[83] Bacher's concerns culminated in the addition of a section titled "Unnecessary Secrecy to Avoid." Further inclusions and changes were made over the spring of 1949.[84]

It would be June 1950 before the commission would agree on a statement about its secrecy policy, some two years after the GAC's request for clarification. By this time, the Soviet detonation of an atomic bomb and revelations about the extent of Soviet penetration into the Manhattan Project would derail the effort for classification reform. Of course, there had already been difficulty articulating a secrecy policy that the AEC commissioners and AEC staff could agree upon. "Analytical" approaches like those employed by Langmuir left room for disagreement, reevaluation, and shifts in emphasis and procedure. Attempts to base secrecy reforms on empirical evidence of the benefits or harms of secrecy proved elusive, as data on Soviet intelligence was not available at the time. Even consensus on whether the primary technological goal of the AEC program should be "positive" (producing more and better US bombs) or "negative" (denying the USSR its own progress) proved elusive.

None of these deliberations over policy was happening in a vacuum, of course. From 1946 to 1949, the Cold War was evolving, from the idea of "containment," to the "Iron Curtain," to the conflict in Greece and the Berlin Airlift and subsequent War Scare. The House Un-American Activities Committee was beginning its period of maximum influence, and any immediate postwar hopes for a collaborative relationship with the Soviet Union had been shattered. The efforts at international control of atomic energy fell flat: the Acheson-Lilienthal Report, adapted into the Baruch Plan, was rejected by the Soviets, and it is not clear that the White House (much less Congress) was ever enthusiastic about it in the first place. The AEC had been born in a mixed atmosphere of hope and fear, but as the 1940s began to transition to the 1950s, fear dominated.[85]

Lilienthal found himself in an unenviable position. The different groups he was trying to satisfy had mutually exclusive visions for AEC policy, and none of his bold re-formulations of the problem ever satisfied any of them. His journals are full of depressive assessments of the situation: the AEC was in charge of the industry that might save or destroy the world but was unable to cement its thinking on key issues and was always vulnerable to political attack. "The world's worst job,"

Lilienthal called the chairmanship in 1947, when an acceptable balance of security and sensibility proved elusive.[86] But as difficult as his first two years were, the last two (1949–1950), would be even worse.

5.3 THREE SHOCKS

Three events, each shocking in its own way, dashed the possibility of secrecy reform in the final two years of Lilienthal's tenure as AEC chairman. First: in September 1949, the US detected the first Soviet test detonation of an atomic bomb. Second: in response to the Soviet test, an acrimonious debate about whether to develop the hydrogen bomb began in classified policy circles, but soon leaked out. Third: only days after Truman affirmed that the US would build the H-bomb, the world learned that the Manhattan Project had been deeply compromised by Soviet espionage. These three shocks, in close succession, rocked American attitudes on the bomb: a new sense of vulnerability was followed by a reinforcement of the idea of the all-powerful "secret," which itself was followed by a reinforced sense of vulnerability and loss.[87]

Even before the shocks of 1949–1950, the tensions around nuclear secrecy were intensifying, notably in the interactions between the AEC and the Congressional Joint Committee on Atomic Energy (JCAE), with AEC Commissioner Strauss and the JCAE frequently favoring harsher responses than Lilienthal or his allies. The areas of disagreement included whether the US should reengage the UK on atomic matters, encouraged by the fact that wartime arrangements had given the UK considerable control of uranium resources that the US desperately coveted. Deep disagreements emerged about whether the US ought to give the UK any atomic assistance, even that of a non-military nature, once it became clear that the UK was pursuing its own weapons program. And a nagging question remained: could the UK be trusted to safeguard US secrets?[88]

There were other troubles. One of the AEC's early initiatives was to set up research fellowships with the National Research Committee in 1947. These were intended to be unclassified grants to further scientific knowledge on topics of AEC interest, with the additional hope that they would nurture new students who might later be interested in working for the AEC. In the summer of 1948, the question arose of whether these students would need security clearances or FBI investigations before accepting the fellowships. For Lilienthal (and most of the other commissioners) the answer was clearly "no": the work they were doing was unclassified, FBI investigations were arduous and potentially damaging, and the process would add unnecessary security and political scrutiny to a program meant to encourage intellectual freedom. Strauss felt otherwise, and the hard-liners on the JCAE sided with Strauss. The issue became a public controversy, sparked intense debate amongst scientists and politicians, and eventually resulted in a hardline approach being adopted in August 1949.[89]

A myriad of minor scandals made Lilienthal feel consistantly under attack. In the spring of 1948, acrimony toward the management at Oak Ridge almost led to a strike by plant workers, raising difficult questions about the role of labor organizations in atomic energy facilities. Investigations from the House Un-American Activities Committee (HUAC) in 1948 focused on the alleged political unreliability of atomic scientists during World War II, especially on Oppenheimer's former Berkeley students. And the JCAE criticized Lilienthal at length about the AEC's Fifth Semiannual Report to Congress in early 1949 for releasing too much information about their program to expand production of fissile material. Lilienthal's response, that the information was all declassified, and that it was his responsibility to inform the people and the government about the AEC's activities, met with a bitter riposte from Senator Tom Connally: "Why is it necessary, because you spend public money, to go out and blah, blah all over the country about these bombs?" Lilienthal responded that their disclosures were about public accountability, given the cost evolved, and elaborated that: "The responsibility for what is and what is not restricted data is one we meet every day, and is a question of balance." But the Congressmen were unswayed: they noted that such openness about nuclear facilities would likely be rare among their adversaries.[90]

Further grief came when a vial of 289 milligrams of uranium lost from the Argonne laboratory in May 1949 prompted headlines and congressional inquiry, despite what the AEC and its scientists considered to be its total lack of security significance. (The missing vial was later recovered.) Finally, in May 1949, Senator Bourke Hickenlooper of the JCAE demanded Lilienthal's resignation, declaring that he had uncovered evidence of "incredible mismanagement" in the AEC. Seemingly endless hearings and reports followed; Lilienthal was largely vindicated, in the end, but the ordeal itself broke his spirit.[91]

The period of 1948–1949 was one of great transformation in the relationship between scientists and the federal government, and in public discourse about secrecy. HUAC and JCAE investigations, capable of creating national scandals, reflected both election-year politics and reactions to a worsening international situation. Scientists, especially physicists, became targets of political attacks, an attitude fostered by growing public ambivalence about US reliance on science and technology for its national security. Throughout this period, public discussions about the bomb shifted subtly from processes and "know-how" to the idea of discrete, inscribable, transmissible "secrets."[92] It was a difficult atmosphere in which to espouse a liberal position on atomic energy. In the spring of 1949, Lilienthal noted in his diary "a growing jitteriness in the country" about atomic security, and that talk about secrecy was becoming "more and more crazy. . . . Let's hope my sense of perspective survives the ordeal."[93]

At the end of August, Lilienthal and his wife left for a month-long vacation at Martha's Vineyard. The events of the year had worn him out. He felt no victory in weathering the "incredible mismanagement" hearings and the calls for his resignation. On September 19, he wrote a long entry in his diary, contrasting the serenity of the island with the trouble of Washington: "[T]he rest of the world seems far, far away."[94] He wondered what the future held for him, after almost twenty years of government service.

But his rest would abruptly end. A few weeks before, a specially outfitted American airplane flying a route between Japan and Alaska had detected the first traces of something that had long been predicted and feared. On the same day Lilienthal wrote his weary journal entry, he and his wife went out to dinner. Driving back to their residence that evening, through heavy fog, they found a figure waiting for them at the entrance to their lane. It was General James McCormack, the AEC director of military applications, "hatless, squinting into the lights, looking bemused, hooking his thumb in a hitchhiker's gesture." They retired inside, huddled around a kerosene lamp. Lilienthal recalled: "Then he gave me the news, rather dead-pan, its unambiguous nature rather fuzzy; some reference to the shock and impact, the recriminations, the whole box of trouble it portended."[95] The next morning, after fretful sleep, Lilienthal headed back to Washington. The Soviets had detonated an atomic bomb.

The most immediate policy question was whether the United States should announce the discovery, referred to as "Joe-1" within classified circles. Many things weighed on President Truman's mind at the moment: Yugoslavia had recently declared its independence from Soviet influence; his new secretary of defense, Louis Johnson, was attempting to make massive cuts to American military forces; the British were devaluing their currency. Despite the diminished importance of these compared to "Joe-1," they led to a situation in which Truman felt that the possibility of Soviet-instigated war was high, that Europe was weak, and that the domestic political situation was shaping up as a potentially tough fight.[96]

There were fears that if Truman announced the end of the American atomic monopoly, it could cause a panic, could have uncertain effects on European confidence and markets, and would furthermore alert the USSR to the fact that the US had the ability to detect nuclear tests at long range (not an easy feat). This could potentially lead the Soviets to change their testing methods to avoid future detection, or they could in turn try to spy on US nuclear testing from afar. Truman's secretary of state, Dean Acheson, was strongly against announcement, as it would have unpredictable effects on his negotiations with the British about sharing atomic information. As one member of the State Department staff noted, "[A]n announcement by the President would dramatize the situation too much; and the American people had about all the bad news they could stand."[97]

But there were strong arguments for release as well. The obvious one was the propaganda coup: wouldn't it be better for Truman to calmly announce the detection, and show that the US was in control, rather than having it announced by a gloating Soviet Union? There were fears that the new Soviet Foreign Minister might do so at a scheduled upcoming public address. And the military, despite its penchant for secrecy, favored "instant release": it might keep its budget from being cut.[98]

On the day he arrived back in Washington, Lilienthal personally lobbied Truman to announce the detection, arguing that it would reassure the American people and demonstrate that in a crisis, they would be kept informed. Truman did not seem interested in Lilienthal's moral arguments about public knowledge. Despite technical reassurances from Lilienthal and AEC scientists, Truman was himself apparently not entirely convinced it was a bomb detonation and not a reactor accident. The president continued to vacillate on the issue over the next two days. Finally, on September 23, he released a statement about the detection, primarily motivated by fear that the information would either leak or be announced by the Soviets. His committee-written statement referenced an "explosion," not a "bomb," but everyone understood the import of the statement: the American atomic monopoly was over.[99]

It is one thing to project calm and certainty, and another thing for people to believe it. Though the end of the American monopoly had been predicted for years, it still came as a shock. Most commentators expressed surprise that it had happened "earlier than expected." Estimates of when the Soviets would get the bomb had remained "in about five years" from 1945 through 1949—the timeline was not updated as time marched on.[100] Despite a constant drumbeat of assurances from scientists since 1945 that the American monopoly would be short-lived, the arrival of that fact was still traumatic.

In the aftermath of the Soviet test, there was some initial hope that new attention might be brought to bear on the secrecy question. After Lilienthal had tried to convince Truman to make an announcement on September 20, he spoke with a "frantic, drawn" Oppenheimer. A Lilienthal diary entry records the nervous energy of their interaction: "J.R.O badly upset: 'We mustn't muff this; chance to end the miasma of secrecy—holding a secret when there is no secret.'"[101]

Lilienthal was relieved that Truman had decided to announce the "explosion." In a speech a few weeks afterward, he amplified his support for public discussion on atomic matters and praised: "Atom or no atom, secrecy or no secrecy, the American people—unless I completely misread their history and their present frame of mind—intend to ask many such questions, and out of public discussion to arrive at some answers."

He further praised Truman for showing that "atomic secrecy" did not require the abandonment of "the right of people to essential information by which they may steer their own course."[102]

That the Soviets had the bomb was for Lilienthal not an excuse for "'Papa-knows-best' nonsense." Furthermore, the Soviets' possession of the bomb could lead to reform in secrecy policy. If there was no more secret, couldn't the burden of classification be relaxed? As William Waymack wrote to another commissioner in mid-October, the Soviet bomb "could shake [the hawks] loose from the monopoly concept and therefore the secrecy obsession."[103] Lilienthal "half-facetiously" brought up the fact that the Soviets had "the secret" as an argument in favor of greater cooperation with the British.[104]

In late October, the commissioners once again considered the latest draft of the AEC's official statement on its secrecy policy that it had been working on for the previous year. The AEC staff were cognizant that their effort, originally directed toward denying the Soviets the bomb, might have changed now that the Soviets already possessed the bomb. Nevertheless, the staff reported that they had reexamined their work "in the light of recent developments in the USSR and had concluded that there was at present no reason for revising it." Lilienthal said the goal of the AEC's policy should be to enable individuals to make bold classification decisions in order to reduce the cumbersome "item-by-item" review process, and that while he didn't agree with everything in the statement, it was better to approve one now that could be later modified rather than to continually tweak drafts as they had been doing for months.[105]

The AEC itself had undergone significant changes since the statement was last up for debate. Both Robert F. Bacher and William W. Waymack had retired and they had been replaced by Henry DeWolf Smyth (of the Smyth Report), and Gordon Dean, a lawyer who had once been a partner at the same firm as Senator Brien McMahon. Smyth was, like Bacher, fairly moderate when it came to declassification and public debate. Dean, on the other hand, was no secrecy extremist, but neither was he a champion for openness.[106]

At the October 1949 meeting on security policy, Dean agreed that the AEC needed a stated policy on secrecy but didn't feel that this was the right one. The statement needed to be written "in light of recent events and cover the whole field" in a post-"Joe-1" world, including the extent of Soviet weapons knowledge. In any case, Dean noted, the current draft likely violated the Atomic Energy Act by proceeding on the assumption "that the Commission has the power to do something which the Act prohibits." The current draft included discussions of what information could be released to the UK as part of their continued cooperation. As far as Dean was concerned, the AEC lacked legal authority to exchange any information with other nations, since the Act specifically outlawed cooperation "with respect to the use of atomic energy for industrial purposes" without congressional approval.[107]

The AEC general counsel suggested that the commission was aware there were potential legal issues there, but that they would discuss them with the JCAE in relation to the UK exchange question more specifically. Commissioner Pike offered that in his opinion, since no significant use of "atomic energy for industrial purposes" had yet been developed, they were not in violation of the statute. Lilienthal suggested that from the point of view of "an administrator," the Atomic Energy Act seemed to be filled with statements supporting public release of research and contradictions about whether the AEC was meant to distribute or control information. Lilienthal felt that "drastic modification of the Atomic Energy Act, rather than revision of isolated sections," was necessary, in order to loosen the restrictions that were "holding back . . . the ingenuity and fertility of the American industrial system" from the atomic energy program. In any case, he explained, if the ban on giving out atomic information to foreign nations applied in the way Dean thought it did, wouldn't that mean that the AEC could not declassify anything? The session ended ambiguously; the AEC officially voted to disseminate the new policy, while at the same time, staff studies were commissioned to determine whether it actually would require a revision of the Atomic Energy Act for the new policy to go into force.[108]

Dean had voiced these objections before. He was not a fanatic, but he was a lawyer, and the law in question could be interpreted conservatively. In the summer of 1949, he had written a long memorandum that evaluated the legal ability of the AEC to disseminate information to other nations and to declassify information in general. The contradictory nature of the Atomic Energy Act had been intentional, to encourage a balanced policy, but Dean was unsure how this was supposed to work in practice. The Act contained a "specific and unqualified limitation," he argued, against exchanging atomic energy information with other nations. That the AEC could not do so without congressional approval was a major argument in favor of the bill during the Senate and House hearings. And yet, now the AEC was attempting to assert wide latitude in information control, including the ability to share it with allies under certain circumstances. Dean emphasized that while the Act was ambiguous, there were strong indications that the information policy contemplated by the AEC was illegal.[109]

The AEC pushed forward, largely ignoring Dean's position. In late September 1949, the Third International Declassification Conference was held at Chalk River, Canada, a part of the joint US-UK-Canadian efforts to update their classification guides. The scientists there used this opportunity to discuss the implications of the Russian test on classification policy. The general conclusion was that "Joe-1" ought to shift the emphasis considerably. If the policy henceforth had been about denying the bomb to the Soviets, it needed a reorientation now that they had the bomb: "The picture has changed from a *monopoly* to *competition* so that in our over-all security outlook the emphasis has shifted toward an acceleration of our own advance as a means of attaining maximum security."[110]

Secrecy should not be eliminated, they argued, but there should be "far less restrictive secrecy rules." More "basic science" should be released, because "if you want a field to develop—open up; if you don't want it to develop—keep it secret." Not all present agreed, noting that "substantial scientific achievement" had occurred under conditions of industrial secrecy. The report concluded that "industrial secrecy" could be used to protect atomic information worth protecting, and the bomb should no longer be a special case. The AEC general manager promulgated the report to its general managers, urging them to consider further declassifications.[111]

The AEC decided to make changes to its draft secrecy statement, emphasizing Dean's restrictions on industrial exchange. At a meeting with the Military Liaison Committee, Lilienthal pressed the Department of Defense (DOD) to understand that the law had charged the AEC with two different responsibilities, dissemination and control, and to work with them in examining how to carry out both. The DOD representatives said they had no trouble with the AEC's statement, but pointed out that in many specific cases, their opinions differed from the AEC. Classification of the atom, they argued, did not affect only the AEC; there were "far-reaching implications of over-all policy . . . [and] implementation of any such policy affected many agencies in the Government and segments of the American people."[112] The AEC finally managed to approve its statement on secrecy, with Dean's changes, in early January 1950.[113] After all of the watering down, there is no evidence that had any effect on the AEC's classification practices. After years of effort, secrecy reform had withered on the vine.

The weeks after the announcement of the Soviet atomic bomb were ones of soul-searching for policymakers. While some hoped "Joe-1" could be used as an opportunity to reform the AEC security system, and potentially even begin to normalize Cold War tensions, others took a different tack. For Lewis Strauss, the Soviet bomb meant it was time to pursue a "quantum jump" in nuclear technology. On October 5, he distributed a memo to the other AEC commissioners arguing that the Soviets would not be deterred by America's *quantitative* nuclear superiority alone. Instead, a *qualitative* change was needed "to stay ahead." He proposed another Manhattan Project, but this time with the goal of producing the H-bomb.[114]

The idea of the hydrogen bomb, then called the "Super," dated back to even before Los Alamos' creation, in the earliest days of the Manhattan Project. Though championed by the physicist Edward Teller, the concept had been kept on a back-burner through World War II and into the postwar period. It was seen as a future possibility, but was never a priority. In theory, the idea was to use the energy of a nuclear fission explosion to start a runaway nuclear fusion reaction, allowing for weapons thousands of times more powerful than the atomic bombs of World War II. In practice, it was hard to develop a workable scheme.

Fusion reactions are harder to set off than fission reactions, and there were vast uncertainties involved at every level. The priority of the postwar nuclear program was expanding and improving the then-meager arsenal of fission weapons. In a world where the US arsenal was still limited by the amount of fissile material available, and where creating fusion fuel (tritium, an isotope of hydrogen) not only would consume time and resources, but also (for technical reasons) would decrease the amount of plutonium produced, there was little enthusiasm for an intensive "Super" program that was not guaranteed success, and could interfere with other nuclear efforts.[115] And even if it did succeed, many scientists were ambivalent about whether it would be a good idea: could a weapon a thousand times more powerful than the Hiroshima bomb be used in a non-genocidal fashion? If not, would it be consistent with American (or scientific) values to produce it?[116]

The Soviet test dramatically changed the political calculation. Strauss, Teller, and Senator McMahon became advocates for developing the H-bomb as soon as possible, along with several key scientists like Ernest Lawrence and Luis Alvarez. Many other American scientists, particularly the members of Oppenheimer's AEC General Advisory Committee, and several other policymakers, notably Lilienthal, bitterly opposed any kind of "crash" effort in favor of the hydrogen bomb. There were different arguments on either side of the debate, all fiercely held and debated. From October through November 1949, this debate took place under tight classification constraints. The existence of the "Super" had never been declassified, though the concept had been either leaked or independently derived several times, but the fact that there was a serious and acrimonious debate over whether it was the proper response to the Soviet test was considered a major programmatic secret regarding American military and atomic strategy.[117]

Many of those involved had been part of the secret discussions regarding the use of the atomic bomb in World War II. The wartime deliberations had been justified by the need to "shock" the Japanese, but now what justified making high-level decisions about nuclear weapons without consulting public opinion? GAC member James Conant remarked to Lilienthal, who doubted Truman would make the debate public, that the whole discussion "makes me feel I was seeing the same film, and a punk one, for the second time."[118] But even the staunchly anti-secrecy *Bulletin of the Atomic Scientists* self-censored on the hydrogen bomb topic, later claiming fears that such discussion "might foster the belief that America was actively engaged in developing thermo-nuclear weapons, and that this might stimulate the arms race and further exacerbate international relations."[119]

The debate went public in November 1949 with a staggering leak. Senator Edwin Johnson of the JCAE appeared on a live television talk show on the subject of "Is There Too Much Secrecy in Our Atomic Program?" Johnson was arguing in favor of secrecy, that increased secrecy was needed to keep scientists from leaking too much information about the atomic program. When asked by another guest on the show about whether the Soviet bomb meant that the US ought not worry so much about such secrets, Johnson's rambling reply released its own secrets:

> I'm glad you asked me that question, because here's the thing that is top secret. Our scientists from the time that the bombs were detonated at Hiroshima and Nagasaki have been trying to make what is known as a superbomb. . . . Now our scientists already—already have created a bomb that has six times the effectiveness of the bomb that was dropped at Nagasaki and they're not satisfied at all; they want one that has a thousand times the effect of that terrible bomb that was dropped at Nagasaki that snuffed out the lives of—of fifty thousand people just like that. And that's the secret, that's the big secret that the scientists in America are so anxious to divulge to the whole scientific world.[120]

Johnson's outburst was both bizarre, and ironic. In advocating the need for tighter controls on security, he apparently self-consciously let out a major "top secret." It is impossible that he thought this information had been declassified, and there cannot be any doubt that he knew he was on live television. The irony of a hardliner senator leaking out "top secrets" as an argument that scientists couldn't be trusted was not lost on observers. It is possible Johnson's leak was a deliberate attempt to put pressure on Truman, by removing the issue from the exclusive provenance of scientific experts, most of whom opposed the "Super" program.

Though the leak occurred on November 1, it was not until November 18 that its impact was felt, when the *Washington Post* carried a column about it on its front page.[121] Lilienthal met with Truman on the day that the *Post* piece came out and later recalled that "the President was mad as hops, [and he] started off by cussing Johnson and the Joint Committee out."[122] On November 26, Truman called Senator McMahon and the Attorney General to the White House and ordered them to "plug" any security leaks.[123] He may have also been trying to keep McMahon from going public: on November 21, McMahon had written the President an impassioned letter in favor of developing the hydrogen bomb, further suggesting that the people of the US and the USSR needed to know that the US was pursuing this new weapon at any cost.[124]

Truman did not want the "Super" debate to become public, but it was unavoidable at this point. Once released, it could not be restrained—all that could be changed was that those with access to official information could be "gagged." But more leaks were making their way out over the next two months. The "Super" had become headline news, to the consternation of the president, the JCAE, and the AEC.[125]

Truman felt his hand was being forced: by making the H-bomb project a public issue, there was hardly any position he could take but to approve it. The "Super" was popular with the American public, and many of the unclassified reasons against it relied on appeals to morality and international control. The strong technical arguments against it were classified and not persuasive to the non-technical members of Congress, who saw the issue solely in terms of getting either "ahead" or "behind," or in the need for bigger weapons. As one senator on the JCAE put it, in a closed session, if the Soviets got the H-bomb first, "they will have a gun at our heart if they do it and we do not."[126]

The debate came to a head with a final report to the president by a subcommittee of the National Security Council on January 31, 1950, which recommended that Truman support building the H-bomb. Truman met with Lilienthal and the secretaries of state and defense in his office that day. Lilienthal voiced his reservations, but Truman interrupted him, telling him that, in Lilienthal's words, "we could have had all this re-examination quietly if Senator Ed Johnson hadn't made that unfortunate remark about the super bomb; since that time there has been so much talk in the Congress and everywhere and people are so excited he [Truman] really hasn't any alternative but to go ahead and that was what he was going to do."[127]

Later that day, Truman issued a public statement that he had "directed the Atomic Energy Commission to continue its work on all forms of atomic weapons, including the so-called hydrogen or superbomb."[128] The news was greeted with cheers in the House of Representatives; public opinion polls showed Americans greatly approved.[129] And so the H-bomb debate was, in a sense, over. It was a debate that started in secrecy between hawks whose opinions on the bomb differed primarily only in degree, and was criticized as being undemocratic and "authoritarian," even after it became public.[130] Those within the administration felt that if the debate was made truly "public," with truly democratic participation, it would alert the USSR to US intentions. There is, again, an irony that this "secret" debate suffered from so many leaks, and that many of them came from those in favor of the weapon's development, the same people who held themselves up as the defenders of secrecy.

Along with his public announcement on the "Super," Truman had also issued a "Top Secret" directive to the AEC. It differed by a single additional clause: "I have also decided to indicate publicly the intention of this Government to continue work to determine the feasibility of a thermonuclear weapon, and I hereby direct that no further official information be made public on it without my approval."[131] In this way, the directive to build the H-bomb was also a directive to stop talking about the H-bomb, a "gag order" passed down directly from the president. The subcommittee of the National Security Council had recommended this clause be included in the directive; its own report noted that the Department of Defense believed "that public discussion once initiated and encouraged is extremely difficult to control and inevitably leads to a greater disclosure than originally intended."[132]

Between "Joe-1" and the H-bomb debate, a major shift had begun to take place within the US atomic regime. In November 1949, Lilienthal submitted his notice that he planned to retire and would remain in office only through mid-February 1950. The events of 1949 had been too much for him, and his growing animosity with Lewis Strauss and the JCAE had soured things irreparably. If "Joe-1" had seemed a possibility for a way out of the trends of the past four years, the H-bomb debate showed otherwise.

By 1950, there was little doubt that *some* espionage had occurred against the Manhattan Project. The Canadian "spy ring" headline had appeared three years earlier, and the House Un-American Activities Committee had been buzzing for months with accounts of alleged spying at Berkeley during the war.[133] These were regarded by those within the AEC and the JCAE as likely minor disclosures, if indeed disclosures they were. The disclosures they knew about may have revealed the existence of the then-secret project and perhaps a few technical details. This kind of espionage seems to have been regarded, even by Manhattan Project security forces, as something to track, but nothing more than one would expect in such a large endeavor, and the general attitude within the Lilienthal AEC was that they had been blown out of proportion.

But two days after Truman announced his decision on the H-bomb crash program, even bigger news broke. On February 2, Klaus Fuchs, a member of the British mission to the Manhattan Project, was arrested in London by officials from Scotland Yard. Days before, he had confessed to British officials that he had been a Soviet spy since 1942. He was caught because US cryptoanalysts had decrypted World War II–era Soviet intelligence communications and identified that the American project had a mole, and with their information, the FBI and British intelligence officers narrowed down the suspect to Fuchs, who, once confronted, soon confessed.[134]

Since September 1949 the FBI had known there was a spy at Los Alamos under the code-names "Rest" and "Charles" and had deep suspicions that "Rest"/"Charles" was Fuchs, based on ancillary information from Soviet cables and certain reports to which he had access. In October 1949, they informed the AEC Intelligence office that Fuchs had been a Communist before the war and had connections with other Communists. The matter was briefly raised at an AEC meeting that November, but nothing more was made of it. The significance of Fuchs' Communism was not appreciated, in part because the FBI was avoiding giving the AEC information that would reveal that their suspicions had come from decrypted Soviet intelligence (even Truman was unaware of the decryption successes). The fact that Fuchs was considered a spy, and that he had "admitted [to] furnishing the Soviets with full atomic 'know-how' from Los Alamos," was not relayed to the AEC until February 1.[135]

Lilienthal would learn about it at seven in the evening on February 2, just as he was leaving the office. "The roof fell in today, you might say," he wrote in a dejected diary entry. "It is a world catastrophe, and a sad day for the human race."[136] He was told that it would be world news the next day, when Fuchs was to be arraigned in London. Lilienthal had never heard of Fuchs before, but he quickly learned that he was no minor scientist. He slept poorly, with visions of "the top blowing off things . . . antagonism increased between US and Britain, witch-hunts, anti-scientist orgies, etc."[137]

The AEC hastily prepared a public statement about Fuchs, so hastily they got his name wrong ("Karl"). The AEC meeting on the morning of February 3 dissolved into acrimonious bickering between the General Manager Carroll Wilson and Lewis Strauss over Wilson's security sensibilities. The press release gave a bare account of Fuchs' association with the project and noted his participation in the 1947 and 1948 international declassification conferences. Confusion over when the British would release their own statement made for an "agitated bunch" in the AEC office that morning. An hour later, they were meeting with the JCAE.[138]

The JCAE emergency executive session with all five AEC commissioners was no calmer.[139] The full details of Fuchs' espionage were not yet known, as the British were cautious about sharing his full confession. But based on knowledge of what Fuchs had worked on while at Los Alamos, it was clear that, as Senator McMahon put it, "we are in a hell of a mess."[140] Lilienthal was similarly visceral, in his attempt to have a "sobering effect" on the congressmen:

> [T]his is a very black day—there is no way of minimizing the extent of knowledge that this man had. He was an intimate part of the development of atomic weapons at Los Alamos for two years. This man was not on the edge of things—he was in the middle.[141]

What exactly was he in the middle of? Or, as Senator Millard Tydings put the question to Smyth, "Was [Fuchs] in a position to go to the Russians and say, 'Here's how you can make an A-bomb, and here's how you can make an H-bomb'?" Smyth's reply was not comforting: "I don't know about the H-bomb but the answer is yes on the A-bomb."[142] For Smyth, the one-time historian of the Manhattan Project, the depth of Fuchs' knowledge could not be minimized:

> [Fuchs] had all the information about the bomb itself that anyone could have asked for. I had hoped that perhaps he didn't know about initiators—somewhere in here it states that he practically designed our initiator; I had hoped he did not know about [core] levitation—but he made efficiency calculations on this.[143]

Fuchs knew all the most sensitive aspects of bomb design, had helped develop the theory behind the gaseous diffusion method for uranium enrichment, had indeed worked intensively on the problem of the hydrogen bomb, had been an editor of a 25-volume "Los Alamos Encyclopedia" that summarized the work of the laboratory, and he had attended every laboratory-wide colloquium available to him. He also had a near-eidetic memory, and had been considered so trustworthy at Los Alamos that he often served as a babysitter.[144] As Norris Bradbury, Oppenheimer's successor as director of Los Alamos, put it later: "He worked very hard; worked very hard for us, for this country. His trouble was that he worked very hard for Russia too."[145]

The congressmen sought to come to terms with the implications of Fuchs' deception and to come up with the next steps. Smyth attempted to point out that secrecy was never considered by scientists to be a permanent strategy: "The strength of our position has always been in the efficiency and rate of our production and the efficiency of our weapons rather than any particular secret that we might have—that is perhaps the reason why I am not quite so much alarmed by this as some may be."[146] The appeal seems to have been ignored. Discussion turned instead to the question of how the JCAE would claim jurisdiction over the Fuchs scandal to avoid the incursions of HUAC. Lilienthal tried to impress upon the congressmen how undesirable it would be for the Fuchs case to spark a "witch-hunt," pointing out that alienating scientists at this moment could be the most disastrous move possible, given the need for their cooperation on the hydrogen bomb development. Representative Charles Elston's attempt to reassure him, the last on-the-record utterance of the meeting, could not have done so: "I don't think any loyal person should be disturbed."[147]

Secret hearings continued, as the JCAE, tasked with atomic energy oversight, struggled to understand what had happened and what to do next. On February 4, they called General Groves to their hot seat.[148] Two years into his forced retirement (he had made one too many enemies in the military), Groves had been in touch with Oppenheimer, who "confirmed what I already had suspected as to Fuchs' importance at Los Alamos, and [had] also tried to describe him to me and I just couldn't remember Fuchs."[149] Groves undoubtedly felt he was in a tough spot; his "best-kept secret" legacy was at risk. He argued, not disingenuously, that the Manhattan Project security apparatus had been primarily focused on preventing leaks and indiscretions, not rooting out disloyalty. But the congressmen were not willing to let him off the hook. "This man was a known member of the Communist Party from his youth," one senator pressed. "Isn't that a fact that could be determined, maybe by a real close search?" Groves insisted, as he would years later in his memoirs, that despite his micromanagement of the lives of those at Los Alamos, pressing the British too hard on the issue of background checks would have been a diplomatic faux pas.[150]

When pressed about what to do next, Groves emphasized the need for a reversion to a wartime secrecy model, that "we should start locking the door" on all secret projects beyond atomic energy. What the AEC should bring back, he argued, were the "fundamental security principles of the Manhattan District during the war," notably extreme compartmentalization.[151]

But there was a fatal flaw to this argument, as McMahon saw it: "Under that system we have Fuchs. That isn't said with any intent to blame you, General, I can see the burden you had, but now you say 'Let us have the system that I had during the war.'"[152] Further questioning brought out more contradictions. When Groves pushed for stricter ideological background checks McMahon and others astutely pointed out that such a policy would have eliminated scores of Groves' own personnel picks.[153] As Groves' hearing drew to a close, the JCAE agreed that they ought to call in FBI Director J. Edgar Hoover next, even though he was scheduled to talk to the House Un-American Activities Committee soon. After all, one representative explained, "There is no use in letting them grab our performances from us."[154]

On February 6, Hoover listed all the inadequacies with the existing security setup to the JCAE.[155] He criticized the Atomic Energy Act for requiring "intent to injure the United States" with disclosures—after all, was Fuchs trying to hurt the US, or just help the USSR, and was there a difference? He considered it to be an "unnecessary restriction," and made it clear that he felt the "existing provisions of the law [were] too cumbersome to permit adequate security measures and quick justice."[156] Lastly, in the words of the summary:

> Chairman McMahon expressed considerable concern over our inability to control such groups as those under Joliot-Curie in France and atomic energy workers in other countries. The discussion indicated that it was

> obviously impossible for us to contain secrets within the western world so long as our security measures were limited to either the United States, or the United States, Great Britain, and Canada alone.[157]

This was not policy, it was pure frustration: the most powerful senator on atomic energy matters lamenting that the US could not extend American security concerns to "atomic energy workers in other countries." There seemed to be no possibility of prosecuting Fuchs in the US, and it was expected (correctly) that the British would be more lenient than American courts.[158]

For weeks, the Fuchs case continued to haunt the secret executive sessions of the JCAE. On February 10, they again met with the AEC to discuss progress on hydrogen bomb development. The spying made the congressmen even more desperate for the H-bomb than before. "We have got to have this weapon," one senator implored. "It seems to me for the time being we could let the peacetime aspects of [atomic energy] lag a little and put that emphasis on the wartime aspects of it to a very good advantage, because if we don't stay alive, it won't make much difference what the peacetime effect is, anyway."[159] And the congressmen were frustrated. They had heard no new information about Fuchs other than what they got from newspapers, and were riled that the newsmen seemed better informed than they were. And the British intelligence services, it would much later come out, had been sheltering Fuchs, believing his talents in working on the British bomb were enough to warrant special security dispensations.[160]

One might wonder whether the Fuchs affair would encourage soul-searching on the issue of secrecy like that sparked by the Soviet bomb. As McMahon ruefully remarked in a classified session, "[M]aybe there is nothing left to give away."[161] But, perhaps predictably, the initial impulse was to re-embrace secrecy despite its obvious failings. A month later, the JCAE asked Lewis Strauss what should be done going forward. Strauss, the H-bomb advocate and most pro-secrecy member of the AEC, had announced his resignation from the AEC two months earlier but was still technically a commissioner until that April. Aside from underscoring the urgent need for a hydrogen bomb, Strauss felt that the Fuchs affair warranted what he called Operation "Stable Door"—as in, the horses were out of the barn, and yet, "we have got to put more bolts and locks on what we discover from this time forward, and give a very thorough screening [to personnel]."[162]

The verbatim transcripts of the JCAE hearings show the Fuchs affair to be perhaps the most intense topic ever recorded by the committee.[163] The Soviets having the atomic bomb was something that many had expected to happen eventually. But the shift after Fuchs was profound, because now the Soviet bomb was seen as a result of perfidy rather than technical achievement. Of all the frustrated outbursts, none were more poignant than Lilienthal's haiku-like outburst on the morning after Fuchs' arrest: "As far as information goes—we do not know—this is a black picture—this traitor."[164] The Fuchs scandal reinforced every idea about the importance of "secrets," and the dark consequences that were associated with their loss. The myths were fraying: the Manhattan Project was clearly not the war's "best-kept secret"; American secrecy was clearly not as all-pervasively successful as had been believed; American nuclear superiority was not a foregone conclusion.

Recent scholarship, benefiting from releases by the Russian Federation, has done much to enlighten us about exactly what information Fuchs passed to the USSR, and how the Soviets actually used the information. Fuchs provided a wealth of information on bomb design, without doubt more complete and more technically informative than any other espionage the Soviets received. But there is much beyond design to acquiring a nuclear weapon. Fuchs knew little about the Hanford reactors, for example, and his espionage could not have contributed to the development of the plutonium necessary for the first Soviet bomb. And his work on the H-bomb was more misleading than it was helpful; it did little to aid either the American or Soviet thermonuclear programs. Archival research in the former USSR has shown that the Soviets did *not* make efficient or optimal use of the information, because they did not trust it. The head of the Soviet bomb project, Lavrenty Beria, was also the Soviet's top spymaster, and as such was not the trusting sort, knowing that espionage information could easily be part of a double-agent campaign to feed the USSR misinformation.

Instead, the Soviets kept the fact of espionage secret, using it as a "check" against the work of Soviet scientists (themselves not fully trusted) rather than as a way to accelerate their program or to avoid "false leads." And while the first Soviet bomb was essentially a clone of the Trinity device (a fact unknown to almost all who worked on it), Soviet scientists had already developed more efficient designs that went untested because of Beria's need for the first test to be a guaranteed success. Ultimately the timetable of the Soviet bomb program was set by the acquisition of uranium ore and the production of fissile material, not matters of weapon design.[165]

So it appears, in retrospect, that Fuchs' espionage did not likely *accelerate* the Soviet acquisition of the atomic bomb by very much, if at all. This is not to say that it wasn't useful—they used it to guide and check their work—but they didn't use it to *speed up*. But throughout the Cold War, the idea of the Soviet bomb as a "stolen" bomb was overwhelmingly prevalent, and it remains so even today in most popular American understanding of "how the Soviets got the bomb." And this narrative not only relies upon, but ultimately reinforces, the concept of "the secret."

The period between late September 1949 and early February 1950 was one of transformation and crisis for American thought about nuclear security and secrecy. It was possible to see both "Joe-1" and the Fuchs case as spurs for a more liberal approach to secrecy: if the Soviets had the bomb, *and* the secrets, then what was the point of further keeping secrets? As a columnist for the *New York Post* opined in the days after Fuchs:

> In our pathetic eagerness to be protected from the dark terrors of the unknown, we turn to the religion of secrecy and to the savior-with-a-dossier. . . . The hopes we cling to are pathetic because they are so illusory. What we seek is security. We think we can surround ourselves by the double wall of the H-bomb and the G-mind. . . . When the secrecy with which we have enveloped our research and policy showed its futility—as it has done in the Fuchs episode—we now move to double and triple that secrecy. But it won't work. There can be no security in dreams.[166]

Within the US atomic empire, some voices expressed similar thoughts. The Committee of Senior Responsible Reviewers felt that the

Fuchs revelations "might properly lead to a reconsideration of the entire question of secrecy."[167] Similarly, acting chairman Sumner T. Pike informed Brien McMahon later in 1950 that the AEC was undergoing a comprehensive review of its security and secrecy program, with the goal of declassifying information that had already been released, in order to relieve some of the "unnecessary and stifling classification."[168]

As we will see in the next chapter, the end of the American monopoly and the knowledge that many "secrets" had been classified did result in a relaxation of classification restrictions over many realms of technical information. But it came with a cost: the security focus was moving from the information to the scientists themselves, now seen as the most unreliable component in the secrecy regime. This shift began especially in the wake of the Fuchs debacle.

One of the difficulties the AEC faced in the Lilienthal years and beyond was that from the outside it appeared to be one of the most opaque and secretive organizations ever created, with powers that could reach well into civilian life and stifle entire fields of science. And in some sense, it was that. At the same time, as we have seen, some within the organization had created thoughtful, careful critiques of their own policies, with an aim toward "rational" and even "enlightened" compromises that would fulfill the greatest hopes of the organization. These activities were largely invisible to the outside world because of classification issues as well as "normal" bureaucratic secrecy. This contradiction (a secret organization secretly attempting to escape from the throes of secrecy) was not lost on its participants. In a February 1948 written response to criticism from a physicist outside of the AEC system, AEC Commissioner William Waymack summed up the dilemma very neatly, which is worth quoting at length for its unusually nuanced perspective on being "the censor":

> I agree that secrecy tends to be destructive of confidence, that it should be avoided where possible, that this is particularly important in areas of "meaningful popular education," and that it would be fine to have a lot of universities engaged in unclassified nuclear research.
>
> If I were out there and you were in here, I'd probably be expressing to you the same kind of dissatisfaction because of the impossibility of deciding myself whether critical parts of policy were "right or wrong, wise or foolish, selfish or altruistic."

> How anybody who is really attached to liberal principles can be happy in the situation is beyond me. But, happy or not, we have to do the best we can. We operate under a law. I think it is a pretty good one. Yet it is not possible to "spell out" fully the reasons for saying even that!
>
> Anyhow, don't get mad at me. Don't assume that the Commission loves to be cryptic, least of all that it enjoys appearing arbitrary.[169]

From the outside, the AEC appeared cryptic, arbitrary, and obsessed with secrecy. From the inside, few members in the early years relished classification, and efforts were made to fight against it. But constraints existed beyond the desire of mere will: there was the law, there was Congress, there was the public, and there was an emerging Cold War.

The experimentation and pushback against secrecy that characterized the Lilienthal years ended when he left the AEC in 1950. Though most of the reform efforts failed, they are indicative of how serious, engaged, and politically astute administrators groped for subtle, practical solutions to what they felt was an ever-expanding mentality of secrecy. Lilienthal's attempt to modify the official mindset on secrecy—to introduce strong anti-secrecy sentiments, as well as truly embracing the civilian-military distinction—were also ultimately unsuccessful. It is worth contemplating why that was the case.

Part of it was the very secrecy of the AEC's operations themselves: Lilienthal's attempts at reforms were effectively invisible to anyone not in his inner circle, and this had two complementary consequences. Additionally, the AEC was regarded as universally secretive even when it was trying not to be, and easily appeared compromised to secrecy critics. Morever, the popular visions of nuclear technology, namely as an existential risk based on "secrets," were totally overwhelming, even if those in the inner circles did not subscribe to them. Lilienthal's AEC was both overmatched and under-armed to change broader sentiments on secrecy, and one can see the three "shocks" as exemplifying the magnitude of what they were up against. The fact that the AEC was a politically weak organization, with no natural constituency and no powerful allies other than a President who was only reluctantly engaged in atomic policymaking, didn't help things. Lilienthal had control of one of the institutions for nuclear secrecy, but he didn't have control over how people thought about secrets.

Lilienthal's successors to the position, as we shall see, at best shared none of his zeal (Gordon Dean), and at worst had zeal in the opposite direction (Lewis Strauss). To see this as a "fall from grace" would be an exaggeration, for the Lilienthal years were themselves marked by compromise and often unsuccessful attempts at "balance." But something had indeed changed: where the postwar approach was still looking for a subtle solution to the "problem of secrecy," a new, thoroughly "Cold War" sensibility was beginning to replace it, and it saw only in extremes—both in terms of dissemination and constraint.

6

PEACEFUL ATOMS, DANGEROUS SCIENTISTS

THE PARADOXES OF COLD WAR SECRECY, 1950–1969

Information once compromised is information broadcast forever.

LEWIS STRAUSS, 1962[1]

The strain of the events of late 1949 and early 1950 spelled the end of the postwar ambivalence toward secrecy. Subtle, quasi-philosophical discussions about the "problem of secrecy" would not cut it in a world where the United States no longer had a monopoly on nuclear weapons, and where thermonuclear weaponry could increase the consequences of the bomb by orders of magnitude. In place of the postwar mindset would arise a new, more expansive framing: a true Cold War secrecy regime, which would combine a newly persuasive way of thinking about nuclear weapons with an ever-expanding infrastructure of government secrecy.

Today we associate the Cold War mindset with McCarthyism, witch-hunts, and spy hysteria, all heightened by a sense of existential dread. But there was another side to this mindset as well: a zeal for free-market, capitalist "solutions" to global and domestic problems. In the nuclear realm, this additional angle would ultimately propel declassification efforts well beyond anything the postwar reformers could have imagined, in the name of generating peaceful and cheap electricity. It is this contradictory dyad—hugely invasive policies to govern everything deemed "secret," with a nearly opposite desire for openness regarding anything deemed "peaceful" or that would promote atomic industry—that makes up the somewhat schizophrenic regime that emerged in the Eisenhower years.

If the institutions of the "problem of secrecy" mindset, such as the

Tolman Committee and the Lilienthal Atomic Energy Commission, were characterized by their obsession with shades of gray, the Cold War mindset was one of extremes. And as with many policies made up of extremes, it would produce its own self-destructive contradictions: a multi-polar nuclear world in which the lines between "good" and "bad" technology became increasingly blurred, despite desperate attempts to force them to act like stable categories. But, as we will see, the Cold War regime was remarkably resilient for approximately a twenty year period, ultimately ending due to its own dissemination of nuclear knowledge, expertise, materials, and technology across the globe, and its discursive power can be still felt today.

6.1 THE H-BOMB'S SILENCE AND ROAR

The hydrogen bomb debate was a painful one for the American community of scientists, and also for the idea of secrecy reform. For many of the physicists and chemists involved in it, it was practically a referendum on the capability of the United States to have democratic, informed discussion about its nuclear weapons policies—and in that it appeared to be answered in the negative. For those who knew about it, the fact that Truman's order to build the hydrogen bomb had contained a secret "gag" order as well only reaffirmed this notion.

The hydrogen bomb would become the focal point for the Cold War reformulation of secrecy. Unlike the atomic (fission) bomb, it could not be easily said to be the product of prior, open research. And it was more plausible to consider it a creation of informational "secrets" than a matter of industrial production (there are, to be sure, industrial requirements for producing thermonuclear weapons, but they do not differ significantly from the investments already needed for a fission weapon, which is its prerequisite anyway). But more importantly, it came to symbolize the transformation of knowledge into power, while also, for reasons that are particular to its context, refocusing attention toward the reliability of the scientists themselves. It played a significant role, both practically and symbolically, in the reshaping and reinforcement of the Cold War regime of secrecy, and was still considered the "ultimate secret" until the early 1980s. The hydrogen bomb's explosive power was orders of magnitude more powerful than the weapons used in World War II, and its rhetorical and political power was similarly disproportionate.

The "gag" order on the hydrogen bomb that Truman secretly issued alongside his decision to produce thermonuclear weapons in late January 1950 appears to have been a hastily formulated reaction to the H-bomb debate and its leaks, without consideration of how long it might be kept in place or by what mechanism, other than direct presidential order, it might be lifted. It was totalizing and absolute: stipulating that "no further official information be made public" on thermonuclear weapons without presidential approval.[2] Such a denial of freedom of speech to the AEC was unusual. The AEC was otherwise free to talk about unclassified topics at its discretion, and the Atomic Energy Act gave it the power to determine what could be removed from the Restricted Data category. With the H-bomb, however, they were totally muzzled, classified or not.

The "gag order" was in stark contrast to the "Information Control" philosophy the commission had adopted since 1947: it forced the people with knowledge to stay quiet, while those without could speak freely. Two days after Truman's directive, Lilienthal fumed to the other commissioners about an article by William Laurence in the *New York Times* regarding the use of tritium as fuel for a potential hydrogen bomb. The fact that tritium might be useful in a hydrogen bomb was fairly common knowledge, but the fact that the AEC was programmatically pursuing tritium was still meant to be a secret. Lilienthal wondered whether there had been a leak, but he was more concerned with the fact that because Laurence had previously had special access to the Manhattan Project, people would assume he had special knowledge of the AEC's current activities, which was false. Furthermore, it would suggest that the boundaries of what the AEC could say about the H-bomb had been relaxed, when in fact the opposite had occurred.[3]

The idea of the "Super," though developed during World War II, had never been formally declassified. It appears Truman himself had no knowledge of the hydrogen bomb until October 1949, when he was informed by Admiral Sidney Souers, who had himself just been informed by Lewis Strauss.[4] Rumors of a vastly more powerful atomic bomb had circulated since the first weeks after the dropping of the atomic bomb on Hiroshima, though the government had never denied or confirmed them.[5] The General Advisory Committee's report of October 1949 had recommended that a public statement about the possibility of the H-bomb be issued, but this was not pursued.[6] By early January 1950, it was common knowledge within the AEC that the President had issued an "explicit injunction against publicity" on the subject, even prior to the formal "gag" directive at the end of the month.[7]

Less than a week after the "gag order" was imposed, the AEC attempted to resist it. These measures consisted of plotting to get the president, or the National Security Council, to alter their directive, as well as considering whether they could be authorized to release a very sanitized statement about the H-bomb that would debunk some myths and rumors. They struggled with the idea that they had been denied control over these matters, and the fact that they were disallowed from telling anyone that the "gag order" itself existed and that it was being imposed on them, not by them.[8] They viewed the order as both impractical and counterproductive—blanket secrecy both made it difficult to deal with public relations while also hindering recruitment to Los Alamos, as newspaper accounts implied that the H-bomb was imminent and merely an "engineering" job. And in the meantime, public speculation on the H-bomb was rampant. People unassociated with the AEC could write freely on the topic, but AEC scientists with good information could not.[9]

In the midst of this came another crisis. In early March, the physicist Hans Bethe had circulated an article he had written on the hydrogen bomb for the April 1950 issue of *Scientific American*. Bethe was an international expert on nuclear fusion, a major participant in the Manhattan Project, and a perennial consultant to the AEC thereafter. His article was to be the second in a four-part *Scientific American* series on the hydrogen bomb; the first had been written by Louis Ridenour, a physicist unconnected with the AEC who was thus free to speculate at will. AEC's director of classification, James G. Beckerley, had reviewed Bethe's article and found certain statements that appeared questionable from a classification standpoint that he would like deleted. Most of what Bethe said was available in other articles, but because of Bethe's long association with atomic energy work, Beckerley argued that he "spoke with an authority which was not matched by writers . . . who had never had access to classified information." Smyth and Pike agreed that they would formally ask for the Bethe article to remain unpublished until it could be determined whether it contained Restricted Data.[10] A further review by classification staff two days later confirmed that, in their eyes, it did contain secrets, and the AEC authorized their general counsel to initiate the necessary legal steps to prevent publication should *Scientific American* decline to accept the request to delete certain portions of Dr. Bethe's article."[11]

The AEC staff contacted the publisher of *Scientific American*, Gerard Piel, and asked that half of the article be deleted. Piel requested they be more specific in their objections, and the AEC narrowed down their deletions to three specific statements, each of which were objectionable *only* because it was Bethe who said them. The statements were by themselves things that had been published previously (and Bethe had assumed were unclassified), but when said by Bethe it was possible to interpret them as confirming the direction and character of the AEC's H-bomb work.[12] Piel agreed to make the changes but reported that many copies of the issue had already been printed. An AEC representative from New York arrived on March 20 to supervise destruction of the existing copies. At the suggestion of *Scientific American*'s general manager, they burned the 3,000 copies and melted down the linotype slugs. The revised article was substituted, and the issue was delayed by only a few days.[13]

Piel was not happy with what he considered blatant government censorship. Because none of the statements were truly secret in the sense of "not known," he believed that the AEC "was not motivated by concern to protect military secrets."[14] In an address to the American Society of Newspaper Editors in April 1950, he argued that the AEC must be using secrecy to avoid public scrutiny and to wield its power arbitrarily. He thoroughly rejected the argument that Bethe's associations with the AEC mattered, and saw the AEC's actions as "clumsy censorship."[15] Piel, of course, could not see the internal, secret discussions that the AEC was having. The main difficulty for the AEC was the Bethe article's poor timing: it came when the AEC itself was seeking to overturn what it considered to be an oppressive directive regarding how much could be said about the hydrogen bomb, and they feared that if they appeared to be running a loose ship it would undermine their cause.

In the end, very little about the article was changed; the article's major political and even technical points remained. Bethe would later say that "it was a considerable loss for the *Scientific American*" in terms of the replacing the destroyed inventory, "but on the other hand they got some advertising out of it."[16] Indeed, Piel was able to turn the issue into a major publicity coup, garnering a front-page story in the *New York Times* that coincided with the release of the April 1950 issue, and he used it to champion freedom of the press, despite his compliance with the secrecy request. The fact that the copies had been burned, and type had been melted down, was used to show the zealousness of the AEC, even though it was *Scientific American* staff who had suggested it.[17] Because Piel was not inside the government, it was his version of the narrative that was allowed to dominate coverage of the story—the AEC was prevented by its own secrecy from giving its side of the story. In May 1950, the American Civil Liberties Union (ACLU) would formally protest the Bethe censorship and the AEC's H-bomb censorship of its staff. The AEC replied that the Bethe issue was separate from whether AEC staff could speak on the H-bomb, and that it had no interest in muzzling scientists.[18]

In late April 1950, Smyth spoke at the annual meeting of the American Society of Newspaper Editors, following Piel's denunciation of the AEC's handling of the Bethe affair. Smyth's talk, on "Secret Weapons and Free Speech," laid out the AEC's attitudes toward secrecy and restriction of public discussion. It was a generally balanced account: it made no great case for the danger of information and argued that nuclear weapons had led to greater government restraint in releasing information. He argued that in principle the AEC desired free dissemination of "basic science" but was required by the Atomic Energy Act to control "weapons information." "Unfortunately," he noted, "these principles are easier to formulate than to apply. There is a twilight zone of information that does not automatically fall into either category."

Smyth put forward the idea that classification of speech must, in part, depend on *who* was talking. If the author of an article had no access to AEC information and was not affiliated with the government, they could generally speculate however they saw fit. Government employees, though, spoke as representatives of government policy. When citing the work of non-government employees, they gave it the air of authenticity. Smyth urged the audience to consider the idea that the AEC had almost no interest in trying to censor the truly private sphere, but that it did not consider current or former AEC employees to be in that sphere. He ended on a somber note: "We do not enjoy making these judgments."[19]

The AEC edited Smyth's speech into a statement about its "Information Control Policy" and circulated it within the agency as its current stance on secrecy.[20] It was a very different approach than the analytical attempts at reform that had been pursued in the late 1940s. It would be one of several such shifts that began to coalesce in the high Cold War, with a stronger emphasis on the control of government *personnel* as a means of controlling official *information*. The *Bulletin of the Atomic Scientists* wrote an editorial against the practice (dubbed "It's Not What's Said, It's Who Says It"), and the Federation of American Scientists issued a strident protest.[21]

Throughout the spring of 1950, the AEC attempted to develop a new policy on what agency staff could say about the H-bomb. Despite getting support from within the AEC, and from the General Advisory Committee (who noted that since the only true secret of the H-bomb was how to make one, it was "presumably secret from everyone," since nobody knew how to do that), they met with strong opposition from the Departments of Defense and State, despite the AEC's argument that the "blackout of official information is hampering the program and having other bad effects."[22] Finally, in June 1950, the National Security Concil approved a relaxation of the "gag" in order to permit testimony and to give the AEC the ability to respond to inquiries, but it stood firm in its desire for no public statement.[23]

The "gag" had been slightly lifted, but the AEC released no statement. The Joint Committee on Atomic Energy, filling the gap, released a primer prepared by its staff on basic hydrogen bomb technical facts, fulfilling much of the function of the proposed AEC statement, and even going a bit further than the AEC had considered.[24] Ironically, the JCAE would appear to the public to be the more forthcoming organization. The AEC's efforts to reduce secrecy were once again hidden by its own secrecy.

In the summer of 1950, Gordon Dean was confirmed as the new chairman of the AEC. As previously noted, Dean was no ideologue. He was by profession and temperament a lawyer, and interested in implementing the mandate he was given, which involved both producing thermonuclear weapons and building up the American nuclear arsenal. His lack of a strong ideological bent on secrecy is notable: he does not seem to have seen the issue, as did Lilienthal, as a fundamental question to be answered, but rather just part of the job, specifically connected with serving the commission's duties as outlined by the Atomic Energy Act. On some issues he favored openness, on others secrecy. It is easy to consider Dean somewhat dull when compared to either his predecessor (Lilienthal) or successor (Strauss). But Dean's lawyerly professionalism and effective lack of partisanship would ultimately be what characterized the AEC as it matured in the Washington ecosystem.[25]

In the spring of 1951, after much work, Los Alamos scientists finally zeroed in on what seemed to be a successful hydrogen bomb design. The original approach taken by Edward Teller since the Manhattan Project, what would become known as the "Classical Super," had been discovered to be a dud. In its place, a new idea had taken hold, the Teller-Ulam design.[26]

This design, a collaborative product by Teller and the Los Alamos mathematician Stanislaw Ulam, was conceptually quite different than what had come before, though its basic idea was deceptively simple. The "Classical" Super concept was to use the heat of an atomic fission bomb to create a self-sustaining, linearly-propagating fusion reaction in a column of deuterium. Calculations had shown this was unlikely to work, as the heat losses would be too great to keep the reaction alive (the fusion reaction would get too cold to self-propagate). The Teller-Ulam idea by contrast involved putting a "primary" fission bomb into a heavy "radiation case" that would, upon detonation, redirect the radiation energy and use it to pre-compress a "secondary" capsule made of fusion fuel and fissile material at the other end of the casing, and the high pressures, followed by high temperatures, would start the fusion reaction. This new idea drew upon the half-decade of research into fission and fusion weapons that had continued after the end of World

War II, and unlike the Classical Super idea, was immediately compelling, even to skeptics like Bethe and Oppenheimer.[27]

A full-scale test was planned for the fall of 1952, held at the AEC's Pacific Proving Grounds site in the Marshall Islands. The device to be tested was not a weapon per se, but a large, conservatively designed proof-of-concept: the fusion fuel was deuterium kept in a liquid state by 80 tons of cryogenic equipment, by no means an air-deliverable device. The timing was awkward, happening almost concurrently with the 1952 presidential election. Hans Bethe had written to Gordon Dean, warning that should news of the test become public, there would be political dangers: it might be used as "campaign materials" by the press should it leak out.[28]

Bethe predicted that news of the test would leak no matter how much secrecy was applied to it for the simple reason that if you detonate a multi-megaton weapon in the Pacific, someone is going to notice:

> If there is no disclosure, the test may still become public knowledge because of large fall-outs, visual observations from Kwajalein, or possibly observations of shock or seismic phenomena. Whichever may be the method of revelation, the evidence of a test with enormous yield combined with a lot of previous discussion in the columns of newspapers will almost undoubtedly lead the public to the right conclusion.[29]

Bethe's recommendation was to delay the test for a week, until after the election. For technical reasons, the Los Alamos scientists working on the test, and the AEC itself, strongly favored holding it on schedule anyway.[30]

There were other reasons to think testing a full-sized H-bomb was a bad idea. Vannevar Bush, who sat on the committees that had analyzed the fallout from the Soviet atomic testing, warned that the Soviets no doubt had the ability to intercept and analyze American fallout as well. By careful analysis of the radioactive debris from a nuclear test, even debris extremely distant from the test itself, one can learn many useful things, such as the types of fission fuel used, the relative ratios of fission and fusion in the explosion, and even aspects of the bomb's design. Bush later noted that this kind of information, combined with press speculation, would "deliver it to them on a platter," to the degree that, as he put it, "I can't understand why they need any spy network in this country."[31]

But the need to confirm the principle, and confirm it at full scale, overwhelmed these concerns. No announcement would be made, and Operation Ivy, as it was known, would be kept as secret as possible. This would be no easy task, since the staging of the operation would require some 10,000 military and scientific personnel to be stationed overseas at the test site, any one of whom could, through a stray rumor or indiscretion, indicate that the US had tested something well beyond the experience of the norm.[32]

The detonation, code-named "Mike" for "megaton," went as planned, releasing an energy equivalent to over 10 million tons of TNT, over 500 times more powerful than the weapons dropped on Hiroshima and Nagasaki. The blast completely destroyed the small island of Elugelab, leaving a massive crater in its place. The fallout from "Mike" indeed circled the world, but the Soviet chemist assigned to analyze it botched the work.[33]

Along with the thousands of direct witnesses, there was an extensive documentary film made about the test, kept highly classified and meant to be shown to government officials back in Washington well after the test series was completed. Physical evidence of the test was available for those who looked for it. Edward Teller did not attend the shot, but he witnessed it nonetheless: availing himself of a seismograph at UC Berkeley, he watched the expected bump at the prearranged shot time.[34]

As Bethe had predicted, despite heavy attempts at secrecy, news of the successful detonation leaked out immediately. Even a day *before* the test had gone off, a reporter from *Time* magazine had called the AEC and inquired about the pending H-bomb explosion: "Is this the big day? . . . We understand that the H-bomb has just been set off." The AEC response was simply: "We have a standard policy of no comment about weapons tests."[35]

After the test, there were more leaks, largely in the form of "eyewitness" stories from servicemen, in violation of security rules, and then reprinted by newspapers. The military tracked at least sixteen of these stories and claimed to have identified and reprimanded the leakers. The AEC responded with a terse, unenlightening press release, noting that there had been nuclear tests that had "included experiments contributing to thermonuclear weapons research. Scientific executives for the tests have expressed satisfaction with the results."[36]

This press release neither confirmed nor denied the H-bomb's existence, leaving it in an epistemic limbo: implied but unconfirmed. The test itself left the hydrogen bomb program in a similar limbo: the AEC had proved it was possible to make multi-megaton thermonuclear explosions, but had not shown how to make them into deliverable weapons. That would come in the spring of 1954, with the Operation Castle series, where the first weapons that would use solid (lithium-deuteride) fusion fuels removed the need for cryogenic equipment.

In the meantime, the question of what could and could not be said about the H-bomb remained. This was not taken up as a political issue in the 1952 elections, to the relief of the AEC. The president-elect, Dwight D. Eisenhower, was informed of the new capability almost immediately after the votes were tallied. At Truman's request, Dean sent Eisenhower a letter explaining that the US had detonated "the first full-scale thermonuclear weapon," but that it would "probably be a year before we will be able to test the first deliverable thermonuclear weapon." Dean acknowledged that despite the goal "to keep the Russians in the dark," the size of the weapon and the number of personnel meant that "it is not likely that we can for long keep from the Russians the fact that there has been a thermonuclear explosion." He similarly warned that seismographs would reveal the tremor, which might be perceived as an underwater earthquake, but said that the AEC was not planning to release much more information on the subject. He attached a much longer memo outlining the US nuclear stockpile's current "position."[37] This would be the first instance of the "handoff" of the nuclear arsenal from one president to another.

But Truman was still president for several months. He approved no further releases on the H-bomb, though in January 1953, Truman's speechwriters considered adding a reference to his final State of the Union Address. The successful completion of the hydrogen bomb might be an excellent way to mark the end of Truman's presidency and could be turned into an argument for the abolition of war and international control. The language of their draft was clear about the accomplishment:

> From now on, man will be able to release forces of a magnitude never before approached on earth. The energy of these new weapons is so tremendous that it will be measured in a new unit—the megaton. . . . The stark fact is that the explosion of one of these weapons could strike a mortal blow to any of the great cities in the world. And there are no scientific bars, no technical obstacles, to keep a nation that is able to invest enough in money and materials from building such weapons in quantity. . . . This new kind of power exists; it will be with us all the days of our lives.[38]

Commissioner Smyth spent considerable time attempting to make Truman's speechwriters understand that the AEC valued greatly the fact that the Soviets probably did not know exactly what had happened in the "Mike" test, and did not know whether the US had thermonuclear weapons in its stockpile yet. Smyth further emphasized that President-elect Eisenhower was strongly opposed to further revelations about the H-bomb and felt that his considerations would need to be taken into account as well. Finally, while the AEC was not entirely opposed to some note about the successes of nuclear testing and the US nuclear program being added to the State of the Union, they thought it should be indefinite and made in passing.[39]

The next day, Commissioner Thomas Murray met with White House Counsel Charles Murphy about the statement. Murray was frustrated with the "inappropriateness" of what the speechwriters wanted to say about the H-bomb. The AEC believed that the "Mike" test, if it was to be discussed, should be used as a diplomatic "counter" to bring the Soviets back to the table to talk about international control. Murphy was apparently unimpressed. Murray realized that the only way they would understand his point of view was if he could apprise them of "the facts of the situation as to our thermonuclear capacity"—namely, that the US didn't have one yet. As Murray related to the other AEC commissioners, "these men were completely non-plussed upon hearing my statement and said that it was contrary to their understanding as given to them by . . . others."[40]

Ultimately, the January 1953, State of the Union message did contain a fleeting, careful message about thermonuclear weapons:

> [R]ecently, in the thermonuclear tests at Eniwetok, we have entered another stage in the world-shaking development of atomic energy. From now on, man moves into a new era of destructive power, capable of creating explosions of a new order of magnitude, dwarfing the mushroom clouds of Hiroshima and Nagasaki.[41]

This coy wink at new American capabilities would be all that the Truman administration would release on the subject. And for the first year of the Eisenhower administration, there was little more released. Newspaper accounts in early February 1954 reported that Eisenhower had authorized a "sanitized" version of the Operation Ivy film to be shown to members of Congress, accompanied by a statement that the "Mike" test had been "the first full-scale thermonuclear explosion in history," and "the first step in the hydrogen weapon program of the United States." That this information was released in a semi-official way—a statement made to Congress that was apparently leaked to the press—decreased the impact such a statement might have made, with most papers relegating the news to their back pages.[42] The Eisenhower administration, like the Truman administration, preferred to keep their progress on the H-bomb close to their chest. But the hydrogen bomb would not be complicit in its own silence.

Manhattan Project security officials had proven capable of keeping the Trinity test's atomic nature secret for the two weeks prior to the bombing of Hiroshima. This was, in part, because atomic bombs were still considered science fiction by most of the people who might have otherwise suspected its nuclear nature. But in a post-Hiroshima world nuclear tests became much harder to conceal. For the first postwar test series, Operation Crossroads, held in the summer of 1946, the Manhattan Engineer District did not even try. Instead, it was turned into a public relations exercise, with members of the press and the United Nations, including scientists from the Soviet Union, being allowed to view the two explosions at the Bikini Atoll. Of course, some aspects were kept secret, but compared to what came before and after, Crossroads was strangely open, and inspired considerable public attention and even kitsch (such as lending the test site name to a new swimsuit, the bikini).[43]

Nuclear testing under the AEC resumed in 1948 as a far more sober and secretive affair. The AEC was not especially interested in drawing attention to its testing activity because there were new weapons concepts being tested, because they feared that the USSR could use any information to understand intercepted fallout, and because they feared adverse publicity. But terse press releases provided some information, always after the fact and never disclosing details about the timing, nature, or power of the shots tested.[44]

The AEC developed two major test sites and by the early 1950s had fallen into a habit of regular test series. The test site at the Marshall Islands had been used since 1946, but its remoteness and rugged conditions meant that test setup and diagnostics were difficult and sometimes even deadly: at least one scientist met an untimely end when the helicopter transporting him from one atoll to another crashed into the sea.[45] While the Pacific site would continue to be used for high-yield tests, like the "Mike" test, the AEC also began using a continental test site in Nevada from 1951 onward. The Nevada Test Site, valued for its easy access and predictable climate, would become the site of most American nuclear detonations in the Cold War.[46] As with Crossroads, the easy detectability of nuclear testing for the American public (the brightness of the tests, and sometimes the resultant mushroom clouds, could be seen from the casinos of Las Vegas), despite attempts at secrecy, produced kitsch as an initial public byproduct, which would eventually be followed by anxiety.[47]

Operation Castle, held in the spring of 1954, was eagerly anticipated by the AEC and the DOD. It was a proof-test of deliverable thermonuclear weapons and would, if successful, pave the way to a new American arsenal, one that would contain the explosive power of literally millions of Hiroshima-equivalents.[48] The AEC released another bland and unrevealing announcement about detonating an "atomic device" that sparked speculation in the press that it might be another hydrogen bomb. The AEC released no further clarification.[49]

But the device for the first test shot, "Bravo," would reveal the real dangers of nuclear testing and thermonuclear weapons in general. The failure of scientists to anticipate a physical process that resulted in far more fusion reactions than expected meant that the explosion was two and a half times more powerful than predicted, detonating with the violence of 15 megatons, the largest US weapon ever set off. Ten of those megatons came from a final fission stage in which a uranium-238 tamper was bombarded by high-energy fusion neutrons that caused even the normally reticent isotope to split. This meant that the "Bravo" test produced 500 times more radioactive fission products than were released by the Trinity test. The "Bravo" fallout plume traveled hundreds of miles downwind over the next few hours, depositing hazardous levels of radiation over tens of thousands of square miles, requiring the emergency evacuation of several hundred Marshallese people and many American servicemen. Many of both groups were exposed to high levels of radiation, with many of the Marshallese developing skin burns and symptoms of radiation sickness.[50]

Even a mishap on this scale might have been containable, if not for the fact that a Japanese fishing boat had accidentally entered the "danger area" established by the AEC. The twenty-three sailors saw a flash of light in the distance and then, after some time, felt a fine white dust fall upon them—vaporized, radioactive coral. The seamen returned to Japan, their holds full of contaminated tuna, before growing ill. Their radiation sickness might have been bad enough (one of them eventually died), but when it became clear that their radioactive tuna had entered the Japanese fish markets, the country panicked. Japan, only recently released from the censorship of the American Occupation, had for the first time since World War II a national opportunity to discuss their status as a "radioactively-exposed" nation. The United States, many argued, had once again subjected Japan to nuclear violence, and the price of tuna plummeted.[51]

The unfavorable publicity was greeted by the AEC not with soul-searching, but with reflexive secrecy, characteristic hostility, and even conspiratorial suspicion. Lewis Strauss, the combative conservative AEC commissioner from the Lilienthal years, had been appointed by Eisenhower as the new chairman of the AEC the summer before. After "Bravo," Strauss released a statement emphasizing the importance of building the H-bomb, its role in US security, and that the Soviets were moving fast as well. He criticized the "exaggerated and mistaken characterizations" of the size of the blast, insisting that "at no time was the testing out of control." He described "Bravo" as an example of due diligence and noted that even the unexpectedly large blast was "a margin of error not incompatible with a totally new weapon." He put the blame on the wind for having "failed to follow the predictions," shifting toward the fishing boat who was "well within the danger area." The exposed Marshallese, he reported, were "well and happy," and no further ill effect was expected. He criticized the Japanese for not letting American doctors inspect the sailors and suggested their health problems would soon heal. Privately, he told Eisenhower's press secretary that the Japanese boat was probably a "Red spy ship."[52]

Most frustrating to Strauss was that the H-bomb was front-page news. The fact that it produced massively contaminating fallout was known not just to the newspapers, but to the Soviets as well. This revealed that the weapon worked, and that it relied heavily on U-238 fission reactions. Even if had not been obvious, scientists outside the United States, notably the nuclear dissenter (and former Manhattan Project scientist) Joseph Rotblat, published articles about it.[53] The hydrogen bomb was now undeniable, and Strauss had to preside over a curious, suspicious, and demanding national and international community.

Operation Castle was still too classified to discuss in detail, but the Operation Ivy film was approved for release to the public on April 1, 1954, giving the first direct information about hydrogen bombs, and confirmation that the United States had them, to the American people. The *New York Times*' television critic skewered the awkward film ("a turning point in history was treated like another installment of 'Racket Squad'") and lamented that it was "talking down the American people"; ironically, the audience of the original film was not the American people at all, but the American president.[54] The fact of the H-bomb's existence, in any case, was finally out, albeit only because of the disaster of "Bravo."

The development of high-yield thermonuclear weapons presented new problems relating to secrecy for the AEC and the US government. The weapon's mere idea came to symbolize the ultimate secret that must be guarded with absolute security in an age in which the American nuclear monopoly was over. Its development appeared to prove that with a little bit of ingenuity, scientists could once again conjure up a new world-threatening weapon out of seemingly basic concepts of nature. And while there was "no secret" to the basic invention of the atomic bomb in the mind of most of the scientists who worked on it, there was a secret to the hydrogen bomb: the Teller-Ulam design. Even a liberal scientist like Hans Bethe, who had thought the AEC's secrecy in discussing thermonuclear matters was overblown in 1950, and who had railed against the idea of "the secret" in the 1940s, changed his tune, writing to a JCAE staffer a few weeks after the "Mike" test that "this time we have a real secret to protect."[55]

The problem was that the Teller-Ulam design, while clever, was not an exceptionally complicated secret. It could be conveyed in a single sentence: "radiation from a fission explosive can be contained and used to transfer energy to compress and ignite a physically separate component containing thermonuclear fuel."[56] Unlike the reams of reports required to explain, in detail, how to construct a gaseous diffusion plant, the Teller-Ulam design could be given away on the back of a napkin. Of course, if a foreign agent did receive such a napkin, how would he or she evaluate its truth? Even the American scientists required a pilot test (Greenhouse "George") to confirm the basic truth of the idea and were unsatisfied until they had tested it at full scale. Presumably a foreign power would be equally skeptical *unless* they knew it had come from someone who was "in" on the secret. Thus, there is a real epistemic difference between a Los Alamos scientist uttering the words "radiation implosion" and the same words being muttered by an outsider.

Such is how the H-bomb helped bend the AEC's approach to secrecy away from the careful balance of the postwar period into the more hardline approach of the Cold War. When secrets are both important and easy to transmit, contortions must take place in order to keep them. All "inside" sources would need to be constantly screened to reaffirm their loyalty, and any utterances they directed outward would need to be scrutinized with a careful eye. The H-bomb would remain the "ultimate secret" well after the Soviet Union, in 1955, demonstrated that it too could make multi-megaton bombs.[57]

6.2 DANGEROUS MINDS

The discovery of a competent mole in the Manhattan Project in 1950 had far-reaching effects inside the American secrecy system. In his confession to the British, Klaus Fuchs had indicated that he had sensed there were other spies in the project unknown to him, setting off a mad rush to mine the same Soviet decrypted cables that had caught Fuchs for evidence of other spies. The search for a "Second Fuchs" would continue for several years, and would eventually identify the young physicist Theodore Hall as another Soviet spy. But unlike with Fuchs, the FBI was not able to build a "clean case" to prosecute Hall, and rather than divulge their secret source of information—the decryption project, code-named Venona—they contented themselves with the fact that Hall had left weapons work.[58]

Probing the edges of the Fuchs case, both in cooperation with the British to extract information from Fuchs and with Venona, soon produced more spies. Fuchs identified several salient details about Fuch's "courier," a middle-man who conveyed Fuchs' information to the Soviet Embassy. The FBI identified him as Harry Gold, a sallow chemist who apparently had fallen into spying not out of strong ideological or monetary affiliation, but because he was lonely and the Soviets provided friendship, appreciation, and comradery. Once he was in FBI lockup, however, this same desire for a human connection led him to happily share as much information as he could. Gold's cooperation in turn led to the arrest of David Greenglass, a Special Engineer Detachment machinist at Los Alamos during the war. Greenglass in turn implicated his sister and brother-in-law: Ethel and Julius Rosenberg.[59]

The Rosenberg trial lasted only the month of March 1951, but it further polarized an already-divided nation, embodying for some the threat of Soviet infiltration, and for others the excesses of McCarthyism. Practically all the material evidence against Julius and Ethel came from the testimony of Gold and Greenglass. Neither were exceptionally reliable witnesses. Gold was an odd, shifty character in apparent psychological distress, and Greenglass had a conflict of interest: he was, the Rosenbergs' defense alleged, trying to save his—or his wife's—own skin. And the overzealous rhetoric of the prosecutor and the judge made it easy for skeptics to see the entire thing as a frame-job. But we know today, through both the declassification of the Venona cables and Greenglass' later admission, that Julius Rosenberg was definitely a spy, but also that Greenglass had perjured himself to enlarge the paltry case against Ethel in order to keep his wife, Ruth, from prosecution.[60]

But Venona's secrecy prevented the unimpeachable evidence against Julius from being presented. Venona was so secret that even Truman was not told about its existence, though, in later years, it was revealed that the Soviets had learned early on about the Venona project through yet another mole.[61] This case presents a curious aspect of secrecy. Typically, secrets are seen as a strength of the government's position. In criminal prosecutions, however, secrets can be a weakness. In the case of Venona, the FBI worried it would lose a valuable "source" of new information. The result was a weaker criminal case, one viewed as a stain on American justice by a substantial portion of the country for decades. It has been argued that if the FBI had revealed the Venona data much earlier, it would have been a net benefit for the American political system: the far-left would have had to accept that there were several Soviet spies within the United States government, entertainment industry, and so on; the far-right would have had to temper their more far-reaching fantasies about the total extent of Soviet espionage, because while it was at times potent, it was nowhere as large as the hardcore anti-Communists believed. Instead, the secrecy led to increased polarization.[62]

Even with a "clean case," the Rosenberg trial presented novel difficulties. How, for example, would the United States prove that Julius Rosenberg aided in the stealing of "secrets of the atomic bomb" without giving away some of those secrets? Secret evidence was not, at the time, admissible in American courts.[63] This was not an entirely new issue. In the fall of 1946, three soldiers attempted to sell photographs of "an exact replica of the atomic bomb" to a Baltimore newspaper, whose staff turned them in to the police. Both Groves and Lilienthal (the case straddled the handoff from the Manhattan Project to the AEC) refused to allow the photographs to be introduced into court, and the cases had to be dropped as having "insufficient evidence" to convict.[64]

The Rosenberg case was viewed as worth giving away some secrets in order to send a message to future spies. AEC Chairman Gordon Dean gave his approval to declassify the basic design of the implosion bomb, since the Soviets already knew all about it. From Dean's perspective, there was little reason to keep it secret, and if releasing it would send a message that caught spies who did not turn state's witness would be met with the full fury of the law, all the better.

But there was some risk in this. Greenglass knew about more than just the basics of implosion, and on a witness stand, being cross-examined by a defense attorney, there was no guaranteee that other secrets would not be inadvertently released. Most worrisome to the AEC classifiers, Greenglass had been made aware of work that had been done on another weapons design concept that was still very secret, known as "levitation." Levitation was a small tweak to the core of an implosion bomb (an air gap is added between the tamper and the pit, allowing the tamper to accelerate, increasing the efficiency dramatically), and had been studied during the war when Greenglass was at Los Alamos, but had been brought to fruition only in the postwar period. If Greenglass was interrogated on his Los Alamos work, especially by technical experts, there was a risk that this still-secret design idea could come out. After lengthy discussions between the AEC, the Justice Department, and the JCAE, an arrangement was decided upon. The prosecution would hew closely to only the technology that dated from the war. The AEC would not declassify new information but would limit its declassifications to what Greenglass said. They could only hope that the defense would not probe too deeply.[65]

On the fifth day of the Rosenberg trial, Greenglass took the stand. He testified that in September 1945, he had given information to Julius Rosenberg about a "newer type of atom bomb" than the weapon used on Hiroshima, including a sketch, a replica of which was entered into testimony as the prosecution's Exhibit 8. After Greenglass presented the sketch, the Rosenbergs' defense attorney requested, to everyone's surprise, that the sketch be impounded "so that it remains secret to the Court, the jury and counsel." The prosecution was happy to join them in this request. And when Greenglass started to describe the sketch, the defense approached the bench for a conversation out of earshot of the jury to discuss whether the courtroom ought to be cleared of the general public. The defense attorneys emphasized that they were concerned that national secrets might be released, even though the AEC had already declassified the information. After ten minutes, Judge Irving Kaufman invited the reporters back in, informing them that the federal prosecution and attending members of the AEC had agreed that the press could hear the bomb-related testimony. Regarding the proliferation of the information, Kaufman put forward a good-faith request for

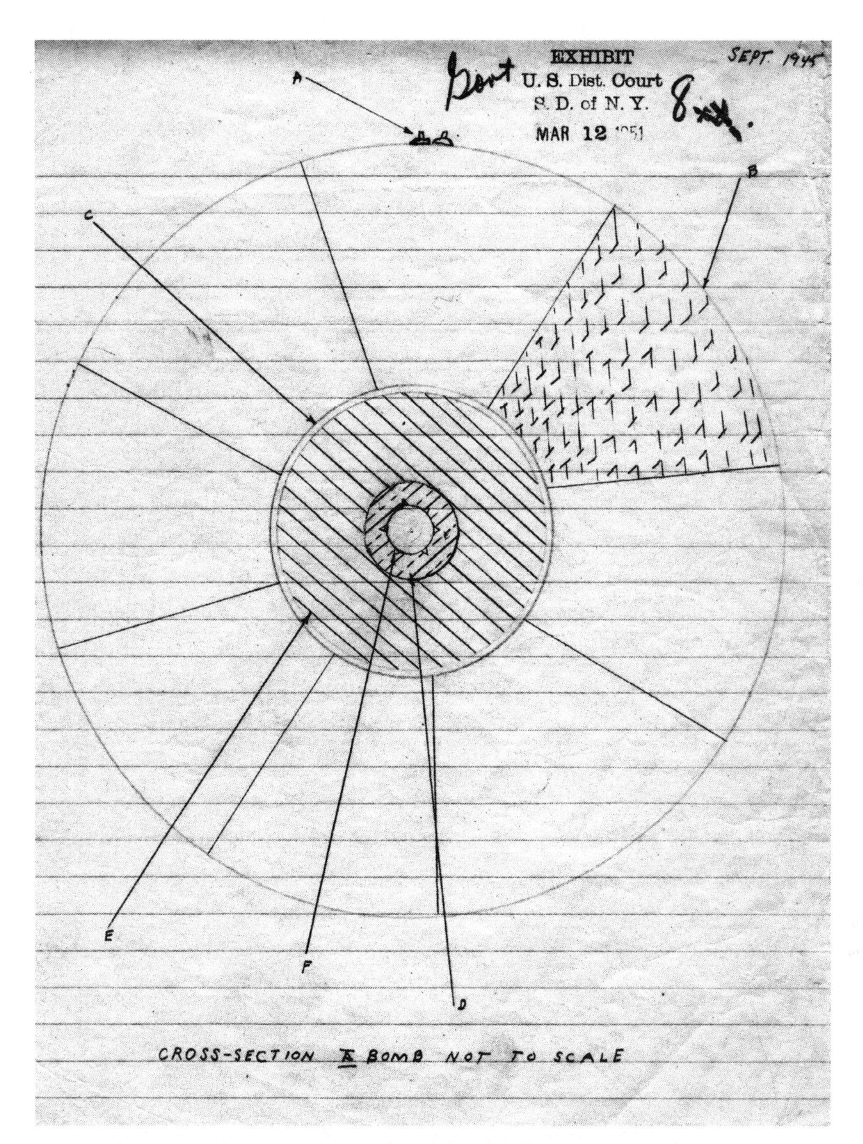

FIGURE 6.1. Exhibit 8 from the Rosenberg trial, a "sketch of the very atomic bomb itself," drawn by David Greenglass while in custody. Source: National Archives and Records Administration Northeast Region (New York City), NRAN-118-SDUSATTY-114868–7(11).

discretion: "We're going to trust to your good taste and judgment as to the publishing of portions of the testimony."[66]

The next day's headline on the front page of the *New York Times* was less-than-demure: "ATOM BOMB SECRET DESCRIBED IN COURT." As the lead declared, "[T]he first public disclosure of the composition and functioning of the super-secret Nagasaki-type atomic bomb came yesterday from the smiling lips of a witness in the spy trial."[67] *Life* magazine presented a pseudo-3-D artists' rendition of the weapon—one full of errors, but the gist was correct.[68] Although the testimony had been limited to members of the press, the information was out: the atomic bomb dropped on Nagasaki was not the "gun-type" design used on Hiroshima, but a more complicated "implosion" design utilizing high explosive charges detonated simultaneously in order to symmetrically compress a sphere of plutonium into a supercritical state. Here, at last, was a visible "nuclear secret."

Contrary to the fears of the AEC, the Rosenbergs' lawyers not only did not challenge or cross-examine the technical evidence, they hid from it. Such was the power of "the secret": in trying to prove that they were not interested in the dissemination of secrets, they gave Greenglass' testimony even more power. Exhibit 8 was not released until fifteen years later, when co-defendant Morton Sobell, who was convicted in the 1951 trial, asked for a new trial after a book by Walter and Miriam Schneir claimed that the sketch evidence was of little value (the Schneirs had not seen Exhibit 8, however, and were attacking the other Greenglass sketches; Judge Kaufman, then a member of the United States Court of Appeals, denied a 1962 request by the Schneirs to release the sketch).[69] Ironically, by 1966 the sketch had become a liability for the government because enough knowledge about the first nuclear weapons had been declassified for serious questions about its accuracy to be undertaken by the defense. Manhattan Project veterans Philip Morrison and Henry Linschitz testified that the sketch was "too incomplete, ambiguous, and even incorrect to be of any service or value to the Russians in shortening the time required to develop their nuclear bombs" (Linschitz), and that it was a "caricature" (Morrison).[70]

But even in 1966 it was ruled accurate enough to constitute a "secret," since it was evocative of a classified idea (implosion). And in 1951, it had been held up as a sketch "of the very atomic bomb itself."[71] The Rosenbergs were convicted and executed, in part based on a crude drawing. General Groves, a few years later, would confide that the information was "of minor value . . . I would never say that publicly. Again that is something while it is not secret, I think should be kept very quiet, because irrespective of the value of that in the overall picture, the Rosenbergs deserved to hang, and I would not like to see anything that would make people say General Groves thinks they didn't do much damage after all."[72]

The Rosenberg trial was only the most visible of a wide net of surveillance that had started to be deployed during World War II and persisted well after it. Well before McCarthyism, Truman had instituted a "Loyalty Program" in March 1947 that authorized "disloyalty" investigations for millions of American governmental employees, with "disloyalty" including membership in communist organizations.[73] The FBI separately sought evidence of subversion and vulnerability in wide swathes of the population. Communism was an obvious area for fear, but so was any kind of sexual activity that was deemed heterodox. Homosexuality, for example, was a disqualifier for a "Q" clearance, included in the same clause that disqualified someone for "demonstrating unreliability . . . abuse of trust, dishonesty, or homosexuality."[74]

This mindset was common in American security agencies during the "lavender scare" of the high Cold War, and the AEC was no exception.[75] In early 1951, Gordon Dean reported to the JCAE, in a closed session, that the AEC had discovered a homosexual Oak Ridge employee who was arrested on a trip to Washington, DC. He was described as "perfectly normal," someone who "is a married man, he engaged in sexual intercourse. When he goes out of town, apparently this other thing comes on him. He got liquored up. It is when he drinks excessively. There is no indication from anybody down there he was even suspected of this sort of activity." Dean assured the JCAE that "we removed him from the payroll immediately, fired him," though there was no evidence that he was a significant security risk, aside from his homosexuality.[76]

It is noteworthy that this single case warranted several minutes of discussion with a congressional oversight committee, and it would hardly be the last. In 1953, the JCAE staff alleged that the AEC had given "Q" clearances to at least seven potential homosexuals; Dean was required to go over each case, showing that the evidence for homosexuality was weak, and pointing out that such an allegation could be difficult to prove, though the AEC took the allegations seriously.[77]

The questions at hand were how to evaluate the "characters, associations, and loyalty" of another complex human being who might be granted access to secrets that could save or destroy a nation and what to do with someone who knows sensitive things but may be unreliable. Another example of this dilemma was the physicist Philip Morrison, a student of Oppenheimer's who had been a key member of the Los Alamos team and one of the people on hand to assemble the atomic bombs at both Trinity and Tinian island. Even during the war, he had a fat security file due to his Communist Party affiliations prior to the war. Despite this, he was cleared and even allowed to be a Responsible Reviewer for years into the postwar period, when he was working at Cornell University. Finally, in 1950, it was decided that a person like Morrison, however brilliant, could not be part of the weapons complex. Morrison was not abruptly severed from the program; he was assigned no further contracts, and when his last contract expired, his clearance was revoked as unnecessary. Freezing him out seemed like the least dangerous way to handle this situation, as he still "had more in his head than [the JCAE] liked for him to have."[78]

Such complex problems became a recurring theme in the security discussions from 1949 through the early 1950s. The Manhattan Project was by Cold War standards full of unreliable and suspicious people, and even after they left the project, voluntarily or involuntarily, they still carried the secrets with them. Thus, the scientist himself or herself became a security risk worth worrying about, and even their movements demanded control: one wouldn't want them to vanish abroad unexpectedly, as happened with Bruno Pontecorvo, a physicist who absconded from the UK to the USSR in 1950, either because he was afraid of the widening net of McCarthyism after the Fuchs arrest, or because he was a spy (the verdict is not in).[79]

As the historian David Kaiser has shown, theoretical physicists became a particular focus for these anxieties and were disproportionately targeted for investigation. These were the creators of the bomb, in the popular imagination, and that made them figures to be both revered and feared. In the Fuchs and Rosenberg cases, the literal bodies of these scientists and spies were scrutinized: reedy theorists like Fuchs were seen as "eggheads" with brains developed at the expense of their frail bodies, while Julius Rosenberg, Harry Gold, and David Greenglass were judged for their sweaty corpulence.[80]

Oppenheimer's students in particular suffered, for reasons that are not mysterious: Oppenheimer's left-wing politics while a professor at Berkeley had attracted similar-minded students, many of whom he brought into the wartime work with him; as the importance of nuclear "secrets" rose in the popular and official imagination, their proximity to him meant that they would always be subjected to scrutiny. Oppenheimer, for his part, did not do them many favors; he was often the first informant to bring their name to the attention of the security offices. In 1943, for example, he told General Groves that Charlotte Serber, the wife of his good friend and Los Alamos colleague Robert Serber, "came from a Communist family in Philadelphia and was at one time herself a member of the Communist Party." Though Oppenheimer reassured Groves that she was probably not currently a Communist and that Robert Serber probably never was one, the damage was done. An investigation into the Serbers' lives began that would last over a decade, involving wiretapped conversations and mail opening, and would ultimately generate some 300 pages of "salacious" material of little substance.[81]

Why did Oppenheimer do this sort of thing, repeatedly? Probably to ingratiate himself to the security officers, to show that he was himself reliable despite his own compromised associations. But therein would lie a contradiction: can a man so associated with dubious characters be trusted, even if he informs on them? And what does it say when he expresses some doubts about someone like Charlotte Serber, and then makes her the head librarian at Los Alamos, the person in charge of making sure the secret reports are properly filed and routed?[82] In retrospect, Oppenheimer's approach to security would increasingly appear naïve: the notion that he would, through informing on his numerous left-leaning colleagues, somehow appear more reliable became harder and harder to sustain.

The war had given the physicists a mystical aura, and a newfound political power. But with power comes suspicion. No one embodied that hazard more than Oppenheimer, who had become famous in the postwar period. Oppenheimer was the "No. 1 thinker on atomic energy," and he was also the "father of the atomic bomb."[83] But the H-bomb debate had exposed his vulnerabilities, and it was well known, both inside and outside the scientific community, that Oppenheimer had opposed its development. And for those who were looking for subversives, his background and associates gave much pause. Beyond his students and colleagues, Oppenheimer's wife had been previously married to a Communist; his beloved brother and his sister-in-law had joined the Party; one of his lovers had been a "fellow traveler." No evidence has ever emerged that Oppenheimer was a spy. But he accrued an FBI file of over 1,300 pages nonetheless.[84]

Even as his brother and his former students suffered, sometimes publicly, for their past political affiliations, Oppenheimer himself remained relatively insulated. The House Un-American Activities Committee (HUAC) had feared to tread too closely to the remit of the powerful Joint Committee on Atomic Energy. When Oppenheimer did once talk to HUAC in a closed session, they made clear he was considered a friendly witness, and they allowed him to avoid questions he felt uncomfortable answering (such as about his brother).[85] But as time went on, Oppenheimer accrued powerful enemies. His role in the General Advisory Committee's report against a hydrogen bomb crash program was the final straw for some more hawkish scientists (like Edward Teller and Ernest Lawrence), and his advocacy of land-based tactical nuclear weapons put him at odds with the US Air Force.[86]

Oppenheimer's most dangerous political foe was Lewis Strauss, who in 1954 became the chairman of the AEC. Strauss was intelligent but thin-skinned, and in many senses an ideological opposite to Oppenheimer's friend Lilienthal. Everything Oppenheimer stood for Strauss seemed to oppose: he was conservative in his politics and hawkish on the military. Where Oppenheimer had championed policies that would decrease secrecy, Strauss was notorious for his obsessive fear that secrets once released were gone forever.

There were deeper personal divisions as well. Where Oppenheimer was an elite, highly educated New Yorker who had abandoned the largely secular Judaism of his wealthy family in pursuit of Far Eastern philosophy and Southwestern American aesthetics, Strauss was the self-made, uneducated son of a shoe salesman from the Southern United States who always went to temple. Even if Oppenheimer had not humiliated the grudge-bearing Strauss in congressional testimony a few years earlier, they probably still would have ended up butting heads, but as it was, Strauss carried a deep resentment of Oppenheimer, and a deep suspicion of his politics and policy recommendations, that he was able to take action upon as the chairman of the AEC.[87]

But Strauss needed a legitimate excuse to take on a figure as publicly well-regarded as Oppenheimer. The circumstances that led to this excuse emphasize the kind of bizarre climate developing around nuclear secrets in the early Eisenhower years. William Borden, the head of the JCAE's staffers, took it upon himself in 1952 to develop an "objective" history of the hydrogen bomb. Borden's goals were hardly academic: he was convinced that Lilienthal and Oppenheimer (among others) had engaged in a conspiracy to suppress the development of thermonuclear weapons and believed that if he and his colleagues could chart out every development that had taken place, the intellectual dishonesty of the H-bomb opponents would be obvious to anyone who read it. By January 1953, he and his staff had compiled a 91-page "Top Secret" history of the H-bomb work, going back to the Manhattan Project, written as a hit-piece against Oppenheimer and others.[88]

But Borden was no scientist, and for several of the technical aspects, he required assistance. In particular, he wanted to make sure he properly described the genesis of the Teller-Ulam design. Borden and his staff had consulted with Teller and others on this subject, but he wanted his work to be checked by another informant of his, the Princeton physicist John A. Wheeler. Wheeler was employed by the AEC to run a thermonuclear research laboratory at Princeton (Project Matterhorn B), and was, in essence, conspiring behind his employer's back with a congressional staff member to create a document that would be used to attack Oppenheimer and other former AEC employees.[89]

One of Borden's staff members had sent Wheeler a six-page extract pertaining to the development of the idea of "radiation implosion," the mechanism by which the fission bomb in a thermonuclear weapon uses its energy to start a fusion reaction. The exact six pages are still mostly classified to this day, but in a later affidavit Wheeler noted that an informed reader who had them would learn a few key secrets: the US was well on its way to acquiring multiple types of thermonuclear weapons, that lithium could be used as a solid fuel, and that "radiation heating provides a way to get compression" in the thermonuclear fuel.[90] As of January 1953, these were the "crown jewels" of the thermonuclear program, and the basis of the Teller-Ulam design.[91]

Early 1953 was a dangerous time to be carrying around such secrets on a few pieces of paper. The US had indeed tested a thermonuclear design in November 1952, but it was only a prototype. The US had no thermonuclear weapons in its arsenal yet, and it would not be until the spring of 1954 that it would identify a viable path to becoming a true thermonuclear superpower. So there was still a feeling of vulnerability and haste, and a worry that the Soviet Union could be nipping at their heels, in particular since the key element of the new design, radiation implosion, had been known to Fuchs.[92]

Wheeler agreed to read the pages and decided to do so on a sleeper train from Philadelphia to Washington, DC. Somewhere along the way, the pages went missing. Wheeler realized this early in the morning, after arriving in Washington, and after a frantic search of the train and Union Station's lost and found office, he sullenly reported the loss to his contacts at the JCAE. They rushed to the train station and put the train car on lockdown, searching every inch to no avail. Finally, Borden himself contacted the FBI, who after initially declining to help a congressional staffer find a lost memo (not their job), became very interested once it became clear it was the "loss of a classified atomic energy document."[93]

Despite a very long investigation, the mobilization of an immense number of Special Agents, and both President Eisenhower's and J. Edgar Hoover's personal investments in the outcome, the FBI never found the document. They did, however, tell the AEC about the loss (something Borden chose not to do), which set off a new investigative question: why were JCAE staffers creating secret documents containing H-bomb secrets and giving them to AEC scientists who, against security protocol, took them onto a sleeper train and either lost them or had them stolen? For once, the AEC was in a position to turn the tables on the JCAE. They alerted Eisenhower, who put pressure on the JCAE congressmen, who were shocked to learn what their staffers had been up to. Wheeler was chastened but otherwise untouched, and Borden was soon fired.[94]

Even prior to being fired, Borden had started to worry that the investigation into the lost pages would steal the thunder of the real show, his scathing history of the H-bomb. After being fired, he turned to drinking and began to wonder if it wasn't all a bit too convenient: his H-bomb attack had been defanged by the coincidental loss of a few pages. Could this somehow have been orchestrated by people like Oppenheimer? In such a state, Borden's thoughts became conspiratorial, culminating in a letter to J. Edgar Hoover essentially accusing Oppenheimer of being a Soviet agent, the sort of thing he could never have written in an official capacity. The world was now wound so tight with atomic anxiety that six lost pages could derail multiple careers through several layers of feverish conspiracy theories.[95]

Hoover forwarded Borden's letter to Strauss, who took it to Eisenhower, using it as the catalyst for the attack he had been longing to begin. Eisenhower, who knew and liked Oppenheimer, deferred to Strauss' judgment and agreed to put a "blank wall" between Oppenheimer and further nuclear secrets. Strauss took pleasure in informing Oppenheimer that his security clearance would be revoked, and Oppenheimer could either accept it (and see it as a forced retirement from the secret community), or he could challenge it, subjecting himself to what might be a humiliating and drawn-out experience. Oppenheimer chose the latter out of principle. Thus began the Oppenheimer affair, a deeply divisive event seen by many as one of the great scientific trials of the ages, indicative of the Cold War's new standards of security.

Oppenheimer's security hearing was not supposed to be a literal trial. The AEC had appointed a permanent Personnel Security Review Board only in 1949. Security hearing procedures had originated from the General Advisory Committee under the Lilienthal AEC, and in Lilienthal's mind would constitute the due process necessary to prevent a "wild nightmare of fear" from compromising the nation's "leadership in science" through "drastic and dumb limitations on scientific men and standards of 'personal clearance' that are impossible."[96] But as with many of the Lilienthal AEC's policies, this one could be turned to different ends in different hands. Oppenheimer's Personnel Security Review Board Hearing would indeed be a trial of sorts—not legally (and not afforded many of the legal niceties that a criminal defendant would be), but in practice it became a formal, adversarial experience with lawyers for the "prosecution" and "defense" involved in lengthy cross-examination of dozens of witnesses.

In principle, the hearing was a referendum on Oppenheimer's "character, associations, and loyalty" as framed by the Atomic Energy Act and the federal "Loyalty Program." "Loyalty" was a tricky term to characterize, and because there was no evidence of true treason, his "associations" could be more easily attacked. It was well known that Oppenheimer had many Communists and "fellow travelers" in his circle of family and friends, though there was no solid evidence that Oppenheimer was himself a "card-carrying" member of the Communist Party. (Nor has any emerged in the decades since; at most, several historians and some of his former colleagues have suggested he may have been considered a "secret," unofficial Party member. Whatever that means in practice, it does not sound like a "card-carrying" member subject to Party discipline.)[97]

As for Oppenheimer's "character," how should one measure a person's "character" from a security standpoint? In practice, the "prosecution" for the hearing spent a good deal of time going over Oppenheimer's interactions with the various security apparatuses of the American nuclear infrastructure, from wartime through the postwar period. Oppenheimer had, as noted already, always tried to maintain a relationship with the security men, offering up derogatory information about his students and colleagues. But his was an inconsistent position: he had appointed the same people he reported to important jobs at Los Alamos, and in one particular episode he gave security agents wildly different accounts of a sensitive situation.

The Chevalier affair, as it was called, involved an alleged approach to Oppenheimer by a Berkeley colleague, Haakon Chevalier. In various versions of the story, either Chevalier himself, or an intermediary, approached Oppenheimer at a social event in Berkeley during the war, offering to make a connection between him and the Soviet Union for the purposes of aiding the latter in their struggle against the Nazis. Oppenheimer, by every account, turned this offer down and reported it to Manhattan Project security officials with the seemingly naive sentiment that this was not a major issue but something to keep an eye on. The security officials did not consider a direct approach to their top scientist by the Soviets as a light matter and repeatedly tried to get Oppenheimer to clarify the story. In the process, Oppenheimer gave several different versions in an effort to avoid having undue scrutiny placed upon his friend and, perhaps, his brother.[98]

At his 1954 hearing, these contradictions became a key element of discussing Oppenheimer's "character." He offered up very little by means of coherent explanation. When asked why he had lied to Manhattan Project security forces about the Chevalier incident, he could only reply: "Because I was an idiot."[99] Such a defense is not much of one at all, not in an ever-heating Cold War and a time in which Oppenheimer himself was no longer essential to any weapon project. Oppenheimer's inconsistency with the truth on matters of espionage could be, in the 1940s, overlooked as an unfortunate flaw, but an acceptable one given his importance and the fact that it did not seem rooted in malice. By the 1950s, things were trickier.

True to the era's concerns, Oppenheimer's sex life also came under scrutiny. His relationship with the ill-fated Jean Tatlock (mentioned in chapter 2) was another probing of "character." When asked whether the director of Los Alamos spending the night with his ex-Communist, ex-girlfriend was "consistent with good security," Oppenheimer glibly asserted that "it was, as a matter of fact." When pushed on whether keeping "social contacts with Communists" while "working on a secret war project was dangerous," Oppenheimer only asserted that he didn't think Tatlock was a Communist.[100]

These exchanges make clear how unsatisfying Oppenheimer's self-account was. Most of the other witnesses called to testify in his favor did a better job: they argued, quite persuasively, that Oppenheimer's judgment on areas of technical expertise had always been informed and educated, and that his priorities had always been justifiable, even if they were controversial. Oppenheimer's policy positions on international control, the hydrogen bomb debate, and the use of tactical nuclear weapons for the defense of Western Europe were all put on display, and his positions, while hardly universal, were all essentially reinforced, at least to a point where no one would call him treasonous.

Even Edward Teller, whose distrust and unhappiness with Oppenheimer could be traced back to the wartime period but had magnified greatly during the H-bomb debate, could manage only a rather tepid objection. Oppenheimer, Teller argued, had never been visibly "disloyal" to the United States. But, he continued:

> In a great number of cases I have seen Dr. Oppenheimer act I understood that Dr. Oppenheimer acted in a way which for me was exceed-

> ingly hard to understand. I thoroughly disagreed with him in numerous issues and his actions frankly appeared to me confused and complicated. To this extent I feel that I would like to see the vital interests of this country in hands which I understand better, and therefore trust more.
>
> In this very limited sense I would like to express a feeling that I would feel personally more secure if public matters would rest in other hands.[101]

Teller was not the only scientist to testify against Oppenheimer, but he drew the most attention. Despite his efforts to qualify his statement, it was taken as a crude denunciation and a betrayal by a former colleague and friend. Teller became a villain in the Oppenheimer affair and an example of what it meant to be a government scientist in the Cold War: deeply embedded in the work of making new weapons (at Teller's new laboratory at Livermore), with more colleagues in the US Air Force than in American academia, operating under the belief that security could come only from greater military strength, whatever the cost. Some of this was caricature, but it served as a symbolic shorthand for the changes that had occurred in the United States over the decade.[102]

In the end, it was Oppenheimer's self-testimony that was most damning: it indicated a willingness to bend the truth when it was personally convenient, was consistent with a lack of attention to security, and cast dark clouds on Oppenheimer's judgment. A majority of the Personnel Security Review board concluded that Oppenheimer's clearance should not be reinstated, and a majority of the Atomic Energy Commission concurred.[103]

That Oppenheimer's judgment was occasionally very poor does not take away from the injustice of the affair. It was clear even at the time, and much more in retrospect, that this was more about Oppenheimer's enemies kicking him off the pedestal than any real security concerns. His clearance was set to expire within days anyway. Strauss' conduct during the affair was atrocious: he obtained wiretaps of Oppenheimer's conversations with his lawyers and fed them to the prosecuting attorneys so they could better anticipate his weaknesses. Such behavior was not only unjust, but illegal.[104]

Oppenheimer would later denounce the hearings as a "farce." But they were also a prolonged examination into the contrasts between the Eisenhower 1950s and the Truman 1940s. The wartime years had allowed shades of gray in terms of security, and those had gradually turned into a harsh black-and-white. Henry DeWolf Smyth, despite viewing Oppenheimer and his transgressions quite severely, dissented from the Strauss AEC's decision, airing a view more consistent with the Lilienthal era:

> With respect to the alleged disregard of the security system, I would suggest that the system itself is nothing to worship. It is a necessary means to an end. Its sole purpose, apart from the prevention of sabotage, is to protect secrets. If a man protects the secrets he has in his hands and his head, he has shown essential regard for the security system.[105]

What is doubly contradictory about the Oppenheimer case is that it was meant to be confidential (but not officially secret); and yet, large portions would play out in the public sphere. In principle, none of the evidence presented was meant to be classified, because at least one member of the audience—Oppenheimer!—had been denied access to classified information. The irony of this was not lost on anyone, given that many of the documents introduced into evidence were by Oppenheimer himself, and no one in the room likely knew as many secrets as he was now not supposed to know. In what is perhaps the perfect graphic depiction of both this irony and the new imagery of the "dangerous minds" of physicists, an editorial cartoonist depicted the situation with an anxious Uncle Sam imploring Oppenheimer, whose head was encased in a box labeled "Top Secret," not to "think up any more atomic secrets."[106]

It was impossible for witnesses to discuss the issues at the core of the hearings (the development of the H-bomb, recommendations on nuclear strategy, security practices, etc.) without some classified information being revealed. So at the end of every day, the AEC director of classification, James Beckerley (and a small number of his staff), would review the day's stenographic transcripts, marking any classified items for deletion, and sanitized versions were delivered to Oppenheimer's counsel for review. All such deletions have only recently been declassified, and reveal the peculiarity of the eye of the redactor. For example, here is a famous quote by the physicist I. I. Rabi, expressing his frustration with the attacks on Oppenheimer, in light of the latter's contributions:

FIGURE 6.2. "And please don't think up any more atomic secrets." Cartoon drawn by Hugh Haynie, *Greensboro Daily News*, reprinted in "Oppenheimer case—five views," *New York Times* (6 June 1954), E5.

> We have an A-bomb and a whole series of it, * * * and what more do you want, mermaids? This is just a tremendous achievement. If the end of that road is this kind of hearing, which can't help but be humiliating, I thought it was a pretty bad show.[107]

Rabi's "mermaids" line has been quoted and requoted over the years, both for its exasperation as well as its New York turn of phrase. But for six decades, the scar of the redaction—the three asterisks—always remained. Something had been removed. In 2014, the Department of Energy fully declassified the Oppenheimer hearing transcripts, and the answer was revealed: "We have an A-bomb and a whole series of it, *and a whole series of Super bombs* and what more do you want, mermaids?"[108]

For Rabi, this seemed like perhaps a safe thing to say: it was no secret, by April 1954, that the United States had tested a variety of different thermonuclear weapons at Operation Castle. But in the eyes of Beckerley, this clearly revealed more than could yet be openly said. The redactions do seem to have been done in good faith. There is nothing that would exonerate Oppenheimer in what was removed; almost all of the redactions concern either technical matters, matters of pressing American nuclear policy (like the deployment of nuclear weapons to Western Europe), or statements that the witnesses explicitly noted should not be made public (such as Groves' declaration that "the Rosenbergs deserved to hang," but, that he "would never say that publicly"). Beckerley, who would leave his AEC post shortly after the Oppenheimer affair, would later report being disillusioned with the security system, and was not wielding it as a weapon. And in any case, he was preparing these transcripts for Oppenheimer himself, not the general public.[109]

But the general public *did* get to read the Oppenheimer hearing transcripts. The idea that they should be made public was first floated during the last day of testimony by Representative Sterling Cole, the then-chairman of the JCAE, who suggested to Strauss:

> It would of course be exceedingly unfortunate, and detrimental to the future of our atomic enterprise, if the notion were to gain currency that Dr. Oppenheimer's suspension resulted from capricious administrative action, or that the findings of the review panel before which he is now appearing were inconsistent with the testimony it developed.[110]

Were its release to occur, he reasoned, "the American people would themselves be able to decide whether this entire matter has been handled in a manner combining maximum security to the United States and maximum fairness to Dr. Oppenheimer." Strauss saw the appeal, and his interest would only grow over time. He learned from the FBI's wiretaps that Oppenheimer and his counsel feared that publication would harm his case, and they were discussing leaking favorable portions of the transcript to the press. Suddenly the normally secretive Strauss was in favor of open publication, but the Personnel Security Board was hesitant, since it had promised the witnesses their comments were off the record.[111]

Unbelievably, lost secrets on a train would once again play a role. Smyth had requested that AEC staff create a summarized version of the testimony for the commissioners' use. A copy was given to AEC Commissioner Eugene Zuckert, who left it behind while on a train to Boston on June 12. Though it was later located by the FBI, Strauss argued that the information had been compromised and thus an authoritative version needed to be published as soon as possible.[112] At a late-night AEC meeting held on the day Zuckert lost his summary, Strauss introduced a motion to publish the full, unclassified version of the transcript. He thought doing so would vindicate the proceedings and the AEC's decision, arguing that "the importance of having as much factual material as possible available to the public because of the conclusions that were being reached in the absence of such material, and because of distortions and misquotations from the report being made available to the press," justified the release.

Strauss seems to have thought that Oppenheimer's counsel was leaking documents and was infuriated by pro-Oppenheimer journalists' assertions that the hearing had been a sham. He was outvoted, 2-to-1, by Smyth and Commissioner Thomas E. Murray. No direct transcript was kept, but from meetings held afterward, it appears Smyth had argued that it was improper to release the transcript prior to the final determination in Oppenheimer's case being made by the AEC, and Murray was hesitant to release anything without permission of the witnesses. Strauss was clearly angered; he made clear that he felt he was within his rights to make public that the motion had been defeated. Three days later, they reviewed the issue again, after receiving assent from the witnesses, and voted, this time 3-to-1 in Strauss' favor (Murray switched his vote, and Zuckert was at the meeting this time).[113]

The transcripts were reviewed again for matters to be cut, literally, with scissors and knives ("physical deletions") prior to being delivered to the Government Printing Office to be typeset.[114] The AEC briefly considered whether "some sections of the transcript relating to the personal life of Dr. Oppenheimer be deleted." It was decided not to do so, "since the information in question was germane to the question of character."[115] The transcript was released to news organizations on June 15th, with an embargo until the next day that was mostly obeyed.[116]

The news that the AEC would uphold the denial of Oppenheimer's security clearance was released on June 29, with Smyth's vigorous dissent accompanying it. The release of the transcript did not bolster Strauss' cases in the eyes of most commentators. Strauss had hoped that the publication would simultaneously air Oppenheimer's dirty laundry and make clear that the hearings had been anything but superficial. But to those who made their way through all 992 pages of small type, the probing into Oppenheimer's personal indiscretions came off as a tawdry witch-hunt. The fact was that Oppenheimer had never been shown to be disloyal. To be sure, the events of the hearing are complicated enough to afford multiple interpretations; had Oppenheimer's clearance not been stripped, perhaps he would be better remembered for the way he testified against his students and friends rather than as a martyr. In any case, Strauss found the sense of Oppenheimer as a victim hard to shake, and his rare interest in openness appears to have backfired, securing his role as one of history's great villains.[117]

The Rosenberg and Oppenheimer cases contain useful insights into how security hardened and became more *personal* in the years after Fuchs. The gray areas and acceptance of idiosyncrasies that had marked the wartime period were largely gone. Information could be released or held back as a judicial and political weapon. And in both cases, only the most superficial appeals to the benefits of public knowledge were made. The Lilienthal AEC's days of worrying about meaningful public debate and close relationships between the AEC and the press appear to have gone by the wayside.

Gordon Dean's tenure as chairman of the AEC (1950–1953) was marked by an attention to the letter of the law, appropriate for a lawyer. As noted, Dean appears to have had no lofty ideals about security. His approach was pragmatic, fairly non-partisan, and arguably professional: he was a bureaucrat, not an ideologue. As an AEC chairman, his tenure would seem more reflective of the sort of temperament that summed up the agency as it matured. Dean's successor, Lewis Strauss, by contrast, was a man of definite ideological commitments and long-standing grudges. On security, he was as conservative as they came, but even he could argue for openness if it accomplished his goals. If relationships between outside scientists and the AEC had become strained in the Lilienthal and Dean years, under Strauss they became abysmal, and the AEC gained the reputation as a secretive, arbitrary, and capricious organization on security.

But even with an increased attention on scientists as potential security problems, there was still some leeway, at least for important scientists. John Wheeler, despite literally losing the secret of the H-bomb on a train, was allowed to keep his clearance because he was useful. Asked how someone with as much security training as Wheeler could lose such a thing, Dean offered up a reasonable answer: "When you put a heavy load of TS's [Top Secrets] and Secrets on the man, the chances are out of a seven-year period maybe one is going to get lost."[118] As the secrets multiplied, and entered into the hands and heads of more people, so did the chances that a few might go astray, one way or another.

6.3 MAKING ATOMS PEACEFUL AND PROFITABLE

While security concerns dominate much of how we view the Cold War approach to nuclear technology, it is only half of the story of its secrecy. At the same time as the Rosenberg trial and execution, and of Oppenheimer's own travails, the AEC under Strauss would preside over a release of secrets several multiples larger than any that had come out under the supposedly liberal Lilienthal regime. This seeming paradox resulted from trying to make the military and peaceful promises of the atom both equally and simultaneously real.

This paradoxical impulse came out of a deep unease with the status quo. Since World War II, scientists had hoped that nuclear technology would be more than just a means of more efficient slaughter. But these dreams had not significantly materialized. The only "peaceful" application available in any quantity was the production of radioisotopes for medical and industrial use, which, while important, was underwhelming when contrasted to the wartime applications.[119] While the AEC was created with the hope that it would spend its energies on both peace and war, by the end of the 1940s, it had produced far less in the former category than the latter. The reason was quite clear: the emerging Cold War had put almost all national priority into the prospect of nuclear war. "I have never had much sympathy with the idea about civilian use of atomic energy," Senator Tom Connally commented in a classified session of the JCAE in 1950. "You will find it will cost three or four times what we are spending for it now and I think we ought to centralize on the military and defensive features of this thing and if anything else interferes with it, let the other thing wait and set it aside."[120]

But by 1952, as the US military was moving from nuclear scarcity and into nuclear plenty, and with an increased experience with reactor development for military propulsion, the AEC was ready to begin expanding its efforts to create power-generating reactors. Within both the AEC and Congress, there had been a long-standing belief that this effort would be improved if private industry was brought into the work. It was an ideological push, one in step with 1950s American politics, and in part an explicit move away from the New Deal policies that had been dominant during nearly two decades of Democratic administration. With Eisenhower's backing, a reorientation of the AEC's mission would take place. Weapons production and improvement would continue apace. But an increasing effort would be made to make nuclear technology more accessible for private industry.[121]

This policy reorientation was strengthened with the appointment of Lewis Strauss to the AEC chairmanship in July 1953. Where Lilienthal had been a New Dealer, Strauss was a successful businessman, enthusiast of industry, and veteran on the atomic policy scene. Initially, however, Strauss was no more inclined to create a private nuclear industry than Lilienthal: if it involved releasing secrets, Strauss preferred to err on the side of restraint. But the promise of privatization of the nuclear industry, which Strauss had long believed in, and Eisenhower's insistence on releasing information, would lead to a reorientation of his worldview.

Ironically, the roots of Eisenhower's own reframing of the atom as a "peaceful" entity in the 1950s came from Oppenheimer, in one of the latter's final acts as a policy influencer, and his last attempt to reform American nuclear secrecy. Oppenheimer was part of a panel assembled by Secretary of State Dean Acheson in 1952 whose job it was to review the matter of "Armaments and American Policy." The panel's final report was a wide-ranging document that urged that the emergence of a nuclear-armed Soviet Union required greater "flexibility" in American nuclear policy. They recommended that the United States reduce its dependence on making nuclear threats as a means of achieving its policy goals and open up channels for political engagement with the USSR on efforts to stem the arms race. But their very first recommendation, central to the entire endeavor, was a call for "candor":

> We think it of critical importance in the development of a national policy which takes full account of the realities of the arms race, that the United States Government should adopt a policy of candor toward the American people—and at least equally toward its own elected representatives and responsible officials—in presenting the meaning of the arms race. The best and wisest government, in this country, is always dependent in large measure upon the support of the American people, and this support, if it is to have the strength and solidity which are necessary in great affairs, must rest upon an adequate basic understanding of the realities of the situation.[122]

If the American people and their elected officials did not understand that the US nuclear stockpile was growing at an exponential rate, and the Soviet Union would likely follow in turn, they were heading toward "great danger." Secrecy, in this view, was creating an existential ignorance, and it was "difficult to overestimate the importance" of a policy of candor. It was a 1940s theme, updated for the new nuclear 1950s.[123]

Eisenhower was receptive to the panel's recommendations. "Candor" seemed to strike a chord with him, though he was dubious about the release of information about nuclear weapons stockpiles. When it became known that Eisenhower was considering implementing some of the panel's recommendations, Strauss (not yet AEC chairman) worked to torpedo any such efforts. Eisenhower pushed back, and Strauss backed down, but only temporarily. "Operation Candor," as it became known, would eventually fail, undone by Washington politics and the announcement in August 1953 of a Soviet thermonuclear test.[124] Instead of increased flexibility, the Eisenhower administration would put even more faith in "massive retaliation" as a means of cutting costs and deterring Soviet activity. Atomic strategy and stockpiles would remain among the most tightly held secrets of the Cold War.[125]

Within the AEC, formal analysis of the expanded possibilities of "candor" was carried out by Commissioner Smyth. His analysis was from within a two-state, bipolar framework, where advantages were things that helped the US directly, and disadvantages were those that might aid the Soviets. Releasing information to the American public or industry would be beneficial in some instances, but there were some areas where "the tightest possible secrecy should be maintained." Overall, Smyth was in favor of greater release of technical information: it would improve democratic deliberation, and since the Soviets already had the bomb, the largest impetus for constraining information had been relieved.[126]

Eisenhower was apparently disappointed by the failure of the "candor" effort. Flexibility was something he desired, and finding a way out of the depressingly apocalyptic arms race was for him a deeply meaningful goal, even if he still maintained an immense distrust of the Soviet Union. In the fall of 1953, he and his staff came up with an idea that would redeem his initial enthusiasm for a safer world. The US and the USSR could divert stocks of their fissionable material toward international "peaceful" applications. The plan, called "Atoms for Peace," was initially opposed by Strauss as pointless, but with Eisenhower's personal interest he carried it forward. A speech by Eisenhower to the UN General Assembly in late 1953 would cement the notion and serve as a vehicle for Eisenhower's hopes for a less "military" atom.[127]

Rhetorically, the rebranding of the atom as "peaceful" was necessary because it was primarily associated with "war"; the opposite of the "peaceful atom," in Eisenhower's speech, was the "fearful atom," a term about once a minute in his UN speech.[128] And to the "fearfulness" of the atom, he linked secrecy:

> But the dread secret and fearful engines of atomic might are not ours alone. In the first place, the secret is possessed by our friends and Allies, Great Britain and Canada, whose scientific genius made a tremendous contribution to our original discoveries and the designs of atomic bombs. The secret is also known by the Soviet Union.[129]

Eisenhower was attempting to provide an alternative framework for thinking about nuclear technology in the Cold War, but neither "secrets" nor "fear" could be abandoned with just a gesture. The main policy thrust of the "Atoms for Peace" plan was relatively modest. The US would distribute fissionable material in order to facilitate international nuclear research and make its library of declassified AEC publications widely available and easily accessible.[130]

Strauss would tell anyone who would listen that "Atoms for Peace" did *not* involve the divulging of "secrets." In a 1954 address, Strauss, striking a discordant note from Eisenhower's appeal for peace, argued at length the wisdom of making the hydrogen bomb, arguing that without it, the "whole world [would have] eventually end[ed] up in the maw of Communism and slavery." "Atoms for Peace," he emphasized, was not a formula for disarmament, and it did "*not* endanger the atomic weapons secrets of any nation that now has or may possess such secrets."[131]

On the tenth anniversary of the end of World War II, the first international "Atoms for Peace" conference was organized in Geneva. Despite being an enterprise founded on candor and openness—or because of it—discussion about secrecy was core to run-up to the conference. At a press conference with US conference officials in May 1955, journalists were eager to know whether new secrets would be released for the conference. The physicist I. I. Rabi, the US representative on the UN Advisory Committee for the conference, responded with enthusiasm while they were not declassifying anything specifically for the conference, they were in the middle of a massive declassification effort, and surely this would be reflected.

The question for the press was whether there were any new "secrets" that were going to be revealed at the conference. As one reporter pushed Rabi: "Approximately how much would have been classified prior to the time they were submitted?" Rabi's response noted that many of the papers on reactors probably would have been classified a year prior. But he strained to emphasize that declassification was safe, methodical, and deliberative. In what may not have been the most appealing visual metaphor, he explained that the declassification process "is a continuing one, like digestion."[132]

At a later press conference, the technical director for the US delegation, George T. Weil, struggled with the fact that journalists were more interested in talking about secrecy than they were the newly unveiled research reactor designs; only the idea of "secrets" was interesting, not the actual technical details. Weil repeatedly tried to dodge questions about whether military secrets were being released. He explained that there was no safe answer he could give:

> Well, if I answer that question one way I will get kicked in the teeth, and if I answer it the other way I will get kicked in the teeth. . . . I don't think there is any point in trying to hide the fact that there is classified information and why there is classified information. We all know the reasons.[133]

The conference itself was part scientific forum, part stage for a display of national nuclear developments. The introductory lecture, given by the Indian physicist Homi J. Bhabha, praised the conference for breaking down the "barriers" that had closeted nuclear research "behind a wall of secrecy." Niels Bohr gave an address calling for a return of international freedom of information, for scientific cooperation that would transcend national borders. In his closing address, Bhabha would argue that "knowledge once given cannot be withdrawn—the free flow of knowledge has been established."[134] Strauss would have agreed, but with more negative implications. As he put it in his later memoirs: "[I]nformation once compromised is information broadcast forever."[135]

"Atoms for Peace" was meant as a way out of secrecy and fear, but secrecy still dominated the narrative. The actual scientific papers were dryly technical and no match for the mystique of nuclear secrecy that had been building for a decade since the bomb had become public. From the media's point of view, there was almost nothing that guaranteed an atomic discovery's dullness than its discussion at an "Atoms for Peace" conference, or its promotion by the UN's International Atomic Energy Agency, which would be formed in 1957. Agricultural genetics, magnetically confined fusion, and endless new research reactor designs, even if they promised a new, high-tech modernity, could not compete with mushroom clouds for attention.

Related to Eisenhower's enthusiasm for newly peaceful atoms was his encouragement of the rapid development of a peaceful nuclear industry. By 1953, Eisenhower, the AEC, and the JCAE were resolved that the development of commercially viable nuclear energy would be a top agenda item of the American nuclear program, but all believed that this would be best facilitated by deepening ties with the private sphere. Here the Atomic Energy Act of 1946 posed a stumbling block, as it was built around the 1940s idea of nuclear technology as an "island of socialism" within the American economy. Eisenhower's interests, and his frustrations with the shortcomings of the original Act, culminated with a push for comprehensive revision. The Atomic Energy Act of 1954, as it was labeled, was eventually moved through congressional committee and passed in August.[136]

The issues with the original Atomic Energy Act of 1946 (referred to below as the Act) were complex, and the changes in the 1954 revision (referred to here as the Revision, for clarity) both reflected and reinforced the new Cold War sensibility of nuclear technology, informed by eight complicated intervening years. The concept of Restricted Data went under revisions reflective of a thermonuclear age: nuclear weapon "design" was now explicitly part of the definition, and "fissionable materials" had been transformed into the more inclusive term "special nuclear materials." What is more interesting, though, is that further clarifications were *not* made: though there were extensive discussions by the 1950s about whether the Restricted Data definition was too broad, more substantive clarifications or attempts to bring it more legally in line with other categories of secrecy in the American classification bureaucracy, were eschewed.[137]

And though the Act gave the AEC the power to remove information from the Restricted Data category (and did not give it the power to add information to the category), the Revision stressed this power. The Revision, for example, noted explicitly that the AEC's responsibility was "to control the dissemination *and declassification* of Restricted Data." Where the Act's provisions for removing information were vague, the Revision outlined declassification procedures at length, mandating the

AEC to continually review whether information in the Restricted Data category still posed a threat to "common defense and security," and, if not, obligated the AEC to release it. Even here, though, there were limits: with weapons information, the AEC was now required to get Department of Defense approval before removing it from the protected category. And if the AEC and the Department of Defense could not come to an agreement about the release of a piece of information, the decision would be made by the president. If the information related to the nuclear programs of other nations, the AEC had to find concurrence with the Central Intelligence Agency. In general, the AEC could formally release information to the Department of Defense as it saw fit. In practice much of this had already become part of AEC protocol, but codifying it emphasized that the AEC was itself now becoming considered more subservient to the broader national defense establishment.

Access to Restricted Data was still limited to those with security clearances, of course, but where the Act had mandated a full FBI investigation for prospective employees, the Revision allowed the Civil Service Commission to make preliminary investigations and send only cases where "questionable loyalty" was involved to the FBI. The president was also explicitly given the ability to mandate an FBI investigation rather than a Civil Service Commission one. The overall purpose of this section was clearly to free up the FBI from the onerous task of personnel investigations, especially for lower-level personnel or "clear cut" cases.[138]

Consequences for mishandling Restricted Data where malicious intent was not evident were treated in a far milder fashion, with a maximum fine of $2,500, some ten times less severe than the Act's prescription. The experience of the previous eight years had shown that in a system with hundreds of thousands of employees, dealing with millions of potential secrets, a few mistakes were bound to occur. The original teeth, including the death penalty, remained for actual spies. The Revision also set a statute of limitations of ten years for all criminal provisions of the law, excepting capital offenses, and created a new category of punishment for the transmission of special nuclear materials, rather than just information. Anyone who attempted to interfere with the AEC's control of enriched uranium or plutonium, for example, could be prosecuted with the same maximum penalties as for giving away secrets, a reflection of the rising quantity of such materials outside of official government control.[139]

The original Act had effectively banned "exchange of information" with all other nations. There had long been frustration with the fact that this provision severely limited the United States' ability to cooperate on nuclear matters with its allies, especially the United Kingdom. The Revision liberalized these conditions. The president was given the ability to authorize the AEC to cooperate with other nations and to communicate Restricted Data regarding refining and purification of source materials (e.g., uranium ore), reactor development, production of special nuclear material, health and safety, industrial atomic energy, and applications of atomic energy "for peaceful purposes," so long as none of this information would communicate Restricted Data that related to the "design or fabrication of atomic weapons." The president was also given the capacity to authorize the Department of Defense to share information with US allies related to its possible military use of atomic weapons, provided that the information would not help said nations acquire their own bombs. This would allow considerably more cooperation between the US and NATO on matters such as the defense of Western Europe.

The original Act had also prevented private industry from taking a strong role in the development of atomic energy. No patents could be granted for "any invention or discovery" useful solely for producing fissionable material or atomic weapons, and any inventions related to atomic energy could be made public property at the whim of the AEC. The AEC's means of compensating private innovation was a Patent Compensation Board, but this was an unwieldy and slow institution, and the compensation doled out was never anything close to the commercial value of the inventions, to the great frustration of the handful of scientists who received any compensation at all.[140]

The Revision restored private patenting of the production of fissionable material, opening up an entire field of research to nongovernmental ownership, and simplified the provisions for private companies to receive private patents for the production of nuclear energy. The AEC reserved the ability to declare patents to be in the "public interest," and thus available for free AEC use or non-exclusive licensing to private companies, for the purpose of preventing patent monopolies (and to avoid giving existing AEC contractors an undue advantage).[141] Beyond patents, the Revision provided a system by which private companies could apply for licenses that would give them access to government-derived Restricted Data and to work in otherwise restricted areas of research.

Today, the Atomic Energy Act of 1954 often has a reputation of being more strict than the Atomic Energy Act of 1946. This is almost surely due to simple confusion, if not an association of the year 1954 with the height of McCarthyism. All Restricted Data stamps from 1954 onward invoke the Revision as its statutory authority, meaning that there are far more invocations of the Revision than the original Act. This is only an artifact of legal precedence; the Revision replaced the Act as the dominant legal authority, and no further comprehensive revisions were ever attempted. In almost every respect, the 1954 Revision was more permissive than the 1946 Act. In the respects that it was not more permissive, it was generally codifying practices that had already been in place.

Despite these changes, nuclear power still took considerable time to become a full-fledged industry, dogged by safety concerns and high capital costs. Encouragement took the form of heavy federal subsidies on research, giving private industry greater access to classified information, and the almost total declassification of documents relating to civilian nuclear power. Ultimately, the US nuclear power industry did not proceed until a tested power reactor design (the Pressurized Water Reactor) was developed and deployed in a military context first (submarine propulsion).[142]

The expansion of technical release necessitated an expansion of the existing information control system as well. In 1956, with plans to begin releasing information accelerating, a new system of coordinating news of classification decisions among the far-flung branches of the AEC was organized as a series of *Monthly Classification Bulletins*.[143] The fact that there were sufficient changes to classification policy to warrant monthly updates was itself a sign of the new regime. The bulletins provide a close view of the mechanics of Cold War declassification. Each bulletin consisted of individual limits on technical facts that were made clear in short statements to AEC employees and contractors. For example, one from March 1957 informed AEC employees thus:

> The fact that 93.4% enriched U-235 is "weapons-grade material" is classified Secret-Restricted Data. In practice we also have been keeping classified any statements on 93.4% enriched material that clearly say it is the top product [final output] of the [enrichment] cascade. Such statements are classified Secret-Restricted Data. We have been *declassifying*, as a necessary part of the civilian reactor program, (without reference to "weapons grade" or "top product"), the assay of the enriched material which is being used in such reactors.[144]

The bulletins were not instructions about what could be *done*, but statements about what could and could not be *said*, indicating what words, phrases, and numbers were or were not considered sensitive. The maintenance of language was central to the bulletins. The replacement of code words, for example, was a common occurrence:

> With respect to the procurement of yttrium by the AEC, the over-all code word for this program is now "Radex." The old code words, "Buckside" and "Calamar" have been compromised. No association of the new word with the old words should be made in an unclassified document. In addition, any association of yttrium with the AEC Reactor Program is classified S-RD [Secret-Restricted Data].[145]

Within these bulletins, classification and declassification reinforced each other as truly two sides of the same coin: the release of some information was used to uphold the importance of not releasing other information. Lines were drawn and boundaries were marked; there were no shades of gray, and definitely no attempts at deep philosophy or reform. The basic assumptions and methods of classification were made routine, if not calcified. Only occasionally did broader questions come up. In 1958, for example, the question of proliferation risks beyond the USSR was raised in a classification context for what appears to be the first time:

> It has been generally assumed that the degree of sensitivity of our information depends on known information about USSR successes in the weapons field. It must be recognized, however, that the declassification of weapons information must be based not only upon USSR accomplishments, but also on the possible effect that such actions may have in disturbing the world balance of power by enabling other nations such as France and Argentina through declassification of weapons information to build atomic weapons.[146]

The sensibility that proliferation applied beyond the bipolar worldview of the Cold War steadily grew over the late 1950s and into the 1960s. It is curious, in retrospect, how slow proliferation was to become recognized as both a crisis and a paradox. The AEC of the 1950s had reoriented to consider weapon designs (like the Teller-Ulam design) to be paramount secrets to be kept from all foreign nations, whether friend or foe. But tried-and-true methods for producing fissile material, like nuclear reactors and several forms of uranium enrichment, were seen as candidates for both private and industrial development, not only the means by which the United States would maximize the peaceful promise of nuclear technology, but also the tools of diplomacy.[147]

In the 1950s, for example, early reactor designs and enrichment technologies like those used in the Manhattan Project were both declassified and widely publicized. AEC Seminannual Reports proudly announced that hundreds of AEC-owned patents had been declassified and released for industrial use between 1950 and 1959, including those for the original "neutronic reactors" of Fermi and Szilard, and Ernest Lawrence's electromagnetic method of enriching uranium.[148] The fact that these technologies had been used for exclusively military purposes during World War II does not seem to have dampened the enthusiasm to actively promote their dissemination. They were "primitive" technologies, after all, the sort that any advanced nation could pursue, and a decade out of date. In a bipolar Cold War world, the boundary between "safe" and "dangerous" technology had been redefined as a relative one: the difference in advancement between the USSR and the US. Technologies that made no difference in that relative distance were no longer "dangerous," even if they could still be used to produce bombs.

The 1950s focus on "weapons information," which was largely a preoccupation with weapon *design* information, is characteristic of the Cold War secrecy regime, one still rooted in the 1940s belief that the primary barrier to achieving a bomb was theoretical engineering information.[149] By the 1960s, the difficulty in this position would become clearer: the US was, in effect, dispersing scientific and technological "know-how" around the world, which could be used to multiple ends. The detonation of an atomic bomb by the People's Republic of China in 1964 indicated that the bipolar Cold War was now considerably more complicated, and that even disorganized and relatively poor countries were capable of nuclear feats. Whether American declassification and distribution policies contributed to global proliferation is a complex debate—one that would become a site for continued reflection on the means and uses of secrecy.[150]

In the second half of the 1950s, Strauss and the AEC came under increased attack for their policies on secrecy. The Oppenheimer case was part of this, as was Strauss' apparent lack of candor regarding hydrogen bomb fallout in the wake of the 1954 "Bravo" thermonuclear test. The attempt to balance the AEC program with "peaceful" interests and industrial development did not dampen the larger public sense that the atom was still "fearful" indeed.[151]

The AEC's reputation of paternalism and excessive secrecy was only partially deserved, and Strauss attempted to dispute it by pointing to his work in encouraging international and industrial collaboration. In a 1958 speech, Strauss argued that while the AEC "has been charged by some critics as 'super secret' in its non-military operations," the actual result of his policies had been a massive declassification effort. "More than three and one-half times as many documents were declassified in 1958 as in 1954," he argued, and noted that "there was little declassification prior to 1953."[152] In terms of raw numbers, Strauss was correct, but his program of technical release was accompanied by both a tightening focus on scientists' "loyalty" and the creation of a strict security culture.

This Cold War approach to nuclear secrecy was a radical move away from the "problem of secrecy" that had evolved out of the "absolute secrecy" of the Manhattan Project. If the "absolute secrecy" approach regarded *everything* as secret, and the "problem of secrecy" approaches worried over the gray areas, the Cold War approach regarded information in a purely binary fashion—either secret or not—but with strong weighting given to specific, industrially-profitable categories of openness, such as fissile material production and nuclear power plant design and operation, that would have previously been considered dangerous. At the same time, the body and mind of the scientist came under increasing scrutiny, with standards of character and loyalty being applied that in some ways were far stricter than under previous secrecy regimes. In the Cold War regime of nuclear secrecy, those secrets still considered "dangerous" needed to be held very close indeed, and anyone in contact with them needed exceptional vetting and close surveillance. But those secrets that could benefit industry or diplomacy would be not only released, but actively spread throughout the world.

This sharp divide between the peaceful and fearful atom was explicitly ideological. It was a vision of what nuclear technology could be used for, one that in many ways was not entirely compatible with the reality of the technology. The Cold War mindset attempted to draw strict lines through technological distinctions that were not always meaningful, and at the same time prioritized the most extreme practices associated with both secrecy and openness. This seemingly contradictory approach made for a somewhat schizophrenic system. The Cold War regime was ultimately long-lived—we still live with a version of it today—but contained inherent, even obvious contradictions. Cracks began forming in the early 1960s, but the real ruptures would not come until the 1970s (as we shall see).

The late 1950s through the 1960s brought radical changes to the composition of not only the US nuclear arsenal, but to the global nuclear situation as well. The United States added compact thermonuclear weapons to its stockpiles and joined the Soviet Union in a race to develop accurate, long-range missiles. Though the US still enjoyed massive nuclear advantages, the Soviet Union gradually gained the capability not only to annihilate American allies, but to threaten the continental United States. The American nuclear command and control systems became more complicated and more automated, all under a heavy veil of secrecy, but with enough discussion and release for a rich cultural narrative about accidental nuclear war and its consequences to develop. Other nations, starting with the United Kingdom but extending to France, the People's Republic of China, and Israel, acquired nuclear arms by the end of the 1960s.[153]

In 1974, when India, regarded as a "developing" nation without industrial or scientific infrastructure comparable to that of the United States or the Soviet Union, detonated its first bomb, the folly of making a simple division between "peaceful" and "military" applications of the atom became self-evident. The Indian atomic infrastructure had been developed over the years, originally under the direction of Homi J. Bhabha before his death in the late 1960s, based largely on "peaceful" Western atomic science and technology. When the country made the decision in the late 1960s to start a bomb production program, much of its nuclear know-how and knowledge had been imported from abroad. It was only fitting that the Indian government labeled its 1974 test as a "peaceful" atomic detonation, a meaningless distinction to its regional rivals.[154]

PART III

CHALLENGES TO NUCLEAR SECRECY

7

UNRESTRICTED DATA

NEW CHALLENGES TO THE COLD WAR SECRECY REGIME, 1964–1978

Where is the alternative to nuclear *laissez-faire* and nuclear monopoly? Our inability to find the answer to that question, in full awareness of the risk of not finding it and in spite of our search for it, constitutes the tragedy of our nuclear policy.
HANS MORGENTHAU, 1964[1]

Though the regime of nuclear secrecy established in the 1950s is still with us, some of its harder edges and extremes have become muted over time. Its persistence in the face of challenges and contradictions has been impressive, enabled by its pretensions of mastery over both the hopeful and fearful aspects of nuclear technology and the absence of strong competing alternatives. It had become so embedded in the fabric of American bureaucracy, and the American security mindset, that it is difficult to imagine anything different at this point.

Challenges and contradictions did arise, however, though they were slow to gain traction. While the Soviet Union had built its own nuclear arsenal by the 1950s and the United States had begun working to distribute the fruits of "peaceful" nuclear technology, in many ways the United States still retained a de facto monopoly over nuclear knowledge in the non-Soviet sphere. The US government was still the largest funder of nuclear technology in the "free world," and anyone who wanted to have a hope at competing in that sector needed to pass through US institutions, which meant security clearances and complicity. Even the opening up of work to private industry was largely a controlled activity, with industry playing a submissive role to government declassification and subsidies. And while parallel work was being done in the USSR, the Soviets were not in the habit of spreading their own produced knowledge and work too liberally, either.[2]

But the de facto monopoly was beginning to loosen. The number of actors, both domestically and abroad, was starting to multiply. And as nuclear technology was becoming more common, the line between the military and civilian threatened to blur in ways US experts found alarming. What had once seemed controlled and relatively safe threatened to unravel in ways that could be catastrophic.

7.1 THE CENTRIFUGE CONUNDRUM

The de facto US monopoly on innovation in nuclear technology lasted for the first decade after the Manhattan Project. Even after the British had joined the "nuclear club" in 1952, the United States remained the primary innovator of new nuclear technology, spending billions of dollars per year on the massive, sprawling industry controlled by the AEC. Other nations who were interested in doing nuclear research or sharing the fruits of said research would most likely need US assistance to make significant strides. But the US, from Eisenhower's "Atoms for Peace" program onward, was willing to grant that assistance.

Starting in 1955, the US also began to enter into dozens of bilateral agreements, most for research but some also for power programs, with friendly or neutral nations around the world. It also encouraged the development of the European Atomic Energy Committee (Euratom), meant to unite the once fractious European allied states around technical cooperation on nuclear issues, and supported the growth of the International Atomic Energy Agency (IAEA) in the late 1950s as a promoter and eventual "watch dog" on peaceful nuclear matters. US policymakers were not truly altruistic on this issue. Rather, they believed that peaceful atomic energy could be a "carrot" that would serve US goals abroad and that encouraging international dependence on the US would prevent such countries from developing too much indigenous nuclear knowledge on their own. If these nations wanted nuclear reactors, it would be better for the US to provide them, because it would allow the US to set the terms and monitor the use of the reactors.[3]

But the US monopoly was largely a function of having gotten an initial advantage through the expenditures of the Manhattan Project and being in a much better economic and political position than most countries in the immediate postwar period. Throughout the 1950s, some policymakers began to suspect that the US lead was not as large as it had been, and many nations were beginning to question whether the restraints that came with US assistance outweighed the benefits.[4]

One particular technology, the gas centrifuge, would challenge the US monopoly and come to embody the rising fear that the "nuclear club" could grow much larger. The gas centrifuge is a means of enriching uranium, separating the fissile U-235 isotope from the more common U-238 isotope by circulating gaseous uranium hexafluoride through a cascade of tubes spun at extremely high velocities.[5]

Along with other methods of uranium enrichment, the gas centrifuge was investigated during the Manhattan Project. The physicist Jesse W. Beams, at the University of Virginia, had been working on centrifuges since the mid-1930s and was tapped to head the original centrifuge effort. Initially, the centrifuge work was allocated a much larger budget than was gaseous diffusion and considered much more promising. But the centrifuge work advanced more slowly than was expected, and Beams' initial designs had disqualifying engineering flaws. It was eventually defunded in favor of other methods, despite some Project scientists believing the centrifuge still had promise. But just because Beams couldn't make it work didn't mean it was unworkable.[6]

The Soviet Union had also considered gas centrifuges as a possible method of enrichment from the beginning of their own atomic project. After the defeat of Germany, the Soviets were able to recruit a number of former Axis scientists to work on the Soviet nuclear program. Among these was the German Max Steenbeck and the Austrian Gernot Zippe, who were assigned the task of investigating gas centrifuge enrichment by the Soviets around 1947. They drew upon Beams' pre-war publications and their own research to debug the centrifuge's engineering problems. The Steenbeck group and other Soviet researchers eventually developed a gas centrifuge that, while not yet competitive with gaseous diffusion in terms of efficiency or capability, had a clear path forward for further development. Furthermore, it was conceptually and practically very simple. Unlike the tall centrifuges that Beams was focused on, whose height increased their separative power but introduced severe engineering hurdles, the Steenbeck group's was short and efficient.[7]

Like the Americans, the Soviets primarily used gaseous diffusion enrichment for their early nuclear program, but they would eventually augment their nuclear program with centrifugal enrichment. This by itself would not disturb the Americans much; the Soviets already had a great supply of enriched uranium through the diffusion method. But Soviet centrifuge design did not stay inside the USSR because the researchers who created it did not. Remarkably, Steenbeck had in 1949 negotiated with Lavrenty Beria himself that if he and his team could produce a workable centrifuge pilot factory, they would be allowed in due time to leave the USSR. Even more remarkably, the deal was honored; in late 1953, having produced their success and having convinced the Soviets that the Germans were no longer necessary for the project, the Steenbeck group was put into a "quarantine" of nonmilitary research so that any knowledge they might give the US would be significantly out of date. They were allowed to emigrate in late July 1956 and even paid tens of thousands of rubles for their trouble.[8] Upon leaving, Steenbeck became a professor in East Germany, but Zippe and another colleague, Rudolf Scheffel, went into the capitalist West.[9]

Zippe's name would become synonymous with centrifuge entrepreneurship. After his release by the Soviets, US intelligence agents found and interviewed him for what knowledge he had about the Soviet nuclear program. His description of his centrifuge expertise piqued their interest, and using a false passport, he was allowed to visit the United States for a longer debriefing.[10] By his own account, Zippe had not given much thought to monetizing his knowledge until he attended an unclassified conference on isotope separation held in Amsterdam in April 1957, at which point he realized that the Steenbeck group had produced work that "far exceeded" what was being done in the West. Zippe not only obtained the permission of his former Soviet colleagues, but he drew up a contract with the West German firm Degussa that would clarify patent rights for himself, Steenbeck, and Scheffel.[11] At the conference, he spent two hours talking to the chief Dutch centrifuge researcher, Jaap Kistemaker, and on the basis of this conversation alone, Kistemaker stopped work on the Beams-style long centrifuges and shifted instead toward Zippe-style short centrifuges.[12]

In the summer of 1958, Zippe was persuaded to join the program at University of Virginia and demonstrate what his team had built for the Soviets. This work, remarkably, was done under an unclassified contract. The unclassified nature of this work stemmed from contractual arrangements made between the AEC and Degussa, and the fact that the US did not have an agreement in place to exchange classified information with West Germany. Had Zippe truly been a "free agent," things might have been different, but due to his affiliation with a major West German manufacturing firm, if the US wanted to see what Zippe knew, they had to agree to the whole world potentially seeing it. Zippe produced several unclassified reports on his work for the AEC, including a 98-page final report filed in July 1960. Fifty-two copies of the report were distributed to US researchers, but anyone who requested one could get one. Not long after, Zippe returned to West Germany.[13]

The US had already restarted its centrifuge research at the University of Virginia around 1953, in response to new publications coming out of West Germany and the Netherlands.[14] The West German and Dutch work did not, by itself, highly concern the AEC. The US viewed these early incarnations of the gas centrifuge as less economical and efficient than gaseous diffusion, and the US had enough enriched uranium to supply not only its own military needs but the foreseeable needs of the free world as well. The AEC's own research program, under Beams, was competing better with gaseous diffusion than the European designs. It was making some progress by the late 1950s, largely because the engineering problems encountered by Beams' long centrifuges were being solved through new breakthroughs in materials that had occurred as part of the US space and missile programs.[15]

But there was another framework through which to look at the gas centrifuge. The US was accustomed to thinking about the development of nuclear weapons in the framework it had developed in the 1940s and early 1950s, which focused on large industrial states with many resources valuing efficiency and bulk above all else. But what if future countries did not follow this template? What if they went in a direction that the US had not anticipated, and thus could not control? What if instead of a massive, detectable, difficult-to-produce, efficient gaseous diffusion plant, future nuclear countries chose the technologically less-advanced, but easier-to-produce Zippe-type centrifuges?

The question of "who's next" to get nuclear weapons (as the satirist Tom Lehrer would famously put it) had gradually arisen since the British built their own atomic bomb in 1952. By 1957 this was called, within the realm of intelligence and diplomacy, the question of the "fourth country." But it was gradually becoming clear how unlikely it was that there would be only four nuclear nations; within technical circles, it became fashionable to refer to the spread of nuclear weapons as the "Nth country" or "Nth power" problem (a change made necessary once France detonated its first bomb in 1960). Only by the late 1960s would the present-day term for the spread of nuclear arms to new countries, nuclear proliferation, become standard.[16]

American interest in the centrifuge's proliferation potential was first piqued by the British. John McCone, a businessman who briefly served as AEC chairman starting in 1958 before leaving to become head of the CIA in 1961, met with Sir William Penney, "father" of the British atomic bomb program and veteran of the Manhattan Project, in London in late 1959 to discuss centrifuges. The British were interested in them as a potential source of low-enriched uranium to use for European nuclear power reactors, but Penney was also the one who convinced McCone that they posed a proliferation threat. Penney feared that the West Germans might decide to make bombs, which might threaten the postwar European alliance and revitalize fears of a revanchist foe from a world war still in living memory. Upon his return to the US, McCone commissioned a series of studies on the risks of the gas centrifuge.[17]

One of the first was completed by analysts at the Union Carbide Nuclear Company in February 1960, which divided nations into three categories (low, medium, high) of industrial development and technological competence. The conclusions were grim: the gas centrifuge might not, in 1960, be competitive with gaseous diffusion, but a working plant could probably be made by a less technically proficient nation. A gas centrifuge plant capable of enriching enough uranium for a small number of bombs could be built relatively cheaply, take up a modest amount of floor space (and thus be hidden in any warehouse-sized building), and have power requirements modest enough not to stand out. The authors reasoned that a high-competency country could develop such a facility without any outside help and perhaps have a weapon within 5 years. A low-competency country would need outside assistance, but with that, even it could manage a weapon within 8 years, albeit at a high price.[18]

Moreover, the gas centrifuge drew attention to a serious problem facing the Cold War secrecy regime. Gas centrifuges were inherently "dual-use": the exact same technology, operated in exactly the same fashion, could produce enriched uranium for a military weapon or for civilian power reactors. Whereas other "peaceful" nuclear technology promoted by the United States, like civilian power reactors, could technically be considered dual-use, they had to be operated somewhat differently for military applications, and such activity either could be easily monitored or might produce plutonium that was less reliable for military purposes. But a centrifuge plant that enriches uranium for nuclear power programs on an industrial scale needs only to be run longer for weapon purposes.

All of which is to say, a nation could develop a civilian centrifuge program and immediately switch it into a military program; indeed, there might not be any real distinction between the two. The result, the AEC concluded, was that a gas centrifuge project would be an ideal choice for an Nth power to run either "covertly" (e.g., as a totally secret project) or "overtly" (as a dual-use project). And there was already some evidence that one of the suspected Nth powers, Brazil, was interested in this route: they had purchased three centrifuge prototypes from West Germany and had sent several of their own people to train with the Germans in their operation, all ostensibly under the rubric of open, peaceful science. Separately, Japan had a research project on long centrifuges of its own that was known to the AEC.[19]

An AEC guide on centrifuge classification developed in the late 1950s had determined that only the long centrifuge work of Beams—i.e., the work they thought might be economically competitive with gaseous diffusion—would be classified. The seemingly less promising short centrifuges were kept completely open. In general, the AEC policy through early 1960 was that centrifuge work was considered unclassified. An internal memorandum makes it clear that this was in part a diplomatic issue: the AEC was monitoring West German and Dutch work on centrifuges and until a "breakthrough" occurred they did not want to muddy the waters with their allies, who were conducting their work without any security classification at all. But by 1960, the AEC judged that the "breakthrough" had happened, and it was time to clamp down.[20]

Several people, notably McCone himself, expressed doubts that secrecy could be used to rein in this emerging threat.[21] After all, the Zippe "breakthrough" did not originate from AEC research. While the AEC could exercise monopolistic control over its own productions and scientists, and even demand compliance from US industry, could it do the same with West Germany or the Dutch? And would it want to? After all, these were allies, and the US had a vested interest in making sure that these countries remained happily within the fold of NATO. Pushing them too far could lead them to sever their ties and perhaps pursue a more independent path as France had done, developing their own nuclear force. As one US diplomat opined confidentially to another, "I'm afraid some of our AEC friends do not realize that the way German industry has a free hand they could just as easily tell both the German Govt and us to go to hell."[22]

Though US officials considered other options, such as price manipulation of reactor fuel to undercut West Germany and the Netherlands, they ultimately chose to try to extend classification powers into the industries of other sovereign states. The AEC knew this would be difficult and possibly dangerous territory. The sheer number of actors was daunting: they were dealing with not merely foreign governments, but foreign industry and scientists, neither of which were used to working under government secrecy. Though the scientists appeared to have truly intended that their technology be used only to generate low-enriched fuel for nuclear power reactors, the AEC worried that their national leaders might eventually develop other ambitions, and that in any case, the open publication and circulation of this work would certainly benefit other Nth powers in waiting. The AEC understood that there were severe political issues. The treaty that had created the Euratom organization required a certain amount of sharing between Euratom members; if West Germany and the Netherlands suddenly put its civilian work under military classification, it could raise the ire and suspicion of their fellow Euratom members, if not constitute a treaty violation.[23]

Through diplomatic channels, the US provided the West Germans and Dutch with a 2-page guide to classification practices, effectively exporting the AEC system in July 1960. Both countries accepted, albeit with some reservations. Both nations treated this agreement as a matter of domestic policy, not international obligation, and reserved the right to modify or abandon the arrangement unilaterally. The AEC perceived this to be a means of retaining autonomy and avoiding some of the political difficulties that might come from seeming unduly beholden to the US, while also giving them some leverage to keep the US from pushing them too hard.[24]

The AEC briefed the Joint Committee on Atomic Energy on the arrangement at a closed session in October 1960. It was a "sobering" presentation for the congressmen. The list of countries the AEC feared getting nuclear weapons was not a comforting one: Cuba, Japan, Israel, Egypt, Argentina. But questions were once again being raised about secrecy as a blanket solution. Demonstrating the JCAE's shifting views on secrecy, its chairman, Senator Clinton P. Anderson, criticized an AEC witness for labeling his charts "Secret," even though they were just lists of US companies doing unclassified research into centrifuges.[25] Later, as the AEC representatives explained that publications on centrifuges were coming out of not only Western European countries, but also East Germany and Poland, Anderson again questioned whether secrecy would matter at all. The Brazilians, after all, had already bought West German centrifuges, legally and uncontrolled. "You don't just make an automobile and try to sell [it] to people and then say, 'this is a secret device,'" Anderson lampooned. He was doubtful the Germans would want to participate at all: "How do you take it out of the market? What are you going to pay them for staying out of the market? What is your proposal in order to get them to classify? Are you offering to pay them so much money so they won't use it?"

The representatives from the AEC, as well as a representative from the State Department, took pains to emphasize that the Germans and Dutch had already agreed to keep future work under classification. McCone still expressed grave doubts: "As to any clear pattern of how we are going to control [centrifuges], we haven't reached that point yet. It may have gone entirely past the point of no return. Maybe we can't do anything about it at all." AEC assurances of cooperation may have mollified the congressmen a bit, but the overall tone of the meeting was one of suspicion and dread. One congressman noted toward the end of the session that "the possibilities are horrifying."[26]

The revisions to the Atomic Energy Act in 1954 had been made in part to open the field of nuclear technology to private industry within the US, while attempting to keep certain controls over the flow of classified nuclear information in place. These two goals were recognized as being difficult, though not impossible, to reconcile. Some of this was accomplished through declassification efforts, but the 1954 Revision also explicitly allowed industrial access to Restricted Data for civilian purposes through AEC licensing. A 1955 program allowed vetted industrial companies to have access to Restricted Data in exchange for the company giving the AEC full access to any technical data or experimental equipment generated, and allowing the AEC to license any deriving technologies for what the AEC considered reasonable compensation.[27] It was, to put it mildly, a somewhat one-sided deal.

But Restricted Data was a tricky concept that was getting even trickier over time. The original conception of Restricted Data, starting with the Atomic Energy Act of 1946 but continued in the 1954 Revision, was that the information was secret by definition. Classifiers didn't tell you that they had determined whether a document was Restricted Data, they told you whether it contained it. This is a subtle distinction, but an important one when it came to the private sector. Could a company that was researching centrifuges without any access to government-produced Restricted Data still generate Restricted Data? A strictly legal reading of the law implied that the answer was *yes*: Restricted Data was not Restricted Data because of its association with the AEC, it was Restricted Data because it was about nuclear weapons. This matter seems not to have been anticipated by the drafters of either the 1946 Act or its 1954 Revision, again, likely because prior to the late 1950s, there were no competitors to the AEC as a generator of Restricted Data.[28]

The gas centrifuge caused the AEC to revisit this policy. In August 1960, it decided that all further work on the gas centrifuge in the United States, even that produced by private industry, would be regarded as Restricted Data. The policy was a stark one: while that which was already unclassified did not change its status, any elaborations were suddenly controlled. This came as a surprise to the few US companies working in the field who suddenly saw the limits of secrecy closing in around them.[29] Could the AEC tell a US company that had put its own funding into gas centrifuge research based on unclassified publications that it was not allowed to pursue the work without first obtaining a security clearance? The official AEC position was now *yes*, so companies that wanted to work on centrifuges now had to join an AEC permit program even if they had no intention of receiving any Restricted Data.[30]

Over the course of the 1960s, these issues got more complicated. In 1967, the AEC proposed that companies would need to get "private restricted data access authorization" to work in areas such as centrifuges that might impinge on weapons design or isotopic enrichment. This was criticized as vague, burdensome, and bureaucratic.[31] More specific regulations designed to clarify the proposal were labeled by one former AEC lawyer as having "produced an ambiguous, hair-splitting, administrative monstrosity" and were likewise abandoned.[32] By 1969, there had been no official policy clarification; indeed, there would not be one before the AEC was abolished in 1975.[33]

Ultimately, US industrial companies who wanted to research centrifuges were allowed to, so long as they got the required permits. This forced several companies out of the field, but a handful stayed in the game. By 1967, only 140 technical employees were working in the US private sector on centrifuges. From a security standpoint, this was a good thing; as an AEC report noted, "[I]n a general sense, the possibility of inadvertent disclosure in any control system is in part a function of the number of people—and particular the number of organizations—who have access to the information being controlled."[34] The AEC record on voluntary domestic compliance with its demands regarding Restricted Data, through the end of the 1960s, was essentially perfect: legal intervention was never required, and the "born classified" interpretation of the Atomic Energy Act was not put to a real test.[35]

Internationally, things were trickier. Ideally, the AEC might have hoped to open up a more formal collaboration with the West Germans and the Dutch, as they had on nuclear matters with the United Kingdom. The US and the UK began collaborating on classified centrifuge technology from the end of 1960 through early 1965, an arrangement that not only afforded the US considerable latitude in monitoring British progress, but also made it difficult for the British to serve as competitors: because the British scientists had been exposed to AEC-generated Restricted Data, they were essentially prohibited from commercializing their work in any way that might expose Restricted Data. Lamenting how much this complicated efforts to develop civilian technology, the UK Minister of Technology wrote in his diary that the UK was "absolutely tied hand and foot to them, and we can't pass any of our nuclear technology over to anybody else without their permission."[36]

The Dutch had scaled back their project to a mere six scientists and ten engineers, but they still made considerable progress.[37] The Dutch scientists were particularly unhappy that classification would inhibit their collaboration with others (they were exploring partnering with the West Germans and perhaps even a US firm) and wondered how a nation that did not have significant scientific secrecy was expected to impose it. How would scientists be screened? Under what criteria? These were hard questions for a scientific community that had largely avoided secrecy regimes.[38] Both the West Germans and Dutch would complain to the US that they wanted the ability to file secret, potentially lucrative centrifuge patents in other NATO countries, a request that the US, in meeting with the UK, Dutch, and West Germans in 1962 and 1964, pointed out created a whole host of problems with regard to secrecy, international treaties (e.g., Euratom), and practicality.[39]

By 1964, the Dutch felt confident enough in their research that they were planning to develop a pilot plant. They requested another meeting of the US, UK, Dutch, and West Germans with the aim of reexamining secrecy arrangements. At the meeting, the Germans and Dutch reaffirmed their commitment, but the AEC representatives noted that they were "reluctant partners to the classification arrangements." They further reported that in West Germany, the Ministry of Scientific Affairs had recommended that the classification of gas centrifuges be ended, and that it was preserved only through political intervention by the Ministry of Foreign Affairs. The Dutch had suggested that classification should be regularly revisited, and secrecy would have to become less rigid as information barriers eroded over time. The AEC representative urged to the rest of the commission that while the policy was being maintained for the present, the internal pressures building against classification within West Germany and the Netherlands were "substantial."[40]

Within the US, new breakthroughs in long centrifuges led, in 1965, to a termination of cooperation with the British. The UK pushed ahead on their own centrifuge research, both for its technological possibilities and also to "wean" West Germany off close collaboration with the French.[41] At the same time, the US began to worry more about the proliferation risks posed by the centrifuge, with AEC Commissioner James Ramey ordering a major study on the subject. The study, not completed until early 1967, reaffirmed that the gas centrifuge could be a game-changer, highlighting that centrifuge work allowed a nation to work toward a bomb with "relatively low capital investment, low electrical power requirements, and easy concealment." The AEC study also concluded that there was now enough information in the open literature for a country "willing to pay a high price for a few weapons" to begin an effective centrifuge program from scratch. The list of countries that the AEC believed could get a weapon within a decade was a long one: Belgium, Denmark, Italy, the Netherlands, Norway, Portugal, Spain, Argentina, Brazil, Czechoslovakia, and East Germany. Furthermore, neither the AEC nor the IAEA had ever attempted to apply international safeguards to any isotope separation facility. Complicating things further was that with a booming nuclear power market, the US was no longer confident it could continue to meet demands for low-enriched uranium in the future. Nevertheless, the AEC was still stuck in its secrecy regime: "[T]here appears to be little choice but to continue the policy of classifying and controlling gas centrifuge technology," the report's authors concluded.[42]

By 1968, the situation had worsened. The West Germans, Dutch, and UK had thrown their lot in together to create a new centrifuge plant operated for profit by an international consortium, known as Urenco. The US had resisted internal pressures to advocate for the prohibition of centrifuge technology as part of the Nuclear Non-Proliferation Treaty because it feared that doing so would give the West Germans an excuse not to sign what was in their country an unpopular agreement.[43] After some initial hiccups created by the "contamination" of UK centrifuge work by AEC Restricted Data during the US-UK period of collaboration, the effort moved forward, with major enrichment plants being built in Capenhurst, UK, and Almelo, the Netherlands.[44] It had taken a little over a decade, but for better or worse, the centrifuge work that had begun in the Soviet Union had finally taken root in the European free market.

The Cold War secrecy regime demanded firm boundaries between what was secret and what was not, between who had access to that information and who did not. But these boundaries began to become "tricky," not only because the number of players, both domestically and internationally, began to expand, but because some of these boundaries were themselves epistemically fraught. Restricted Data was a concept that, in its "born secret" interpretation, crossed boundaries by definition: it did not care who made the data or for what purposes. And the categories of "private," "AEC," and even "American" could be very complicated in practice: Zippe, for example, was an Austrian physicist who had worked in the USSR and then emigrated, who had come to the US to do unclassified research for the AEC on a subject that would later be regarded as Restricted Data, after he had told the Dutch about it and had entered into a contract to develop it for West German industry. Real people were more complicated than neat Cold War idealized categories.

And yet, while the fears of nuclear proliferation, particularly the role of the gas centrifuge, grew over the course of the 1960s, there is little to suggest that the AEC fully understood the difficulties they were creating. Even today, there are questions about how much the centrifuge has changed the nature of proliferation.[45] On the whole, the risk of "Nth powers" acquiring centrifuge technology appeared looming but not imminent. The AEC had gotten what it demanded from both private industry and foreign partners. But the proliferation problems that would become evident in the longer arc of the centrifuge's history would demonstrate the AEC's weaknesses as its own Cold War regime demanded rigid and clear boundaries in an increasingly interconnected and complicated world.

7.2 THE PERILS OF "PEACEFUL" FUSION

Nuclear fusion has proven much less amenable to being used for peaceful applications than nuclear fission. Despite its promise of virtually limitless, clean energy, controlled thermonuclear reactions—which is to say, *not* the kind that you get from hydrogen bombs—have remained relatively elusive to this day. AEC policy on fusion was caught between its hopes and fears in the 1950s, as it attempted to demonstrate a commitment to the peaceful applications of a field of science that at some level was at the heart of what they believed was their most important technical secrets. In the late 1960s and early 1970s, however, the problem became especially acute, when private entrepreneurs attempted to commercialize a type of "peaceful" fusion that had decidedly "military" origins, calling into question not just the Cold War dichotomies of "safe" and "dangerous," but also throwing into fundamental relief the difficulties of the AEC's job as regulator of Restricted Data.

Peaceful nuclear fusion technology had captivated scientists and statesmen since the early 1950s. Unlike nuclear fission, fusion would not produce significant amounts of nuclear waste, and there was no risk of a reaction getting out of control. Pound for pound, fusion fuel produces more energy than fission fuel, and the main fusion fuel, deuterium, was an isotope of hydrogen that was far more abundant and easier to manufacture than enriched uranium. Fusion as a source of electricity would be "too cheap to meter," as AEC Chairman Lewis Strauss infamously prophesied in 1954. In reality, as stimulating to the human imagination as peaceful fusion might be, it was, and has been so far, immensely difficult to realize in practice.[46]

The AEC did not begin to officially work on peaceful fusion energy until 1951. Bizarrely, the impetus for the work was an apparent hoax by a German scientist, Ronald Richter, working for the Argentine dictator Juan Perón. Perón announced to the world that Richter had built a compact nuclear fusion reactor, and while American scientists doubted the truth of these claims (for good reason), they also inspired them to consider what a real reactor might look like. One so inspired was the Princeton astrophysicist Lyman Spitzer Jr. Through his work on the Project Matterhorn H-bomb project, Spitzer had ample exposure to nuclear fusion research, and was an expert on plasmas. Spitzer combined his interests into an idea that would be known as magnetic confinement fusion, which worked by trying to contain a fusion plasma in a "bottle" made of a magnetic field. It was not easy going, because plasmas are not at all easy to contain, and magnetic "bottles" are difficult to construct, but it seemed like a new opportunity.[47]

Spitzer was able to convince the AEC to fund his work, inaugurating peaceful fusion research in the United States. The work was entirely classified, born as it was out of the work on the hydrogen bomb. The Stellerator work at Princeton even shared the same project name as the weapons work: Spitzer's Stellerator work was Matterhorn S, while John Wheeler's bomb-related work was Matterhorn B (for bomb).[48] While magnetic confinement fusion has only superficial similarities to the physics of thermonuclear weapons, the core reactions are the same, and in the early work on the H-bomb, that was enough to keep them secret.

But the idea for controlled fusion reactions occurred to many different scientists in many different countries. By 1955, the idea was sufficiently diffused that, at the first "Atoms for Peace" conference in Geneva, Indian physicist Homi Bhabha devoted a significant amount of time in his welcoming address to the prospect of peaceful fusion. Bhabha's overall address was about the role of energy in human history, with fusion presented as the next step to a new kind of era of plenty: "I venture to predict that a method will be found for liberating fusion energy in a controlled manner within the next two decades. When that happens, the energy problems of the world will truly have been solved forever, for the fuel will be as plentiful as the heavy hydrogen in the oceans."[49]

As the US still considered fusion research classified, the AEC delegation was caught off-guard by Bhabha's address. When asked to comment on it at a press conference afterward, Chairman Strauss punted, saying he would answer questions at a later session. The journalists interpreted this as a sign that Strauss was planning a big announcement. A few days later, he held another press conference to inform the press that the AEC had been working on controlled fusion reactions for some time, but there were no spectacular breakthroughs yet.[50]

All of this drew far more attention to fusion than the AEC had been prepared for. Bhabha suggested utopia was just around the corner, and while the AEC scientists knew how unrealistic this was, they were prohibited from saying so. Thus did the secrecy around controlled fusion set up, not for the first time, a dangerous dynamic: it appeared to outsiders that the AEC was stifling work of immense importance, when in reality classification made it difficult to express just how nascent the effort really was. AEC Commissioner Smyth attempted to express some of this at a speech a few weeks after the Geneva conference, using the kind of cryptic language that secrecy demanded:

> Because the work on controlled thermonuclear reactions is still classified, I cannot say anything more about the problems on which work is being done or the methods used to explore them. . . . Nor should I venture any prediction as to the probability of success or the time estimated to be required. For this last prohibition, I am grateful. . . . Let us agree that we have here an extremely difficult scientific and technical problem of great eventual economic importance, but of no direct military value. It is a long range problem. Even when the technical problems are solved, it may be a long time before its economic importance is significant in this country.[51]

For several years prior to the Geneva conference, there had been pushes to declassify the AEC's peaceful fusion work at least partially, and perhaps even completely. In early 1956, JCAE member Representative Carl Hinshaw wrote to Strauss asking why it was still classified, despite his having talked to several scientists who, in his account, "do not seem to have any convincing reasons for classifying this project." Strauss' response was that the work was classified not because it had any relevance to thermonuclear weapons, but because a full-scale nuclear fusion reactor would be a powerful neutron source, and could breed fissile material from uranium-238 or thorium-232. He downplayed the possibility that declassification would accelerate the research.[52]

But over time, pressure would build for declassification. Announcements about British fusion work appeared in the press, and there were fears among AEC scientists that it might appear that the United States was lagging. Norris Bradbury worried, as he wrote to Strauss in late 1957, that the US might look like it had been "Sputniked" again.[53] This fear, and the British disclosures, finally overcame the resistance to declassification, and in January 1958, the AEC moved to open the field of magnetic confinement fusion and announce it several months before the next Geneva "Atoms for Peace" conference. At the press conference, several of the questions were about whether the previous secrecy had held back the work; Strauss denied it had any effect.[54] In any case, peaceful fusion was now public—at least, magnetic confinement fusion.

But there was another form of peaceful fusion research that was emerging at the same time, and it was a more complex situation. At Edward Teller's Livermore laboratory, which had been set up in 1952 as a "competitor" to Los Alamos in the wake of the H-bomb debate, a young Teller-protégé physicist, John Nuckolls, was exploring a very different concept for controlled fusion.[55] The lab was intensely working on Project Plowshare, a research program that sought "peaceful" application of nuclear explosions, such as using them to excavate canals and harbors. Nuckolls was asked by his division leader to explore whether the detonation of an underground nuclear weapon, suspended in a large cavity filled with steam, would generate enough superheated vapor to run through a turbine to generate electricity. In principle, it could be done, but it would cost a lot of money and the radiation problem was not insignificant.[56]

Nuckolls focused on reducing the size of the bomb needed (the original plan was a massive 500 kiloton weapon) and even looking into whether the bomb could be a pure-fusion detonation, which would remove most of the problem of radioactive byproducts. This led Nuckolls to considering whether a "non-nuclear primary" could be used to channel the energy to compress the fusion "secondary" of the Teller-Ulam-style. Nuckolls proposed using plasma jets, charged particle beams, and even hypervelocity pellet guns to implode a very small volume of fusion material. Under some conditions, his calculations were encouraging; under others, they were not. He continued his work into early 1960, focusing on developing small, efficient ("high-gain") fusion capsules that would be able to optimally use whatever energy his non-fission primary (later known as the "driver") could deliver. This work was entirely theoretical, but benefited from his experience in thermonuclear weapons design and from access to the latest data and models for how thermonuclear reactions worked in hydrogen bombs.[57]

Nuckolls' approach began to look very different from the initial idea of dropping H-bombs into steam-filled holes. The amount of fusion fuel he was trying to burn was very small, on the order of 10 milligrams of deuterium and tritium gas (as opposed to the kilogram quantities used by H-bombs), which might be compressible with the non-nuclear "driver." He also incorporated existing H-bomb design concepts, like the highly important radiation case. He had created a new kind of reactor, one he would later dub the "thermonuclear engine" that, if it operated, would be able to fire one tiny H-bomb after another ("microexplosions"), pulsing with neutrons and heat. But it was all on paper; none of the non-nuclear "drivers" seemed especially promising, and it wasn't clear any of it would work.[58]

The first real breakthrough came in July 1960, when Theodore Maiman reported the development of the first laser. Lasers allow for efficient transformation of electrical energy into optical energy, and it was quickly realized by many physicists at Livermore that although the technology was still embryonic, a sufficiently powerful laser could be used as a non-nuclear "driver" to initiate fusion reactions in exactly the kind of arrangement Nuckolls had been contemplating. By 1961, a group of physicists at Livermore were working seriously on the question of laser-initiated thermonuclear reactions, even though nobody yet knew how to produce lasers powerful enough to work in this way.[59]

Theoretical work on this new type of fusion reactor continued at Livermore and Los Alamos into the 1960s. The scientists—all nuclear-weapons designers—found that by fine-tuning a laser so that the energy was pulsed in an optimal way, they could affect exactly how the fusion capsule was imploded. They explored schemes that involved both "direct drive" compression of the fuel pellet, in which the laser pulse would be shaped by mirrors so that it impacted all sides of the pellet sphere simultaneously, as well as the "indirect drive" that involved firing the laser into a radiation case that would re-radiate a uniform temperature of X-rays all around the pellet. The latter was a direct adaptation of the Teller-Ulam idea, with its only downside being that a considerable amount of laser energy was lost when converted into X-rays. Direct drive approaches promised better compression, but they required extreme simultaneity and near-perfect geometry to avoid asymmetries. Various kinds of fuel pellets were explored, from the very simple and cheap (frozen droplets of deuterium-tritium gas) to the highly complicated (pellets that involved layers of materials, even fissile materials). Even the more powerful lasers that the scientists dreamed would be possible down the line paled in comparison to the power of an explod-

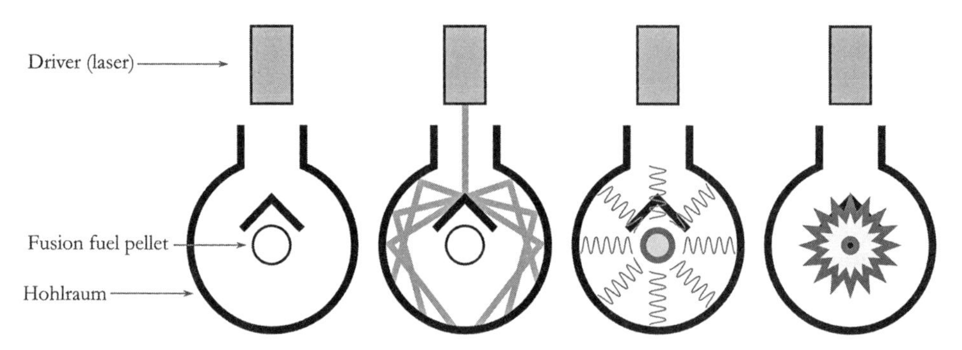

FIGURE 7.1. Nuckoll's conception of laser-driven, indirect-drive ICF, around 1961, shown in a time sequence. From left to right: 1) The basic setup with a spherical droplet of deuterium-tritium fuel within a hohlraum, 2) the laser fires and is directed to multiple spots within the hohlraum, 3) stimulated by the laser pulse, the hohlraum inner surface produces X-rays that ablate the surface of the fusion pellet, beginning its implosion, 4) the pellet implodes to extremely high densities and temperatures, beginning thermonuclear burn. Derived from Nuckolls, "Contributions to the genesis and progress of ICF," figure 6, and National Ignition Facility, Lawrence Livermore National Laboratory, "How ICF works," online at https://lasers.llnl.gov/science/icf/how-icf-works, accessed 10 December 2018.

ing atomic bomb, so every bit of efficiency counted. A key insight borrowed from thermonuclear weapon design was that the laser was good only if it could achieve extremely high (i.e., thousand-fold) compression of the fusion fuel: it is the compression that would start the fusion reaction, not the heat of the laser. The high compression at the center of the fuel pellet would both start the fusion reaction and confine it. Hence this approach to fusion was referred to as laser-driven inertial confinement fusion, frequently called just "laser fusion" or ICF.[60]

One of the things that had afforded Nuckolls and other weapon designers the luxury of researching non-weapons ideas was that the US and the USSR had entered into a nuclear test moratorium beginning in late 1958, which lasted until fall 1961. During the moratorium, the US had slackened in their weapons work and were caught flat-footed when new testing resumed. In 1962, the US launched one of their largest nuclear test series, Operation Dominic, at the Pacific Proving Grounds. Nuckolls put laser fusion aside, and that April proposed an unusual new thermonuclear weapon design, code-named "RIPPLE," which used a "highly optimized pulse" to implode a "high performance" thermonuclear secondary.[61]

After its success, he put forward an "even more radical" design, one in which "we optimized the pulse shape to achieve practically isentropic fuel compression."[62] The "RIPPLE" design details are still very classified, but while Nuckolls does not publicly state that this work was informed by his previous work on laser fusion, he appears to have incorporated the breakthroughs of his earlier work: high-gain, optimized pulse shapes, practically isentropic fuel compression. Put plainly, it seems like the dual-use nature of laser fusion research worked in *both* directions, from military to civilian and back again. Laser fusion's resemblance to H-bombs is not superficial: it is derived from, and can apparently contribute further to, the design of thermonuclear weapons.

But while Nuckolls had been among the first to consider the applications of the laser to fusion, he was certainly not the last. The mathematical physicist Ray Kidder had been put in charge of coordinating Livermore's laser fusion work in 1962, and by 1963 he had discovered that there had already been at least three separate "inventions" of laser fusion outside of the AEC: one by a professor at the University of Michigan, another by a researcher at Hughes Aircraft, and another by a researcher at the New Hampshire defense contractor Sanders Associates. They all appear to have been variations on the idea of achieving fusion by using a laser to *heat*, not *implode*, thermonuclear fuel. This doesn't work by itself, but anyone who followed the idea through to its conclusion would find that using a laser to heat fusion fuel will cause the fuel to expand. Confinement would be necessary to keep the plasma density high enough for fusion, and if that confinement was inertial, then they will have gotten close to the secret to the H-bomb. In the case of the apparently independent "reinventions," Kidder judged that none of them had yet found this, but Kidder considered that soon this would be a problem:

> The more difficult question relates to security, which of course is no problem if the work is done at a weapons laboratory. That is, if Sanders Associates or Hughes or General Motors should decide to go ahead and investigate laser heating of thermonuclear fuels with their own money, then what? The problem arises because the distribution of fusionable materials, including tritium, is not restricted; because many laboratories are experimenting with high-power lasers; and because everyone knows that things can get very hot at the focus of a high-power laser.[63]

The AEC's stopgap measure was to decide to classify any work with what were then considered extremely powerful lasers (with power output over 10 kJ, a thousand times more powerful than the largest lasers at the time), or any approaches that would definitely lead to lots of fusion reactions. It was hoped such guidelines would not interfere with laser research, but would nonetheless "deter private groups or individuals from performing research in areas of military interest."[64]

In the meantime, others across the nation and globe continued to work on both lasers and fusion, and sometimes both together. Over the course of the early 1960s, researchers in France, Israel, Japan, the Soviet Union, Spain, and West Germany all began laser fusion investigations. Kidder's fear that laser fusion would be hard to control because "everyone knows that things can get very hot at the focus of a high-power laser," was proving true.[65] But the most difficult case would be much closer to home.

Keith Brueckner was a physicist, but when people talked about him they tended to emphasize that he was also an "aggressive guy," a literal mountain climber.[66] In 1959, he had been hired to found the Physics Department at the University of California's new San Diego campus, but he was also a government scientist, working for Los Alamos, the AEC, the US Air Force, and the Department of Defense, where he applied high-powered lasers to military problems. For the AEC, he had worked on, among other things, magnetic confinement fusion, nuclear-powered rockets, and high-altitude nuclear weapons effects. He had been acquainted with the basic design principles of fission and fusion bombs since at least 1953.[67] He was, in other words, well known within the Cold War world of defense scientists.

Brueckner was also a member of the AEC Standing Committee on Controlled Thermonuclear Research (CTR). The CTR had been created in 1966 to explore the peaceful applications of fusion technology. The CTR Standing Committee was composed of "prominent US scientists" along with national laboratory heads, with the charter to evaluate the priorities of the American effort for peaceful fusion.[68] It was in this capacity that Brueckner was asked by the AEC in August 1968 to be

CTR's emissary to an International Atomic Energy Agency conference in Novosibirsk, USSR, on plasma physics.[69] The AEC wanted Brueckner to see what was on display regarding magnetic confinement fusion and to observe the work on laser heating of deuterium and tritium that would be presented by the Soviet Union, who had claimed to achieve fusion neutrons earlier that year. Brueckner was encouraged by what he saw and recommended that the AEC pursue research in this area immediately. Though the AEC expressed interest, CTR ultimately declined to fund the work, and its head wrote to Brueckner that the "micro-explosion" approach to fusion was "of questionable CTR interest."[70]

Brueckner seems to have had no knowledge that the AEC had been pursuing laser fusion at both Livermore and Los Alamos since the early 1960s, and they made no moves to tell him this. Brueckner inquired with the AEC to see whether they would fund his own research. They declined. Instead, he got funding for a classified, theoretical study on the laser heating of deuterium from the Department of Defense. Because his home institution did not handle classified contracts, Brueckner instead did the work under the auspices of a private company he did consulting for, KMS Industries of Ann Arbor, Michigan. KMS Industries was run by and named after its CEO, Keeve M. Siegel, an irrepressible entrepreneur who had formerly been a professor of electrical engineering at the University of Michigan before heading into the private sector. Siegel had run a host of private, physics-based companies during the 1960s and had invested millions of his own dollars in KMS, which by 1969 had over 3,000 employees and was making almost $60 million in annual net sales. Brueckner later called him a "fascinating man, full of ideas . . . very ambitious, very intelligent, but very speculative. A gambler."[71]

Brueckner started his work in April 1969, developing a one-dimensional computer model for how a laser would react with a small sphere filled with deuterium and tritium, and by the time the contract lapsed that August, he had come up with intriguing results. Brueckner convinced Siegel to finance further investigations out of KMS's private budget. By September, Brueckner had discovered that with relatively modest laser power, the fusion pellet would not only heat, but have its outer surface vaporized, causing a spherical implosion that would reach sufficient pressures at its core for fusion reactions to take place.[72]

In other words, Brueckner had independently discovered laser-driven inertial confinement fusion in 1969, and in both his and Siegel's eyes, this was an incredible science-based business opportunity.

Siegel and Brueckner decided they ought to approach the AEC with their results. They were put into contact with the AEC director of research, who told them that before disclosing any proprietary information to the AEC, they should first file a claim to any inventions with the US Patent Office. This was a fairly standard approach: the AEC did not want to be accused of stealing their ideas. Brueckner filed three patent applications for laser fusion, covering the basic concepts of imploding deuterium-tritium fuel pellets with lasers. On the same day, Siegel and the KMS board of directors announced that they were going to devote all the company's resources to the development of laser fusion for power purposes based on Brueckner's research.[73]

But Brueckner's patent applications were now making waves within the AEC. In November 1969, while Brueckner was attending a DOD meeting about the use of lasers as defensive weapons in West Palm Beach, Florida, the AEC assistant director of security hand-carried a letter to him. Brueckner was allowed to read it but not keep a copy, because it was classified. The letter said the under the authority granted to it by the Atomic Energy Act, the AEC demanded he "stop discussion and computing work" on laser-driven fusion, on the grounds that it was weapons work that Brueckner was not authorized to perform. "The AEC descended on me, and said, your work is classified, stop," Brueckner recalled. A few days later, he was informed that his patent applications were to be indefinitely considered secret by the US Patent Office, at request of the AEC.[74]

The AEC held that the technology involved was still highly classified because of its connection to thermonuclear weapons design, and because of its potential relevance to creating a pure-fusion weapon.[75] But there was an even trickier issue. As the AEC saw it, Brueckner had been exposed to the essential idea of the Teller-Ulam design while he had been a consultant working for the AEC. As a result, they felt that his insights surely were derivative of AEC secrets.[76] If this were the case, then the AEC would have whole or partial claim to any patents involved, and they felt he was trying to privatize and monetize the H-bomb. While the AEC was happy to work *with* private industry on nuclear technology, they found the idea of private profit off a "stolen" AEC concept—much less a classified one—highly distasteful.[77]

Over the next few years, the AEC and KMS would joust over the question of private laser fusion work. The main surprise for the AEC was that neither Siegel nor Brueckner could be cowed by threats. The AEC had run into difficulties with private industry getting interested in classified fields before, but in every case said industry had been willing to either accept AEC safeguards on their work or abandon the effort. In the laser fusion field, the most evocative example of this came from the Israeli-born engineer Moshe J. Lubin, who had been working at the University of Rochester on laser development since 1964. In 1970, Lubin and a colleague met with AEC representatives with the hope of getting official sanction for a series of laser tests he wanted to perform at Rochester related to fusion problems. Because Rochester didn't support classified contracts, he wanted to see whether there was a way to do the research openly. An AEC scientist told Lubin that the work would, for a number of reasons, definitely be considered classified under current AEC rules. He suggested an alternative: if Lubin modified his experiment so that the observational results were only "bulk measurements," then he would get results that would "have a large variety of explanations," which "seemingly would make it difficult to extract reliable values for individual parameters," then it could be considered unclassified. So the only way Lubin could do his work openly would be to modify the experiment to the point that he wouldn't be able to determine what was happening during the experiment. Lubin was understandably not enthusiastic, but complied.[78] This was the sort of experience the AEC was used to, at least from domestic scientists.

By contrast, when Brueckner was told to stop working on laser fusion, he responded by filing more patent applications—over a dozen. As he later wrote, "I felt that the AEC had been unnecessarily severe in stopping our work and [I] responded by making as wide a range of applications and claims as I could conceive."[79] These patent applications were not only deliberately antagonizing, they were also legal documents that the AEC would now need to treat seriously.

Brueckner and Siegel saw no reason to be cooperative. They truly believed that laser fusion had started with Brueckner's work. They didn't know that Livermore had been looking at the same idea for nearly a decade, or about the many other firms and scientists who had agreed to work with the AEC in preserving secrecy. All of that previous work was classified, and the secrecy surrounding the topic meant that someone like Brueckner could argue that his own invention was not only independent, but arguably had been first.

By late 1970, the AEC had become convinced that the best approach was to firmly discourage KMS from continuing their work. AEC representatives told Siegel that the scientific results were poor and not going to succeed; they explained that the KMS business plan was impossible; and finally, told KMS that even if they did somehow create a successful fusion reactor, the AEC would classify it as a nuclear weapon, and private ownership of nuclear weapons was banned by the Atomic Energy Act. Even if what they made couldn't explode, the AEC elaborated, "the fact remains that the concepts that KMS would need to exploit are essentially weapon concepts" and would probably never be eligible for civilian use. Should KMS still not be persuaded, the AEC would offer only extremely prohibitive contract terms, whereby KMS would pay for all security demanded by the AEC, would allow the AEC to inspect their work at any time, would have any patentable discoveries declared "in the public interest," and would give the AEC the right to use KMS's scientific and proprietary information for any purposes they deemed fit. KMS's staff would be given security clearances, but only if they had never previously had access to Restricted Data, and they would be authorized to work only on information created at KMS. At the same time, the AEC also internally proposed a press statement meant to indicate in no uncertain terms how technically infeasible the KMS approach was, and that they intended to contest any and all KMS patents on the technology, no doubt with the goal of making it difficult for KMS to secure further investment.[80]

But Siegel was not easily discouraged. That September, he wrote to the AEC chairman, Glenn Seaborg, that he thought the meeting had gone well. When the AEC technical experts had said that none of the KMS work was novel, Siegel chose to interpret them as saying that the work was thus "based on well-known and well-tested physics and that new concepts in physics are not required to achieve the results of our calculations." When the AEC said they thought that KMS's estimates for success were off by years, Siegel reported that both the KMS and

AEC estimates were within the same "order of magnitude." Siegel took the threat of the work being classified as a nuclear weapon in stride; he was confident that no one would mistake a laser fusion power plant for a hydrogen bomb. He was willing to accept classification controls and one-way information sharing with the AEC. Above all, the private effort needed to be supported, he argued, for the greater good of the nation: "We believe this to be particularly true at this time when the prognosis for adequate power for the nation in the '80s, with minimal impact on the environment, is becoming increasingly alarming."[81]

The AEC and KMS would eventually work out a compromise: an "unusual" (as the AEC put it) no-fund contract where KMS would agree to work under AEC security constraints without receiving AEC funding or data, but would also maintain private ownership over its ideas. That the AEC was willing to accept this, after all of its tough talk, is reflective of the ways in which Siegel was a deft political operator. To the AEC's annoyance, KMS actively promoted the contract; KMS was desperately seeking funding and attention, even if the subject of its research was still classified. The AEC felt in turn that it needed to make clear that it thought KMS was unlikely to succeed.[82]

But unbeknownst to Siegel and Brueckner, the AEC had been revisiting the classification status of laser fusion since early 1970. KMS was part of the motive for this, as were other researchers both inside the US (like Lubin) and outside (notably the USSR and France). There was a growing sense within the AEC's fusion community that the efforts to keep laser fusion under wraps were close to failing. The head of the AEC's Controlled Thermonuclear Research branch, Robert L. Hirsch, noted that concepts close to inertial confinement laser fusion were being discussed quite openly at a recent European conference on plasma physics. "Many of the key physics concepts which have resulted in the recent wave of optimism at LRL [Livermore], LASL [Los Alamos], and KMS are known to others and are being openly discussed," he wrote to another AEC colleague. "These people have not put all of the elements together in the proper order as yet, but they are nevertheless calculating attractively low input energy requirements. It is anyone's guess as to how soon this matter could crystallize but it could occur at any time."[83]

A new panel of experts had been assembled in late 1970 to reconsider classification guidelines, and concluded in early 1971 that the peaceful benefits of fusion might compel declassifications even though there were direct connections between the work of laser fusion and the design of thermonuclear weapons.[84] It was a controversial position. While some parts of the AEC, including its own Division of Research, General Advisory Committee, and representatives from Sandia National Laboratories, agreed, representatives from Livermore, Los Alamos, and the Department of Defense opposed them vigorously, arguing that the risks outweighed the benefits.[85] Carl Haussmann, who would in a year lead a new, revamped laser fusion effort at Livermore, went so far as to argue that a full understanding of the AEC's laser fusion classification policy would challenge the basic assumptions behind *all* nuclear secrecy policies: "Until the country is willing to relax its policy concerning dissemination of nuclear explosives technology and capability—a charge which clearly transcends the prerogatives of the AEC—very little unclassified latitude is permissible indeed!"[86] KMS was not a party to these discussions, though they were aware they were going on, and that information about KMS's progress had been part of the panel's considerations.

The consequences of these deliberations became clear in late 1971, when Nuckolls and another Teller protégé, Lowell Wood, presented a surprising paper on "Prospects for Unconventional Approaches to Controlled Fusion" at a meeting of the American Association for the Advancement of Science. In this paper, they not only made a case for laser fusion, but for the first time they made public the claim that Nuckolls had invented the laser field at Livermore in 1960. Not only did they talk about past work, they predicted that the Livermore program might see fusion "break-even" within three to five years.[87] The only reference to KMS was sidelong and mocking:

> Incidentally, that private capital is currently pouring into the unconventional CTR [Controlled Thermonuclear Reactions] field on the multimillion dollar scale can presumably be taken to indicate either that fools and their money may be parted much more readily today than in the recent past, or that the light of promise is burning relatively strong in these areas. Either conclusion is remarkable, but the latter seems more likely, a priori.[88]

Brueckner would hear about the talk. In a memo, he tried to make sense of the situation: "The importance of the information [in the 1971

Nuckolls-Wood paper] derives from its source in an AEC weapons laboratory. The release also indicates that [either] a major security violation has occurred or that the AEC is declassifying a highly sensitive area of work with pure-fusion weapon application."[89]

In May 1972, at the International Quantum Electronics Conference in Montreal, Nuckolls and Wood presented a new series of papers on laser fusion, going into the technical details of many basic concepts of laser fusion, the same ones developed by Brueckner, all of which had until then been highly classified.[90] By September, they had published a landmark paper in *Nature* with two other Livermore collaborators covering the direct-drive implosion of fusion targets and laser pulse shaping as a key element of achieving high compressions—the first major paper on laser fusion, and the most cited one to this day.[91] Though the paper did not indicate when this work at Livermore had been done, the fact it was the first published ended any possibility of Brueckner's work being accepted in "first-to-file" (i.e., most non-US) patent regimes, and presented an alternative claim to priority. Brueckner was shocked. It was hard for him not to see this as the AEC playing dirty: they were the gatekeepers of nuclear information, and they were being selective about who was allowed to speak.[92]

The archival record shows *how* the Nuckolls-Wood paper was declassified but is silent on *why*. The AEC's record shows it was aware that the release of the paper would "represent the first open presentation" of the laser fusion work, and that "such a proposed scheme could attract considerable press coverage."[93] Much in the field was still held back, to be sure: only direct-drive techniques against simple fusion pellets were declassified, without any discussions of the more complicated target designs, much less the indirect-drive concept (usually dubbed the *hohlraum* in a laser fusion context, or a radiation case in a weapons context) that had dominated Livermore work from the start.

Nuckolls insists he followed standard procedures when it came to requesting declassification of his paper, and that it was just a happy coincidence that he happened to ask for it just as the AEC was revising its guidelines.[94] But the timing does seem suspiciously convenient. The AEC clearly had it out for KMS, not only because they doubted their technical abilities, but also because they felt that Brueckner was trying to unjustly monetize the secret of the H-bomb, which they abhorred on institutional, legal, and ethical grounds.

The AEC would continue to revisit the classification of laser fusion repeatedly in the early 1970s, with the KMS work at the forefront of their mind. Eventually a more moderate policy would be adopted, one based on the "gradual erosion" of classification: a few specific concepts would be declassified up front, and further concepts would be declassified only when directly implied by the public literature.[95]

After the Nuckolls-Wood papers, Brueckner and Siegel were informed that they could now publicly talk about things that had been discussed at the Montreal conference, but that was it. Even if other AEC employees said something publicly, it could not be taken as a sign that the information was actually declassified; all KMS statements and publications of a technical nature had to be cleared by the AEC Classification Division.[96] In the meantime, KMS continued to push for a final decision on the patent question, enlisting congressmen to appeal on their behalf, arguing that if the AEC would not agree to a speedy resolution on the patents, it was incumbent upon them to start providing some of KMS's funding, since the hope for patents was their sole source of private capital, and the company was getting deeper into debt.[97]

KMS had one brief but important success after this point. In May 1974, KMS announced that they had "for the first time in the US obtained high energy neutrons unambiguously from a process of laser fusion."[98] This was a massive achievement. For all the millions the AEC had invested in its own laser fusion programs, it had not yet accomplished this technical goal, and it got KMS considerable press attention.[99] If the neutrons had been observed, it proved that fusion was taking place, although it was still a long way from generating as much energy from the fusion reactions as went into producing them ("breakeven").

Big claims demand big evidence, but KMS was slow in giving them. Siegel personally brushed away the public calls for more technical details, citing the fact that KMS Fusion had proprietary technical information to protect. "General Motors wouldn't give out more information than we have," he told *Laser Focus Magazine*.[100] The AEC experts were skeptical, but eventually KMS released enough data to confirm that indeed, neutrons had been produced, but they were still a long way away from a viable reactor.[101]

And indeed, by that point Brueckner had become convinced that "there was something badly missing in our theoretical work, in the computational work we'd done."[102] The problem was that symmetrically compressing a fusion pellet with a laser is easy only for a theorist. If one optimistically assumes a high level of symmetry, this reduces much of the requirement for laser power, but it had been known within the AEC for some time that symmetry was very difficult.[103] This was why the labs had spent so much of their time just trying to develop sufficiently powerful lasers and ways of circumventing the asymmetries that Brueckner had never really considered, and why they knew he would ultimately fail.[104] The indirect-drive approach, with the *hohlraum*, for example, had been a centerpiece of the Livermore program since the 1960s, but was apparently never pursued by Brueckner. Brueckner told me that he knew about radiation cases in the context of hydrogen bomb designs, but they never considered them for laser fusion because they didn't think they'd be necessary.[105] The requirements for laser power turned out to be orders of magnitude greater than Brueckner had calculated. There was no shortcut to fusion. Brueckner had been wrong all along. KMS was doomed.

Brueckner left KMS in 1974 and went back to his academic job. He never returned to the laser fusion field. Some of Brueckner's patents on laser fusion would eventually be declassified and granted, but not until the mid-1980s, and he didn't make any money on them. Back in Ann Arbor, KMS pressed on despite increasing uncertainty. The company's goals were receding: instead of making laser fusion power plants, they were focused on making fusion pellets for the AEC, and desperately looking for ways to make their work pay off even though they knew it would never achieve "break-even."[106] One KMS executive offered a grim assessment to a journalist: "Everything is a cliffhanger. We practically live from month to month."[107]

In March 1975, Siegel was testifying before an open session of the JCAE, asking for financial assistance from the government. In a dramatic turn of events, Siegel suffered a stroke while testifying, midsentence, and was rushed to the hospital. He was declared dead the next morning. For the newspapers, his death before Congress was the last of his sensational activities and a fitting literary ending for someone willing to bet it all.[108] KMS continued in his absence, but eventually became just another government contractor, performing ancillary tasks in support of the fusion programs of the national laboratories, a far cry from the dreams of Siegel and Brueckner in the summer of 1969.[109]

Peaceful fusion technology has never taken off in the way that peaceful fission technology has, though there are many who still believe it is a few decades away. Separate from the technical difficulties (which are substantial), the problems for drawing a firm line between the classified and open aspects of the field proved exceedingly difficult in the mid-to-late Cold War. Laser fusion in particular proved a serious challenge to the Cold War mindset, embodying both the extremes of utopian energy production and the proliferation of thermonuclear weapons designs.

The case of laser fusion demonstrated some of the deepest complexities of the period. It wasn't just that the technology was inherently dual-use, or that the secrecy created a context for acrimonious and high-stakes priority disputes. The complexity comes from the ways in which the categories of people no longer fit into the neat boxes that the Cold War secrecy mindset demanded. It wasn't just the AEC versus KMS; the weapons designers at Livermore, for example, had a different agenda than the AEC commissioners, and were a potent force in pushing for declassification, contrary to what one might expect given their jobs. And Brueckner in particular embodied the complex identity of the late Cold War defense scientist. Over the course of a long career in national security science, he had worn so many different hats (academic, government, industry) that he eluded any simple categorization.

The AEC attempted to leverage the power of secrecy to its advantage, and was caught off guard by the brash publicity of the private sector. In the end, the private sector effort was undone in part by the AEC's embrace of publicity: by declassifying Nuckolls' laser fusion work, the AEC was able to deal a hard blow to KMS's position. KMS at times used its own secrecy (proprietary trade secrets) to advance its own position. This muddling of tactics and roles is emblematic of the state of defense science in the late Cold War, and was not easily assimilated into the Cold War nuclear secrecy mindset, as some of the government practitioners seemed to realize: it was not just a case of a specific scientific field that was at stake, but potentially the entire approach to scientific secrecy.

The KMS episode came at a difficult moment for the AEC, when it was facing broad criticism for its paternalistic attitudes as a promoter and regulator of nuclear power. The commission would ultimately be dissolved and reorganized into, first, the Energy Research and Development Administration in 1975, and in 1977 the Department of Energy (DOE). The same forces that led to these massive changes are arguably the same ones that complicated the AEC's position during the KMS incident: the early postwar assumption that nuclear energy mandated unlimited power was no longer as rhetorically or legally convincing as it had once been, and nuclear governance was gradually becoming more "normalized" within the American political ecosystem.[110] By the late Cold War, even the AEC was hardly a monolith, with laboratory directors, scientists, and the central administrators disagreeing on the interlocking questions of secrecy and security and using the mechanisms of the system to their own institutional and personal ends.

7.3 ATOMS FOR TERROR

At the core of the Cold War approach to nuclear secrecy was a fundamental belief about the nature of who would misuse "dangerous" knowledge. Initially, the enemy was straightforwardly the USSR. By the 1960s, as we have seen, this expanded to include "Nth powers" and nascent nuclear states. Even this change was a dramatic one, because the sort of threat posed by the USSR was different than the one posed by, say, North Korea. The "Atoms for Peace" effort of the early 1950s had been enabled in part because of its focus on the Soviets, which allowed the US to judge it prudent to declassify "crude" technologies that the Soviets had already acquired, and to instead focus on the protection of new, "sophisticated" nuclear applications. While expanding the threat to less-developed nations changed this calculus a bit, it was still an approach that focused exclusively on nations.

In the late 1960s and early 1970s, however, a new threat was postulated: the non-state actor, or nuclear terrorist. The question of how seriously to take this threat was controversial then, and is still controversial today. But the rhetoric of domestic and international terrorism, which itself rose in the 1960s and 1970s, was found easily compatible with the existential risks of nuclear weapons. And, curiously, discussions of nuclear terrorism often focused on the futility of secrecy in a late Cold War age, identifying the "peaceful" policies of the early 1950s as the origin of this new fear.

The idea that information about nuclear weapons in the public domain could be mined by nefarious, non-state actors goes back as far as the Smyth Report and the early debates about the nature of the atomic "secret." "Any books on Atomic Power?," a pair of shifty characters (one in wingtips and a Zoot suit) ask a librarian, in a November 1945 *New York Times Book Review* cartoon.[111] Other examples of the "secret is out there" trope centered around similarly non-scientific actors, notably children, figuring out how to build a bomb. Perhaps the earliest example dated to January 1946, when the United Press syndicated the story of a nine-year-old child from Jackson, Mississippi, named Jimmy who had written a "25-page thesis on the atomic bomb which he wanted published 'for kids like me,'" and furthermore, "only nuclear physicists and Jimmy fully understood the second chapter of his work," which dealt with the construction and operation of the bomb.[112] Such stories were a curious byproduct of the early Scientists' Movement's insistence that there were no secrets to be kept—if a child could figure it out, who couldn't?

Further examples of such tropes continued into the 1950s and 1960s as part of the discussions around the "Nth power" problem. In satirist Tom Lehrer's song, "Who's Next?" (1965), Alabama joins the nuclear club; in a *New Yorker* cartoon from 1963, a pair of worried parents, watching their demented-looking "Junior" create a mushroom cloud with his chemistry set, remark: "I certainly hope we have controls before *he* gets the bomb!" A *Saturday Evening Post* short story from 1967 explored what happened when "Albie Watkins," a 42-year-old schlub, builds a hydrogen bomb in his basement.[113] Such notions were not confined to the popular press. From 1964 through 1966, the Livermore weapons laboratory ran an "experiment" to see whether several physics postdocs "unfamiliar with nuclear weapons and with access only to the unclassified technology, could produce a credible [nuclear] weapon design." The study's conclusions are hard to parse, as the declassified version is heavily redacted, but the gist is that while their design would not have met the standards of a nuclear weapons state, it might have been able to yield several kilotons of explosive energy. Even in this study, the official interest was not about terrorism but proliferation: it was dubbed the "Nth Country Experiment," and the students were modeled on scientists working for a state.[114]

There were also discussions through the 1940s and 1950s about the threat of a "smuggled" atomic bomb, although it was always assumed that the bomb itself would still be created by a state actor (the USSR) and the smuggling was conceived of as just an unconventional delivery mechanism.[115] But a realistic threat of nuclear terrorism—that of a non-state actor acquiring and using a nuclear weapon—was an idea that only later arose from these tropes, along with a growing appreciation for terrorism as a form of violent activity.

The origin of nearly all discussions of nuclear terrorism from the 1960s onward was the weapons designer Theodore B. Taylor. Taylor had worked at Los Alamos from 1948 until 1956 and had become one of the primary designers of fission weapons. In 1965, he began to worry about the "easiness" of designing a fission weapon. How hard would it be, he wondered, to design a fission weapon that would detonate with at least a tenth of a kiloton of explosive yield? Such a weapon would still have a devastating effect on densely populated areas. Taylor concluded that it would not be very difficult for someone (or some small group) to accomplish such a thing. The only guarantee against it was that fissile material still required massive investment in the form of uranium enrichment or reactor development. But what would happen if someone *stole* fissile material, or otherwise acquired it clandestinely? Looking at the state of the American civilian nuclear industry at the end of the 1960s, Taylor concluded it would not be hard at all for a dedicated and well-organized terrorist group to pull off such a theft.[116]

Starting in 1969, Taylor began to publicize his fear, and American media found the idea titillating. That year, Taylor was the subject of an article in *Esquire* provocatively titled, "Please Don't Steal the Atomic Bomb." There he argued that "the whole job [of making an atomic bomb] could be done in someone's basement," so long as they stole the fissile material first.[117] In December 1973 he was the subject of a series of *New Yorker* articles by John McPhee, which were then compiled and republished as *The Curve of Binding Energy*.[118] McPhee's book was widely reviewed and praised. Reviewers for the *Wall Street Journal*, the *New York Times*, and the *Chicago Tribune* were shocked and intrigued by the "easiness" of bomb building.[119] The *Washington Post* reviewer, however, questioned whether Taylor's book did more harm than good:

> You must assume that Taylor and McPhee decided that the best way of alerting people to the danger was to put this kind of detail in writing, but still . . . any hoodlum or political loonie [*sic*] whose special madness

is nuclear blackmail could make unhappily good use of the book. Let's hope none of them do.[120]

Even today, discussions of nuclear terrorism are fraught with an uneasiness about whether talking openly about the topic might turn it into a self-fulfilling prophecy. But Taylor felt that the AEC was insufficiently disturbed by the possibility, that it was treating the security of fissile material too laxly, despite his attempts to raise the issue internally. How could this issue be rectified without public pressure? And if there really were no significant "secrets" left to be kept about crude weapons designs, then who would the silence be fooling? Presumably terrorists are not reliant on *Esquire* and the *New Yorker* for their ambition.

Taylor wrote a book of his own on the topic, coauthored with a legal scholar, Mason Willrich. Their 1974 work, *Nuclear Theft: Risks and Safeguards*, was meant for expert use, but written with a simplicity and straightforwardness that ensured it would be easily understood by people without technical training. It attempted to assess with cool reason whether an "illicit bomb maker" could divert American fissile material and use it to kill tens of thousands of people. Unsurprisingly, they found it plausible. They predicted that if the American civilian nuclear power industry expanded as expected, the amount of loose or unaccountable fissile material would expand by an order of magnitude, especially if plutonium reprocessing (which would chemically separate plutonium from spent fuel from civilian reactors) expanded.[121]

Core to Taylor's argument was that things had changed from the 1940s to the 1970s. At one point, the atomic bomb required the geniuses of Los Alamos and the might of the entire Manhattan Project to make only a handful of weapons. But that was because they were doing it for the first time, and there was so much uncertainty involved, and fissile material was so scarce.

By the 1970s, Taylor argued, there were two major shifts. The first was that "Atoms for Peace" and other declassification efforts had made crude weapons designable from the public domain alone. In its obsessive focus on "sophisticated" weapons like the hydrogen bomb, the US had mistakenly declassified information of relevance for designing "crude" weapons, like the bomb that destroyed Hiroshima.

The second was the development of the civilian nuclear power industry, which resulted in the creation of many metric tons of spent nuclear fuel, as well as the circulation of highly enriched uranium for research reactors. There were now tons of separated plutonium and enriched uranium in the world, often in non-military facilities, and only kilogram quantities were necessary to make a bomb. Taylor worried that workers in these facilities globally ("insider threats") might be willing and able to smuggle out small amounts of them without being noticed, and sell them on the black market. He also feared that a cunning terrorist group could find ways to intercept shipments of fissile materials between sites. Scariest of all, because the amount of fissile material needed for a bomb is so small relative to the total fissile material in existence, it might be impossible to detect the missing material until it was too late.

Taylor anticipated that other weapons designers would be skeptical of the idea that an atomic bomb was now "easy" to build. Taylor told such critics to ask themselves: "What is the easiest way I can think of to make a fission bomb, given enough fission explosive material to assemble more than one normal density critical mass?" Or, as he put it to McPhee: "Lay off any sophistication altogether. Try and see what is the simpleminded way to make something that could knock over the World Trade Center."[122] He reminded critics that while a nuclear state would want a highly reliable design that could be produced in large numbers and "mated" to a missile or bomber, a terrorist might be happy with an unreliable, one-off weapon that fit into the back of a van.

J. Carson Mark, another veteran Los Alamos weapons designer, made a strong counterargument against Taylor at a congressional hearing a few years later. He agreed with Taylor that basic, workable weapons designs were easy to come up with, just based on readings of the public literature. The issue, though, was actually making the thing:

> The problems about the design are in realizing the design, in having a configuration, or in having an actual apparatus which will do what you have specified it is supposed to do. Here one runs into a tremendously wide range of complexities and possible difficulties about which details are not written down.[123]

Even if one had step-by-step instructions for, say, reducing plutonium oxide (the most common form used in reactors) to plutonium metal (which is necessary for bombs), the difficulties of doing so were non-trivial, even for the experienced plutonium chemist, and the chances of ruining the plutonium for bomb-purposes, or in accidentally killing oneself, or somehow getting caught in the act, were quite high.

But the idea of homemade atomic bombs resonated with preexisting fears about the dangers of knowledge and the hopelessness of the modern condition. "Atoms for Peace" had become "Atoms for Terror": the "civilian" atom, in this view, was just more rope by which to hang ourselves.[124] Aiding this view was the coincidental detonation of a "peaceful" nuclear bomb by India in May 1974, which neatly merged the threat of a "poor" nation acquiring the bomb with the hypocrisy of labeling anything related to nuclear weapons as "peaceful."[125] Combining popular tropes about the accessibility of dangerous knowledge with the argument that the Cold War nuclear regime had created a novel problem through its own "peaceful" actions made for a potent popular message.

Within the US government, the threat of specifically nuclear terrorism did not really emerge as a force in policy until the late 1960s and early 1970s. This was a direct byproduct of awareness of domestic and international terrorism as significant forces more generally. Within the United States, acts of domestic terrorism, perpetrated primarily by extreme political groups on both the right and the left, had been rising since the 1960s. Bombings in particular had been on the rise; according to the FBI, there were 110 actual (non-hoax, successful) bombings for a political purpose on US soil in 1971 alone.[126] It was not until the hostage-taking and killing at the Munich Olympics in 1972, however, that US domestic policy made terrorism a major concern. In September 1972, President Nixon ordered the creation of a Cabinet Committee to Combat Terrorism in order to coordinate the efforts of all departments and agencies of the US government "to be fully responsive to the efforts of the Secretary of State and assist him in every way in his efforts to coordinate Government-wide actions against terrorism."[127]

As a cabinet-level agency, the AEC participated closely in this work, using it as an opportunity to review its physical controls on nuclear materials and to change its declassification criteria significantly. From 1972 onward, when evaluating the hazards of declassifying nuclear information, the AEC would take into account not merely whether it would aid a foreign power, but whether it would aid terrorists.[128] In the eyes of Taylor and those who agreed with him, however, secrecy had already ceased to exist as a buffer against the threat of home-grown atomic weapons, and the AEC itself considered "reclassification"—the re-designating of declassified information as "secret"—entirely fruitless, if not illegal.[129]

The AEC had been discussing safeguards internally since 1965, when a large mass of uranium—some enriched to bomb-grade levels—went unaccounted for at a reprocessing plant in Apollo, Pennsylvania. Though the AEC was satisfied, after investigation, that the missing uranium had not been diverted, they could offer no absolute proof. (Their theory was essentially that the uranium was "lost" as part of inevitable inefficiencies in the chemistry of reprocessing. Even if less than 1% of any amount of material is lost in every operation, for a large plant that adds up very quickly to many kilograms of material.) There were suspicions, notably by CIA officials, that the uranium had been stolen by another aspirational nuclear power: Israel. If a small country could steal highly enriched uranium from a US civilian plant, that wasn't a good sign.[130]

They had also been taken aback by an incident in October 1970, when the Police Department in the city of Orlando, Florida, received a letter saying that the writer had possession of "a nuclear fusion device, more commonly called a hydrogen bomb," and demanding $1 million dollars in small bills and safe passage out of the United States. "This is no bluff," the letter claimed. "If you think it is ask the Atomic Energy Commission what happened to the shipments of U235 that never got to their destinations." The next day, a further letter arrived, written in a careful, cursive script, threatening that "Orlando will be in ruins" if the demands were not soon met.

As proof of the authenticity of the threat, the letter-writer included a drawing of the alleged hydrogen bomb and noted that the police would probably require "an expert in nuclear weapons to tell you if it is genuine, but believe me, it is." The police contacted the FBI, who in turn contacted the AEC to see if any of their bomb-grade material had been reported stolen or lost. But the AEC's record-keeping, even for bomb-grade fissile material, could not be searched quickly enough to ascertain whether the threat was credible. In the meantime, the FBI had taken the drawing to an expert at the McCoy Air Force Base, who indicated that the proposed design "would probably work."

Four days later, after staking out the "drop site" for the money, the police apprehended the perpetrator: a 14-year-old boy who was interested in science and had cobbled together his "bomb" drawing from public sources and sent the letters off as a hoax. Had the police not identified him when they did, the city was preparing to pay the ransom.[131] A senator, hearing the story a year later, remarked that the student "got closer to the million dollars than he did to the actual bomb," but then quickly admitted that "he did rather well in both." The fact that the AEC could not easily dispute either the perpetrator's ability to make such a weapon or the claims to lost fissile material was understood as a severe problem.[132]

Over the course of the 1970s, the AEC adopted new measures to better secure and account for its fissile materials, but these were measured against the cost to the nuclear power industry. This meant "safeguards" including physical security (fences, bunkers, guards), better procedures for transporting fissile material (reducing the possibility of interception between facilities), and better inventory tracking capabilities (including the ability to better recognize when materials were unaccounted for). There was, however, no way for the AEC to assess the effectiveness of its security measures other than demonstrated failure. The consequence of a worst-case scenario (a terrorist nuclear attack) was so unacceptably high in social terms that it precluded any effective resolution of the issue: the government could never do "enough," as historian J. Samuel Walker has pointed out.[133] By the end of the decade, President Carter had indefinitely suspended reprocessing of nuclear waste, in part because of the security problems in enlarging the volume of separated plutonium in civilian hands.[134]

Taylor's gospel of "safeguards" reached a wide audience through McPhee's bestseller. Attention to the safeguards issue waxed and waned

through the decade, but Taylor doggedly continued to press his point whenever possible, tying his agenda to the large nuclear debate of the day. He was aided by a recurring phenomenon: college students drawing nuclear weapons to prove his thesis, an update to the old "children making the atomic bomb" trope.

The first such case was a television special produced by NOVA that attempted to prove Taylor's thesis by hiring a 20-year-old chemistry major at a "famous university in Boston" (the Massachusetts Institute of Technology) to design a "crude fission explosive." The student, who was kept anonymous (he apparently feared kidnapping), wrote up his report for a plutonium implosion weapon in five weeks. The basic design was shown onscreen, though all numbers were redacted. The actor playing the student even made a theatrical drawing of the bomb for the camera, a set of concentric circles: "Really, it's that simple," he explained. An expert from the Swedish Ministry of Defense reviewed the design, and concluded that while there was a fair chance it wouldn't go off at all, there was also a fair chance that it might have a yield of around 100 tons of TNT equivalent, which, though small by nuclear standards, would still kill thousands of people. "The Plutonium Connection," which aired in March 1975, simultaneously flirted with the argument that information should be reclassified and kept even more secret and the idea that nuclear weapons information had long since escaped into the public domain. An insert included with the VHS tape provided a provocative assignment for teachers to use in their classes: "Using the materials in your own school library, try to replicate the design for an atomic bomb."[135]

But the most famous episode of this sort would come in 1976, when John Aristotle Phillips, a Princeton junior who had taken a seminar on arms control, read *The Curve of Binding Energy* for class. In Phillips' later recollection, the book provoked intense discussion about whether atomic bombs truly were easy to design. Phillips decided to test Taylor's contention by designing an atomic bomb for his junior independent project in the Physics Department:

> Suppose an average—or below-average in my case—physics student at a university could design a workable atomic bomb on paper. That would prove the point dramatically and show the federal government that

stronger safeguards have to be placed on the manufacturing and use of plutonium. In short, if I could design a bomb, almost any intelligent person could.[136]

Phillips asked the physicist Freeman Dyson to be his supervisor for the project. Dyson, a colleague and friend of Taylor's who had been exposed to nuclear weapons and reactor designs as part of his long career as an advisor to the US government, agreed, with the stipulation that he would give Phillips only basic, unclassified information.[137]

The result was a 37-page report titled "The Fundamentals of Atomic Bomb Design." Phillips combed through the public literature on the bomb, and combined it with a search of declassified documents at the National Technical Information Service in Washington, DC, including the *Los Alamos Primer*, a 1943 treatise on bomb design from the first meetings at Los Alamos.[138] Toward the end of the project, Phillips even called up an explosives expert at the DuPont Corporation who shared details with Phillips about the high explosives used in modern implosion weapons. The paper concludes with the assertion that the implosion design he developed would have an explosive power about half the power of the Hiroshima bomb. Dyson gave Phillips an "A" on his paper, and also quietly had it removed it from circulation.[139]

Phillips did not look for public attention. Publicity came instead through a reporter for the *Trenton Times* who had talked to another student in the course.[140] Taylor, who was now working at Princeton, advised Phillips that "going public" with his work would entail a loss of privacy but it would also be extremely important for the cause of safeguards, especially in light of an impending sale of a reactor by France to Pakistan.[141] Phillip's story soon spread to other major newspapers; there was excitement over the idea of a Princeton student designing a bomb, and a gleeful willingness to conflate "design" with "build." Stories about Phillips' work and his political motivations ran from October 1976 onward in a number of US and world periodicals, notably the *New York Times* and the *Los Angeles Times*.[142] The story resurged in February 1977, when Phillips alleged (and the FBI corroborated) that someone from the Pakistani embassy had contacted Phillips seeking a copy of his bomb design. This bizarre and fantastical twist led to more news stories, features on the evening news, a book deal, and even an attempt to produce a made-for-television movie, starring Phillips as himself.[143]

In all of this, everyone, including Phillips, assumed that the bomb design would have worked. This, however, was not clearly the case. As Dyson later explained:

> [Phillips] had mastered quickly and competently the principles of shock-wave dynamics. But his sketch of the bomb was far too sketchy for the question "Would it actually explode?" to have any meaning. To me the impressive and frightening part of his paper was the first part [in which he described how he got the information]. The fact that a twenty-year-old kid could collect such information so quickly and with so little effort gave me the shivers.[144]

The hardest part of *any* of these claims is knowing whether the design would actually work. Anyone can draw several concentric circles and proclaim it to be an implosion bomb. But knowing whether any given design would work as predicted is difficult; this is why nuclear testing was done. This is not to say that designing a weapon "from scratch" is impossible, but it is *very* difficult to look at any weapon design "from scratch" and be confident of its yield, at least without extensive experience in weapons design, testing, and simulation.[145]

Next in the line of students-drawing-bombs was Dmitri A. Rotow, an economics major at Harvard. Rotow had seen the publicity surrounding Phillips and saw bomb-drawing "as a means of gaining credibility and, say, future funding or what have you for studying public policy issues." He also thought that "it would make good fodder for magazine articles, possibly even a book."[146] Rotow would claim to be similarly motivated by the safeguards issue and Taylor's thesis. Starting in late 1977, he researched and wrote up eight chapters of a book on fission weapon design, presenting over two dozen different design variations. By the end of March 1978, he had brought this to the attention of the Department of Energy (DOE), the AEC's successor agency, which confiscated all copies and declared them classified. For his trouble, Rotow got to be a star witness at a hearing on safeguards convened by Senator John Glenn and got to hear his work publicly evaluated by none other than Ted Taylor. On Rotow's work, Taylor was laudatory:

> Mr. Rotow's manuscript is the most extensive and detailed exposition of things to think about and how to think about them in the design of nu-

> clear weapons, nuclear fission weapons, that I have seen outside of classified literature. . . . All in all, however, I was neither shocked nor surprised that an intelligent and innovative person, without extensive training in nuclear physics, could produce such a document, thought I must say I was surprised it took so little time.[147]

Senator Glenn himself clearly saw the import of Rotow's work and the impossibility of using secrecy to stem a terrorist threat. Further, it emerged that in 1977 alone, the DOE received four submissions from private researchers "of the nature comparable to the Rotow document," out of an even larger number of submissions that "reflected a less sophisticated approach by the author."[148]

Encouraged by further witnesses, including a representative from the DOE, Glenn put the blame solely at the feet of the Eisenhower era's attempts to tame the atom.[149] The declassification efforts of the 1950s were now being blamed for giving terrorists the means to kill thousands of Americans in a single moment. The supreme irony of "peaceful" atoms potentially transmuting into "terrorist" atoms was exactly what made the act of "amateur bomb design" so attractive at a time when the anti–nuclear power movement was gaining in mainstream appeal and nuclear expertise was being equated with out-of-touch paternalism.[150]

Designing homemade atomic bombs had become an easy way to make a statement, to claim possession of a secret power. What started as a way to support Taylor's calls for safeguards had begun to morph into its own separate phenomenon, rooted in a consistently blurry distinction between "designing," "making," and "possessing." The theme of college kids not only designing, but also building, nuclear weapons was even the inspiration for a widely reviewed novel published just months before the Phillips case, in which Princeton undergrads got involved in a harebrained scheme to extort the US government into helping them build more bombs for poor African nations who would use them to extort aid from wealthy Arab nations. "Taylor was practically begging someone to make a bomb," one of the characters says to another, after giving him a copy of *The Curve of Binding Energy*.[151]

Others also made the connection between amateur bomb designs and social movements. In the summer of 1978, a "radical" women's liberation underground newspaper, *Majority Report*, printed an elaborate diagram for "How to Build Your Atomic Bomb And Strike a Balance of Power with the Patriarchy." The article was largely tongue-in-cheek, but it took pains to present apparently precise technical details about the weapon's construction. The unnamed author sought to defuse some of the banal technicality and morbid engineering surrounding such a "how to" with a parody of female domesticity (explosive lenses were molded with a copy of *The Joy of Cooking*). But the article still took seriously the connection between this kind of technical knowledge and claims to political power.[152] What had started as a means to advertise that there were *no* secrets (only material safeguards), quickly morphed into a way for those with anti-establishment political views to show that *they* had the secrets.

Taylor's call for increased safeguards as a method for controlling the bomb was predicated on the "loss" of secrecy. Despite this, the most frequent response to the demonstration of a lack of secrecy, especially in newspaper reviews and Senate testimony, was a desire for *more* secrecy. When Phillips and Rotow showed the world that a college student could design a crude atomic bomb, their designs were immediately put under wraps, something both students went along with, perhaps because it fed into their argument that their work was dangerous and valid. Stories about the ease of bomb design in mainstream media often contained disclaimers explaining that "key elements" had been left out, lest anyone get too worried.[153] In a sense, then, the message that there were no longer any "secrets" was a hard sell. The students were spoken of as if they had the expert knowledge of bomb designers, and, in the case of Phillips, were even sought after by aspiring nuclear nations. Yet this narrative contradicts their own theses that anyone could design such bombs in a matter of weeks and with access to public libraries. The entire point of the exercise was to show that bombs could be designed by people who were not privy to expert knowledge, but ironically this frequently reinforced the secrecy mystique.[154]

Over the course of the 1960s and 1970s, the Cold War secrecy regime began to show its strains. The bipolar assumptions inherent to its way of thinking—that information and people could be divided into neat categories of safety and danger, and that the proper administration of this divide would lead to the twin goals of technological superiority and security—became increasingly questioned.

The case of gas centrifuges showed the ways in which these discursive contradictions could translate into dangerous results. Uranium enrichment technology was becoming far more diffuse than anyone had predicted, and even getting American allies to comply with secrecy restrictions became increasingly difficult. Ultimately, the US acquiesced in the pursuit of civilian uranium enrichment by European allies, with the formation of Urenco, the tripartite enrichment consortium by the UK, West Germany, and the Netherlands.

In 1972, a Pakistani-born, Dutch-educated metallurgist, Abdul Qadeer (A. Q.) Khan, began working at Physical Dynamic Research Laboratory, a subcontractor of the Dutch partner in the Urenco consortium. Despite not being cleared to do so, he visited the ultracentrifuge facility in Almelo many times, with the full awareness and consent of his employers. In 1974, a few months after India's first atomic test, Khan was given the job of translating German centrifuge designs into Dutch. For two weeks he had totally unsupervised access to the material. Though the Dutch became suspicious enough of his behavior that they transferred him into other work by 1975, it was too late. In December, he left the Netherlands for Pakistan, carrying with him blueprints and contact information for companies that manufactured centrifuge components. After helping with the Pakistani atomic bomb program (they likely became nuclear-capable in 1986 but did not test a weapon until 1998), he also, possibly with the knowledge of the Pakistani government, created an international network of black market centrifuge suppliers and facilitated the spreading of the enrichment technology to Iran, Libya, and North Korea, among possibly several others.[155]

Laser fusion initially looked like a happier story. There is no evidence that any nation has used laser fusion research to develop a thermonuclear capability where one did not already exist, though the absence of evidence is not evidence of absence. One pioneer of laser fusion told me, confidentially, that he hoped laser fusion would ultimately lead to a new form of energy for humankind—because if it didn't, he would regret that so much information had been released on these subjects.

Today, hopes for laser fusion providing a way toward peaceful energy production are not high. Despite large investments by the US and France in the technology, including the construction of several massive facilities at the cost of many billions of dollars, the scientific "break-even" point has remained elusive. The American and French facilities, instead, are primarily used for research to maintain thermonuclear weapon capability in a world where nuclear testing has become taboo.

And the threat of nuclear terrorism contained within it a new sense of how control could be obtained: a control based not on secrecy (which it argued had been compromised already, by "Atoms for Peace"), but on materiality (fissile material safeguards). This callback to Oppenheimer's gambits of the late 1940s makes for an interesting repetition, focused this time not on nonproliferation, but on non-state actors. The performative quality of the amateur "bomb designers" proved to be extremely successful in the context of American media in part because it seemed to attack so directly the discursive assumptions of the Cold War. But beyond its public relations aspect, the fear of low-tech nuclear threats itself began to rearrange assumptions within the US government as well: if the enemy was not merely states, then entire categories of "safe" information could be suddenly rendered "dangerous."

The three issues embodied by the cases in this chapter—low-tech state-level nuclear proliferation, high-tech "erosion" of thermonuclear knowledge, and the fears that "anyone" could cobble together nuclear secrets—would culminate in a spectacular revelation at the end of the decade, as the "one remaining secret," the Teller-Ulam design, escaped under circumstances that got at the very core of the ideological and legal foundation of the Cold War secrecy regime. Every aspect would be simultaneously challenged: the discursive framework that invested such secrecy with intonations of death; the practices that could be manifested to control such information; and the institutions that positioned themselves as the ultimate arbiters of dangerous knowledge.

8

SECRET SEEKING

ANTI-SECRECY AT THE END OF THE COLD WAR, 1978–1991

I used to think that secrecy is incompatible with freedom of the press. But now it seems our press thrives on secrecy.
EDWARD TELLER, 1971[1]

By the end of the 1970s, the politics of secrecy had undergone a transformation in the United States. This had less to do with the atomic bomb than other factors: the disillusion of the Vietnam War, the official lying revealed by the Pentagon Papers, Watergate and the resignation of Richard Nixon, and a host of other revelations about prior and current misdeeds committed largely in the name of national security had created a cynical view of official secrecy with a large segment of the American population.

"Anti-secrecy" is what I call this emerging public discourse. Its origins have already been described, in a way: it came out of the concerns and push-backs that emerged alongside the different secrecy regimes, whether it was the scientific idealism of the 1940s, the push for "candor" and anti-McCarthyism in the 1950s, and the attempts to emphasize materiality over information, first in the 1940s with Acheson-Lilienthal, and later in the 1960s with fissile material safeguards. But it was only in the 1970s that these critiques became synthesized into a coherent political worldview that began to motivate both individual actors and competing institutions (like the press) to directly challenge not only the *ideas* of the secrecy regime, but its practices and institutions.[2]

This "anti-secrecy" should be distinguished from the politics of "transparency" or "openness." Both transparency and openness imply a mode of doing normal business, explicitly repudiating secrecy by embracing its opposite. Anti-secrecy, by contrast, is a deliberately antagonistic, oppositional stance: it is about tearing down the existing regimes of secrecy. In political terms, we might think of anti-secrecy as the revolution necessary before contemplating what the new order might be. And anti-secrecy is not exactly the same thing as secrecy *reform*. Reform implies fixing a problem in the system, whereas the anti-secrecy of the 1970s and beyond is about a rejection of the system almost as a whole. In political terms, anti-secrecy is radical where secrecy reform is liberal. They can coexist, especially expediently, but a true anti-secrecy proponent ultimately thinks reform is not enough.

In the wake of the Pentagon Papers and Watergate scandals, the government did attempt secrecy reform, including in the area of nuclear weapons. The Pentagon Papers case spurred the AEC into undertaking a massive declassification program.[3] A five-year "Declassification Drive," running from 1971 through 1976, reflected a new swing of the "classification pendulum," as one of the agency's official press releases put it, arguing that information "should be declassified unless a strong justification is demonstrated to keep it classified." Millions of documents were rapidly declassified. The AEC proudly announced its actions in a series of press releases, even distributing glossy photographs of documents being performatively declassified.[4]

The program was further encouraged by President Nixon's Executive Order 11652 (March 1972), which revised federal classification guidelines in ways that would have long-term impacts, as advised by an interagency study headed by William H. Rehnquist. Documents were now given a declassification "schedule," meaning they were to be reviewed for release after a set period of time, and officials were required to "portion-mark" documents, indicating within them the varying levels of classification afforded to each paragraph. And for the first time, the use of secrecy to conceal errors or avoid embarrassment was explicitly disallowed. This, coupled with revisions to the Freedom of Information Act (which had been passed in the 1960s after a long fight), suggested that the government was creating conditions that would not lead to overclassification or misuse of secrecy practices.[5]

But these approaches would not only fail to satisfy the critics of secrecy; they would also provide ammunition for a new form of anti-

FIGURE 8.1. Charles L. Marshall, Director of Classification, declassifying a document as part of the AEC's 1971–1976 "declassification drive." Source: AEC Press Release, "AEC in Midst of Declassification Drive," (n.d., ca. 1973), NTA, document NV0148015.

secrecy practice, what I call "secret seeking," in which the deliberate exposure of self-described secrets became a form of political action. When this activity moved into the nuclear realm at the end of the 1970s, it proved a potent form of attack on the Cold War secrecy regime. Discursively, it tried to show the failures of said regime to correctly identify and protect secrets. From the perspective of practice, it used new methods (including the Freedom of Information Act) to pry out supposedly well-kept "dangerous" information from government monopoly. And it utilized the institution of the free press as a potent weapon against the institutions of legal secrecy. While not successful in dismantling the

Cold War secrecy regime, the anti-secrecy attacks did manage to show the cracks in its façade.

8.1 DRAWING THE H-BOMB

Criticisms of government secrecy and of the infrastructure of government nuclear expertise had existed since the 1940s. Yet, it was only in the 1970s that the challenges to nuclear expertise from outside of the nuclear establishment and nuclear industry took root and were to various degrees abetted by a mainstream press that was willing to make its own challenges to the institutions of secrecy.

Some of this was simply a confluence of several independent factors, one of which was the Pentagon Papers case. In 1971, a former RAND Corporation employee, Daniel Ellsberg, deliberately leaked an extensive classified history of early US involvement in the Vietnam War to the *New York Times*. The Pentagon Papers, as the document became known as, contained numerous revelations about the origins of the war and revealed that the Johnson administration in particular had systematically lied to both Congress and the public about the justifications for the conflict's escalation. The Nixon administration sought an injunction against the *Times*, arguing that to publish the document would cause grave and immediate harm to national security. The case went to the Supreme Court, which ruled in the *Times*' favor. The documents were of such import to deliberations about the war, and the oppositions to publishing it were so vague, the Court held, that in this case the traditional classification regime could not trump First Amendment protections. The Department of Justice responded by filing criminal charges against Ellsberg for leaking the documents, but these were dropped when it came out in court that the government had severely mismanaged the case against Ellsberg (such as breaking into the offices of his psychiatrist without a warrant).[6]

The public repercussions of the case were broad: the US government was widely seen as using classification to cover up its own embarrassments in order to advance the cause of an unpopular war, now commonly seen as unjustified from the very beginning. Interestingly, while nuclear weapons did not play a role in the Pentagon Papers discussions, Ellsberg's earlier career at RAND had been related to problems of nuclear war planning. By his own account, Ellsberg considered the Pentagon Papers his *first* leak, and was planning to make an even larger, more portentous leak of material about nuclear weapons policy, but a bizarre circumstance of events led to his nuclear war files being lost, and they were never leaked.[7]

The Pentagon Papers scandal was followed in 1972 by the Nixon administration's break-in to the Watergate hotel, its attempted cover up, and Nixon's eventual resignation.[8] These events set a new tone for relations between the media and the government: investigative journalists were now dashing heroes defending the First Amendment, whereas government officials were duplicitous villains attempting to use "national security" claims to cover up their own mistakes and misdeeds. It also marked the creation of a new breed of activist: those who saw government secrecy as an evil to be challenged and uprooted.[9]

None of these public cases had any direct connections to nuclear weapons. Indeed, even in this environment, the US Supreme Court had no great difficulty ruling that Congress could be excluded from sensitive nuclear weapons information under certain circumstances.[10] But there were confluences between these changing attitudes about the roles of the government and the press and the declining respect for the Atomic Energy Commission. Much of the "mystery" of nuclear technology had dissolved during the 1960s, ironically aided by the AEC's success in "taming" the atom. As early as 1963, even David Lilienthal, who had once seen atomic energy as a turning point for civilization, was arguing that perhaps the AEC should be abolished: "The reality is the Atom has not justified the separate and unique status which Congress understandably assigned it in 1946."[11]

By the 1970s, attacks against the AEC had led to a significant reform. In 1974, it was split into two agencies, the Nuclear Regulatory Commission, which took over all civilian nuclear power oversight functions, and the Energy Research and Development Administration (ERDA), which took over all other functions, including weapons development. In 1977, ERDA was itself abolished and its functions absorbed into the newly created cabinet-level Department of Energy (DOE). If the AEC had grown less "special" as its organizational procedures calcified and nuclear technology had become less unusual, the DOE was the apotheosis of these processes: a vast bureaucracy that had little ideological force motivating it other than its own continued existence and such a wide docket of pursuits under the heading of "energy" that over time the public would frequently forget its historical origins in the atomic bomb.

Ted Taylor's safeguards advocacy in the late 1960s and early 1970s had pushed the notion that there were no essential secrets not already in the public domain that would prevent a skilled amateur from constructing a crude nuclear fission weapon. But that did not mean that there were no weapons design secrets not yet known. The specifics of the Teller-Ulam design for the hydrogen bomb were, despite many near disclosures, still essentially under wraps. There were, to be sure, many attempts to guess at how the bomb might work in unclassified literature. As far back as the H-bomb debate, authors had speculated as to what such a weapon would look like. One might imagine it was a fission bomb wrapped in fusion fuel as an artist for *Time* magazine did in 1950. Or one might imagine that a large mass of fusion material was simply bolted onto the side of a fission bomb, as an artist for the *Illustrated London News* conceived of that same year.[12] Neither of these drawings appear to be based on any secret knowledge, yet they do both embody the basic concepts of two actual weapons designs: the "layer-cake"/ "Alarm clock" design conceived of in both the US and Soviet Union independently (and actually detonated by the USSR in 1953), and the "Classical Super" idea conceived of in the US and brought to the USSR via espionage (which turned out not to work).[13] Some commentators proposed even more complex weapon designs. In late 1955, after the "Bravo" accident had revealed that the hydrogen bomb likely used a final "dirty" fission stage of uranium-238, *Life* magazine published a full-page diagram of the "3-F bomb" design. This unusual nomenclature, which never caught on, was meant to indicate that it wasn't really a hydrogen fusion bomb, but a fission-fusion-fission bomb, in which a fission reaction would ignite fusion reactions that would ignite more fission reactions. This idea was accurate, but the diagram itself was not: it depicted the internal arrangement of a mass of fusion fuel hugged by several small fission bombs, all within a uranium-238 casing.[14]

These hypothetical H-bomb drawings did not claim to reveal "secrets," purport to be "leaks," or suggest any privileged access to classified information. They were all part of speculative articles aimed at a general audience: none of the drawings adopted the graphical terminology of "real" engineering diagrams (e.g. thin, perfectly straight lines, precise measurements, blueprint-mimicking) and were clearly labeled as "hypothetical" and "theoretical." They garnered no condemnation from the AEC, whose "no comment" approach to private speculation led them to neither confirm or deny in any situation where doing so might plausibly give away real information. These drawings were also not intended as a form of activism, unlike the hypothetical terrorist bomb drawings, which often did adopt the graphical tropes of "real" engineering devices as a part of their claims to knowledge.[15]

But how would one know, in the late 1970s, whether the H-bomb worked in the way depicted by *Time*, *Illustrated London News*, *Life*, or any other speculative source? One could try to reason through the physics of it, but the actual science was difficult to pin down without access to a lot of other information not easily accessible in the public sphere. The easiest way to be sure that one speculative design or another was "real" would be to know that it came from an "official" source or to have an "official" source demand censorship. After the incident with Hans Bethe and *Scientific American* in 1950, the AEC and its successors had been careful both to impress upon those who knew the "secret" the penalties for its revelation as well as to avoid "validating" public domain information where possible, but over the decades, there would be limits to this approach.

Howard Morland was a journalist who had majored in economics at Emory University after a "false start" in physics. A former pilot in the US Air Force, Morland had become a self-described "peace activist" due to his disillusionment with the Vietnam War, and after that war's end had pivoted his anti-war efforts to the nuclear weapon complex. Beginning in 1976, he had developed a slide show on the American nuclear complex, working to identify production sites and civilian contractors who profited off weapons of mass destruction, and to convey the threat of impending Armageddon. Morland was one of many activists in the post-Vietnam era who thought that the spirit of the anti-war movement needed to pivot if it were going to continue, and he took inspiration in part from the rising anti-nuclear power movement, which was based on a similar transition (e.g., the Union of Concerned Scientists had similarly made such a pivot).[16] Morland wanted to demystify the complexity and secrecy of the nuclear complex in order to show that the issue was not as abstract, hidden, or remote as he felt most Americans considered it to be.[17]

As he attempted to get other anti-war activists to focus on the issue of nuclear war, Morland began to feel that secrecy was itself part of the problem. Not merely the actual effect of it, which was bad enough, but the *perception* of secrecy led even well-informed people to feel that the issue of nuclear weapons was inherently non-participatory: they couldn't act on it because they couldn't be adequately informed on it to act. As Morland later put it:

> Most people were prepared to consider armaments questions only in the most general terms. This hesitancy had its source, I thought, in a general sense that they were not qualified to discuss matters about which nothing was known, or could be known, because it was a requisite of our national security that these matters could be revealed only to a select few.[18]

Morland was directly inspired by a 1976 book about the H-bomb debate by Herbert York, the first director of the Livermore weapons laboratory, that described the "Teller-Ulam invention" as the "one truly central technological fact in all this that still remains secret."[19] Morland took this statement as an activist challenge: "Apart from my personal pique at the government for denying me this information, I felt that the H-bomb Secret stood symbolically for all secrets, and that its revelation by an outsider would puncture the bloated sanctity of the weapons priesthood." He believed that if he could figure out the Teller-Ulam "secret," he could manufacture a traveling model of an actual H-bomb, usable as an attention-getter that would simultaneously make nuclear weapons tangible while disabusing the notion of their secrecy.[20] And unlike the student "bomb designers," he didn't just want people to know *he* could do it—he wanted *everybody* to know the H-bomb secret itself. It was not enough to learn it; he also had to *tell* it.

So from the "outside" of the metaphorical "security fence," with no advanced engineering or scientific training, Morland worked with the assumption that enough information existed in the public domain by 1978 that with some research he could come up with a hydrogen bomb design. He began with what he considered the most authoritative source: photographs reproduced in a visitor's brochure from the Y-12 facility at Oak Ridge, Tennessee, a massive complex created during the Manhattan Project but whose role in nuclear weapon production in the 1970s was still not fully known. According to Morland, the brochure "contained over a hundred pictures of machines and work spaces, together with captions explaining the uses to which they are put . . . I could infer that certain types of parts might be present in the bombs, as well as certain materials."[21] His logic was sound: "official" releases could definitely contain "official" information.

But Morland's next step was more curious: he began to collect accounts of H-bombs from encyclopedias. Encyclopedias are many things, but they are generally not seen as places to look for secret information. But their entries on nuclear weapons were frequently written by scientists, including current or former weapons scientists, so the idea is not as far-fetched as it seems. Morland noted that the 1974 *Encyclopedia Americana* entry on "Hydrogen bomb" was credited to none other than Edward Teller, for example, so while it might not contain secrets, what it did have was probably accurate. Accompanying the article was a provocative illustration of a potential H-bomb design, with a bullet-shaped casing that contained a spherical fission bomb at one end and an oval of fusion fuel at the other. However, Morland did not originally know that the illustrations of such publications are rarely the responsibility of the article authors. In the case of the *Encyclopedia Americana* diagram, it was later revealed that not only had Teller had nothing to do with it, it was essentially a plagiarism of another encyclopedia from a decade earlier, itself attached to an article written by Hans Bethe.[22]

In canvassing the public domain, Morland soon encountered patterns in the diagrams that accompanied articles. The "Teller" diagram from *Encyclopedia Americana* was one approach, but he also soon encountered a different approach, one published in *World Book* by former Manhattan Project physicist Ralph Lapp. The "Lapp" diagram was essentially a reworking of the 1955 *Life* magazine "3F" design, but Morland didn't know this. The friction caused by putting these mutually incompatible diagrams next to each other was productive in his mind: "They couldn't all be right, but I knew they might not all be wrong, either."[23] He realized that the H-bomb secret must be, in part, a *sequence* of events, one frame separated from another by nanoseconds, each showing how you go from a "before" setup to the "after" of megatons of explosive energy released. Thinking of these graphics sequentially led Morland to conclude the key to the H-bomb secret was less about special materials and more about a core arrangement that would allow a fission bomb to ignite fusion material that then would produce further fission reactions.[24]

As someone who lacked the technical knowledge to engage deeply with theoretical models of weapon performance, Morland had to adopt a peculiar investigative approach. He weighed the relative contributions of each diagram by their presumed authorship (which turned out to be a red herring), their visual style, and their apparent logical consistency to tell sense from nonsense:

> Teller's [images in sequence] were presumably the more authoritative, but they were deliberately vague and seemed more like pictures of the *idea* of an atom bomb igniting the *idea* of a mass of hydrogen inside a trash can than of real elements. Lapp's were more detailed, but didn't seem to me to make much sense. I had once asked Samuel Glasstone, semi-official science writer for the bomb industry and the author of *The Effects of Nuclear Weapons*, about Lapp's diagrams, and he had remarked that they showed how active people's imaginations could get when they thought about H-bombs.[25]

And as Morland later told me: "I figured it would be difficult to make any drawing that contained no information at all, and all the drawings were advertised as conceptually accurate."[26]

Morland spent a year working on the question. He expanded beyond his encyclopedia-based methodology, including looking at more brochures from weapons plants, reviewing the published literature on atomic weapons (often focusing, again, on the diagrams), and, importantly, talking to people. The people were both connected with the weapons complex and not, and he told them about his project and his ideas on how the weapons worked. There is a sense in Morland's own account that many of these people felt they were humoring him, amazed at the chutzpah involved in a non-scientist attempting to divine something as technical as the H-bomb based on encyclopedia diagrams. He was even able to talk to Teller and Ulam about their work, though they were not cooperative. Morland developed techniques for talking to people who knew "the secret" that are, in retrospect, classic espionage techniques, like "baiting" them by offering up speculation. If they confirmed his guess, he learned something. If they corrected him, he learned something. He found scientists especially susceptible to the need to correct the misconceptions of others.[27]

In later accounts of his efforts, Morland was unembarrassed about the fact that he tried to trick scientists and the government to "inadvertently reveal more than they meant to." For him, it was all something of a game, and he did not think that it could cause national security damage. After all, if a curious but non-technical person like himself could learn something real about the H-bomb, then obviously the information was accessible to a state with serious resources to expend, and at no point did anyone seriously consider that H-bombs could be built by a terrorist.[28]

Eventually, Morland assembled what he considered to be the "secret" of the hydrogen bomb into an annotated diagram describing the firing of a hydrogen bomb in seven frames of a visual sequence. The first frame depicted a cut-away similar to the "Teller" diagram, a bullet-shaped casing of natural uranium containing a fission bomb "trigger" on one end (the "primary") and a coffin-shaped container at the other end representing a cylinder of fusion fuel (the "secondary"). In the detonation sequence, the fission bomb fires (frame 2), is irradiated with neutrons from an external emitter (frame 3), sending out gamma rays and X-rays into the casing (frame 4), which reflects them inward onto the "secondary," compressing it inward (frame 5), resulting in a fusion reaction which itself releases neutrons (frame 6), which causes the external case of uranium to itself fission (frame 7), and then "a fireball begins to develop . . ."[29]

This scheme fit with what he had learned from talking with people, and also satisfied two important criteria he had developed based on what he had read about the history of the H-bomb's development: that the H-bomb "secret" had some non-obvious element to it (because it

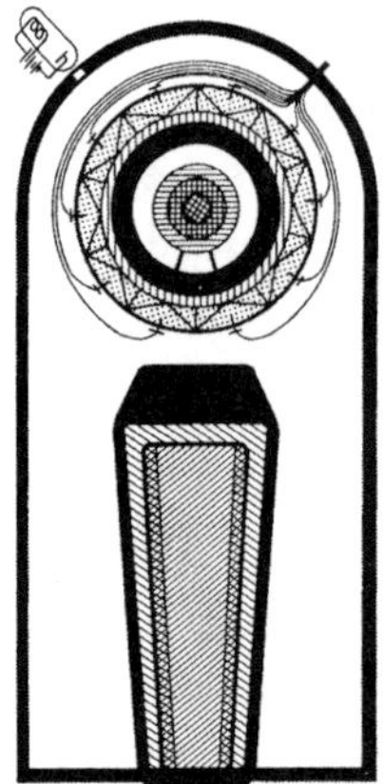

Figure 1. Schematic diagram of a 300-kiloton thermonuclear weapon before detonation. Concentric spheres near the top make up the primary system, or fission trigger. The rest is the secondary system.

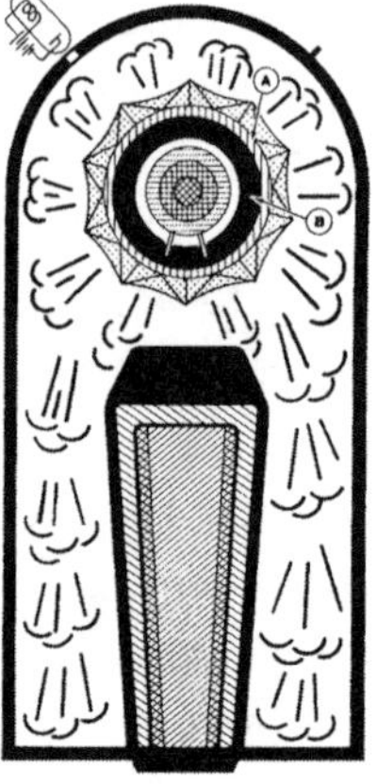

Figure 2. High explosives in the primary system begin to burn, driving beryllium neutron reflector (A) and heavy Uranium-238 tamper (B) inward toward the fissile core. The space between the tamper and the core allows the tamper to develop momentum before hitting the core.

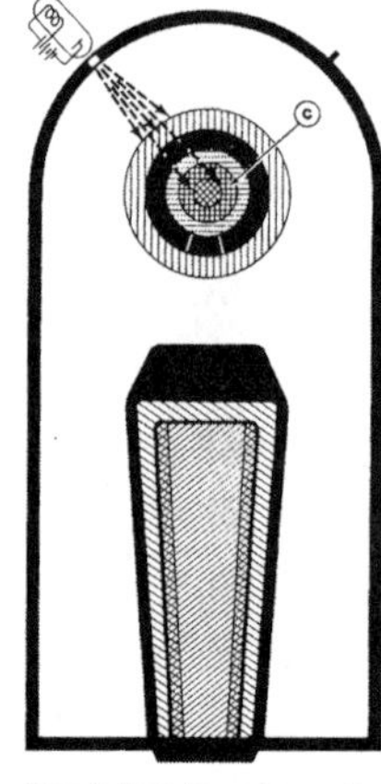

Figure 3. The fissile core is squeezed to more than double its normal density, going supercritical. Neutrons fired from a high-voltage vacuum tube start a chain reaction in the fissile material. The chain reaction concentrates first in the fast-fissioning Plutonium-239 (C).

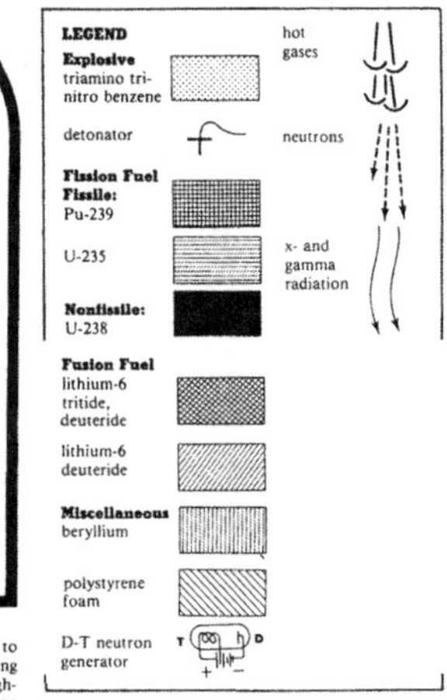

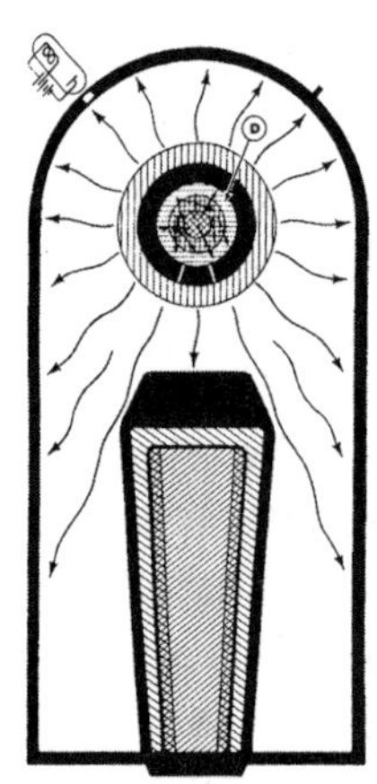

Figure 4. The chain reaction spreads to slow-fissioning Uranium-235 (D). Fusion fuel at the center of the core showers the core with neutrons, "boosting" fission efficiency. As the core expands to its original size, reaction stops, completing the first stage of the detonation. Energy release so far: forty kilotons. Prompt gamma rays and x-rays travel outward at the speed of light.

Figure 5. The weapon casing (E) reflects radiation pressure around the thick radiation shield (F) and onto the sides of the fusion tamper (G), collapsing the tamper inward. Heat and pressure of the impact start fusion in the tritiated portion (H) of the fusion fuel "pencil." The precise location of the tritium within the pencil depends on where the designer intends the fusion reaction to begin. Neutrons from this fusion activity breed tritium throughout the pencil.

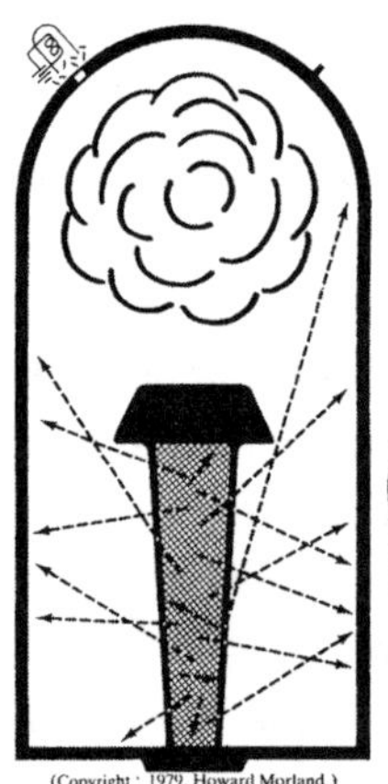

(Copyright © 1979, Howard Morland.)

Figure 6. Fusion fuel reacts virtually simultaneously throughout the pencil, releasing 130 kilotons of energy to complete the second stage. High-energy neutrons from fusion are absorbed by Uranium-238, which has so far served as a fission tamper, radiation shield, radiation reflector, and fusion tamper. Now it serves as fission fuel.

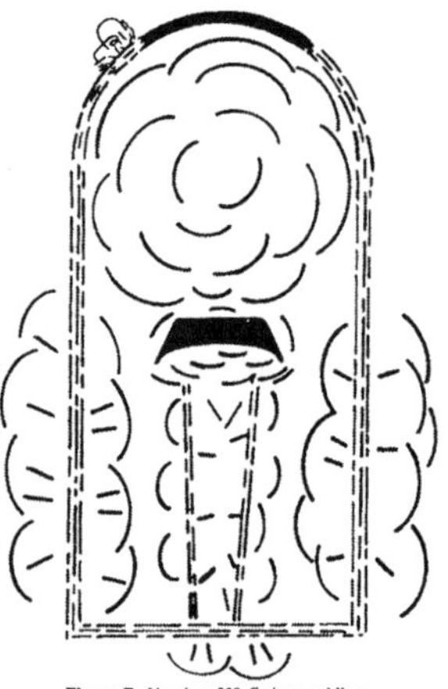

Figure 7. Uranium-238 fissions, adding another 130 kilotons of energy to the explosion and generating enough fission products to kill everyone within 150 square miles with fallout. This is the end of the third stage. A fireball begins to develop....

FIGURE 8.2. Howard Morland's original formulation of the "H-bomb secret." Source: Howard Morland, "The H-bomb secret," *Progressive* (November 1979), 3–12.

took some years to develop), and that it could not be overly complicated because skeptics like Oppenheimer had admired its technical "sweetness" and elegance. A weapon that was "staged" (had its fusion and fission components physically separated) and used a radiation case to effect "radiation implosion" seemed to fit the bill.[30]

After creating the "final" version of his diagrams and explanatory captions, Morland circulated drafts to activists, scientists, and other parties with whom he had started communicating during the project, with the idea of having them published as part of a larger article on the dangers of nuclear weapons. He rarely received encouragement. A science writer at the *Washington Post* told him that "right now you haven't got a story, you've just got these cartoons," while a physicist at Berkeley told him that he thought "all this information has been published before." He also received occasional warnings that he would make a fool of himself and might even suffer the dire legal consequences spelled out for disclosing Restricted Data.[31]

After several rejections, Morland pitched it to a left-wing magazine, the *Progressive*. The editors were dubious about the merit of the article as well. Managing editor Samuel H. Day Jr. recalled years later that in its original form it was a "schoolboy recitation of nuclear arms race history" with an "impenetrable admixture of atomic bomb technology."[32] It was only when others began to tell Day it might be *dangerous* that he became interested in publishing the piece. These "others" were not government scientists or people who were "pro-nuke"; they were anti-nuclear activists with technical backgrounds, who agreed that de-mystifying nuclear secrecy was a noble goal, but they didn't see how encouraging the spread of thermonuclear weapons design information could contribute to a nuclear-free world.[33] One of Day's contacts encouraged him to send a copy of the diagrams to a graduate student at MIT, who in turn passed it along to MIT political scientist George W. Rathjens.[34]

Rathjens, a respected member of the arms control community, had regularly challenged his graduate students to discover the secret of the hydrogen bomb during his courses, but none had pulled it off.[35] Upon receiving a copy of Morland's diagrams, Rathjens called the *Progressive*'s editors on February 15, 1979. According to Day, he expressed agreement with the anti-secrecy thrust of the article but was not convinced that the technical information would further that cause. "I have the impression that the information could be used very mischievously," he explained, "with possibly catastrophic effect."[36] Day was unconvinced, and despite the fact that he admitted to not understanding the technical details, he had already concluded that they "could not conceivably be of more than fleeting value to a nation bent on developing H-bombs."[37] Rathjens, "as a matter of conscience," had submitted the Morland diagrams to the Department of Energy for review, and filed an affidavit to federal court arguing that its publication ought to be blocked.[38]

But there was no immediate response from the DOE. Day asked another scientist who had read the article for his opinion. Theodore Postol, of Argonne National Laboratory, had never had access to classified data but thought Morland's design was probably accurate and not too hard for a trained physicist to develop. Nevertheless, he also advised against publishing it. After four days, the editors at the *Progressive* concluded that the DOE must not have agreed with Rathjens' assessment and worried that "our blockbuster might be a dud."[39] According to Morland, Erwin Knoll, editor of the *Progressive*, had relished the idea of having the article declared secret: "That would put us on the front page of every newspaper in the country!"[40] The *Progressive* sent another copy of the article to the DOE on February 21, asking for help in "verifying the accuracy of the material."[41] Having received no reply by February 26, Day called the DOE's director of public affairs, who had not heard anything about the article. They sent in yet another copy of the article by registered mail.[42]

The *Progressive*'s editors badgered the DOE, telling them that publication was imminent, even though they had actually begun preparations for a substitute story. Finally, the DOE took the bait. The DOE's general counsel called the editors and asked them to refrain from printing the story as written because the DOE had determined that it contained Restricted Data, but said that the DOE would be happy to help the *Progressive* develop a sanitized version.[43] The DOE saw this as a generous offer, an exception to their "no comment" policy, and they anticipated that the *Progressive* would acquiesce, as many a publication had in the past. As the Secretary of Energy at the time, James Schlesinger, later put it, "the same political points about the dangers of nuclear secrecy could have been made" without inclusion of specific thermonuclear design information. Should the *Progressive* press forward, the DOE had decided it would seek a court injunction to prohibit the publication in a rare instance of American "prior restraint."[44]

The *Progressive*'s editors were excited: the fight they had been spoiling for had arrived, and they had no intention of changing the article. "Whatever happened from this point on," Day reflected later, "whenever and however it was published, the Morland article would serve its principal purpose—to draw attention to the problem of nuclear secrecy and its impact on public policy. . . . We were in high spirits."[45]

In hindsight, it is clear the DOE misjudged the *Progressive*. Neither they nor their AEC predecessors had ever encountered a case of journalists soliciting censorship merely so they could defy it. Though they had encountered publications willing to turn censorship into good publicity, they had never been "baited" quite so brazenly. If they had ignored the *Progressive*'s editors, the article would have likely never come out, and even if it had, it would be yet another speculative H-bomb design. Official censorship was the worst possible option the DOE could have pursued: it not only gave the *Progressive* a newsworthy "cause," it also signaled that there was something correct about Morland's work. Even if the DOE won, the possibilities of information being released were high, if only by inspiring "copycats" to retrace Morland's steps. And if the DOE lost, there was the possibility that the legal authority to regulate non-government Restricted Data might be put into doubt: it had never been tested in the courts, and there were, as lawyers for the Department of Justice had determined by the 1960s, severe questions of Constitutional law involved when assuming the government had the power to enjoin private speech.[46]

Schlesinger later judged that he had a "greater, naïve trust in the law at that time," and they assumed that if they pursued an injunction, it would be granted. They were not ready for anti-secrecy activism of this sort, and did not think that either the courts or the other members of the press would fight to release the "secret" of the hydrogen bomb.[47] Schlesinger felt that the *Progressive*'s goals were "scarcely responsible positions," and while he had reservations about the Atomic Energy

Act, he felt it was his duty to enforce the law as written, which to his eyes meant seeking an injunction. It is important to emphasize that the Carter administration and its officials did *not* think of themselves as agents of secrecy: they considered themselves liberals on matters of nuclear policy, weapons, and public trust.[48]

On March 7, the *Progressive*'s editors told the DOE that they would not accept any proposed revisions to the article. On March 9, DOE lawyers filed an application in the Seventh Circuit of the US District Court for a temporary injunction against publication. The application included an affidavit by the DOE director of classification, John A. Griffin, that stated that Morland's article contained Restricted Data, and that its circulation would materially add to the proliferation of thermonuclear weapons, causing "serious and irreparable injury to the security of the nation."[49]

Court action against the *Progressive* was not taken frivolously or without serious consideration for the many possible consequences. The decision to move forward was made at the highest of levels: along with Schlesinger, the attorney general, Griffin B. Bell, was personally involved. Bell had written a memo to President Carter regarding the possible consequences of trying to censor the *Progressive* and seeking the President's approval to go ahead. The memo's conclusions were stark, both about the possible dangers of publication, and the possibility of the government losing a potential lawsuit: "[W]hile we cannot assure you that we will prevail in this suit, the potentially grave consequences to the security of the US and the world itself resulting from disclosure of the data are obvious and frightening." Carter wrote a hand-written response on the memo: "Good move. Proceed. J."[50]

Thomas S. Martin, the deputy assistant attorney general, was the one who had initially gotten the call from the DOE asking about the possibility of an injunction. Martin's immediate personal reaction was: "It can't be done." The government had *never* won a case of "prior restraint," because of the strength of the First Amendment. It was not necessarily impossible a priori: the two major Supreme Court cases that gave precedent on such attempts, *Near v. Minnesota* (1931) and the Pentagon Papers case (1971), had both been denied, but their denials had contained recommendations on what a successful prior restraint request might look like. In Martin's recollection, the Department of Justice (DOJ) was interested in winning a prior restraint case, especially after its loss with the Pentagon Papers, in order to set the boundaries on its powers. But it didn't want another "loser." Martin knew nothing about the technical details of the hydrogen bomb but told the DOE that if they could get affidavits from a Nobel Prize–winning scientist and all of the top cabinet officials testifying that the information from the *Progressive* article was not already in the public domain and that it would threaten national security if published, that the DOJ would support their case. When the DOE expressed confidence in their ability to do this, Martin recalled, "I was convinced."[51]

8.2 THE "DREAM CASE": THE *PROGRESSIVE* V. THE UNITED STATES

The case against the *Progressive* promised to, for the first time, test whether Restricted Data was a legal concept with teeth. Though the DOE and DOJ lawyers were initially optimistic—after all, what could prior restraint be used for, if not to preserve the crown jewel of American nuclear secrets?—they would quickly find that despite decades of experience with the Atomic Energy Act, the legal foundation for nuclear secrecy was shakier than it appeared.

The legal argument made by the government lawyers in favor of enjoining the *Progressive* from publishing Morland's article rested on distinctions between what the DOE was requesting and what the Supreme Court had rejected in the Pentagon Papers case. Specifically, in the Pentagon Papers case, the Court had been critical of the Executive Branch's attempt to restrain publication by the *New York Times* in the absence of legislation that explicitly gave the Executive Branch this power. In this case, the DOE argued, the Atomic Energy Act gave them this power quite explicitly. And indeed, in his opinion on the Pentagon Papers case, Justice Thurgood Marshall had pointed to the Atomic Energy Act as an example of congressionally sanctioned "statutory provisions prohibiting and punishing the dissemination of information."[52]

On March 9, 1979, District Court Judge Robert W. Warren heard statements from lawyers for the government and for the *Progressive* at a hearing on the government's request for a temporary restraining order. He granted the request and ordered that hearings on a preliminary injunction would be held in a little over a week. In delivering his ruling, he remarked at length on his considerations, knitting together an informal argument based on the fears of proliferation and terrorism:

> I'd like . . . to think a long hard time before I gave the hydrogen bomb to Idi Amin. . . . I can't help feeling that somehow or other to put together the recipe for a do-it-yourself hydrogen bomb is somewhat different than revealing that certain members of our military establishment have very poor ideas about how to conduct a national effort in Vietnam.[53]

Warren's remarks would come under criticism as the rest of the trial unfolded (Morland's article was not a "recipe for a do-it-yourself hydrogen bomb"), but it is hard to see how he could have responsibly ruled otherwise. To ignore the government's claim at this early phase of the proceedings would be tantamount to arguing that the DOE lacked the expertise to identify threatening nuclear weapons information.[54]

Over the next few weeks, both the prosecution and the defense prepared for a fight. Numerous affidavits streamed into the court for both sides. The ones prepared by government officials all testified along roughly the same lines: 1. Morland's article contained Restricted Data; 2. said Restricted Data was not in the public domain in as correct or suggestive a form as it was in Morland's article; 3. publication of this Restricted Data would greatly decrease the time required for a nation to develop thermonuclear weapons after it had already acquired fission weapons, and thus would pose a threat to national security.[55]

These affidavits came from scientists, weapons laboratory heads, and government officials. In a sense, these were highly flattering to Morland's work, testifying to its power, importance, and relative accuracy, albeit as a means of justifying its suppression. The director of Lawrence Livermore Laboratory argued that, "in spite of some minor technical errors, [the article] contains or strongly suggests key concepts for the functioning of the hydrogen bomb . . . Previous publications contain some correct hints mixed with incorrect ones, but in no way come so near to describing the operation of thermonuclear weapons." The acting director of Los Alamos went so far as to say that the article "is perhaps as suggestive of the process used in thermonuclear weapons as the original outline on the subject by Teller and Ulam." The secretary of state argued that its publication would "substantially contribute to the ability of foreign nations to develop thermonuclear weapons, and to develop them in a shorter time than would otherwise be possible." Secretary of Energy Schlesinger wrote that publication would "irreparably impair the national security of the United States."[56]

The primary technical analysis for the government was provided by Jack W. Rosengren, a longtime AEC consultant on classification and a thermonuclear weapons designer himself, who had been a major participant in the AEC's internal decisions about declassifying thermonuclear Restricted Data related to laser fusion. Rosengren argued that while there were hints in the open literature as to the Teller-Ulam design, "nowhere is there a correct description" of it, one that had sorted out the "good and bad ideas" present in the open literature. He concluded that Morland's article "goes far beyond any other publication in identifying the nature of the particular design used in the thermonuclear weapons in the US stockpile."[57]

Similarly, the DOE director of classification, John A. Griffin, argued that the government had been continuously reviewing technical data for declassification since 1947, and despite this, "virtually all significant information regarding thermonuclear weapon design has been determined to require continued classification in order to insure the common defense and security."[58]

The affidavits filed by the *Progressive*'s legal team told a different story. They argued that all the information in Morland's article was indeed already "out there" in the public domain, and if someone like Morland could piece it together, starting from the diagrams in encyclopedia articles, then it wasn't much of a secret anyway. They also disputed that Morland's article would cause any direct harm: it would require many chains of events (including the acquisition of nuclear fission weapons) to imagine Morland's article causing any actual damage. Directness of harm was one of the primary legal questions raised by previous attempts at prior restraint, and they argued that a foreign nation saving a year or two in developing a thermonuclear weapon was too indirectly connected to the act of publication to justify its legal censorship.

Supporting this argument was an amici curiae ("friends of the court") brief filed on behalf of the *Progressive* by the American Civil Liberties Union (ACLU). The ACLU brief emphasized that previous court cases had made clear that the First Amendment overruled all but the most extraordinary and carefully worded of exceptions and alleged that the vague allusions to national security damage in the government briefs were unsupported by hard evidence.[59] The ACLU's public position was that this was first and foremost a major First Amendment issue, not simply a question about bombs and Restricted Data. Within the ACLU, agreement was not unanimous. During the Pentagon Papers case, nobody within the organization questioned whether they were on the right side in arguing against censorship. But during the *Progressive* case, because of the contradicting concerns about proliferation, there was a lot of doubt about whether this was something the ACLU should be supporting.[60]

The defense also had its own technical affidavits provided by sympathetic scientists. Theodore Postol, the physicist who had been an early contact of Morland's, had advised the editors not to publish the article, not because it was dangerous on a technical level, but because he feared that the government's panicked response could create a dangerous legal precedent for censorship. But once the censorship attempt had been made, he worked to subvert it, arguing that the Teller-Ulam design as described by Morland was really quite pedestrian. The basics of the thing had already been shown in the illustration accompanying Teller's *Encyclopedia Americana* article. To Postol's chagrin, the DOE declared his affidavit also classified.[61]

Another group of physicists from Argonne filed additional affidavits in support of the *Progressive*. One of them, Alexander De Volpi, alleged that the government's efforts to classify Morland's article potentially stemmed from a fear that to release the H-bomb secret would demonstrate that their "shallow structure of secrecy and technological denial" was inadequate, and that they were trying to stifle discussion about arms control policies.[62] For the Argonne scientists, it was not just a question of whether the government was correct in asserting that the information was not already easily available in the public domain. It was also a rejection of the government monopoly in expertise on nuclear matters.

The government submitted several of its own affidavits arguing directly against these scientific briefs. The most prominent expert was Hans Bethe, a veteran of H-bomb secrecy issues. Bethe's submission, like many of the previous government affidavits, argued that "all the essential principles" of the Morland drawings were correct, and were not already public. As with the other technical affidavits, it was extremely flattering to Morland's work: "The concepts described in the manuscript are as fundamental and necessary to the design of a thermonuclear weapon as those originally formulated by Dr. Edward Teller and Dr. Stanislaw Ulam." Bethe asserted that in his experience as an expert on nonproliferation policy and someone who had chaired government panels directly relating to it, the publication of Morland's article would "substantially hasten the development of thermonuclear weapon capabilities by nations not now having such capabilities."[63]

On March 26, Judge Warren upheld the injunction against publication, arguing that even if aspects of the Morland design were available in the public domain, Morland's own synthesis and compilation of them still constituted the first instance of a "correct description" of the weapon design. He corrected his own earlier misapprehension that the article was a "do-it-yourself" guide for making a hydrogen bomb, but stressed that it still could, for a nuclear nation, "provide a ticket to by-pass blind alleys." He argued that even if secrecy could just slow down such an acquisition, not prevent it outright, "there are times in the court of human history when time itself may be very important." He rejected the argument that the political points of the defendants could not be made without disclosing such technical information: "[T]his Court can find no plausible reason why the public needs to know the technical details about hydrogen bomb construction to carry on an informed debate on [government secrecy]." And he again reiterated his view that this case was, in many respects, very different from the Pentagon Papers.[64]

Not only had the *Progressive* lost, but Warren's argument was fairly expansive. It seemed to endorse a philosophy of classification known as "mosaic theory," in which the combination of several pieces of unclassified information could constitute classified information.[65] At its most expansive, mosaic theory can argue against the release of a considerable amount of information. Combined with the Restricted Data definition, mosaic theory potentially gave the government a wide latitude in preventing discussion about nuclear matters. This is not to imply that such powers were being sought. The DOE and the DOJ saw themselves as the "good guys," opposing the dissemination of information they sincerely believed was dangerous and did nothing to promote an informed democracy. But one can see how someone deep into the anti-secrecy mindset, like Morland, would view such a ruling: not only as a personal failure, but as an affront to the ideal of informed democracy.

The lawyers for the *Progressive* immediately began to prepare for a hearing to vacate the injunction, but this was denied. The case then entered the appeals phase, and unlike the relatively speedy pace it followed so far, the subsequent trial would continue for six months. The warring affidavits had painted the various positions well, but they had not put the government's case under prolonged legal scrutiny.

The response of the American media to the initial case was wary, but leaning toward the government's arguments. Immediately after the temporary restraining order had been granted, the *New York Times* ran an editorial that neither fully trusted the government's claims of harm nor defended the publication of "the design for a highly dangerous secret weapon." The editorial tilted in the government's favor, despite the *Times*' own experience with the Pentagon Papers. Perhaps this was the *Times*' attempt to show that it was, as it has always claimed, a responsible party when it came to anti-secrecy. Coincidentally, the executive vice president of the *Times* had actually defended censorship along similar grounds five years earlier. In August 1974, James Goodale had been giving testimony about the limits of the free press before a skeptical congressional committee, and has been asked whether they would publish "the plan for the atomic bomb" if it was given to them as a leak. "Well, that is what I would call the classic case of where you draw the line between information that is within the scope of the Government to protect and that which is not, the latter being the Pentagon Papers case," Goodale replied.[66]

By contrast, an editorial in the *Washington Post* considered the whole thing a threatening sham: it was a "dream case" for the government to push through tighter controls over the press in the wake of the Pentagon Papers. The *Post* argued in no uncertain terms that it could not see any public interest to be served "in making available to all information on how to build nuclear weapons," which made the case "a real First Amendment loser," a fight the press was "almost certain to lose."[67]

But by the time of the second court hearing, things had shifted. The *New York Times* had concluded from the affidavits filed that the *Progressive*, not the government, was in the right. The article was not dangerous in the immediate sense that would require censorship, and the government's efforts were simultaneously heavy-handed and too vague about the potential harm.[68] An article in the *Times* portrayed *Progressive* editor Knoll as a "bearded and rumpled" intellectual who had bumbled into the case unaware that it would cause a stir, and only wanted to publish the article because it contained "some information we think [people] ought to know." They also reproduced the full "Teller" *Encyclopedia Americana* illustration, as provided to them by Postol. They were buying what the *Progressive* was selling: that this was a clear-cut case of government infringement of speech imposed on an innocent and well-meaning press.[69]

To hold a trial about a secret that one does not want to release is difficult. Because a prior restraint case is not a criminal case, the trial in principle could be held behind closed doors (*in camera*), with a sealed record. This would have required, however, that the defendants submit to security clearances. Morland, Knoll, and Day all refused, noting that even if they succeeded in winning such a case, it would be, as they put it, a "permanent gag order," compromising their ability to speak freely in the future.[70] Instead, two parallel trials were held: one in public for which the defendants could be present, and one *in camera*, in which their lawyers alone would operate, having agreed to secure the necessary clearances. In theory, this compromise would balance the needs of security and open argument. In practice, it was a disaster: clunky and inconsistent at best, compromising of information at worst.

A crucial part of the defense's argument was that the information in the Morland article was already easily available in public, and thus was not truly "secret" at all, despite its official classification. In order to argue this, the defense had to venture what they thought the government was alleging to be the "secret," and then explain how Morland had derived it. A lawyer for the *Progressive* suggested that there were three main concepts involved in the Teller-Ulam design that were under dispute: 1. "Reflection" (the X-rays from the fission primary are re-radiated by the heavy casing); 2. "Radiation pressure" (the force of said re-radiated X-rays compresses the fusion secondary); and 3. "Compression" (the fusion secondary must be compressed before ignition). The next day, the DOE offered up three alternative concepts for "the secret": 1. "Separation of stages" (the fact that the fission primary and fusion secondary are physically separated, and that the energy from one stage is used to ignite the next); 2. "Radiation coupling" (a more precise combination of "Reflection" and "Radiation pressure" into one physical action describing the transfer of energy from one stage to the next); and 3. "Compression" (as before). These clarifications were filed *in camera*, and Morland was not supposed to have ever seen them. These technical clarifications opened the door to greater discussion about what exactly was and was not already in the public domain. "Separation of stages," for example, was plainly indicated in the "Teller" diagram, with its physically separate fission and fusion components.[71]

The most significant expert testifying on behalf of the *Progressive* was Ray Kidder, who had run the Livermore laboratory's laser fusion effort in the 1960s. No scientist in the entire US weapons establishment who knew thermonuclear weapons design information was as informed on what was and was not publicly known about concepts like "radiation coupling" and "compression" than Kidder. Since the early 1960s, Kidder had been reviewing nongovernmental work done on laser fusion; he had been intimately involved in decisions relating to the declassification of related concepts in the early 1970s, and had been actively lobbying for declassification of some of these same general principles. Kidder had, as he put it, a "big stake" in the success of the case: if it succeeded, it would lead to the declassification of more laser fusion concepts.[72]

Kidder was able to point to dozens of places where the key concepts in Morland's article had been discussed in the scientific and even popular literature. As he argued in an *in camera* affidavit: "The concept of radiation implosion that is identified in the Morland article as the key secret of the hydrogen bomb is appearing with increasing frequency in the open scientific literature."[73] He could even cite direct evidence that the DOE director of classification had been informed in 1977 that laser fusion researchers in West Germany, Canada, and Japan had all independently derived the concept through their own work.[74] Kidder's participation gave the defendants the ability to make substantive and authoritative technical arguments with experts at least as experienced in weapons development as those deployed by the prosecution.

And while the prosecution presented a unified front, internally there was dissent. Frank Tuerkheimer, one of the US attorneys on the case, twice petitioned Attorney General Bell to drop the suit. The evidence, Tuerkheimer argued, was wrong: the DOE had made clear statements that the Morland information was not already in the public domain, and this was demonstrably false. According to Tuerkheimer, Bell chastised him for his position the first time, arguing that he was "reacting to pressure from others in what he called the most liberal community in the country." (Which "community" he had in mind is unclear.) The second time, Tuerkheimer says he managed to convince "everyone else" in the DOJ that the case was unraveling, but Bell was adamant because "he had made a commitment to Secretary of Energy Schlesinger that the case would go ahead and felt he had to live up to that commitment."[75]

Attorney Thomas S. Martin, who had initially been the one at DOJ to urge moving forward with the prior restraint request, found the evolution of the case "enormously distressing." As the defense affidavits piled up, especially those from scientists who argued that the information was already publicly known, Martin became convinced that the DOE experts the prosecution had been relying on for their technical arguments were themselves unclear as to what had already been published. "Nobody really knew, and nobody could know, what was out there," he recalled. He felt that the defense affidavits had "put us in a much more murky world than we expected when we started," but that the DOJ had committed themselves to seeing the case through, since to abandon it would be to draw even more attention to the information in the article.[76]

From the beginning, officials were aware that bringing the case to court risked focusing attention on something they were trying to keep secret. But they did not anticipate the extent to which this would occur. The community of private citizens interested in ferreting out bomb secrets had grown since the early 1970s, following the example of the high-profile "amateur bomb designers" discussed in chapter 7. The *Progressive* case provided a national focal point in the mainstream news cycle to bring these people together. Morland began receiving articles and theories and diagrams sent in by others, and while the case progressed, he continued to work out the kinks in his own bomb design, based on pieces of information that filtered through the clumsily deployed *in camera* screen.[77]

In this way, the *Progressive* case became a magnet for anti-secrecy activism of a new sort. Activists like Morland were not interested in deferring to the opinions of scientists, or even in acknowledging them as necessarily more "learned" in nuclear policy. Morland viewed a scientific degree as "no more than a license to practice science," and saw the questions of the use to which science was put as belonging to "an informed citizenry."[78] His suspicion seems to have been mutual: scientists and arms control experts were largely dubious about Morland's project. Jeremy Stone, director of the Federation of American Scientists, explained at an event in 1979 that he felt any argument about the public interest of thermonuclear weapon secrets was "just malarkey . . . comparable to saying that a study of the environmental implications of the automobile industry requires the public to know just how the sparkplugs are inserted."[79]

One of the "amateur bomb designers" from the safeguards debate (discussed in the previous chapter), Dmitri Rotow, also played a key role in the *Progressive* case. Rotow, who was now working for an antinuclear, pro-environment advocacy group, the Natural Resources Defense Council, had been in touch with the staff of the *Progressive*. Morland and Rotow met up, and the latter displayed his own take on the Teller-Ulam design to an unimpressed Morland. Rotow noted that while he had been researching his own book on nuclear weapons designs, he had found some interesting and revealing documents regarding fission weapons at the public technical library at the Los Alamos laboratory. Morland suggested he go back to Los Alamos and poke around, and the ACLU agreed to pay for his trip.[80]

The technical library at Los Alamos, the National Security and Resources Center, was open to the public, a resource provided to the PhD-heavy populace of the isolated town.[81] It contained declassified reports received from other DOE nuclear facilities. Rotow went to its card file and looked up "H-bomb." There he found a cross-listing that said, "see Weapons." He looked up "Weapons," got a list of two dozen declassified reports. Ten minutes later, he had a stack in hand, within which was a progress report from the Livermore laboratory with the designation UCRL-4725.[82]

UCRL-4725 was a report on the Operation Redwing nuclear test series, held in the Marshall Islands in June 1956. The report assumed a high level of preexisting understanding of weapons design, yet even to Rotow and Morland it clearly revealed a good deal of design information they had not otherwise seen.[83] It contained copious amounts of classified weapons data, including masses and densities of material used, specific design calculations, and even information specific to the Teller-Ulam design that Morland had not before seen. Weapons physicist Ted Taylor later testified before Congress that the public accessibility of UCRL-4725 was "the most serious breach of security I am aware of in this country's post-World War II nuclear weapon development programs."[84] It would later come out that UCRL-4725 had been one of the millions of documents hastily processed during the AEC "Declassification Drive" of 1971–1976. Only a short section of the report was intended to be accessible on the open shelves. But the change in classification had been entered incorrectly: whoever had filed it had neglected the note the "(EX)" next to the report's title, indicating that only an "extract" was to be released. And so the report had been erroneously declassified and readily available since 1977 at the latest.[85]

The discovery of UCRL-4725 attracted a lot of press attention, as the defendants had hoped. The government immediately withdrew the document from circulation, but the damage was done.[86] Senator John Glenn, who had held hearings in 1978 relating to Rotow and the accessibility of nuclear weapons information, convened additional hearings to review what had happened and its significance. He was not happy to have Rotow in front of him again, and he informed the DOE representatives that the "Rotow II" session would be enough for him: "Gentlemen, I don't want to be sitting here one year from today having Rotow III. I tell you that."[87] Throughout the hearings, Glenn expressed frustration with the DOE representatives because they were trying to handle the matter carefully (and "legally") in the face of the ongoing *Progressive* case, adhering to "niceties of law" when thermonuclear weapons information was involved.

The UCRL-4725 incident was disastrous to the government's case; their claims that H-bomb secrets were well kept were becoming hard to swallow. The *Los Angeles Times* reported that most of the lawyers on the government case were in favor of dropping it rather than potentially losing it at a higher level.[88] As the case was heard before the Seventh

Circuit Court of Appeals, the judges were indicating suspicion toward the government's claims. In one notable instance, Judge Wilbur Pell Jr. challenged the entire interpretation of the "born classified" provision: "In candor, I'd be more impressed if you [the government] were just trying to keep secret our secret. But you're keeping secret the whole world's secrets, aren't you, under this Act?"[89]

One of the like-minded individuals drawn to Morland's cause was Charles "Chuck" Hansen, a computer programmer living in Mountain View, California. Hansen was another "amateur bomb designer" who had made a hobby of digging for nuclear weapons secrets and had been in contact with Morland. Hansen focused primarily on fission bombs and reconstructing detailed technical histories of specific US weapons development; he was interested in nuclear history more than modern nuclear politics and had gotten into the hobby as a weapons "buff" who wrote articles for hobbyists who liked to build scale models.[90] But when the case of the *Progressive* made national headlines, he was drawn to it as a way to show off his own skills at nuclear secret sleuthing. He told a reporter later that he "never gave a good goddamn what the secret of the H-bomb was" until the case began but decided to prove Morland right by deriving the "secret" himself, starting by going over in detail all of the public affidavits filed for the case, including ones that had been sloppily declassified by the DOE.[91]

In April 1979, Hansen took out advertisements in several college newspapers for a "Hydrogen Bomb Collegiate Design Contest," offering $200 to the first person who drew a hydrogen bomb diagram that was declared "classified" by the DOE. Hansen received several submissions, including one from a UC Berkeley graduate student in physics who, after seeing Hansen's advertisement, reasoned that "if Howard Morland could do it without my physics background, then maybe I can [too]," and submitted three of them to the DOE. The DOE, "ever humorless" (in Morland's assessment), told Hansen that he was under investigation for conspiracy to violate the Atomic Energy Act.[92]

In August 1979, Hansen wrote an 18-page letter to Senator Charles Percy outlining his version of the "secret," replete with detailed diagrams. He also sent a copy to multiple newspapers, including student newspapers. Morland was disappointed by Hansen's letter, as the diagrams were clearly wrong: "I saw in it the same old bunch of errors and some new ones, and thought: . . . Poor Chuck Hansen, whose life's ambition has been to generate a piece of classified information, blew it. They'll just ignore it."[93]

Instead, two weeks later, the government declared the letter to be secret, though by then it had acquired a large circulation. They requested from Hansen a list of all parties he had sent it to and then demanded that no newspapers publish it. This shifted the sentiment of the press corps: multiple editors indicated they would not comply with the censorship order. The editors of the UC Berkeley student newspaper, the *Daily Californian*, which had already defied a previous order not to publish another letter by Theodore Postol, threatened to publish the letter, though the editors didn't think it contained anything interesting. The appeal of Hansen's letter was entirely the clumsy attempt to censor it, not its largely impenetrable technical content of dubious validity. The DOE filed a temporary restraining order against the newspaper as a response.[94] But September 16, the Madison *Press-Connection* published the "Hansen letter" in a special edition. The *Press-Connection* was not one of those that Hansen had sent it to directly, and thus had not been enjoined against publishing by the government.[95]

In what came as a surprise to both the defense and the appeals court judges, the government abruptly dropped the case against the *Progressive* the next day, declaring it "moot" after the wide publication of Hansen's letter.[96] They argued that it would be pointless to continue to enjoin the *Progressive* from publication if the same information had been published in roughly the same form by Hansen, but they added a warning that this decision was not an indication of their backing down from enforcing the Atomic Energy Act. Indeed, on the same day they announced that they would be investigating whether information had been leaked to Morland from people with access to government-derived Restricted Data.[97] The *Progressive* could now publish the article, and so in the November 1979 issue, the front page of the magazine was emblazoned with Morland's bomb design and the text: "The H-bomb secret: How we got it—why we're telling it."[98]

Ironically, by this point, Morland had decided that his original

H-bomb design was *not* correct. He had taken to heart the comments about his "errors" in the various DOE affidavits and had scoured the contributions sent to him from other "amateur bomb designers." One that would profoundly influence his thinking on the bomb mechanism came from a fusion physicist named Friedwardt Winterberg containing a news article clipped from a 1976 issue of the *New Solidarity* newspaper.[99] *New Solidarity* was published by an organ of the United States Labor Party, which, starting in 1976, began running the cultish and conspiratorial Lyndon LaRouche for president. Among the LaRouche group's diverse and idiosyncratic policy positions was support for nuclear fusion, and Winterberg had been an active participant in some of the work conducted by the LaRouche-funded Fusion Energy Foundation.[100] The 1976 article, by FEF's research director, Uwe Parpart, was about supposed new breakthroughs by the USSR in thermonuclear weapons that would allow it to construct "gigaton" (a thousand megatons) strength hydrogen bombs.[101] The article included a single-frame diagram of a hydrogen bomb that looked somewhat different than Morland's but utilized "staging" and the X-rays from the "primary."

Morland did not view the article uncritically, but found it interesting.[102] He took from it the idea that at the center of the fusion mass was more fissile material, called a "sparkplug," that would exert pressure on the fusion material from within while it was being compressed from without.[103] And during his trial, Morland concluded that his original mechanism of compressing the "secondary"—radiation pressure alone—was incorrect as well. In his new understanding, the lower half of the bomb casing was filled with a polystyrene foam that the X-rays from the "primary" would turn into a hot plasma, which would in turn do the compressing. He justified it in part from a new source of information: declassified documents from the *in camera* side of the *Progressive* trial that hinted at secrets he was not supposed to know.[104]

For "historical" reasons, his article in the November 1979 issue of the *Progressive* was published in the form that the DOE had tried to censor. His revised thoughts on the bomb mechanism were included as errata, complete with a new cut-away diagram with "exploding foam" and a plutonium "sparkplug."[105]

Before the article was published, Morland presented his H-bomb theory at a press conference. He wore a t-shirt he had made early in the trial that contained the essence of the idea drawn across it, and ex-

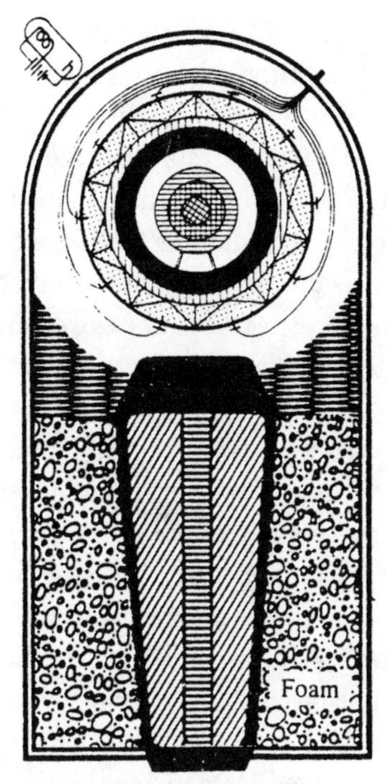

FIGURE 8.3. Morland's final, revised bomb diagram, printed as errata along with the original article in the November 1979 *Progressive* issue. Note the plutonium "sparkplug" (in the lower center), and the "exploding foam." Source: Howard Morland, "Errata," *Progressive* (November 1979), 35.

plained his new discoveries about the foam and "sparkplug." The journalists' eyes glazed over. "This illumination, over which I had struggled for months, was received with polite uninterest," Morland recalled.[106] The design of the hydrogen bomb was exciting mostly in the abstract, for its mystery and controversy. Like a magician's trick, the details of how it worked were banal once revealed—the fun was not in knowing, but in desiring to know.

All parties involved saw the *Progressive* case as important, but its meaning was unclear. The *Progressive* had won, in a sense, but only because the government had voluntarily dropped out. By being mooted prior to any ruling, the case had not established a legal precedent. The government had not lost its ability to regulate privately generated Restricted Data, though the limits of enforcing such rules had been put on display. Almost everyone involved agreed that the government's interests would have been better served if it had simply ignored the article. As the DOE and DOJ had known from the outset, challenging the article "validated" it and made it the subject of media spectacle, and this attention had also highlighted many errors the government itself made in its information control. It is clear in retrospect that the government underestimated how deleterious these effects would ultimately be, both to the case and to its ability to control thermonuclear information.

The case had several long-term effects on how secrecy was treated in the United States. Most importantly, it reconfigured the question of nuclear secrecy from one about freedom of *research* into one about freedom of *speech*. "Born secret" became an issue for the press and civil liberties, not just private industry or private scientists. Legal scholars would pick up this thread, arguing that the Atomic Energy Act did give the government authoritarian powers that threatened civil society.[107] After the case, the DOE largely hewed to its original "no comment" policy, and treated Restricted Data in a far less "special" manner.

It is easier to see today how many different motives the case encompassed for all the parties involved. Morland saw himself as a nuclear Daniel Ellsberg, a figurehead for an anti–nuclear weapons movement who saw the possibility of deriving legitimacy and strength from the fact that he could show that nuclear knowledge was not restricted to a limited "priesthood."[108] Knoll and Day, the editors at the *Progressive*, clearly saw the case as a way to boost their magazine's profile and as a complement to their long-standing anti-nuclear journalism. The ACLU, and later the press corps, saw the case as important in avoiding a devastating legal loss that would have given the government far more leeway to enjoin publication in the name of "national security." The Argonne scientists saw the case as a means to argue for less government secrecy, even if they were not in favor of distributing technical information about the bomb. Ray Kidder at Livermore saw this as an opportunity to reduce classification in an area in which he worked (laser fusion). All the anti-restraint factions were motivated by slightly different motivations and desired outcomes.

The internal motives of the DOE and the DOJ are the hardest to derive, and were the ones most speculated about. In retrospect, the idea

TABLE 8.1. A rough taxonomy of the historical actors, the respective motivations behind their (direct or indirect) participation in the *Progressive* case on the side of the defendants, and their desired outcomes.

Actor(s)	Motivation(s)	Desired outcome
Howard Morland	Anti-nuclear and anti-secrecy activism, publicity	Publication; end of secrecy
The *Progressive* editors	Anti-nuclear activism, freedom of press, publicity	Publication; sales
The Argonne scientists	Anti-secrecy activism, freedom of research	Lack of censorship
Fusion researchers	Specific declassification goals, freedom of research	Declassification of science
ACLU, *New York Times*	Freedom of press	Lack of censorship

that the government was making a mad grab for power appears less likely than their having made a serious miscalculation. Neither Secretary of Energy Schlesinger or Attorney General Bell seems to have been making a grand "political" statement when they decided to enjoin the publication of the article. Schlesinger seems never to have realized that the *Progressive* editors would relish the fight and that Morland would totally reject government experts' assertions of harm. Schlesinger also, as he later admitted, greatly overestimated the ability of the government to control public information, not realizing that the assumptions Congress made about control of information in 1946 did not hold in 1979, if they ever had.[109]

Within the DOJ, there was never unanimity about the wisdom of prosecuting the case or the strength of the legal arguments involved. Reliant on the DOE for technical assessments, the DOJ lawyers found themselves in a tough situation when it became clear that the DOE did not have a monopoly on technical expertise, even regarding nuclear secrets. US Attorney Thomas Martin recalled feeling additionally frustrated with the way that both the DOE and DOJ were demonized by the press corps and accused of political machinations. "None of these people were ideologues," he insisted to me. They were largely Carter Democrats, with the DOE seriously interested in nonproliferation and the DOJ seriously interested in cleaning up its post-Nixonian image.[110]

The DOE classification experts consulted present an additional consideration. What was their motivation in repeatedly asserting the value of the Morland article and its uniqueness, despite many claims to the contrary? Paradoxically, classification experts are not generally as informed about what information is in the public domain as they believe they are. They know what is *supposed* to be legally secret, but that is very different from what is actually known in the world. Intimate knowledge of a secrecy regime may be an obstacle to understanding exactly where the limits of that regime are because the experts internalize the structure of the regime to such a degree that they are unable to see outside of it. Something appearing in the public domain does not automatically lead to a formal declassification (to have that policy would be to incentivize leaking, and the fact that something is "out" in the world does not mean that its truth has been publicly established), but in a prior restraint case, where the onus is on the government to prove that grave harm would come from publication, the fact that the information is already easy to access is undermining.

The *Progressive* case was, in many respects, the inheritor of the debates about laser fusion, safeguards, secrecy, and proliferation from the early and middle years of the decade. It fused these concerns with a strident form of anti-secrecy activism that was itself an outgrowth of the Pentagon Papers and Watergate. This was an activism that respected no expert opinions, saw the government as a hegemonic and monolithic expression of state control, and was clever about "gaming the system." In this context, what had started as a "dream case" for the government's ability to censor information in the private sphere had transformed into a "dream case" for those wanting to show that such secrecy was impossible to practicably enforce.

8.3 OPEN-SOURCE INTELLIGENCE IN A SUSPICIOUS AGE

The US government's failure to successfully prosecute the *Progressive* and its effective "validation" of the Teller-Ulam design as described by Howard Morland had effects both short- and long-term. Most immediately, it resulted in the formal declassification of the basic concepts of thermonuclear weaponry. In late 1979, the AEC promulgated new guidance on what was an unclassified statement about thermonuclear weapons: "In thermonuclear weapons, radiation from a fission explosive can be contained and used to transfer energy to compress

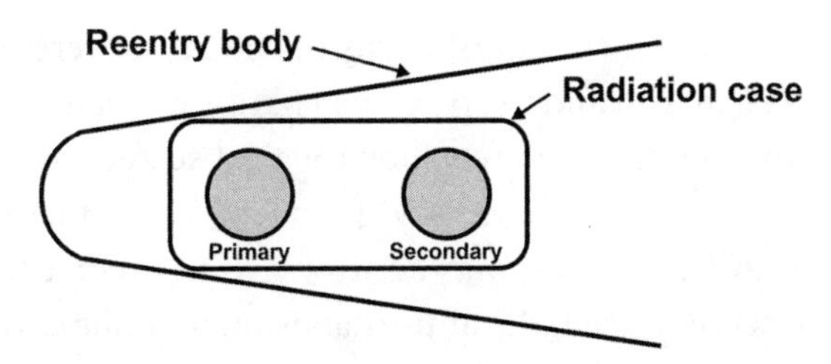

FIGURE 8.4. The maximum graphical schema US government employees are allowed to use to illustrate about the Teller-Ulam design idea, since at least the 1990s. This particular version was redrawn from John J. Vandenkieboom, "Nuclear weapon fundamentals," Los Alamos National Laboratory Report LA-UR-11-03126 (June 2011), online at https://permalink.lanl.gov/object/tr?what=info:lanl-repo/lareport/LA-UR-11-03126, accessed 10 December 2018.

and ignite a physically separate component containing thermonuclear fuel."[111] They warned, however, that "any elaboration of this statement will be classified."

Such was the limited nature of the concession, a confirmation of Morland's essential idea without elaboration, but still no doubt a painful one for the DOE after some 30 years of thermonuclear designs being the "crown jewels" of the nuclear secrecy regime. It would eventually become possible for DOE publications to illustrate the basic schema, represented as two circles (one labeled "primary," the other "secondary") within a square box, sometimes within the generic shroud of a reentry vehicle. A singular, barely descript frame of the H-bomb detonation sequence, it contrasted strongly with the hyper-detailed, faux engineering diagrams of Morland and other "secret seekers." Simultaneous with these disclosures, the DOE also declassified the related laser fusion concept of indirect-drive compression, the herald of what would be many important declassifications for the field in the 1980s and 1990s, effectively opening almost all of it up to unclassified research.[112]

The longer-term effects were more subtle. By the early 1980s, US public attitudes toward secrecy were undergoing a sea-change. The *Progressive* was assimilated into a new anti-secrecy politics as a lesson of the power of free speech and the fallibility of American nuclear secrecy. The anti-secrecy worldview that motivated it not only had triumphed but had been, despite initial misgivings, validated: even the secrets of the H-bomb were "silly," as the "secrets" were hardly better than what you could find in children's encyclopedias.

Critiques of secrecy had been present over the course of the entire

Cold War. But the trenchant criticisms of the 1950s were more about the ills of McCarthyism and the perceived harms to science by the security systems than genericized rebellions against secrecy as a practice.[113] There were criticisms of "silly secrecy" in the 1960s, but these were more about the absurdities of keeping the wrong things secret than against the notion of secrecy itself; the anti-establishment politics of the 1960s had rarely focused on secrecy as a particular target of its ills.[114] And while the Pentagon Papers and Watergate scandals of the 1970s had created the sense of distrust that anti-secrecy politics thrived on, and had created anti-secrecy heroes such as Daniel Ellsberg, Bob Woodward, and Carl Bernstein, even they had been focused on the kinds of secrets that were worth exposing: secrets that concealed lies, or embarrassments, or crimes. By the late 1970s, and especially into the 1980s and beyond, this had coalesced into a true anti-secrecy politics that saw secrecy of all forms as being a social and governmental ill, and which saw the only possible remedy as a total rejection of secrecy regimes.

The early Reagan years were a return to a Cold War approach to nuclear arms, a period some scholars have called "Cold War II." Détente was declared a failure, and new arms systems, deployments, and global strategies were put into effect.[115] Most famous of these new nuclear initiatives was the Strategic Defense Initiative, an ambitious plan for space-based ballistic missile defense, inspired by supposed breakthroughs at the Livermore laboratory championed by Edward Teller, but there were other more mundane throw-backs as well, such as the deployment of new nuclear weapons systems and the return to more active "covert measures" abroad. Unlike the heyday of the Cold War, though, these build-ups were met with extreme skepticism and disdain, with the space-based defense, lampooned as "Star Wars," facing scrutiny in the open literature about its feasibility. In the case of the Strategic Defense Initiative, a whistleblower at Livermore alleged that the classification of the program had been used to hide the fact that its technical foundation had always been shaky, fitting into what was by then a well-worn narrative about secrecy's use as a tool to push for funding without proper oversight.[116]

To anti-secrecy activists, all official statements were probably lies and most policies were deceptions. Secrecy was a rot that had infested the US government and was at the core of all its problems. The answer was a cleansing via the much-championed antiseptic qualities of sunshine and openness—by force, if necessary. In such an environment, the "secret seekers" could transform themselves from threats into heroes, even if they still embraced the cultural archetype of the trickster.

Accompanying the public loss of faith in secrecy was a rise in what would later become known as Open Source Intelligence (OSINT). Journalists had been engaged in investigations of classified matters since the beginning of their profession, frequently working on stories relating to the military and nuclear complexes, with or without government sanction. But in the 1980s, the act of "secret seeking," where private individuals outside of secrecy regimes attempted to learn what they were not supposed to know, moved from the fringe into the mainstream. What had once been the territory of amateur undergraduate "bomb designers," self-described activists like Morland, and fringe political groups like the LaRouche organization, was now being done by relatively respectable advocacy groups and nongovernmental organizations.

Organizations engaging in open-source intelligence collection were looking at unclassified information but collating and collecting it in ways that created "products" that looked similar to those created by state intelligence agencies. As with most things, this was not an entirely new endeavor, but it flourished in the 1980s. Jane's Information Group, for example, had been producing encyclopedic compendiums of military hardware since the late nineteenth century, but their business expanded in the 1980s and began to provide weekly assessments and even dedicated coverage of nuclear matters for the first time. At the Natural Resources Defense Council (NRDC), an environmental advocacy group founded in 1970, coverage of nuclear weapons matters increased dramatically due to the fears of the Reagan era. The NRDC had already been involved, since the 1970s, in lawsuits to stop the development of certain types of nuclear reactors and to halt proposed US nuclear testing in Alaska, but in the early 1980s they started a project to provide "basic facts" about US nuclear weapons: "how many there were, what they looked like, where they were deployed, and how and where they were made."[117]

The impetus for this work came from the efforts of a dedicated and accomplished "secret seeker" who had applied considerable attention to acquiring a different brand of "nuclear secret" than had preoccupied previous "seekers." William M. Arkin was a former US Army intelligence analyst who, starting in 1978, decided to explore the question of where the physical nuclear weapons of the United States actually were. The problem was that "they weren't anywhere . . . the corporal nature of nuclear weapons seemed opaque." While nuclear weapons occupied a central place in the rhetoric of security, strategy, and secrecy, classification policies kept the number of warheads in the US arsenal and where they had been dispersed throughout the world over the course of the Cold War extremely secret. Arkin started this work without any "thesis" of what it would tell him about the state of the world. It was just sheer intellectual curiosity that drove him, coupled with the realization that, as an intelligence analyst, he knew more about the Soviet nuclear system than he did about that of his own country.[118]

Arkin worked entirely in the public domain, patching together obscure reports, declassified documents, congressional testimony, and even his own impressions of suspected nuclear weapons sites developed by driving past them. Though his approach was unclassified, it looked an awful lot like what used to be considered espionage; for example, one way Arkin located bases with nuclear weapons was by collecting phone books of US military commands and comparing their listings with abbreviations he had obtained from a Freedom of Information Act request that indicated there were nuclear arms stationed on the base.[119] Through this approach, Arkin was able to piece together specific details to corroborate a picture that had previously only been gestured at: that the United States had dispersed tens of thousands of nuclear weapons into dozens of countries, sometimes without the knowledge of their high commands, and almost entirely without the knowledge of their inhabitants. Arkin found this to be "surprising and alarming," with broad political implications about the nature of democratic consent.

Arkin's first major article written using this research was published in February 1981, in the West German magazine *Stern*, three pages describing US deployments in the country, along with a map with around a hundred dots indicating the presence of weapons at various bases. The sites were not identified by name, but it didn't take much effort to discern some of them, and the group that Arkin worked for, the Center for Defense Information, teased that they would reveal the names soon. However, pressure came swiftly to not publish further details, and Arkin was himself fired for his insistence on the necessity of releasing them.[120]

At the same time, the US secretary of state, Alexander Haig, asked the legal arm of the State Department to look into the possibility of criminal charges against Arkin for publication of classified information. Upon being contacted by State Department lawyers, Arkin obtained his own counsel, and a deal was worked out where he would present on his methods to top nuclear security experts in the military and Department of Energy. If they were satisfied that Arkin had published only information available in the open literature, he would be free to continue; if it seemed likely he had obtained controlled secrets, then charges might follow. Arkin arrived at the State Department with his notepads, index cards, and reference materials. In Arkin's account, a colonel shouted the names of bases, and Arkin would then display the tangled web of documents proving that each base contained nuclear weapons. This continued for maybe twenty or thirty possibilities, and with each demonstration, the colonel "slumped deeper and deeper into his seat." Finally, he announced he had seen enough, and Arkin was given a clean bill of legal health.[121]

Arkin saw his work as being of a different character than Morland's, which he characterized as merely "mischief," without concrete, measurable, positive political consequences. He viewed open-source secret seeking as a lever for policy. By rendering tangible what had been made abstract, he sought to make it targetable for policy analysis and change. His open-source "secrets" were weapons against the nuclear industry and secrecy itself. The initial result of this work was the *Nuclear Weapons Databook, Volume 1*, the first of a multi-volume series covering the nuclear capabilities and facilities of the United States, the Soviet Union, the United Kingdom, France, and the People's Republic of China. Published at a time when Reagan's secretary of defense, Casper Weinberger, was alleging vast increases in the Soviet nuclear arsenal as a justification for increases in American defense spending, the *Databook* contained some of the closest-kept secrets in the US: the sizes of the nuclear stockpile and the locations and types of deployed nuclear weapons, what the authors considered "basic facts" necessary for any serious policy discussion.[122]

The work was heralded by many commentators and given wide press coverage, much of which noted that the information came from "public sources." And unlike the work of Ted Taylor or Morland, there was little to no hand-wringing in the press about whether it was inappropriate to publish such information. McGeorge Bundy, the former national security advisor to President Kennedy, praised the work's usefulness, while making a general lament: "no administration has had an unwavering attachment to nuclear candor, but things have been worse than usual in the last three years." A wider view of history makes that a hard statement to support—there does not appear to be anything unusually secret about the Reagan administration's handling of these matters compared to previous administrations—but it was true inasmuch as the *expectation* of candor had changed.[123]

In 1987, the NRDC group began publishing a regular "Nuclear Notebook" column in the *Bulletin of the Atomic Scientists*, first acting as updates to their books and short news pieces on nuclear weapons matters, but giving annual updates on nuclear deployments and stockpile sizes. These were not based on classified information, but on scouring reports, statements, official speeches, and information obtained through the Freedom of Information Act to create a synthetic picture. If you believed their results, then you had information that was meant to be classified in not just the United States, but in countries that were even more close-lipped: the Soviet Union, the People's Republic of China, and even the elusive Israel.

Arkin's 1985 book, *Nuclear Battlefields*, applied his methodology to the entire world, documenting nuclear deployments and storage in long tables and copious maps. He contacted journalists throughout the world, ensuring that every jurisdiction had an opportunity to report on these revelations. The book even made it into the *Tonight Show* monologue of Johnny Carson, who was impressed that California led the country in nuclear warheads: "Now most of them I think are on military bases. But if you happen to go out in front of your house on the street where you live, and you see a manhole cover about twenty-five feet wide—there could be a problem."[124]

Though the public response was positive, internally, the NRDC was conflicted. Antagonizing the powers-that-be made for good press, but it potentially exposed the organization to legal threats or difficulties with fundraising. Though he continued working with the group, Arkin increasingly saw his agenda as more radical than the "liberal institutions supported by peacenik foundations" he collaborated with. As another former NRDC employee put it to me, there were constant conflicts between the "strait-laced" New York branch of the NRDC, which handled fundraising and agenda-setting, and the "bad boys" in the DC office, where the analysts worked. Increased centralization of NRDC's work in the New York office would eventually lead to the elimination of their nuclear program, which would be taken up by the Federation of American Scientists.[125]

Arkin's most successful campaign, in his mind, was with an even more radical organization, Greenpeace, in their "Nuclear Free Seas" campaign that ran from 1986 until the end of the Cold War. Arkin judges it "probably the most successful activist campaign that has ever been run, ever." The work involved identifying which US ships carried nuclear weapons and showing that said ships were docking in foreign ports, including those of nations whose domestic politics were antagonistic to nuclear weapons, like Japan, New Zealand, Iceland, and Sweden. Greenpeace would then publicize this information, generating unwanted controversy. President George H. W. Bush, supported by the Navy, removed the nuclear arms from the ships to avoid diplomatic headaches, accomplishing exactly what Arkin and Greenpeace had desired. No doubt there were other factors involved with the decision, but clearly the negative publicity caused considerable angst between the US and its allies.[126]

In the long run, the techniques and activism pioneered by Arkin and his collaborators in this period would become more routinely used by activists (both anti-nuclear and otherwise), journalists, and academic researchers. To merely rely upon the word of intelligence agencies and government sources for key matters of policy would mark one as dangerously naive in a suspicious age.

One of the "secret seekers" of the 1970s who continued his work well into the 1980s was Chuck Hansen, the computer programmer who had unexpectedly led to the mooting of the *Progressive* case. Prior to the case, Hansen had been working on what he would later characterize as "a comprehensive unclassified technical history of the US postwar nuclear weapons program."[127] Through such extreme use of the Freedom of Information Act that he supposedly was known on a first-name basis by FOIA officers, Hansen managed to extract far more details than had ever been compiled in one place, published in a richly illustrated oversized volume as *US Nuclear Weapons: The Secret History* in 1988. There were rumors afterward that men in gray suits had been seen scrutinizing bookstores carrying the book and that the government had bought out many stores' stocks in order to restrict circulation, but by the 1980s such stories had become badges of honor and legitimacy, not something to be feared. Friedwardt Winterberg, the physicist who had contributed clippings from the LaRouche newspapers to Howard Morland's work, published his own book in 1981 on H-bomb "secrets" that proudly carried on its back cover a DOE refusal to evaluate its contents: "Our policy is not to comment on things like this for publication, because if it does contain something classified, then we are revealing what is classified."[128] For Winterberg's publisher, even a banal statement of the "no comment" policy hinted at the possible secrecy of the work, and thus its accuracy and power.

The illustrations in Hansen's book were a large part of its claim to secret knowledge and power. The photographs of nuclear explosions and the weapons themselves made a claim to authenticity, but nothing went so far as the elaborate diagrams describing how the weapons worked. They mimicked blueprints with their fine-lined style and numerous component labels. The most impressive diagrammatic tour de force was the half-page "Fat Man Assembly," showing an isometric, exploded view of Hansen's understanding of the Fat Man bomb, complete with twenty-six component labels drawn with enough detail and clarity to make one believe that Hansen knew exactly how it was created. And if Morland had demonstrated his own mastery of thermonuclear processes through a seven-frame filmstrip, Hansen pushed it to the limits, showing twelve stages of an exploding bomb.[129]

These detailed bomb diagrams were made for Hansen by a technical artist named Mike Wagnon, who worked for Hansen's publisher, Aerofax, Inc. Wagnon worked directly with Hansen, drawing the devices "partly from descriptions, partly from some photographs and drawings scrounged together." To achieve the desired hyper-realism, they were originally drawn very large—four-feet across in the case of the Fat Man diagram—and then scaled down. Utilizing his technical training and deductive skills, Wagnon created diagrams that, in his words, "ad-

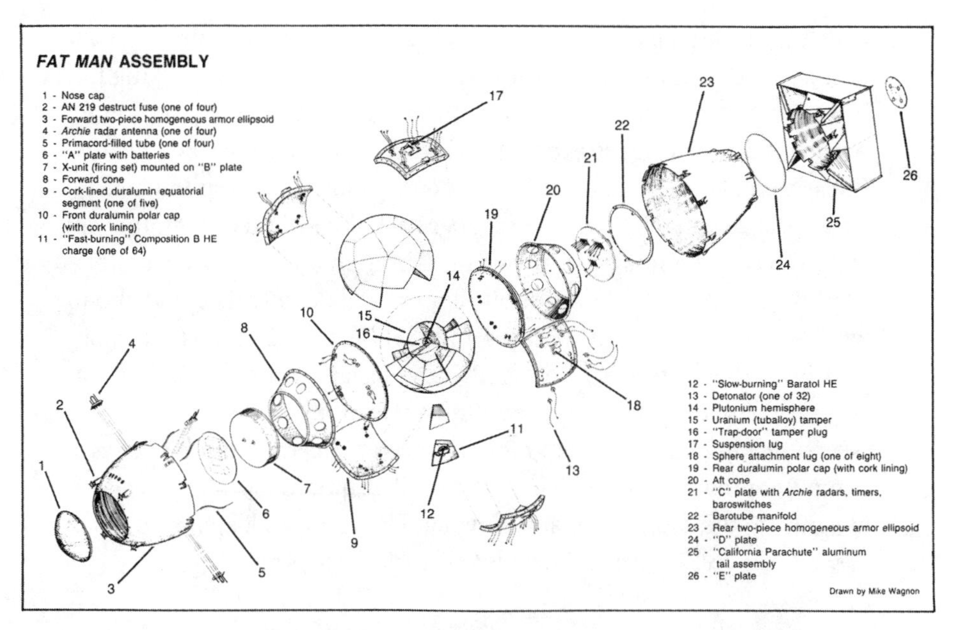

FIGURE 8.5. Isometric, detail-heavy drawing of the Fat Man atomic bomb from Chuck Hansen's *U.S. nuclear weapons: The secret history*, created by technical artist Mike Wagnon. Source: Hansen, *U.S. nuclear weapons*, 22.

vertise an accuracy they do not have," using graphical tropes borrowed from engineering to suggest access to "secrets."[130]

Hansen's book was lavished with praise for revealing a hidden history, and Hansen was lionized as an amateur made good. His obsession with nuclear secrets, which left his suburban house "cramped, cluttered . . . its floorboards sagging with nuclear documents," was viewed as a curiosity but not a threat. In interviews, Hansen described his work in terms of anti-secrecy, but his real motivation seemed more personal: "I like to write about things that are not well documented, because information about them is difficult to obtain." His anti-secrecy sentiments were rooted in his own attempts to extract information via the Freedom of Information Act, and the many ways that the FOIA officers could make that difficult with their exemptions, fees, and slow bureaucracy.[131]

Hansen does not appear to have come under any serious scrutiny by the FBI in the 1980s. In 1993, there was a small investigation into his work. In one of his FOIA requests to the Defense Nuclear Agency, Hansen had asked for very specific pages out of a classified document, which made a DNA employee suspicious. The employee met with an FBI agent, who told them that the information Hansen was asking for "could only come from within his agency" and that he believed someone had been leaking information to Hansen. Hansen responded to the FBI's attempts to speak with him by demanding all correspondence in writing, and then telling the newspapers that the FBI was attempting an "intimidation" campaign against him. Through a lawyer, Hansen told the FBI that the DNA was simply sloppy in its record-keeping about what they had already declassified and that he had gotten his information from the footnotes of other already-declassified reports. The FBI seems to have dropped the case within a few months. Despite Hansen's "secret seeking," he was later granted a "Secret" clearance for his job as a computer programmer, which he had until he quit in 1991. An FBI agent assigned to the case noted favorably that Hansen's book appeared to be "a very comprehensive and detailed historical account."[132]

The early Reagan administration was, arguably, in pursuit of a "return" to some measure of secrecy, in the same way that they advocated a "return" to an Eisenhower-era stance toward the military and Cold War confrontation. The early years of the administration were wracked by internal leaks to the press, apparently to the personal frustration of Reagan himself. Attempts to combat the leaks generated even more press attention. National Security Decision Directive 84, drafted by Reagan's national security advisor, aimed to crack down on the leak epidemic in ways that were widely decried as heavy-handed: it required all federal employees with access to classified and sensitive information to sign non-disclosure agreements that would remain in effect even after the end of their employment, allowed for the use of a polygraph in investigating leaks, and called for the sanctioning of employees who refused. Newspaper articles and op-eds routinely decried the "obsession" with secrecy in the Reagan White House, saying that such measures had "Orwellian implications" and could be used to silence critics without regard to national security. Many of these changes would be, over the course of the administration, scaled back or eliminated as a result of both internal and external criticism.[133]

The Reagan administration was also imbued with the suspicion born out of the late 1970s. It was not, despite some critics' fears and allegations, a complete return to the Eisenhower period. It was Eisenhower by way of Ellsberg, a period of leaks and suspicions and conspiracies, in which secrecy was criticized at every turn. Secrecy was there, to be sure, as it had been under every presidency of the postwar period. But public attitudes toward it had become radically suspicious: in every scandal, a conspiracy was seen, whether it was the war in Grenada, the Iran-Contra affair, or the new security directives that seemed like they would lead to further concealment. When the Reagan administration labeled a Canadian anti-nuclear film as "foreign propaganda," it was given an Oscar in response (despite its being essentially a university lecture by Helen Caldicott, filled with enough inaccuracies to perhaps warrant the "propaganda" designation). What may have been a series of isolated incidents became an overarching narrative about a "cult of secrecy" that was attempting, one policy at a time, to manipulate and control the American people. The laundry-list of sins is somewhat humorous: increased fees for photocopying Freedom of Information Act requests, the denial of visas to a smattering of explicitly anti-American speakers, and a new push to enforce existing export control requirements added up to, as one op-ed put it, "peacetime censorship of a scope unparalleled in this country since the adoption of the Bill of Rights."[134]

As with the rest of the government, so in the nuclear realm. The Department of Energy's attempt to create a new category of Unclassified Controlled Nuclear Information, intended for information that was not formally classified but might be useful to terrorists interested in sabotaging or attacking sensitive facilities, was greeted with fear and contempt. DOE officials attempted to reassure congressmen and watchdog groups that they had no such intention, that their hope was to make the world a safer place, but such assurances fell largely on deaf ears.[135]

Those within the Reagan administration continued to invoke nuclear weapons as the ultimate justification for secrecy policies. At a debate hosted by the American Bar Association, a DOJ official labeled as a "chief architect" of the Reagan anti-leaking policies argued that secrecy was necessary because it was "a dangerous world . . . we face adversaries who have military power greater than any enemy this country has ever faced in time of war. We live with the ever-present threat of nuclear war, which can destroy the world as we know it."[136]

This is not to say that the Reagan administration was *not* moving in a direction of increased secrecy. But attempting to crack down on internal leaks is a perennial obsession, and there are always ebbs and flows in how classified information is handled. The anti-secrecy critique of Reagan was rooted less in the specifics of his policies than in an imagined "falling back" (which presumes previous administrations had been moving in a progression toward less secrecy). To say that these assertions are a bit dubious is not the point. What is interesting about the 1980s, in this respect, is that anything that touched on secrecy was fair game for criticism. This is the essence of the emerging and evolving anti-secrecy politics, a mainstreaming of a new kind of critique that, unlike in the 1970s, barely acknowledged that there were "some things that national security requires that shouldn't be talked about" and lacked faith in the government to make that determination.[137] The notion that the Reagan administration was uniquely secretive—the worst since the invention of the atomic bomb, declared one watchdog group—seems hard to sustain in light of what came before it, but the fact that such a criticism could exist and have mainstream impact is a sign that something profound had occurred with respect to the public discourse on secrecy.[138]

Perhaps chastened by the incidents in the 1970s that made it clear that trying to regulate outsiders was a recipe for very public failure, the Reagan brand of secrecy drew careful lines between insiders and outsiders. Insiders were people who had worked for the government, signed secrecy orders, or signed non-disclosure agreements: people who had, for the purposes of employment, entered into obligations that left them legally vulnerable. So when *Jane's Defence Weekly*, an open-source intelligence news source that had started in 1984, published leaked photographs from American spy satellites, the Reagan administration moved to identify and prosecute the leaker (an employee of the Naval Intelligence Support Center), successfully having him convicted of espionage and theft of government property, which came with a two-year prison sentence. Censoring or censuring *Jane's* was out of the question, but the leaker was fair game.[139]

Ironically, even Reagan could embrace anti-secrecy when it was applied to the USSR. During a summit in Geneva shortly after the Chernobyl accident in 1986, he declared that secrecy was a main difference between the two countries: "The contrast between the leaders of free nations meeting at the summit to deal openly with common concerns and the Soviet government, with its secrecy and stubborn refusal to inform the international community of the common danger from this disaster, is stark and clear." Aides reported to the press that Reagan had personally added these lines to the speech as part of his "tough" stance toward the Soviets. And to be sure, the Soviet secrecy over Chernobyl, both internally and externally and in contradiction with Mikhail Gorbachev's policy of *glasnost*, was a major factor in the accident and its consequences. But for the United States to portray itself as free of secrecy while attempting an aggressive campaign to shore up its own information control was more than a bit opportunistic and hypocritical, especially when American disclosures about accidents and releases in the nuclear weapons complex have always been less than full.[140]

The late Cold War, starting in the 1970s and moving toward the fall of the Berlin Wall in 1989 and the dissolution of the Soviet Union in 1991, saw a rising tide of anti-secrecy activism and a mainstreaming of the core tenets of anti-secrecy. With the increased size of the national security state and the growing perception of nuclear weapons as a unique existential threat, secrecy became not merely a *symptom*, but the *cause* of many national and global political ills. Secrecy became seen as perhaps the defining factor of the Cold War United States, and any Cold War fears and ills could be partially laid at its feet. What had been previously seen as an evil that might need to be tolerated as a *result* of an increasingly dangerous world was now being understood and cast as the *cause* of that danger.

Secrecy and anti-secrecy politics did not align themselves along strictly partisan lines, but they did tend to "clump." All administrations during the Cold War embraced secrecy when it served their aims, just as all embraced the national security state. But anti-secrecy tended to present itself as a liberal critique, perhaps because its core contention—that the state could not be trusted in its claims to security—tended to be part of a liberal critique. There were a few conservative critics of secrecy, but they tended to be the surprising exceptions to the rule. Attempts, especially in the post–Cold War period, to rebrand secrecy as a form of "government regulation," in an attempt to align it with a familiar conservative target, were not especially successful.[141]

Surprisingly, the most consistent conservative anti-secrecy advocate was Edward Teller, the "father" of the hydrogen bomb. As early as 1945, he had declared secrecy an evil to be overthrown, and throughout the Cold War he wrote editorials and op-eds opposing secrecy in the weapons complex. But Teller's critique of secrecy was not that of other champions of anti-secrecy. Secrecy, in Teller's mind, was a problem because it got in the way of scientific development, which was necessary for further technological development. As he somewhat coyly put it in a 1986 editorial comparing the development of nuclear weapons and computers:

> Nuclear weapons have been developed under a maximum amount of secrecy. The Soviet Union is probably ahead of the US. Practically no governmental secrecy, only mild proprietary limitations, has been introduced in the development of computer. The US is undoubtedly ahead of the Soviets in this field. Secrecy does not lead to security.[142]

Putting aside the contradiction of the Soviet Union's nuclear program flourishing under even more secrecy than the US had put into place, Teller's argument was a simple and consistent one: the US would have more, and better, nuclear weapons if it had less secrecy. Herbert York, a longtime colleague (and antagonist) of Teller's, interpreted Teller's anti-secrecy efforts more suspiciously:

> I can't get over the idea [that] the reason he's against classification is he believe[d]—and this is a serious thought—that the way to keep ahead of the Russians is to have everybody working on weapons design. Not just Los Alamos—everybody! So he want[ed] to declassify so [that] you can get every department of applied science in America working on nuclear weapons, and other weapons, not just nuclear.[143]

Teller's approach here is not so different from the common anti-secrecy critique during the Manhattan Project that secrecy would hamper US scientific developments in weapons areas as well as peaceful ones. More unusual is that Teller clung to this idea well after US predominance in weapons was demonstrated, and long after any of his colleagues had shifted their efforts into arms control. In any event, Teller's conservative argument for declassification, aside from being couched within the context of weapons development, was always a vague and blunted one, never really challenging the power structures or incurring any real wrath from them.

Nations are not homogeneous entities, and the US is nothing if not capable of sustaining large populations of people who subscribe to different worldviews. Within the US government, the Cold War secrecy regime continued even in the face of attacks from anti-secrecy activists. The *Progressive* case opened up some information about thermonuclear weapon design (to the delight of the laser fusion scientists), but it didn't fundamentally change the government's underlying assumptions. The only mode of operation it seems to have changed was to restore the "no comment" policy that had been dominant since the 1940s and had been violated only by the *Progressive* case itself. Whatever one thinks of Morland's case and its success (opinions are still divided today among the people involved), it did not, in the end, dismantle American secrecy, even if it fueled a new form of Cold War critique.

But the Cold War would not last forever; its end came more swiftly than anyone expected. One of the precipitating factors was secrecy reform, but not in the United States: Gorbachev's policies of *glasnost* and *perestroika* were part of the potent brew that led the Soviet Union to undergo a rapid implosion. In 1991, the Soviet Union would cease to exist—but American secrecy, nuclear and otherwise, would persist.

9

NUCLEAR SECRECY AND OPENNESS AFTER THE COLD WAR

It is time for a new way of thinking about secrecy.
MOYNIHAN REPORT, 1997[1]

For most Americans, the end of the Cold War was greeted with a sigh of relief. Almost overnight, the existential dangers of the previous decades had seemingly dissipated. The world was still a complex one, still held its threats and fears, but for an all-too-brief moment it seemed like the possibility of nuclear destruction had abated. Perhaps, many also hoped, the overwrought American national security state might itself also dissipate, or at least lessen.

Initially, there were promising signs. President George H. W. Bush did make some dramatic changes, nearly halving the nuclear stockpile in 1991–1992 through deep cuts in the numbers of both strategic and tactical weapons. The Department of Energy began a comprehensive review of its guidelines regarding Restricted Data, concluding in a report finalized in the summer of 1992 that the new international context allowed for some fundamental reforms of DOE classification practices. As the *Classification Policy Study* put it, the existing classification system was an outgrowth of the Cold War, and "much has changed since then." After 45 years of the Atomic Energy Act, "in view of the rapidly changing world situation," it was "time for a fundamental review of classification policy for nuclear weapons and nuclear-weapon related information." The overall conclusions were that while the primary goal of classification should be to deter proliferation and nuclear terrorism, the way forward would require greater international cooperation and greater attention to the impact of classification policy on "other US national objectives such as environmental cleanup, technology commercialization, and cost reduction."[2]

Though a contrast with Cold War priorities was part of newly elected President Clinton's campaign pitch, secrecy reform was not a major component of it. Within months of his election, though, public calls began for such reforms. "Someone should tell the CIA and FBI that the Cold War's over," one columnist put it.[3] Within the first few months of his administration, Clinton issued a presidential directive beginning a two-year process of complete review of national classification provisions, with the aim of drafting "a new executive order that reflects the need to classify and safeguard national security information in the post-Cold War period."[4] His 1993 appointment of Hazel O'Leary as the first post–Cold War Secretary of Energy signaled further changes. O'Leary was a lawyer who had worked in the Carter administration and had most recently been the executive vice president of the Northern States Power Company. Her style was non-technical and her interests were in modernizing the DOE, both in practice and outlook. She even looked different, a point not missed by commentators at the time: in a field of policy and science dominated by white males, she was an African-American woman.[5]

In December 1993, O'Leary launched what would become the hallmark plan of her administration. The Openness Initiative would attempt to implement many of the recommendations of the 1992 study, emphasizing broader reform of classification and declassification procedures, with a special emphasis on the release of historical information of relevance to "stakeholders": environmentalists, anti-nuclear activists, communities surrounding government nuclear sites, historians, other activists, nongovernmental organizations, and think-tankers, all of whom had been for secrecy reform since the 1980s. "We are starting with a simple piece to say that the Cold War is over," O'Leary announced at the press conference held in DOE headquarters. Her language was that of atonement, and of distancing the present attitude from the bad days of the past: "We were shrouded and clouded in an atmosphere of secrecy. I would even take it a step further. I would call it repression."[6]

To accompany the launch of the Openness Initiative, O'Leary had prepared a set of facts about the historical atomic energy program for release. First was the fact that 20% of all US nuclear tests were done without official announcement so that the USSR would not be able to monitor them. Second was the total amount of separated plutonium produced by the United States up until 1988 (89 metric tons). Third, she announced massive declassifications in the field of laser fusion, and that only approximately 20% of US research in the field would remain classified by the end of their review. Fourth, she released information about the use of mercury around the Y-12 complex at Oak Ridge, an issue of local environmental and health concern. Lastly, she announced that the DOE would be releasing information about the hundreds of "human radiation experiments" that had taken place over the Cold War, an act of unusual and extraordinary revelation about past ills.[7]

The human radiation experiments had been exposed somewhat earlier. An extensive Pulitzer-Prize winning account of the human plutonium injection experiments had recently appeared in the *Albuquerque Tribune*, documenting how between April 1945 and July 1947, eighteen Americans who had previously been diagnosed with terminal diseases had been injected with solutions of plutonium, with minimal standards of informed consent, to better understand the human absorption of plutonium for the purpose of establishing plant safety standards. The lack of consent made it scandalous even in 1947, when the AEC had opted to bury it with classification rather than worry about the "adverse effect" of such publicity.[8] When O'Leary heard about the experiments, she was "appalled, shocked, and deeply saddened" and announced that there were some eight hundred total human radiation experiments over the course of the Cold War, most conducted according to the standards of consent in their time but woefully inadequate by the research standards of the 1990s.[9]

O'Leary worked to reframe nuclear rhetoric away from secrecy and instead around trust and "stakeholders." All classification policy going forward, she explained, would be determined only after extensive discussion with these outside groups. The DOE was in the process of reviewing some 32 million pages of documents for declassification and was planning, she announced, to substantially revise the Atomic Energy Act. It was paramount, she argued, to embrace government transparency in the post–Cold War world and to rebuild public trust in the DOE. During the following question and answer session with journalists, O'Leary was pushed for more details on the human radiation experiments. The Secretary of Energy explained that if it was up to her, she'd release everything immediately, but the lawyers were holding her back until privacy and liability issues had been fully worked out.[10]

O'Leary's DOE struggled to rebrand itself as something different from the notoriously secretive and paternalistic AEC. "We have listened to everyone," she claimed at a May 1994 event, "from Edward Teller to Greenpeace."[11] For years, Openness and its revelations would get considerable press attention, serving as an exemplar, whether deserving or not, of the post–Cold War focus of the Democratic administration.[12] Whether the focus on scandals was actually productive is questionable; it may have simply focused more attention on the possibility of government misdeeds, fueling an anti-governmental anti-secrecy rhetoric that targeted O'Leary's organization as well as all previous.[13]

In June 1994, at the second Openness press conference, O'Leary said that they had declassified "119 separate formerly secret facts" that year alone, without elaborating on exactly what that might mean. She said they were now going to release information about the long-term environmental damage done to the Marshall Islands during the years of nuclear testing and the contamination of the Rocky Flats fabrication plant in Colorado. Declassifications, she claimed, were happening at a greater rate than ever before: what would have previously taken years to review was now being done in months.[14] Over the course of the 1990s, Openness became an opportunity for greater liberalization of classification policy, increased outreach by the DOE in circulating information, and publicity. Symbolically, in 1993, the DOE Office of Classification was renamed as the Office of Declassification, reflecting the new priority and mission. The goal of the secrecy apparatus was to release information, not hold it back.[15] Taking this to perhaps its ultimate end, in 1996, the DOE Fundamental Classification Policy Group recommended that the government should jettison the "born classified" concept completely, removing the "special" status that had been given to nuclear weapons secrets since 1946, rendering it just another form of classified national defense information.[16]

But Openness eventually ground to a halt. There are indications that within the government, past the friendly press releases and the appearance of a unified front, feelings about Openness were mixed. In the US nuclear weapons laboratories, anxieties over the end of nuclear testing (the last American nuclear test took place in September 1992) had produced uncertainty over the future of the nuclear establishment. Within the classified rooms of the Los Alamos and Livermore laboratories, discussions began over how the labs might stay relevant by shifting more of their work into unclassified areas. But those who worked on nuclear weapons design appear to have been largely in favor of continued classification for their work, believing it was essential for stemming proliferation.[17] At a January 1994 stakeholder meeting, the director of the Office of Declassification, A. Bryan Siebert, who had been pushing for the revision of policy since at least 1992, told those in attendance that there had been resistance within the DOE to the new outlook. The old guard of the DOE, as Siebert characterized it, had gotten used to the DOD being the only "stakeholder" of interest, their sole "customer." They had been trained to think in terms of keeping things secret and did so both as a compulsion and because they enjoyed the exclusivity it gave them. They had been told that they had to get with the "different philosophy."[18] Over the course of O'Leary's tenure, according to Siebert in 1996, the DOE declassified more information than it had done in the entirety of the previous efforts of it or of its predecessor organizations combined, having reviewed some 300 million pages of material, which didn't "even approach the size of the problem."[19]

The push against Cold War secrecy was not restricted to the bomb, of course. In March 1997, a voluminous "Report of the Commission on Protecting and Reducing Government Secrecy," the product of a Senate committee chaired by Daniel Patrick Moynihan, was released to the public, following several years of investigation into the history, practice, and even philosophy of secrecy. The Moynihan report ultimately concluded that secrecy had produced a negative effect by warping American politics in ways that were harmful to the nation, and argued that secrecy itself needed rethinking and, perhaps, reframing. "If the present report is to serve any large purpose," Moynihan argued in its introduction, "it is to introduce the public to the thought that *secrecy is a mode of regulation*." In an era where attacking regulation was a common conservative talking point, this framing can be seen as an attempt to make secrecy reform a bipartisan issue.[20]

But the politics of openness and secrecy were still politics, and the post–Cold War period was highly partisan. In 1994, the Republican party took control of both houses of Congress and began sustained attacks on the Clinton presidency. One area they focused on was an alleged weakness on national security, especially with respect to the People's Republic of China. In the mid-1990s, allegations emerged about China funneling money to the Democratic National Committee, and further reports alleged illegal technology transfer deals between the Chinese and the United States, in possible violation of export control laws. In 1998, a new House select committee was created to investigate these interactions, chaired by Representative Christopher Cox.[21] The Cox Committee would report in early 1999 that the Chinese had somehow managed to steal classified design information on "all of the United States' most advanced thermonuclear warheads," revealing that while some of this espionage had taken place over the course of two decades, some was as recent as the mid-1990s.[22]

Less than a decade after the Cold War ended, the fear of nuclear secrets and their loss was back, and with a vengeance. Clinton-era Openness was equated with laxity and sloppiness. Even seemingly innocuous attempts to explain basic weapons concepts were seen as weakness: "Visitors to Los Alamos National Laboratory are provided a 72-page publication that provides, among other things, a primer on the design of thermonuclear weapons," the Cox Report explained, showing the simplistic (two spheres in a square) version of the Teller-Ulam design that had been declassified after the *Progressive* case. The substance of the allegations was that the Chinese had stolen US weapons "codes," computer data that correlated theoretical bomb knowledge with decades of nuclear testing. Using said codes, the Cox Report argued, the Chinese had been able to modernize and miniaturize their warhead designs without the extensive nuclear testing the US had required to develop the same technology.[23]

By that point, O'Leary was already out, having resigned due to an unrelated scandal, and the idea of Openness was increasingly seen as a political liability that had mollified few critics and drawn lots of attention to past governmental misdeeds. Fears of Chinese weapons had stepped in to fill the gap left by the Soviet Union, nuclear terrorism fears had been rekindled, and the new nuclear powers of Pakistan and India (both of whom tested warheads in 1998) made it clear that any relief at the end of the Cold War was short-lived. In late 1998, Congress passed a new law requiring the DOE to review all declassified documents for possible "inadvertently released" Restricted Data, setting off a laborious and costly process that threatened to effectively "re-classify" innumerable documents that had previously been available to the public.[24] The 1990s even experienced its own genuine nuclear spy scandal. The case against Wen Ho Lee, a Taiwanese-American computer programmer, would eventually collapse—he was found to have mishandled classified data but not to have participated in espionage—but it indicated to many a return to Cold War norms, fears, and hysteria.[25] Due to these factors, the reforms ultimately ground to a halt.

Another new factor was the cheap availability and accessibility of networked digital technologies. Though publicly framed in positive terms, the free movement of information on scales that would be almost unimaginable only a few decades before, had profound implications for information control. Information freely flowing tends to flow to undesirable places, as the massive leaks by Chelsea Manning and Edward Snowden would make unnervingly clear. Well before these leaks, however, weapons laboratories struggled with the fact that a single hard drive could store an incredible amount of classified information and was physically quite small. In mid-2000, two hard drives were lost at Los Alamos, making national headlines and provoking an FBI investigation that involved polygraphs and interviews. They were later found behind a photocopier.[26]

The terrorist attacks against the United States on September 11, 2001, triggered new waves of fear and new fears of secrecy. As before, these went hand in hand: some new restrictions of information, each accompanied by worries about whether we were returning to the Cold War, or something even worse.[27] Much of this anxiety in the nuclear realm expressed itself as fears of foreign secrets, but among countries different than the traditional enemies of the United States (e.g., Russia and China).

Nuclear terrorism fears resurfaced, but this time with a new locus: the collapse of the Soviet Union had left the Russian Federation in a decade of disarray, including its nuclear weapons infrastructure. American efforts had been made to funnel money into the country to help the state keep control over its nuclear stockpile and fissile materials, and efforts were made to keep its nuclear experts fed and employed, lest they decide to work for the highest bidder. Fears that these efforts had come too late or had failed fueled concerns that enriched uranium or even "loose nukes" had made it onto the black markets of the world. What had largely been a domestic discourse of secrecy was now focused almost exclusively on threats from abroad.

And the fact that there actually was a nuclear black market of sorts also emerged: A.Q. Khan, the Pakistani metallurgist who had been employed at a Dutch commercial centrifuge enrichment plant in the 1970s, was identified as a key agent in the selling of nuclear expertise, technology, and weapons designs to pariah states (Iran, Libya, and North Korea, among potentially others). Many of Khan's offerings were material: cast-off centrifuge designs from the Pakistani nuclear program. Some were informational: nuclear weapon designs. Much appears to have been ultimately logistical: international contacts for a web of producers necessary to manufacture the specialized components.[28]

In 2003, the US began a war against Iraq, ostensibly to find weapons of mass destruction. Though none were found, the remains of a long-aborted Iraqi nuclear program were analyzed at length, part of a secrecy regime that had been dismantled along with the country that had created and maintained it. (Many Iraqi records were even briefly put onto the internet by the US government, until Western experts noted that the files contained what the US government considered nuclear secrets.[29]) The Iraqi program seems to have been heavily reliant on open-source information about nuclear matters, even choosing the by-then quite outdated electromagnetic method of enrichment because it was the most easily accessed, having been declassified in the 1950s.[30]

In 2003, North Korea withdrew from the Nuclear Non-Proliferation Treaty and began reprocessing plutonium from its Yongbyon nuclear reactor facility. By 2006, it had detonated the first of what would be half a dozen nuclear warheads, including one plausibly claimed to be a thermonuclear warhead in 2017. Unlike any of the other thermonuclear nations, North Korea showcased its bombs, releasing warhead casings to the international press, including an intriguing peanut-shaped hydrogen bomb that, it claimed, could be fit into the nose of their long-range missiles. This performative openness of course served a purpose: to convince the world, and especially the United States, that its nuclear capability was real and dangerous, in order to ward off any thoughts of military action against the state. This deployment of publicity should, of course, be understood not as a true striving toward openness (North Korea still deploys secrecy reflexively and routinely), but as exactly the same kind of selective "information control" honed in the earlier years of the Cold War by the United States.[31]

The secrecy surrounding Iran's nuclear program and intentions has driven international concern and speculation for decades. The language of secrets, and the inference that secrecy indicates nefarious intent, surrounds political discussions of the subject. Iran has declared that it had a secret weapons program in the past, but that it stopped, but suspicion led to demands to open up clandestine sites. Iran in turn has claimed that while the sites are secret, they are not *nuclear* secrets: the Nuclear Non-Proliferation Treaty allows only inspection of nuclear sites, and "regular" military secrets and sites are allowed to remain sacrosanct. Scientific measurements can help contest such claims (such as soil samples that reveal the presence of uranium enrichment), but ultimately they appear to remain essentially contested, as Iran either finds itself in the position of trying to prove a negative (if you are sympathetic to their claims) or deploys secrecy to further cloud its goals (if you are suspicious). Time will tell which, if either, of these is the correct interpretation.[32]

The state that has gone the furthest in publicizing alleged Iranian secrets is that of its regional rival Israel, who has gone to great lengths to publicize the "laptop of death," an alleged Iranian computer recovered that was full of "secrets." Some of these alleged secrets look somewhat banal: power curves showing how a nuclear weapon releases its energy over time; a "spheres within spheres" diagram of an implosion nuclear weapon; and other forms of representing nuclear knowledge that are declassified in the US. Such is the strangeness of this modern state of secrecy: declassified drawings can be used to establish dangerous work under the allegation that they are classified in the country that they were created.[33]

Israel itself remains a conundrum of secrecy and publicity: it is well known to be a nuclear weapons state, despite its having never acknowledged it publicly. This policy of "nuclear opacity" looks, from a distance, like an attempt to have it both ways: by allowing the world to know, obliquely, that the state has such weapons, it gains the benefits from nuclear deterrence. By not officially acknowledging that it has them, it gains the diplomatic benefits of being an American ally despite not being a member of the Nuclear Non-Proliferation Treaty. Israel's bomb is, as the scholar Avner Cohen has called it, "the worst-kept secret"—

an inversion of the trope of the Manhattan Project's secrecy glory—but that, ironically, may make it the most *useful* secret for Israel's specific international political context.[34]

But as secrecy returned to the political fray, so at times did openness. As the idea of secrecy had dominated the criticisms of the George W. Bush presidency, the campaign and early presidency of Barack Obama focused explicitly on creating an "unprecedented level of openness in Government," as a memorandum from the president to all executive departments and agencies put it.[35] By 2010, this led to the release, for the first time ever, of the current stockpile size of the US nuclear arsenal. To great press attention, the Department of Defense revealed a poorly rendered computer graph of the US nuclear stockpile, and a number (5,113) that surprised nobody—a fitting anticlimax, perhaps.[36] But as with previous pushes for "openness," the political realities pushed back: plagued by leaks, notably those enabled by the ease of transferring data via digital and networked technologies (such as the massive leak of State Department cables through the organization WikiLeaks in 2010 and the leaks relating to the National Security Agency provided by the contractor Edward Snowden), the Obama administration developed the dubious reputation of being one of the most litigious when it came to prosecuting leakers and whistleblowers, to the dismay of many of its supporters.[37]

I am hesitant to try to extend this narrative beyond this point, for as we have seen, what is visible on the surface of America's secrecy regime is often but a pale reflection of what is happening in the layers underneath. The secrecy regime of the early twenty-first century appears to be a combination of several things, some new, and some old. In the category of "old": the laws and restrictions appear to be essentially the same as those that were developed for controlling information (and people, and spaces) during the Cold War. The changes made to them, such as the PATRIOT Act, seem primarily to have reinforced these modes of operation. The practices of enforcing secrecy have evolved only slightly, to cover some new cases brought on by a digital, networked age. The institutions have also changed slightly, but not in ways that feel especially meaningful: the DOE split off some of its weapons functions into the National Nuclear Security Agency, but its general approach seems the same as before. If anything, it is the fact that so little has changed, despite the now many decades separating the end of the Cold War from the present, that is most striking.

In the "new" category, there are some discursive shifts. The rhetoric of anti-secrecy has become so mainstreamed that major political figures across the spectrum have invoked it, even if their engagement with it is unclear. On both sides of the political spectrum, anti-secrecy has become merged with a form of paranoid politics, imagining conspiracies (which does not mean they do not exist) and fueled by leaks. Nuclear secrecy itself seems primarily to rear its discursive head when talking about foreign concerns (e.g., North Korea or Iran), and is rarely reapplied domestically. As I write this, current discussions of secrecy broadly seem less concerned with the national security state (except when a political figure can be accused of "mishandling" secrets as a form of political attack), and with broader fears of economic misdeeds. Fear of Chinese scientific and industrial espionage are dominated by whether or not secrecy could be used to increase the "economic competitiveness" of the US, and may signal a new shift in this dialogue within universities—there have been increasing calls as of late to restrict Chinese students from participating in American university programs, not because the information they might learn is classified (it is not), but because of fears they will gain access to proprietary commercial information that would be useful to their home country. Even for a historian who frequently points out that the ethos of scientific openness has always been more of an ambition than a true "norm," it is shocking how quickly policymakers and analysts have accepted the premise that they should attempt to restrict the exchange of unclassified knowledge along national lines, and in universities specifically.

The idea of nuclear weapons as the ultimate secrets still exists, but it manifests in strange ways: talk about anything relating to nuclear secrecy on social media, and someone will inevitably comment that we're all "on a list now," or imply that the National Security Agency, FBI, or CIA is also silent party to the conversation. This is all done in jest, but it is telling in the ways in which the early twenty-first century is a time in which secrecy and privacy have become linked in their mutual attention from a state that is considered (with good reason) to have newfound powers of scrutiny and observation.

CONCLUSION

THE PAST AND FUTURE OF NUCLEAR SECRECY

World War II is over, but the Battle of the Atom still goes on.

CONGRESSMAN WALTER ANDREWS, 1946[1]

When Congressman Andrews wrote of the continuing Battle of the Atom, he meant only the legislative battle over domestic atomic energy legislation. But as we have seen, the Atomic Energy Act of 1946, and its "Restricted Data" concept, did not stop the battles. If anything, it entrenched them. We live in a world vexed by the same sorts of fears and hopes over nuclear weapons and their control (and lack thereof) that began shortly after the bomb itself was conceived and continued through the decades that followed. World War II is over, and so is even the Cold War, but indeed, the Battle of the Atom still goes on. And the problems of secrecy still sit squarely at the center of that battle.

Over the course of this book, we have looked at some eighty years of this history. We have seen how the idea of nuclear secrecy was initially born out of a very specific fear—that the world's worst nation could achieve nation-destroying power—that, over the course of many transformations, morphed into something more generalized. More than totalitarian regimes were the target of this secrecy: allied nations, private industry, democratic institutions, and "the public" more broadly became understood as targets as well, for reasons both justifiable and not. The secrecy problem moved wherever the problem of the bomb went, which was far indeed. The desirability of control over nuclear knowledge became totalizing because the threat of loss of control was tied to the almost unimaginable power of the bomb itself. These two forces—the desire for control, the fear of its loss—became locked in a vicious cycle, each driving the other.

But despite this, the institutions of nuclear secrecy, while at times quite powerful, could also prove quite brittle in the American context. Turning that fear into reality provided truly difficult, for the forces mitigating against total secrecy were also powerful. The ideal of freedom of expression, enshrined into the US Constitution and core among its stated civic values, has never been absolute, but it persists nonetheless. The ideal of scientific openness, while never as powerful or all-encompassing as scientists sometimes claimed they were, does exist, and does embody another counteracting power against absolute secrecy. From the earliest days of nuclear secrecy, it provided a rhetorical answer to those in favor of secrets: too much secrecy would stifle the creation of knowledge, dangerous or otherwise. And the desire for peaceful and profitable nuclear technology also mitigated against the secrecy, for better and worse, and worked to produce an enduring-if-flawed regime that attempted to strike a wary compromise between these competing impulses.

And so the discourses of secrecy and anti-secrecy swirled around each other from the very beginning. The tension between them appears to be the most definitive and generative component of the American context of secrecy. There have been nations with greater standards of openness, and certainly there are nations with more powerful and totalizing secrecy regimes. In the United States, one finds both strands of thought in excess: two powerful, compelling ideas that are constantly warring, each vying for both rhetorical and political power, producing controversies, conflicts, and bizarre "hybrids" of their interactions, like an atomic-bomb-drawing trucker who both embodies and rejects the idea of the secret simultaneously.[2] These are the parts of this history that feel distinctly American, because they embody a conflict of values, and America is nothing if not a place built upon values in conflict: a simmering mix of high-minded idealism and ugly, fearful power.

In the years after the Cold War, the United States has found itself in an uneasy position. Nuclear weapons did not vanish, and some nuclear threats are arguably as bad as they had ever been, if not worse. We still live in a world where nuclear proliferation is a major concern, where superpower rivalries and arms races are still playing out between the United States, the Russian Federation, and the People's Republic of China, and where the risk of nuclear terrorism seems simultaneously unknowable and too plausible. And yet, the secrecy regimes that were developed and deployed over the course of the Cold War have been, since the late Cold War onward, viewed by a large segment of the population and political classes as unaccountable, unworkable, expensive, and ineffective. There is a pervasive, conspiratorial, bipartisan suspicion of secrecy in American political thought, and there are deep questions about the legitimacy of government claims to control information in the name of national security. This suspicion allows even mundane information controls to be construed as the presaging of an Orwellian dystopia.

And yet, the secrecy has persisted, despite all attempts to undertake systematic reform. There is no sign that it will be going away, despite the fact that at some level it is acknowledged by all involved that it is impossible to maintain a permanent, effective secrecy regime over technical information in a world where there are multiple technically-competent entities and a lack of totalizing control. The number of potentially existential technologies has increased: increasingly, there are questions from people in communities working on emerging scientific fields as varied as synthetic biology, artificial intelligence, and climate geoengineering who are looking to the history of nuclear technology for suggestions to alleviate the fears of technological misuse.

So this is the point where we ask, as people living through the first decades of the twenty-first century, *what should we make of all of this*? It is easy enough to point out that the past informs the present: the references are quite literal ones, with the same questions, themes, sites, technologies, and even historical figures (such as the ever-persistent Edward Teller) looping back again and again on the question of secrecy and nuclear weapons. It has become clear to all at this point that the end of the Cold War was not the panacea that it might have been hoped to be: some things changed, some things reconfigured, but ultimately many of the structures of power were able to adapt, persist, and multiply. Among them, nuclear secrecy has also persisted.

Why? Part of it is that there has never been a simple, singular thing called nuclear secrecy. What we call "nuclear secrecy" is really a bundle of many different ideas, desires, fears, hopes, activities, and institutional relationships that have changed over time, at times dramatically. What began as a conversation between two scientists (Szilard and Fermi) has since expanded into a system that is now routine for literally millions of Americans, and has become so taken for granted that many cannot imagine how it could have ever been any other way.

These sorts of deeply rooted regimes cannot be simply waved away, any more than an economic system can be waved away. It is of note that for some states, the only remedy for massive secrecy systems was the total collapse of the state (East Germany, Hussein's Iraq), and in some cases, even that was insufficient (post-Soviet Russia). Secrecy systems are not merely policies that can be switched on and off; they became deeply embedded with the functioning of the state. It is easier to imagine the elimination of nuclear weapons than an elimination of the secrecy surrounding them.[3]

And while there have been attempts at reform, the reformers, once on the "inside" of the system, have tended to find themselves faced with a much more difficult task than they realized. There is no more acute example of this than the transformation of David Lilienthal. In 1945, Lilienthal was thoroughly convinced of the foolishness of secrecy and the need for transparency. By 1946, once he had been exposed to "secrets," he was no longer so sure how that would work. By 1947, he was engaged in activities his previous self would have reviled. By 1949, he had resigned in anguish. Is this the inevitable path of the idealist? It was not merely the "system" that warped Lilienthal: it was the change in perspective that accompanied the transformation from an outsider with limited knowledge and responsibility to an insider who was suddenly confronted with the high stakes of the nuclear world, and tasked with taking personal responsibility for it.

This is not to say that reforms are impossible or that the "classification pendulum" is doomed to swing back and forth ineffectually forever. But anyone who seeks to enact serious, lasting change to how we think about or act upon nuclear weapons secrecy should heed the lessons of this history. The first and foremost is that the idea of nuclear weapons has been, and may always be, tightly bound to the idea of knowledge-as-power. Whether it should be or not is somewhat beside the question: both popular and expert understandings of nuclear weapons are tightly linked to discussions about science and engineering, and both expert and lay understandings of science and engineering revolve around knowledge at their core. The result of this is that any approach to nuclear information, whether an advocacy for restraint or release, is going to be inherently controversial and political, because it will be always construed (again, rightly or wrongly) as potentially enabling terrible consequences to come from it.

To put it bluntly, if knowledge is power, then nuclear knowledge is quite a lot of power. There are those who have pushed against this, such as Oppenheimer's attempts in 1946 to reframe the control of technology as the control of materiality, or later sociologists of science who have emphasized that knowledge, devoid of a social context necessary to make it operationally useful, poses no threat.[4] But these approaches are far more counterintuitive, and potentially incomplete. While the exact role of "information" in nuclear proliferation, as opposed to other factors like organizational context and political will, is an area of considerable scholarly disagreement, practically nobody disputes that there is some amount of "dangerous information" that ought to be controlled in the name of protecting against nuclear proliferation, innovation, or terrorism.[5]

And even the mere perception of information control or release can become an effective weapon in a political dispute. There is no position that is truly unassailable in an age that still fears such weapons *and* which is dramatically suspicious of government overreach. The political world we live in has both of these elements well-represented, and routinely weaponizes both.

The anti-secrecy politics that gained significant mainstream support in the 1970s has similarly not created deep structural changes, though it has occasionally had important victories, like the Freedom of Information Act, which while flawed, provides some institutional structure for forcing a review of specific instances of secrecy (but does not challenge the ability of the government to classify information as secret). Ultimately the claims of anti-secrecy have tended to devolve down to questions of competing expertise, which make them hard to resolve. As an example, nearly every anti-secrecy advocate that I have interviewed has stated that *of course* there are "some secrets that need to be kept"— to insist otherwise is to appear dangerously naïve. But who makes that determination? The answer is always unsatisfying: someone other than whomever is currently doing it, because presumably they are doing it poorly or for ideological reasons.

But when we open the files and look at these matters through the eye of the redactor, we rarely find things so straightforward. One can, to be sure, find many cases of abuse of classification and "silly secrecy" (the latter being a broad category including "things that are actually well known," "things that were never secret to begin with," and "ridiculous policies, like keeping people from 'secrets' they themselves created"). But in most cases, the line between what is safe and what is not is tricky, dependent on projections of harm and benefit, and dogged by a lack of clairvoyance. And though much of Lewis Strauss' philosophy of secrecy is odious and paranoid, his notion that once "out," a secret cannot be easily returned to its unknown state appears frustratingly true, and even more so in our hyper-archived, highly scrutinized, networked moment. Information is notoriously difficult to control; this can be both a statement about the futility of secrecy as well as one about the caution to be taken in releasing information.

Critics of secrecy who do manage to get "inside" such systems have tended to find that not only are there significant institutional forces that maintain the status quo, but also that while it is easy to say that some secrets "need to be kept," when one looks at practical cases of secrecy it quickly becomes apparent that secrecy is "sticky." Let us imagine we all agree that, in principle, information that would materially aid a terrorist in acquiring a nuclear weapon should be kept under wraps. What information would fall into this category? Would the location of fissile material stockpiles and information about how securely they are being kept fall into this definition? How about information about how a terrorist would shield stolen uranium-235 from detection? If fundamental information about uranium-235 revealed the latter, would it too be swept up in the secrecy, if it was plausible it could save lives? How far down the "rabbit hole" are we willing to go in terms of indirect connections to the threat we fear? Would we instead get pulled into the safeguards vs. secrecy debate, asking whether "information" is really the most important vector to focus upon? How likely are we to wage the fate of thousands—potentially millions, in some cases—of future lives on a decision about releasing information? How likely are we to wage our nation itself on such decisions? Is it at all possible that the positive value of releasing the information could compete with the hypothetical, worst-case scenario negative outcomes? Individually one can imagine people being intellectual about these points; as a nation, it is clear where we have thrown in our lot. We would need to have a *very* compelling replacement for secrecy as a form of control ready to deploy if we were trying to overturn the status quo.

We can easily see that even a group of well-informed and well-meaning people is going to find this a difficult problem, even assuming they have the same values and understanding. As Oppenheimer put it impoliticly behind closed doors in 1949: "The fiction that some supreme intelligence exists that can paint red for danger, and green for caution—that's not so. . . . I don't believe a disembodied agency has a Chinaman's chance of wisely deciding what to make public."[6] But where does that leave us, if we believe that some of these lines must be drawn? Such is the quandary the reformers find themselves in, once they are forced to translate their rhetoric into practice, and why they are assimilated by the secrecy regime into yet another form of classification officer.

Any attempt at reframing this issue around "balance" is going to be anemic and ineffectual. There is nothing inherently wrong with the idea that in a democratic society there needs to be a trade-off between security and transparency. But the ultimately toothless nature of such an idea is exactly what dooms it: it is a statement that everyone, from the ACLU through the National Security Agency, can endorse, because it does not give any real policy determinations.[7] In its blandness, the "balance" idea assumes that secrecy or openness is a simple property that can be weighed and considered. And as we have seen, secrecy and openness both take many forms. What, for example, does a "balanced" approach to clearance investigations look like? Does it mean not probing too deeply into an individual's personal life? How would such a thing be made into regulations and rules that could be implemented by the thousands of FBI agents who would need to carry out this work?

A better approach is to avoid the "balance" and instead try to address the harm directly. So one might say, sexual preferences should not be a consideration in the granting of security clearances (currently, sexual orientation is not a criteria for clearance denial, but sexual behavior can be, as it is seen as indicative of character, stability, and reliability). The major problem with this approach is it is largely piecemeal, and frequently ineffective. It is one thing to pass a law that says that "information shall not be classified in order to conceal inefficiency, violations of law, or administrative error."[8] It is another thing to actually enforce it. To my knowledge, nobody has ever gone to prison for engaging in overclassification. If the harm inflicted by secrecy is systemic and massive, then a systemic and massive answer is required.

The difficulty of real reform is also a consequence of the historical evolution of the nuclear classification system in the United States. As we've seen, the legislative aspects are relatively limited: they define (both through the Atomic Energy and Espionage Acts) the importance of regulation of information, give some guidance as to the type of information that is regulated, and allow for the application of criminal and civil penalties should the laws be violated. But this "high-level" approach ultimately devolves downward to the regulations promulgated by the executive branch, both in the "high-level" requirements of the Presidential Executive Orders that periodically change procedures and in the more "low-level" regulations developed by individual agencies. What this means in practice is that any reform is easy to undo: they require far less consensus than the writing of laws, and the judicial branch has typically given the executive a very wide leeway in exercising its judgment on these matters. This is one of the reasons the "classification pendulum" seems to swing so freely at times, administration to administration, and no amount of swinging seems to get rid of the "pendulum" itself.

Of course, it is worth asking: does nuclear secrecy actually *work*? As we have seen when looking at Groves' motivations for secrecy during the Manhattan Project, it is clear that the question of what "working" means is itself a complicated one. The stated reasons for secrecy in the Cold War period and beyond have been about differential advantage (over the Soviet Union, China, etc.), nuclear proliferation, and nuclear terrorism. As noted, it is not entirely easy to judge whether the secrecy has been effectual in any of these cases: the Soviet Union and China managed to develop their own nuclear capabilities apparently unhindered by American secrecy efforts, and at times in spite of them (due to successful espionage attempts). Whether secrecy deserves any credit for any slowed or averted nuclear proliferation is unclear—it may be that sensitive information is less important than many other factors (such as export control over difficult-to-fabricate technological components, diplomatic interventions and treaties, and other matters). And with nuclear terrorism, material safeguards seem likely to be the major reason for lack of non-state actor acquisition of these weapons, not secrecy.

Which lends credence to the anti-secrecy position: that the regime of nuclear secrecy is at best a form of "security theater," meant to make its practitioners, and the American populace, feel safe, while providing no benefit; and at worst, it "works" by prohibiting engaged public deliberation around national nuclear policies. But the stakes of being wrong on this question are immensely—impossibly—high. What if nuclear secrecy has slowed things down, or avoided nuclear acquisition by hostile forces altogether? Would it have been worth all of the harms done to American democracy, science, and the lives it painfully touched? How could we know? As someone who has thought about this for a very long time now, I admit: I still don't know, and I am not sure I ever will. And in the absence of certainty, what are our options?

It is easy, as someone who does research with secret sources, to end up in an oppositional, antagonistic relationship with the government. Wanting to know something and encountering a "DELETED" stamp is not the same as simply not finding information one hopes to find. The stamp screams a different message: the information *is known*, but *you are not allowed to know it*. Why? Because it is *important and dangerous*. And nothing feeds an outsider's hunger for knowledge like thinking it might be important and dangerous. To know that somewhere, at some point in time, some bureaucrat decided that he or she could know the information but *I cannot*, is frustrating. To be told that the government *knows best* on such matters, in the face of many examples of their clearly *not* handling classified information correctly or drawing the line between classified and unclassified justly, is also frustrating. And so it is easy to imagine an evil redactor, high behind his or her wall of secrecy, keeping me from doing my work, learning my truths, and sharing the stories I think are important.

But those on the outside of the secrecy regimes should resist this characterization. Careful attention to how the secrecy system worked in the past, and talking with participants in the system today, make it clear that the reality is far less dramatic. Our modern-day classification system is banal. There is no interesting ideology behind it anymore: there is simply the enforcement of laws and regulations long-since written. This is, of course, an ideology of its own—an acceptance as status quo of ideologies that were baked into the system's origins—but it is a far cry from the radical and imaginative battles over secrecy that took place during the early Cold War, when there still seemed to be many possible futures available.

Today we have an overgrown bureaucracy attempting to manage an unfathomable amount of information, some of which they truly believe actually could cause grave harm in the wrong hands, sometimes with good reason. And for everyone outside of a security clearance system, the "wrong hands" and "our hands" are unfortunately synonymous, and to try to differentiate between the two would require a clearance system, which is to say, a secrecy regime. We might disagree on whether a given fact, idea, or document should still be classified, but the essential setup—that some will have access, and some will not—means that such disagreements will outlast anything short of a complete disintegration of the secrecy regime, which seems unlikely without either a radical restructuring or destruction of the state in which it is embedded. This is not to say that some regimes are "better" or "worse" than others, or that they cannot change. The US nuclear secrecy regime has changed in ways both subtle and dramatic over time, for better and worse. And if the past is an indication of the future, these changes will continue, again for better or worse, both mirroring other changes in the historical context, and affecting them.

Secrecy is not a wall, though secrecy regimes can feel like one to those on the outside. If anything, secrecy is a door, albeit one with a lock on it, and with keys that are only selectively distributed. For what is a door but a temporary wall, one where access can be controlled and modified over time? There must always be some kind of access to secrets, and some kind of movement of the information, if the secrets are to be made "useful" in any way. For what is a room with all walls and no doors? Functionally, it's no different from a tomb. Looking at how historians have written about the bomb over the last few decades makes it clear that while the door doesn't open as widely as perhaps we'd like, it has indeed let a lot out over the years. This is not a call to embrace the status quo, or complacency in the face of government denial of information—but it is an acknowledgment that over the long run, despite their apparent solidity, secrecy regimes do wither away, eventually.

In recent years, the history of nuclear secrecy has been invoked in thinking about emergent technologies that once again seem to hold the secrets of life and death. Emergent areas of biology (such as synthetic biology, which sits at the intersection of nanotechnology and cellular biology) have been particularly prone to such soul-searching, as scientists and non-scientists alike have wondered whether the ability to build viruses out of pure information will auger new horrors. In 2006, after virologists created living, synthetic polio virus out of a digitized genetic code and published their methodology in *Nature*, such calls went from being theoretical to very practical, and questions about the self-censorship regime of Leo Szilard seemed newly relevant.[9] Similar questions have emerged around other fields that are still relatively nascent but whose potential could be immense, such as artificial intelligence.

I am asked occasionally whether the history of nuclear secrecy gives lessons as to the benefits and pitfalls of using secrecy to control potentially dangerous science in non-nuclear fields. The answer to whether secrecy *could* be used (not *should*) seems to depend a lot on the technical specifics of the field in question. In the case of nuclear weapons, secrecy appears to have had very little effect on preventing their spread: the chief difficulties in acquiring nuclear weapons have not been informational, but rather material and political. Information certainly plays a role—you cannot build a centrifuge from spare parts alone, much less an intercontinental ballistic missile—but ultimately the value of "secrets" appears overblown. The line between a nuclear power and a non-nuclear one is not determined merely by whether one knows a given number of facts that the other does not; it is determined by the infrastructures that have been built up to develop the technology, and the political choices that have created them. This is not to say that information cannot be dangerous. It is just to say that information itself is not sufficient, and appears far less important than the other factors.

Attempting to use secrecy to stop proliferation has not been successful on the whole; at best, it might be credited with increasing the time and cost of nuclear acquisition, but even that is hard to know. One cannot really know, for example, what the value of Fuchs' espionage information was in "speeding up" the Soviet nuclear program—all estimates are uncertain hypotheticals. The sort of information that any single Fuchs could provide, such as weapon design details, ultimately is only a tiny component of the larger effort necessary to make such a weapon, and frequently the kind of thing that could be relatively easily acquired by foreign scientists. Much harder to develop are the industrial techniques, tacit knowledge, experience, organizational infrastructure, delivery vehicles, and so on. It is not at all clear that any nuclear aspirant has *ever* found that a lack of secret knowledge was a significant hurdle toward their acquiring a weapon, or that it caused them to fail at it.[10]

In non-nuclear fields of science and technology, however, the translation of information-about-a-threat into the-threat-itself is more direct. The ultimate examples of this are cyberattacks and digital viruses, where the distinction between the information of how the threat works and the threat itself readily dissolves away: a computer virus is an informational entity, and transferring it from one party to another, even in the form of an attack, is in essence a transfer of capability. So cyberexperts warned in the wake of the Stuxnet attack on Iran in 2010: now everyone in the world has access to the "weapons-grade" code that had gone into writing a virus to attack physical infrastructure and a clear model for adapting it to new purposes. Recent devastating cyberattacks were made with code that had been leaked out of the National Security Agency, as well as "proofs of concept" offered up by academic researchers. The line between "the information" and "the weapon" is, in this case, nil.[11]

It is of note that in this field, many cybersecurity researchers do not believe secrecy is an effective countermeasure, and tend to prefer radical openness, because these informational "weapons" can in fact be countered by other informational countermeasures. Keeping such techniques secret also keeps the vulnerabilities secret—which only works if nobody else discovers the same problem, or your own secrets are completely secure. And in the field of cryptography, openness is not just a social benefit but a true virtue: a truly secure cryptographic algorithm succeeds because it has been peer-reviewed and attacked unsuccessfully by a large community of researchers, and does not rely at all on any "secret" to function (other than the "secret keys" that are used to decrypt a message), but on a secure mathematical foundation.[12]

Where does biology fit in this rubric? At the moment, the kinds of virological feats that produced synthetic poliovirus still appear to be achievable only by a small group of experts and are limited by tacit knowledge (experience, "know-how") more than by access to materials (which for biological work are fairly "open") and information (the genetic codes of polio, smallpox, bubonic plague, etc., are all easily available online). Secrecy might indeed be useful to slow the diffusion of these capabilities or keep them within a small number of hands and heads, but ultimately these developments seem to be just the beginning. Many non-nefarious developments in biology could be applicable to nefarious purposes, which may lead to an increased risk of misuse over time. Should these same sorts of researchers produce instruments and tools that simplify the conversion of information into output (minimizing the tacit knowledge requirements) and circulate these instruments and tools in ways that allow nearly anyone to manufacture, say, modified influenza at will, then we may be heading for a scary future indeed.

To put it another way: if tacit knowledge is the only "barrier" preventing biological information from being weaponized, then either the diffusion of this tacit knowledge, or the development of tools that remove or reduce it, will increase the risk. Could secrecy have prevented this? Maybe, maybe not. But information is not *currently* the vector by which these techniques are weaponized. What this points to is the idea that any technique that allows information to be easily weaponized should be regulated fairly tightly, if the information is not or cannot be so regulated. Whether these things can or will occur remains to be seen: my point is that one cannot see this as simply a case of "secrecy or not"; secrecy is merely one form of technology control among many, and the technical differences between nuclear physics and synthetic biology are significant enough that the lessons to be drawn between one and the other will be more complex than simple one-to-one relationships.

Some have asked whether, instead of secrecy, we might just make sure that only the "right people" are allowed to use said tools or information.[13] I should hope that the reader of this book will see what folly this is. Once we ask, "who shall determine whose hands are right?" and "what sort of consequences will motivate those with access to not violate the pact?" it is clear that this is, in fact, a secrecy regime, not an alternative to one. And this is, perhaps, the real application for the history of nuclear secrecy to these fields: once the controls come in, they don't go away fast, and they may not even work well to prevent the proliferation of technology. But they will do other kinds of work in their effort to partition the world into multiple parts: creating in-communities and out-communities, drawing scrutiny to those who practice in these arts, and monopolizing patrons. There may be good reasons for other scientific communities to embrace secrecy—if the information in question truly was unlikely to be independently discoverable, had potentially large negative applications relative to the possible positive applications, and could be effectively controlled, then it might be a candidate—but if they took my advice, they would think long and hard about what types of secrecy activities they wanted to adopt and how to make sure that their attempts at secrecy did not outstrip their other values.

From a technical standpoint, nuclear weapons should have been very easy to control. As Oppenheimer understood in 1945, the material pipelines to acquiring nuclear weapons are relatively large, and controlling them (the uranium, the enrichment facilities, the reactors) meant controlling the spread of the bomb, even in the face of possibly incomplete or non-existent secrecy. That nuclear control has been elusive should give us pause. The problem of controlling nuclear weapons is not and has never been a truly technical one—it is, rather, a political problem. And technical solutions to political problems are rarely adequate.

What if the nuclear secrecy regime had never been created? It is hard to imagine the Manhattan Project succeeding without the secrecy it imposed (their fears of congressional cancellation were not unjustified), but one can ask whether the decisions of the immediate postwar period could have been different, when those in favor of limited or non-existent scientific secrecy for a moment seemed to have considerable influence. If there is any junction point in this history where another road could have been taken, it was probably then: in the wake of

World War II, but before a Cold War mindset had calcified, where the questions of secrecy, arms policy, and international control were still taken seriously by people who had the power to do something about them. What if the visions of secrecy espoused by Vannevar Bush, James Conant, and J. Robert Oppenheimer had been made more real, in the period of late 1945 through the summer of 1946?

Obviously, it is hard to say: counterfactual, hypothetical historical questions of this magnitude are hard to answer. We cannot re-run history like a physics simulation with a few tweaked variables. But we can make informed guesses. Would the Cold War have been averted? Probably not. Secrecy on the side of the US likely played a role in that, but the forces of distrust and fear between the US and the Soviet Union were broad, and our knowledge of Soviet leaders at the time does not suggest that they would have treated more nuclear openness as anything more than a potential intelligence source. It takes a very active imagination to conceive of a US-led and Soviet-accepted policy that would have convinced the Soviet Union not to develop its own nuclear weapon and avoided a superpower stand-off.

Would it have increased the speed of nuclear proliferation? This also seems unlikely. There is little evidence from the histories of other national nuclear programs that lack of access to information played a dramatic role in the timing of such programs or the decisions to pursue nuclear ambitions.[14] The political decisions that launched nuclear weapons development programs have been complicated, and not simply related to technical capabilities (countries that could do it easily have not done it, while countries for whom it was a struggle did it). The pace of such programs seems to have been tied largely to the engineering difficulties of running full-scale fissile material production facilities and the trade-offs between cost, speed, and secrecy from other nations that might try to stop them, as well as administrative arrangements that can slow the work down.[15] So I see little reason to think that the timeline of nuclear acquisition would have been affected much either way. (In this analysis, I am assuming no workable system of international control was developed, which is a separate issue. Had international control somehow been implemented much earlier, it is possible to imagine a very different twentieth century.)

Could the United States have avoided the domestic political spasms that have been associated with secrecy and nuclear fears, like McCarthyism? Nuclear secrecy and the increasingly powerful sense of "national security" were interlocking even if they were, in some respects, quite separate. Nuclear secrecy emerged in parallel alongside broader changes to American sensibilities of national security, and while it exported some practices (like declassification guides) to the broader system, its largest contribution seems to have been as a rhetorical underwriting of the needs for expansive and permanent classification. The imagery of nuclear weapons underwrote the conspiratorial and paranoid politics of the 1950s, and the idea of "the secret" and its loss gave amplification to the new Red Scare.[16] But the Red Scare of 1919 showed that that existential fear of the Soviet Union could be powerful without the threat being technological in nature. So it is hard to know.

While the national security state, and the nuclear secrecy regime, overlapped in important ways, they also maintained some separate features. The AEC and its successor agencies, for example, did have very different "cultures of secrecy" than the military, the Department of Defense, and the CIA. The AEC approach appears to have been generally more bureaucratized and formalized, and reflected the "scientific" origins of its mandate (more inclined to evaluate each secrecy release individually and earnestly).[17] But as we have seen, over the course of the twentieth century, the centrality of the specifically *nuclear* secrecy regime to the national security regime seems to have moved. By the twenty-first century, nuclear secrecy is a subset of a generalized secrecy state rather than the driving force behind any of it. Nuclear weapons as an "ultimate" form of power have persisted, and "nuclear secrets" as a trope still exist (though often in foreign, not domestic, contexts), but the size of the national security state has outstripped the size of the nuclear infrastructure considerably.

I suspect that removing fear of "the secret of the atomic bomb" being "released" or "stolen" and returning to pre-war attitudes toward scientific information might have decreased some of the "heat" of the Cold War controversies and certainly might have influenced the shape they took (it is hard to imagine physicists being so specifically targeted, for example). But the pessimist in me wonders whether some other fear would fill the breach, and generalized appeals to "national security" would have still found plenty of fertile ground onto which to spread.

While the "secret" of the atomic bomb was a convenient image to draw upon, it was far from the only one. Classical McCarthyism was not about nuclear secrets alone, but the fear of "infiltrators" and "corruptors" subtly undermining the American state to the advantage of the USSR. In any Cold War atmosphere, this would still be a potent accusation, and we have multiple examples from history of persecutory politics being waged, even in the United States, well before the advent of nuclear weapons.

So what difference might there have been if nuclear secrecy had been smothered out at the end of World War II? This is perhaps the most useful and fruitful of questions, however speculative our answers. I can imagine, for example, that the development of civilian nuclear power might have been sped up if an increase in the number of people working on these questions hadn't been delayed by a decade.[18] But it isn't clear that this would have radically altered the "timeline." Would serviceable nuclear power stations have been available by 1952 rather than 1958? If so, so what? Would that have been better, worse, or neither? Would, as Edward Teller hoped, there have been rapid innovation in US nuclear weapons designs if every PhD student could study such ideas without a clearance? Maybe, but with the exception of Teller himself, this is not generally the goal of people who have been against secrecy. Would the impact of Soviet spy cases be undercut? Again, this seems dependent on the broader Cold War fears. Would Americans be more likely to participate in nuclear policy decisions without the secrecy? Perhaps yes, though in which direction that would drive the policy is hardly clear; the notion that a more informed American public would vote in one direction or another on this seems rather unfounded and, frankly, optimistic.

Could such a system of non-secrecy have been sustained in the face of revelations about the Soviet Union's own atomic program and the extent of Soviet spying? Would the invention of thermonuclear weapons have returned us to a reification of "the secret"? Without a very firm reaffirmation of the value of scientific and governmental openness, it is hard to imagine a system of this sort surviving such "scandals." And therein is perhaps the rub: the United States, despite its Enlightenment attestations to the benefits of transparency and openness, found itself extremely willing to go down the path of secrecy, a path it had been on well before the atomic bomb was made a reality, but a path that was paved even more deeply with the existential threat offered by nuclear weapons. Despite the apparent openness of American society, and despite the increased suspicion of secrecy over the years, secrecy regimes have proven remarkably adaptable to the American context, and the bomb in particular has provided an endless source of motivating imagery.

This all sounds very pessimistic, as if I am about to conclude that "nuclear secrecy is here to stay" and maybe, in some form, was always likely to have occurred. It may have been, to some degree, inevitable. The exact regime, of course, as we have seen, was extremely contingent: we would not have the Restricted Data clause, for example, if Groves had not leaked a spy scandal while McMahon's committee was deliberating on their legislation. Without Oppenheimer's input, we might not have quite the same declassification system we have today. And so on. But it is hard to avoid seeing the temptation of conceiving of the atomic bomb as dangerous knowledge (and not just dangerous materials) being very persuasive in the American Cold War political context.

So let me end on a somewhat different note. The reason that American nuclear secrecy is interesting to study is because it has *not* always been an "easy fit." The United States contains a multitude of voices, values, and political forces. It was founded on principles that lent themselves to openness and transparency, even if such ideals were not made absolute mandates and, like many founding American principles, were at times deeply subverted and undermined. Secrecy was not present in the United States since the beginning: it had to develop, over the years, and has met with challenges in a wide variety of venues ever since. This is in great contrast to, for example, the Soviet Union, where secrecy went essentially unchallenged, and so while its effects may still be interesting to study, the question of whether secrecy *ought* to exist was nowhere near as persistent and driving as in the American case. In other nuclear countries (Russia, China, France, Israel, Pakistan, and North Korea, in particular), the national nuclear programs appear to exist in near-total secrecy without any truly powerful challengers.[19] The United States, despite its reputation for secrecy, is also the state that has released the most information about its nuclear activities to its public and the world and responds (however sluggishly) to the demands of its citizens for more information.

Secrecy has always been at tension within the United States, and nuclear secrecy, with its connections both to existential risk and to the generative power of science and technology, has always been especially fraught. And that, I hope, will always continue: it is a productive tension, one that creates political pressures from a variety of directions (not always positive), one that forces politicians, bureaucrats, and citizens to make hard decisions. They will not always make the best decisions, as we have seen. These tensions will produce scandals, mistakes, and polemics, but they also occasionally produce moments of great revelation and understanding, and even sometimes produce policy that moves in the direction of increased peace and justice. Nuclear secrecy and anti-secrecy both seem here to stay, and we should expect them to continue to writhe and struggle.

ACKNOWLEDGMENTS

Over 15 years ago, I was sitting on the floor of UC Berkeley's Moffitt Library, looking over a trove of books on nuclear weapons, when I opened Chuck Hansen's *U.S. Nuclear Weapons* (1988) for the first time. It wasn't his text, or even the photographs, that grabbed me: it was the meticulous depiction of the Fat Man nuclear bomb (figure 8.5 in this book) and its expanded, exploded, hyper-annotated, isometric view that I found provoking. How much of it was true? How did he get the information? When was it released? Could you trace, piece by piece, exactly the conditions that were necessary to bring this diagram into existence? And, more reflexively: why does this drawing feel so powerful?

These questions hooked me good, and my life has been a muddle of Freedom of Information Act requests, document databases, PDF files, "secret" diagrams, and concentric "circles of death" drawn over city maps ever since.

It's been a long journey to finish this book, and there are so many people who have aided me in this endeavor, directly and indirectly, that I cannot hope to possibly acknowledge them all.

First, I need to thank the mentors who encouraged me, pushed me, and enabled me. I would never have gotten this far without the faith and hard work of these people. Cathryn L. Carson is responsible perhaps more than anyone for my interest in the history of science and in nuclear history, and I credit her for making me into a source-obsessed scholar. Peter Galison always pushed me to think big about the epistemological implications of secrecy; David Kaiser taught me the practice of doing Cold War history of science; and Mario Biagioli encouraged me to think beyond the specifics of nuclear weapons in thinking about the practices of concealment of knowledge. And lastly, but surely not least, Sheila Jasanoff, along with consistently pushing me to further develop my thinking and writing, also taught me how to blend a nuanced political sensibility into questions about claims to knowledge and power.

While at Harvard University, I had the fortune to work and talk over the years with an almost impossibly wonderful group of people whose conversations and interactions have indelibly marked my ways of thinking: Tal Arbel, Jeremy Blatter, Alex Boxer, Janet Browne, Luis Campos, Jimena Canales, Lisa Crystal, Alex Csiszar, Stephanie Dick, Paul Doty (d), Sam Weiss Evans, Tope Fadiran, Megan Formato, Jean-François Gauvin, Adam Green, Jeremy Greene, Orit Halpern, Anne Harrington, Daniela Helbig, Benjamin Hurlbut, Louis Hyman, Andrew Leifer, Rebecca Lemov, Deborah Levine, Andrew Mamo, Daniel Margocsy, Aaron Mauck, Ernest May (d), Everett Mendelsohn, Matthew Meselson, Grischa Metlay, Latif Nasser, Cormac O'Raifeartaigh, Katherine Park, Sharrona Pearl, Christopher Phillips, Chitra Ramalingam, William Rankin, Lukas Rieppel, Hanna Rose Shell, Steven Shapin, Alistair Sponsel, Hallam Stevens, Judith Surkis, Leandra Swanner, Will Thomas, Jenna Tonn, Elly Truitt, Matthew Underwood, David Unger, and Heidi Voskuhl. And always alphabetically last, but never the least, I want to especially thank Nasser Zakariya for his apparently boundless patience for reading my work, and talking with me about the ideas I struggle to express.

Scholars, historians, and people who shared things with me (document, thoughts, insights, ideas, comments, etc.) who were of great help over the years in ways both large and small in my thinking about this project and accomplishing my goals include: Steven Aftergood, Elena Aronova, Emma Belcher, Barton Bernstein, Jeremy Bernstein, Kai Bird, Ellen Bradbury, Marisa Brandt, Joan Bromberg (d), Thomas Burnett, William Burr, Lila Byock, Alan Carr, Eugene Cittadino, John Cloud, Avner Cohen, Matthew Connelly, Campbell Craig, Angela Creager, Gene Dannen, Michael Dennis, Paul Edwards, Terrence Fehner, Ann Finkbeiner, Paul Forman, Francis Gavin, Edward Geist, Slava Gerovitch, Lisa Gitelman, Alexander Glaser, Skip Gosling, Susan Groppi, Benjamin Gross, Hugh Gusterson, Barton Hacker, Jacob Hamblin, Tsuyoshi Hasegawa, Gabrielle Hecht, Evan Hepler-Smith, Gregg Herken, Toshihiro Higuchi, Stephen Hilgartner, David Holloway, Ann Johnson (d), Matt Jones, Cindy Kelly, R. Scott Kemp, Daniel Kevles, Timothy Koeth, Alexei Kojevnikov, John Krige, Hans Kristensen, Robert Krulwich, William Lanouette, Milton Leitenberg, Jeffrey Lewis, Allison MacFarlane, Kristie Macrakis, Sean Malloy, Glenn McDuff, Priscilla McMillan, Zia Mian, Alexander Mikhalchenko, Erika Milam, Mary X. Mitchell, Alexander Montgomery, Teasel Muir-Harmony, Allan Needell, Ingrid Ockert, Kathryn Olesko, Benoît Pelopidas, Martin Pfeiffer, Pavel Podvig, Martha Poon, Rebecca Press Schwartz, Joanna Radin, B. Cameron Reed, Richard Rhodes, Carl Robichaud, Cheryl Rofer, Robert S. Norris, Silvan S. Schweber (d), Scott Sagan, Eric Schlosser, Sonja Schmid, Stephen Schwartz, Suman Seth, Sam Shaw, Martin Sherwin, Leo Slater, Matt Stanley, Abel Streefland, Nina Tannenwald, Kathleen Vogel, Frank von Hippel, J. Samuel Walker, Jessica Wang, Zuoyue Wang, Spencer Weart, Anna Weichselbraun, Stephen Weldon, Peter Westwick, Benjamin Wilson, Audra Wolfe, and Dan Zak. I would like to give special appreciation to Michael D. Gordin and Patrick McCray in particular as scholars who, along with their productive mixtures of support, encouragement, and constructive criticism, have also deigned to read and listen to my work in far excess of their quota.

I had the good fortune to spend a year at the Harvard Kennedy School's Managing the Atom and International Security Programs in 2011–2012, which in retrospect was absolutely essential in my learning to communicate (however awkwardly) with people in the world of "policy." For their insights and camaraderie, I am grateful to Tom Bielefeld, Matthew Bunn, Neal Doyle, Martin Malin (d), Rolf Mowatt-Larssen, Nicholas Roth, Karthika Sasikumar, and William H. Tobey.

My three years at the American Institute of Physics allowed me considerable time to work on this manuscript, and the good fortune of working with Joe Anderson, Charles Day, Gregory Good, and Melanie Mueller. They also not only tolerated, but encouraged, my creation of *Restricted Data: The Nuclear Secrecy Blog* (http://blog.nuclearsecrecy.com), and for that I am very grateful.

At my present institutional home, the College of Arts and Letters at the Stevens Institute of Technology, I have been graced with unusually pleasant and interesting colleagues. For giving me support, friendship, and much food for thought, I want to thank Lindsey Cormack, David

Farber, Bradley Fidler, Edward Friedman, Hamed Ghoddusi, John Horgan, Kristyn Karl, Ashley Lytle, Theresa MacPhail, Christopher Manzione, Jen McBryan, James McClellan III, Billy Middleton (who also ably helped me edit this manuscript), Gregory Morgan, Samantha Muka, Nicholas O'Brien, Anthony Pennino, Julie Pullen, Andrew Russell, Yu Tao, Kelland Thomas, Jeff Thompson, and Lee Vinsel.

I've had the honor of working with many talented undergraduates (and a few graduate students) at a number of institutions over the years. I am grateful for my students, and consider my teaching to be an essential part of my practice. In particular, several able undergraduates, at both Harvard and Stevens, have assisted me as researchers who contributed to this work in fairly direct ways: Emma Benintende, Max Rizzuto, and Benjamin Sakarin.

In the course of this work, I also spoke and worked with several "historical actors," probing them for knowledge, document, and experiences. Sometimes these were in the form of formal oral histories, but often they were informal communications. My gratitude is immense for their giving me their time: William Arkin, Benjamin Bederson, Keith Brueckner (d), Alan Carr, Robert Christy (d), John Coster-Mullen, Tony deBrum (d), Hugh DeWitt (d), Freeman Dyson (d), Daniel Ellsberg, Erik Erpelding, Bill Graham (d), Michael Hayden, Donald Hornig (d), "Jimmy" the child prodigy, Lawrence Johnson (d), George Keyworth (d), Ray Kidder (d), Peter Kuran, Howard Morland, Philip Morrison (d), John Nuckolls, John Aristotle Phillips, Theodore Postol, Thomas S. Martin, Carey Sublette, Wendy Teller, Kenneth W. Ford, Mike Wagnon, Herbert York (d), and Peter Zimmerman.

I also would like to thank the many archivists who aided me in my search (notably William Davis at the Center for Legislative Archives at the National Archives and Records Administration, who went above and beyond in helping me find newly released materials, and Martha DeMarre at the Nuclear Testing Archive, who has quite a stack of my requests at this point), and the many FOIA officers (especially at the Department of Energy, the Federal Bureau of Investigation, and the National Archives and Records Administration) who put up with, occasionally responded generously to, my many requests, appeals, demands, etc.

Karen Darling of the University of Chicago Press gave me extraordinary patience, input, and editorial suggestions. And the anonymous scholars who reviewed my manuscript gave me the most constructive comments I have ever received from peer review.

There is not a day that goes by that I am not aware and appreciative of the efforts that my family has made for me, and the absolutely essential role this has played in my successes in life so far. My ever-supportive parents (Jeff and Wanda) and my sister (Robyn) mean the world to me.

The animal companions in my life will never read this, even if they were all here, but I still feel the need to acknowledge their contributions as well: difficult Zeno (d), dear Shelley (d), and odd little Lyndon, you're contributors, too, in your own ways.

Last, but certainly not least, my wife, co-conspirator, and fellow-traveler Ellen Bales requires acknowledgment not just for the love she has provided, or the material assistance in keeping a household together, or even the intellectual companionship of another historian of science who has been willing to talk to me about nuclear weapons on practically a daily basis for over 15 years now, but also for pushing me to be a better historian and teacher, essentially teaching me how to write, and inspiring me daily with her erudition, intellectualism, and engagement. This book would not exist without her, and it is dedicated to her (thank you, dear).

NOTES

INTRODUCTION

1 Testimony of Leslie R. Groves, Executive Hearing, JCAE (24 February 1948), 25.
2 "Statement by the President Announcing the Use of the A-Bomb at Hiroshima," (6 August 1945), MDH, Book 1, Volume 4, Chapter 8, "Press Releases," item 1.
3 There has been much written about the history of the atomic bomb, especially during World War II, and into the Cold War. Secrecy has always been a theme in this work, though there has been little that has focused on the history of nuclear secrecy specifically as their subjects of study. Two of the rare offerings in this area are Seidel, "Secret scientific communities," and Westwick, "In the beginning." A useful resource on tropes of secrecy historically is Shattuck, *Forbidden knowledge*.
4 See esp. Ezrahi, *The descent of Icarus*.
5 J. Robert Oppenheimer to David E. Lilienthal (22 October 1947), JRO, Box 46, "Lilienthal, David E., Correspondence from Oppenheimer 1946–1950."
6 This usage of "secrecy regime" is my own, and is about as much jargon as I will use in this book. This conceptualization of a "secrecy regime" as a bundle of discourse, practice, and institutions is deliberately vague, and serves primarily as an effort to redirect the reader's attention to how secrecy is transformed from desire into a physical reality. My thoughts on these matters have been strongly affected by Certeau, *The practice of everyday life*, Lenoir, *Instituting science*, and Foucault, *Discipline and punish*.
7 Hewlett and Anderson, *The new world*, 657.
8 And Congress itself made its own records exempt from the Freedom of Information Act, so any congressional records that were once secret—such as the Executive Sessions of the Joint Committee of Atomic Energy that I use extensively—are released essentially whenever they choose to do so, without any possibility of expediting or appeal.
9 I did briefly serve on a fellowship for the Department of Energy while I was a PhD student, as the amusingly named Edward Teller Graduate Security Fellow

in Science and Security Studies. I did not require a clearance for my job, and I was not given privileged access to any material.

10 See, e.g., Hewlett and Quatannens, "Richard G. Hewlett," and Hacker, "Writing the history of a controversial program," as two accounts of nuclear historians who struggled to publish after obtaining clearances. For a more mundane account of the difficulties of engaging with once-secret sources, see Burton et al., "Following the paper trail west."

11 The whistleblower/leaker Daniel Ellsberg, in his book *Secrets*, has an excellent account of his (failed) attempt to dissuade Henry Kissinger from the smug, self-defeating epistemology that can come with a security clearance: Ellsberg, *Secrets*, 237–239.

12 Testimony of Leslie R. Groves, Executive Hearing, JCAE (24 February 1948), 25.

CHAPTER ONE

1 Ralph Lapp quoting Leo Szilard, in Ralph Lapp to William Lanouette (9 April 1987), private communication, provided by William Lanouette to the author. A variation is quoted by Lapp in De Volpi et al., *Born secret*, 30.

2 Kragh, *Quantum generations*, ch. 18; Rhodes, *The making of the atomic bomb*, ch. 1–9.

3 See Pearson, "On the belated discovery of fission," and Segré, "The discovery of fission."

4 Segrè, "The discovery of fission," 42. See also Sime, *Lise Meitner*, ch. 10.

5 There are many histories of the early Atomic Age, and for the sake of brevity I am omitting many fascinating narrative details. Those wishing for more detail about the discovery of fission should turn to sources such as Rhodes, *The making of the atomic bomb*; Weart, *Nuclear fear*; and Kragh, *Quantum generations*. A trillion trillion (10^{24}) uranium atoms splitting would release half of the energy as the bomb that destroyed Hiroshima.

6 Weart and Szilard, *Leo Szilard*, 14.

7 Szilard has been credited with having several "first" ideas, but he rarely brought them to fruition first, as he was largely uninterested in investing the labor required. On Szilard, see esp. Lanouette, *Genius in the shadows*.

8 John Jenkins has argued that Rutherford was at times quite bullish on the liberation of atomic energy from radioactivity, and may, then, have been deliberately misleading regarding such pronouncements, fearful that they might lead to commercial or military exploitation of his physics. The hard evidence for this is very slim, but it is an interesting possibility. See Jenkins, "Atomic energy is 'moonshine.'"

9 Kragh, *Quantum generations*, ch. 12.

10 Weart, *Nuclear fear*, ch. 1; Campos, *Radium and the secret of life*, ch. 1; Weart and Szilard, *Leo Szilard*, 16–18; Smith, *Doomsday men*, prologue; Rhodes, *The making of the atomic bomb*, ch. 1.

11 Jenkins, "Atomic energy is 'moonshine,'" 136; Lanouette, *Genius in the shadows*, 147.

12 For the state of Szilard's research, see his patents of 1934: Leo Szilard, "Improvements in or relating to the Transmutation of Chemical Elements," GB630726, filed 28 June 1934, and issued 28 September 1949.

13 Weart and Szilard, *Leo Szilard*, 18–19, 45–48; Rhodes, *The making of the atomic bomb*, 223–225, 254.

14 Weart and Szilard, *Leo Szilard*, 53.

15 Bernal, *The social function of science*, 150, and 107–110.

16 Merton, "The normative structure of science [1942]," esp. 273–275. See also Hollinger, "The defense of democracy and Robert K. Merton's formulation of the scientific ethos," and Gieryn, "Boundaries of science." On the attempts within Germany to politicize the discipline of physics ("Deutsche Physik"), which both Bernal and Merton had in mind when making these pronouncements, see Walker, *Nazi science*, esp. ch. 2–6.

17 The literature on secrecy and science generally speaking is quite large, and has been lately growing dramatically. For a historical/analytical overview, see Hull, "Openness and secrecy in science." On secrecy and the development of ancient and medieval technical arts, see Long, *Openness, secrecy, authorship*, and Eamon, *Science and the secrets of nature*. On the early modern period (including Galileo, Newton, and their contemporaries), see Biagioli, *Galileo's instruments of credit*; Westfall, *Never at rest*; and Iliffe, "In the warehouse." On the example of Darwin, see Browne, *Charles Darwin*. For a critical overview of secrecy and science, see Dennis, "Secrecy and science revisited." On the complicated situation regarding industrial secrecy, see Shapin, *Never pure*. See also Vermeir and Margocsy, "States of secrecy," and other articles in the same issue. On alchemy's relationship to secrecy and science, see, e.g., Newman, *Atoms and alchemy*.

18 Turchetti, "The invisible businessman"; Heilbron and Seidel, *Lawrence and his laboratory*, 103–116.

19 The public disclosure requirement of patents—whereby the exchange of temporary government monopoly of an idea is given in exchange for full public disclosure of its workings—was not always part of patenting practices, but had become the norm by the late nineteenth century. For the American case, see Walterscheid, *To promote the progress of useful arts*, esp. ch. 10.

20 Weart, "Scientists with a secret"; Weart and Szilard, *Leo Szilard*, 54.

21 On the German nuclear program, see Walker, *German national socialism and the quest for nuclear power*, and Walker, *Nazi science*. The Germans had a modest reactor research program by 1945; they did not have an atomic bomb production program.

22 Weart, *Scientists in power*, 43–45, 56–57, 63, 69; Norman Feather, "Fission of Heavy Nuclei: A new type of nuclear disintegration," *Nature* 143, no. 3630 (27 May 1939): 877–879.

23 Weart, "Scientists with a secret," 24.
24 Weart and Szilard, *Leo Szilard*, 55.
25 Weart and Szilard, *Leo Szilard*, 56–57.
26 Weart and Szilard, *Leo Szilard*, 56–57.
27 Weart, "Scientists with a secret," 25.
28 Weart, "Scientists with a secret," 27–28.
29 Weart, "Scientists with a secret," 27–28; Weart and Szilard, *Leo Szilard*, 69–78.
30 Weart, *Scientists in power*, 90–91.
31 Weart, *Scientists in power*, 91–92. In fact, Joliot's number was overly optimistic: the accepted value today is around 2.5.
32 Holloway, *Stalin and the bomb*, 50. See also, Badash et al., "Nuclear fission." A detailed bibliography of publications on fission at the time can be found in: Louis A. Turner, "Nuclear fission," *Reviews of Modern Physics* 12, no. 1 (January 1940), 1–29.
33 Weart and Szilard, *Leo Szilard*, 77. Joliot, for his part, later wrote to Szilard explaining that he would not, in principle, be against a *future* self-censorship regime, but only if it were "extended to *all* the laboratories which could concern themselves with this question." Emphasis in original. Weart and Szilard, *Leo Szilard*, 78–79.
34 Ernest O. Lawrence memo (27 November 1945), EOL, box 29, folder 4, "Historical Reports 1942–1945," microfilm reel 43, frame 095985. The work in question was being done by the physicist Phil Abelson.
35 Hewlett and Anderson, *The new world*, 19–21; Rhodes, *The making of the atomic bomb*, 304–309; Isaacson, *Einstein*, 471–476.
36 Niels Bohr and John Archibald Wheeler, "The mechanism of nuclear fission," *Physical Review* 56, no. 5 (1 September 1939), 426–450; Rhodes, *The making of the atomic bomb*, 283–288.
37 Weart, "Scientists with a secret," 29–30.
38 Walker, *German national socialism and the quest for nuclear power*, 24.
39 Weart, "Scientists with a secret," 30; Gregory Breit to Leo Szilard (5 June 1940), LSP, Box 5, Folder 6, "Breit, Gregory."
40 On the evolution of internal classified scientific practices, like classified journals, see esp. Westwick, "Secret science," and Seidel, "Secret scientific communities."
41 Szilard to Breit (6 July 1940), LSP, Box 5, Folder 6, "Breit, Gregory."
42 Breit to Szilard (16 July 1940), LSP, Box 5, Folder 6, "Breit, Gregory."
43 The State Secrets Privilege, for example, was not upheld by the US Supreme Court until 1953, in the highly controversial case of *United States v. Reynolds*. On *Reynolds*, see Siegel, *Claim of privilege*. For two good overviews of the early years of American secrecy, see esp. Relyea, "The evolution of government information security," Relyea, "Freedom of information, privacy, and official secrecy," and Quist, *Security classification of information* (both volumes).
44 Quist, *Security classification of information*, vol. 1, esp. 13–14, 18–22.
45 Hershberg, *James B. Conant*, ch. 3.

46 Quist, *Security classification of information*, vol. 1, 37.
47 Quist, *Security classification of information*, vol. 1, 26–37.
48 Galison, "Secrecy in three acts."
49 Lee, "Protecting the private inventor under peacetime provisions of the Invention Secrecy Act," 348–349.
50 Hearing before the Committee on Patents, US House of Reps., 65th Cong., 1st sess. on H.R. 5269 and H.R. 5287 (13 July 1917), 6–8. On the use of patent secrecy into the nuclear age, see Wellerstein, "Patenting the bomb," and Turchetti, "'For slow neutrons, slow pay.'"
51 Fox, "Unique unto itself," 477.
52 Quist, *Security classification of information*, vol. 1, chapter 3.
53 Hewlett and Anderson, *The new world*, 17–19; Dupree, *Science in the federal government*; Roland, "Science and war"; Hershberg, *James B. Conant*, ch. 3.
54 Hewlett and Anderson, *The new world*, 25; Goldberg, "Inventing a climate of opinion"; Stewart, *Organizing scientific research for war*; Owens, "The counterproductive management of science in the Second World War."
55 Hewlett and Anderson, *The new world*, 25; Stewart, *Organizing scientific research for war*, 12.
56 Karl T. Compton to Vannevar Bush (17 March 1941), BCF, Roll 1, Target 2, Folder 1, "S-1 Historical File, Section A." Every reference to "Compton" in this book is for Karl Compton's brother, the physicist Arthur Compton, unless otherwise indicated.
57 James B. Conant to Vannevar Bush (16 April 1941), BCF, Roll 2, Target 2, Folder 7, "Miscellaneous S-1 JBC Materials." For further discussion of the pessimism of this period, see Hewlett and Anderson, *The new world*, 32–39.
58 These same concerns, it now appears, were present in the German program as well. See esp. Walker, *Nazi science*, 195–197.
59 Hewlett and Anderson, *The new world*, 41; Stewart, *Organizing scientific research for war*, 35–40.
60 The name MAUD was due to an interesting misunderstanding, one predicated on an assumption of attempted secrecy. After the occupation of Denmark, Lise Meitner sent a telegram telling an English friend that she had heard from Niels Bohr that he was doing well, and asked to pass the message on to the physicist John Cockcroft and "MAUD RAY KENT." Cockcroft assumed a secret message was being sent, and interpreted the latter name to be a near anagram for "radium taken." In reality, Bohr was trying to assure the governess of Bohr's sons, Maud Ray, that the Bohrs were doing fine, and she happened to live in Kent. The truth of this would not come out until Bohr made it out of Denmark in 1943. Rhodes, *The making of the atomic bomb*, 340–341. On the British work, see in particular Gowing, *Britain and atomic energy, 1939–1945*.
61 Hewlett and Anderson, *The new world*, 41–44.
62 E.g., James B. Conant to Robert F. Bacher (4 September 1941), RFB, Box 16, Folder 1.

63 Vannevar Bush to James B. Conant (9 October 1941), BCF, Roll 1, Target 3, Folder 2, "S-1 Historical File, Section B (1941–42)." Cf. Rhodes, *The making of the atomic bomb*, 378; Hewlett and Anderson, *The new world*, 45–46.

64 Bush, *Pieces of the action*, 134.

65 Hewlett and Anderson, *The new world*, 48–52; Jones, *Manhattan*, 33–34.

66 Laurence followed up his prophetic statement with an assertion that has held up less well: that "such a substance would not likely be wasted on explosives," since it could instead be used as a power source. William L. Laurence, "The atom gives up," *Saturday Evening Post* 213, no. 10 (7 September 1940), 12–13, 60–63.

67 See Weart, *Nuclear fear*, ch. 1–5; see also Campos, *Radium and the secret of life*.

68 "Writer charges U.S. with curbs on science," *New York Times* (14 August 1941), 15; "New explosive reported censored," *Los Angeles Times* (14 August 1941), 8.

69 James B. Conant to J. D. Ratcliff (15 August 1941), BCF, Roll 5, Target 9, Folder 39, "C (1942–45)."

70 Vannevar Bush to Charles G. Darwin (20 September 1941), BCF, Roll 1, Target 2, Folder 1, "S-1 Historical File, Section A."

71 Irvin Stewart to Lyman J. Briggs (27 November 1941), BCF, Roll 9, Target 13, Folder 129, "Miscellaneous (L. J. Briggs' Materials, 1940–42)."

72 Gordin, *Five days in August*, 40. See also MDH, Book 1, Volume 1, Part 1, "General," for quite a long list of examples with the "special" adjective (materials, pumps, work, documents, committees, accounts, groups, and chemicals). By May 1945, it became quite common to refer to the weapon as a "special bomb" when communicating about it to the military. See, e.g., Lauris Norstad to Curtis LeMay (29 May 1945), Document 10 in Burr, "The atomic bomb and the end of World War II," or J. Robert Oppenheimer to Gen. Thomas Farrell (11 May 1945), NTA, document NV0103574.

73 Priest and Arkin, *Top secret America*, 56–57.

74 E.g., Vannevar Bush to Arthur H. Compton (9 October 1941); and Vannevar Bush to Harold Urey (13 December 1941), both in BCF, Roll 1, Target 3, Folder 2, "S-1 Historical File, Section B (1941–1942)."

75 Vannevar Bush to Charles G. Darwin (23 December 1941), BCF, Roll 3, Target 2, Folder 15, "S-1 OSRD Research Program, Executive Planning Committee (Fldr.) No. 1 (1941–42)."

76 James B. Conant to Ernest O. Lawrence (20 December 1941), BCF, Roll 1, Target 3, Folder 2, "S-1 Historical File, Section B (1941–1942)"; James B. Conant to Ernest O. Lawrence (2 January 1942), BCF, Roll 5, Target 14, Folder 44, "Classified Material, Handling of (1942)"; Irvin Stewart to James B. Conant (26 January 1942), BCF, Roll 5, Target 14, Folder 44, "Classified Material, Handling of (1942)"; Robert F. Bacher to Arthur H. Compton (2 April 1942), RFB, Box 16, Folder 1.

77 "Pledge of Secrecy" (24 November 1941), RFB, Box 16, Folder 1; Vannevar Bush

to Ernest Lawrence (5 December 1942), EOL, Box 27, Folder 30, "OSRD, Section S-1; Personnel," Roll 41, Frame 091662.

78 The MIT technician in question had worked at the Chicago Metallurgical Laboratory, and had discussed some of the uranium work with another company when trying to get another job. John W. M. Bunker to J. R. Killian (6 October 1942), BCF, Roll 7, Target 12, Folder 77, "G (1941–44)."

79 Vannevar Bush to James B. Conant (30 December 1941), BCF, Roll 7, Target 10, Folder 75, "Espionage."

80 Arthur H. Compton to James B. Conant (22 January 1942), BCF, Roll 1, Target 3, Folder 2, "S-1 Historical File, Section B (1941–42)."

81 Hewlett and Anderson, *The new world*, 55.

82 James B. Conant to Ernest O. Lawrence (30 January 1942), BCF, Roll 5, Target 14, Folder 44, "Classified Material, Handling of (1942)." The OSRD generally adopted its classification grades (Secret, Confidential, and Restricted) and guidelines from the Army and Navy regulations at the time. The organization did not have its own investigative body for personnel clearances, and instead utilized either the Army or the Navy, depending on which was the most relevant agency for the particular project the individual in question was working on. In the spring of 1942, these requirements were revised somewhat, so that the Army would do the majority of the investigative work. Personnel working for OSRD contractors, though, were generally not officially investigated, and it was left to the contractor to determine their trustworthiness. Stewart, *Organizing scientific research for war*, 247–255. For the development of World War II classification guidelines generally, see Quist, *Security classification of information*, vol 1, ch. 2 and 3.

83 C. P. Baker to Robert F. Bacher (13 May 1942), RFB, Box 16, Folder 1.

84 Vannevar Bush to James B. Conant (24 February 1942), BCF, Roll 6, Target 2, Folder 54, "Compton, A. H. 1941–42."

85 Vannevar Bush to R. H. Thayer (28 February 1942), BCF, Roll 3, Target 2, Folder 15, "S-1 OSRD Research Program, Executive Planning Committee (Fldr.) No. 1 (1941–42)."

86 E.g., OSRD Administrative Circular 4.02, "Classified Information, Documents, or Materials, Revision No. 1," (7 February 1942), EOL, Box 27, Folder 13, "OSRD; Administrative Circulars 1941–April 1944," microfilm reel 40, frame 089483.

87 James B. Conant to Arthur H. Compton (30 January 1942), BCF, Roll 1, Target 3, Folder 2, "S-1 Historical File, Section B (1941–42)."

88 For a detailed treatment of patent secrecy in the OSRD and Manhattan Project, see Wellerstein, "Patenting the bomb."

89 Vannevar Bush, "Report to the President—Status of Tubealloy Development," (9 March 1942), HBF, Roll 4, Target 4, Folder 58, "Vannevar Bush Report—March 1942."

90 Franklin D. Roosevelt to Vannevar Bush (11 March 1942), BCF, Roll 1, Target 3, Folder 2, "S-1 Historical File, Section B (1941–42)."

91 On the construction of the "special" nature of the bomb, and the stakes involved in seeing it as "special," see Gordin, *Five days in August.*

92 Roosevelt employed secrecy throughout the war in regard to diplomatic arrangements, and was not above deceiving the American people if it played toward "greater good" political ends. On this last point, see Schuessler, "The deception dividend." See also Craig, "The atom bomb as policy maker," which explores more closely than any other publication so far the possible role of the atomic bomb in Roosevelt's thinking, and Bernstein, "Roosevelt, Truman, and the atomic bomb, 1941–1945."

93 Vannevar Bush to Frank B. Jewett (8 July 1941), BCF, Roll 8, Target 7, Folder 91, "J (1941–44)." See also Goldberg, "Inventing a climate of opinion," and Kevles, "The National Science Foundation and the debate over postwar research policy, 1942–1945," for context.

94 Frank B. Jewett to Vannevar Bush (3 November 1941), BCF, Roll 1, Target 2, Folder 1, "S-1 Historical File, Section A."

95 "Long-range planning" is an unusual phrase to use in 1941—is it a reference to the *use* of the still-far-off bombs, or is it about the *even-further-off* notion of a postwar situation, before the US was even a party to the war? Vannevar Bush to Frank B. Jewett (4 November 1941), BCF, Roll 1, Target 2, Folder 1, "S-1 Historical File, Section A."

96 Gregory Breit to Lyman J. Briggs (18 May 1942), BCF, Roll 1, Target 3, Folder 2, "S-1 Historical File, Section B (1941–42)."

97 James B. Conant to Vannevar Bush (n.d., ca. May 1942), BCF, Roll 1, Target 3, Folder 2, "S-1 Historical File, Section B (1941–42)."

98 Emphasis in original. Vannevar Bush to James B. Conant (n.d., ca. May 1942), BCF, Roll 1, Target 3, Folder 2, "S-1 Historical File, Section B (1941–42)."

99 Frank Parker Stockbridge, "War inventions that came too late," *Harper's Monthly* (November 1919), 828–835. Conant's deadly product never saw action—the war ended before it could be completed, and the remaining stockpile was sunk off of the coast of Baltimore. For more on Conant's chemical warfare work, see Hershberg, *James B. Conant*, ch. 3.

100 Ernest O. Lawrence to James B. Conant (26 March 1942), BCF, Roll 1, Target 3, Folder 2, "S-1 Historical File, Section B (1941–42)"; Hewlett and Anderson, *The new world*, 103.

101 Hewlett and Anderson, *The new world*, 104; Herken, *Brotherhood of the bomb*, 65–68.

102 Vannevar Bush and James B. Conant, Report on "Atomic Fission Bombs," (17 June 1942), HBF, Roll 4, Target 4, Folder 58, "Vannevar Bush Report—March 1942."

103 Vannevar Bush to Lyman J. Briggs (19 June 1942), BCF, Roll 2, Target 2, Folder 7, "Miscellaneous S-1 JBC Materials."

104 OSRD Administrative Circular 4.02, "Classified Information, Documents, or Materials, Revision No. 1" (7 February 1942), EOL, Box 27, Folder 13, "OSRD; Administrative Circulars 1941-April 1944," microfilm reel 40, frame 089483.

105 Stewart, *Organizing scientific research for war*, 12.

106 Ralph Lapp quoting Leo Szilard, in Ralph Lapp to William Lanouette (9 April 1987), private communication, provided by William Lanouette to the author. A variation is quoted by Lapp in De Volpi et al., *Born secret*, 30.

CHAPTER TWO

1 Ernest O. Lawrence to James C. Marshall (24 February 1943), EOL, Box 29, Folder 18, "Marshall, James C.," microfilm reel 44, frame 096799.

2 This is a summary of considerable bureaucratic complexity over the course of a number of months. See Jones, *Manhattan*, ch. 1–3; on the naming and formal creation, see 43–44. The best overall biography of Groves is Norris, *Racing for the bomb*. For general overviews of the Manhattan Project, and its secrecy, see esp. Rhodes, *The making of the atomic bomb*; Herken, *Brotherhood of the bomb*; Hales, *Atomic spaces*; and Groves, *Now it can be told*.

3 Groves, *Now it can be told*, 5.

4 Jones, *Manhattan*, ch. 4, Norris, *Racing for the bomb*, 10–14, and see esp. Goldberg, "General Groves and the atomic West," for discussions of Groves' clashes with other military brass.

5 On said books, films, and so on, see Hecht, *Storytelling and science*.

6 "All theoretical physicists are prima donnas," Groves remarked to Lise Meitner at a postwar cocktail party, apparently with Niels Bohr primarily in mind. Quoted from Meitner's notes in Sime, *Lise Meitner*, 333.

7 On Oppenheimer and Groves, see Norris, *Racing for the bomb*; Bird and Sherwin, *American Prometheus*; Herken, *Brotherhood of the bomb*, Groueff, *Manhattan Project*; Cassidy, *J. Robert Oppenheimer and the American century*; and McMillan, *The ruin of J. Robert Oppenheimer and the birth of the modern arms race*; Thorpe, *Oppenheimer*; and Bernstein, "Reconsidering the 'Atomic General,'" among many other books and articles that have been written on the pair.

8 See, e.g., Norris, *Racing for the bomb*, ch. 12; Bird and Sherwin, *American Prometheus*, ch. 15–17.

9 Hales, *Atomic spaces*; Brown, *Plutopia*; Jones, *Manhattan*, ch. 5–9, 13; Rhodes, *The making of the atomic bomb*, ch. 14.

10 James B. Conant to Vannevar Bush (23 October 1943), BCF, Target 5, Folder 35, "Bush, Dr. V. (Fldr.) 1 (1942–43)."

11 See, e.g., Ulam, *Adventures of a mathematician*, 144.

12 Groves, *Now it can be told*, 140.

13 On turnover, losses, and numbers, see MDH, Book 1, Volume 8, "Personnel."

14 Report from the Recreation and Welfare Association of Oak Ridge (n.d., ca. September 1945), ARC, identifier 281580.

15 Ibid.

16 "We the People," radio show (9 February 1947), ARC, identifier 281583.

17 See esp. Kiernan, *The girls of Atomic City*. It is of note that women played a larger role in the development of the atomic bomb than they did during the later Project Apollo, in a case of expediency winning over misogyny. Howes and Herzenberg, *Their day in the sun*, 11.

18 See esp. Hunner, *Inventing Los Alamos*, ch. 1.

19 Los Alamos Special Bulletin, "Badge Classification Form," (2 August 1945), NTA, document NV0319479; J. Robert Oppenheimer to Los Alamos division leaders, "Dissemination of Information within Los Alamos," (21 June 1943), MEDR, Box 65, "Publications."

20 MDH, Book 1, Volume 14, "Intelligence and Security—Supplement," Appendix CS-8, ("Security Manual, Manhattan District, 26 November 1945"), Exhibit II.

21 Hewlett and Anderson, *The new world*, 238–239. The list of colloquia topics comes from the FBI file of Klaus Fuchs, as he attended all of them. List of Los Alamos colloquia, 24 October 1944–3 June 1946, (n.d., ca. February 1950), included in KFFBI, part 49, as well as quoted in Special Agent in Charge, Albuquerque to J. Edgar Hoover (6 February 1950), in KFFBI, part 4.

22 Edward U. Condon to J. Robert Oppenheimer (26 April 1943), MEDR, Box 101, "Investigation Files."

23 On this theme, see also Westwick, "In the beginning."

24 Henry DeWolf Smyth quoted in "Talk of the Town," *New Yorker* (19 January 1946), 15.

25 For a vivid example of the laughing approach to secrecy, see Feynman, *"Surely you're joking, Mr. Feynman!"* For bitterness on compartmentalization, see esp. the testimony of Leo Szilard at the hearings of the Special Committee on Atomic Energy in late 1945 and early 1946.

26 These regulations were updated at various points during the war. See, e.g., Leslie R. Groves to Arthur H. Compton and J. Robert Oppenheimer, "Interchange of information between Chicago and Los Alamos," MEDR, Box 31, "Releasing information."

27 Leslie R. Groves to A. V. Peterson, "Interchange of information," (16 October 1943), MEDR, Box 31, "Releasing information"; Herbert F. York, oral history interview with author (24 April 2008), OHCNBL.

28 Hewlett and Anderson, *The new world*, ch. 8. Groves, a notorious Francophobe, had issues with several of the specific scientists there, notably Hans van Halban, whom he distrusted both because of his connections to known-Communist Frédéric Joliot-Curie and because of his attempts to prosecute French patent interests with the British. See Weart, *Scientists in power*, ch. 12–14.

29 Norris, *Racing for the bomb*, 232–236; Lanouette, *Genius in the shadows*, 236–237, and 311–313.

30 The most comprehensive discussion of Tatlock affair and her death is to be

found in Bird and Sherwin, *American Prometheus*, ch. 18. On the surveillance of Tatlock, see Herken, *Brotherhood of the bomb*, 101–102.

31 MDH, Book 1, Volume 14, "Intelligence and Security," 2.3, 3.7, 7.1–7.2, and 7.8. The "espionage attempts" in question likely were the Soviet Union's alleged recruitment attempts in Berkeley, which were known to Groves during the war.

32 MDH, Book 1, Volume 14, "Intelligence and Security," 1.6 and Appendix A2; Goudsmit, *Alsos*.

33 Bird and Sherwin, *American Prometheus*, 186–187; Leslie Groves to the District Engineer (20 July 1943), MEDR, Box 99, "Investigation Files."

34 Groves, *Now it can be told*, 140–141.

35 See, e.g., Leslie R. Groves, "Atomic Fission Bombs," (23 April 1945), included as Document 3a in Burr, "The atomic bomb and the end of World War II."

36 Bird and Sherwin, *American Prometheus*, 193–194; Herken, *Brotherhood of the bomb*, 107–125; Mullet, "Little man."

37 Jones, *Manhattan*, ch. 9.

38 Quist, *Security classification of information*, vol. 1, 46–49.

39 Leslie R. Groves to J. Robert Oppenheimer (14 August 1944); J. Robert Oppenheimer to Leslie R. Groves (19 August 1944), copies of both in CHNSA, Box 11, "1945–1949," Folder 1.

40 See esp. Roberts, *The brother*; Albright and Kunstel, *Bombshell*; Haynes, Klehr, and Vassiliev, *Spies*. On the usage of the information, see Holloway, *Stalin and the bomb*; Gordin, *Red cloud at dawn*; Kojevnikov, *Stalin's great science*. In recent months, another alleged Soviet spy at Los Alamos was revealed: Oscar Seborer, an electrical engineer who, like Greenglass, worked in the Special Engineer Detachment. Seborer and his brother defected to the USSR after Fuchs was revealed, and lived out their lives there. The extent of his knowledge, and his alleged espionage, is as of this writing still unknown. See William J. Broad, "Fourth spy at Los Alamos knew A-bomb's inner secrets," *New York Times* (27 January 2020), D1.

41 By my count, there were about ten serious spies within the Manhattan Project, with "serious" meaning that they had a more than passing acquaintance with it (in other words, they would have secrets to give up) and were directly offering up the information themselves (not just being a courier for other agents). *Index and Concordance to Alexander Vassiliev's Notebooks and Soviet Cables Deciphered by the National Security Agency's Venona Project* (Revised 1 November 2014), hosted by the Cold War History Project, Woodrow Wilson Center, online at https://www.wilsoncenter.org/sites/default/files/Vassiliev-Notebooks-and-Venona-Index-Concordance_update-2014-11-01.pdf, accessed 11 December 2018.

42 Leslie R. Groves, "Memorandum," (30 December 1944) and Leslie R. Groves, "Extract from notes made after conference with the President," (31 December 1944), both in CTS, Roll 3, Target 7, Folder 24, "Memorandums to (Gen.) L. R.

Groves Covering Two Meetings with the President (Dec. 30, 1944, and Apr. 25, 1945)."

43 Leslie R. Groves, "Extract from notes made after conference with the President," (31 December 1944), CTS, Roll 3, Target 7, Folder 24, "Memorandums to (Gen.) L. R. Groves Covering Two Meetings with the President (Dec. 30, 1944, and Apr. 25, 1945)." See also Craig, "The atom bomb as policy maker."

44 Testimony of Leslie R. Groves, Executive Hearing, JCAE, "Study of the British Disclosure," (4 February 1950), 454.

45 Vannevar Bush to R. H. Thayer (28 February 1942), BCF, Roll 3, Target 2, Folder 15, "S-1 OSRD Research Program, Executive Planning Committee (Fldr.) No. 1 (1941–42)."

46 Emphasis in original. H. T. Wensel to James B. Conant (23 July 1942), BCF, Roll 4, Target 12, Folder 30, "Breit, Gregory (1940–44)." See also James B. Conant to Gregory Breit (13 December 1942), BCF, Roll 4, Target 12, Folder 30, "Breit, Gregory (1940–44)."

47 "Science hush-hushed," *Time* (11 May 1942).

48 Holloway, *Stalin and the bomb*, 77–79; Zaloga, *Target America*, 1–3, and 11–13.

49 For an in-depth discussion of the wartime Office of Censorship practices regarding the atomic bomb, see esp. Washburn, "The Office of Censorship's attempt to control press coverage of the atomic bomb during World War II." On the Office of Censorship more broadly, see Sweeney, *Secrets of victory*.

50 Vannevar Bush to General W. D. Styer (13 February 1942), BCF, Roll 5, Target 5, Folder 35, "Bush, Dr. V. (Fldr.) 1 (1942–43)."

51 Franklin T. Matthias to Leslie R. Groves (9 April 1943), MEDR, Box 65, "Publications" (emphasis in original).

52 Leslie R. Groves to N. R. Howard (2 June 1943), MEDR, Box 65, "Publications."

53 N.R. Howard to Leslie R. Groves (3 June 1943), MEDR, Box 65, "Publications."

54 Washburn, "The Office of Censorship's attempt to control press coverage of the atomic bomb during World War II," 7.

55 Washburn, "The Office of Censorship's attempt to control press coverage of the atomic bomb during World War II."

56 "Work of scientists hold key to victory," *Chicago Sun Times* (9 February 1943), quoted in James C. Marshall to Ernest O. Lawrence (16 February 1943), EOL, Box 29, Folder 18, "Marshall, James C.," microfilm reel 44, frame 096729.

57 The only member of the University of California cabinet who knew about the goals of the Manhattan Project was its treasurer and secretary, Robert M. Underhill, who had been told by Ernest Lawrence (and been made to dramatically hide under a table while being told) in order to expedite purchase orders. See Herken, "The University of California, the federal weapons labs, and the founding of the atomic West."

58 James C. Marshall to Ernest O. Lawrence (16 February 1943), EOL, Box 29, Folder 18, "Marshall, James C.," microfilm reel 44, frame 096729. Similar letters were also sent to Arthur Compton and Harold Urey. James C. Marshall to

James B. Conant (22 February 1943), BCF, Roll 8, Target 19, Folder 103, "Manhattan District (New York) (1943)."

59 Ernest O. Lawrence to James C. Marshall (24 February 1943), EOL, Box 29, Folder 18, "Marshall, James C.," microfilm reel 44, frame 096799.

60 On the tensions that were created by Groves within the Army due to his demands for secrecy and high priority of the Manhattan Project, see esp. Goldberg, "General Groves and the atomic West."

61 John Lansdale, Jr., "Publicity concerning Clinton Engineering Works, Knoxville, Tennessee," (3 January 1944), and Leslie R. Groves to Henry L. Stimson, "Violation of Vital Security Provisions by Brigadier General Thomas A. Frazier," (10 January 1944), both in HBF, Roll 4, Target 8, Folder 62, "Security (Manhattan Project)."

62 Franklin T. Matthias to Leslie R. Groves (9 April 1943), MEDR, Box 65, "Publications."

63 John Lansdale to Leslie R. Groves (4 January 1944), HBF, Roll 4, Target 8, Folder 62: "Security (Manhattan Project)."

64 John Lansdale, Jr., "Publicity concerning Clinton Engineering Works, Knoxville, Tennessee," (3 January 1944), HBF, Roll 4, Target 8, Folder 62, "Security (Manhattan Project)."

65 Translation of article from *Svenska Dagbladet* (14 March 1943), CTS, Roll 2, Target 1, Folder 7, "War Department Special Operations," Subfile 7E, "Norsk Hydro Incident—Bombing of Norwegian Heavy Water Plant, 1943."

66 "Nazi 'heavy water' looms as weapon," *New York Times* (4 April 1943), 18.

67 Washburn, "The Office of Censorship's attempt to control press coverage of the atomic bomb during World War II," 6–7; Leslie R. Groves to Henry Wilson (22 April 1945), CTS, Roll 2, Target 6, Folder 12, "Intelligence and Security."

68 Vannevar Bush to R. C. Jacobs (6 April 1943), BCF, Roll 1, Target 6, Folder 5, "S-1 Intelligence (1943–45)."

69 R. C. Jacobs to Vannevar Bush (8 April 1943), BCF, Roll 1, Target 6, Folder 5, "S-1 Intelligence (1943–45)."

70 George V. Strong to George C. Marshall, "Control of Dangerous Publicity," (14 June 1943), MEDR, Box 65, "Publications."

71 Harold C. Urey to Watson David (8 April 1943), CTS, Roll 2, Target 1, Folder 7, "War Department Special Operations," Subfile 7E, "Norsk Hydro Incident—Bombing of Norwegian Heavy Water Plant, 1943."

72 "Discoverer doubts use of 'heavy water' in war," *New York Herald Tribune* (4 April 1943).

73 It is not clear whether Urey knew, at this point, that deuterium *was* being considered as a possible source of fusion reactions for the "Super" or hydrogen bomb by theorists at Los Alamos.

74 See J. Edgar Hoover to Harry Hopkins (9 February 1945), in HBF, Folder 62: "Security (Manhattan Project)," Roll 4, Target 8. Leslie Groves, "Notes on the Military Policy Committee of June 21, 1944," (undated, but prior to 1964), LRG,

Entry 7530M, Box 4, “Working Papers.” I thank Robert S. Norris for bringing to my attention this latter document, in which Groves talks about the time in which he “overruled” his “creeps.”

75 The writer of the strip, Alvin Schwartz, of course got his cyclotron information from a public source: pre-war issues of *Popular Mechanics*. Cronin, *Was Superman a spy?*, 12–14; Washburn, “The Office of Censorship’s attempt to control press coverage of the atomic bomb during World War II,” 20–22. Someone—perhaps Lansdale—attempted to dispute this story in an odd way in the post-war period, releasing a memo to *Harper’s* that appeared to recommend against censorship. However the evidence that the storyline *was* changed is conclusive, which only makes this effort even stranger. See John Lansdale Jr., “Superman and the atomic bomb,” *Harper’s Magazine* (April 1948), 355.

76 Berger, “The *Astounding* investigation.”

77 MDH, Book 1, Volume 14, “Intelligence and Security,” Appendix E.

78 The scientists were Dr. Nazir Ahmad, Col. S. L. Bhatia, Sir Shanti Swaroop Bhatnagar, Sir Jnan Chandra Ghosh, Prof. Sisir Kumar Mitra, Prof. Meghnad Saha, and Prof. Jnanendra Nath Mukherjee. Bhatnagar and Saha were singled out for special scrutiny by the security forces. The story of their interactions with Manhattan Project security is told through the following memos: Wallace Murray to James Clement Dunn (Assistant Secretary of State), “Visit of seven Indian scientists to the United States,” (10 February 1945), CTR, Roll 2, Target 6, Folder 12: “Intelligence and Security”; Walter A. Parish, Jr., “Indian Scientists’ visit to the United States,” (26 February 1945), CTR, Roll 2, Target 6, Folder 12: “Intelligence and Security”; and Leslie R. Groves to Harvey H. Bundy (22 March 1945), HBF, Roll 4, Target 8, Folder 62: “Security (Manhattan Project).”

79 John W. Raper, “Forbidden city,” *Cleveland Press* (13 March 1944). Photostat copy in MEDR, Box 99, “Investigation Files.”

80 Washburn, “The Office of Censorship’s attempt to control press coverage of the atomic bomb during World War II,” 11–12, and 37, fn. 43; Norris, *Racing for the bomb*, 275–276.

81 Interestingly, the Japanese did notice an article from a German technical magazine called *Nitrocelluose* published in November 1940 about “America’s super-explosive U-235,” which claimed that the US was working on a bomb that would detonate a single gram of uranium-235 with slow neutrons. It was read with interest by Japanese scientists and technicians. Rumors of US and German atomic efforts did make their way to Prime Minister Tojo Hideki in early 1943, but it is unclear whether these rumors were based in any actual leaks. See Grunden, *Secret weapons and World War II*, 63 and 69.

82 “Writer charges U.S. with curbs on science,” *New York Times* (14 August 1941), 15; “New explosive reported censored,” *Los Angeles Times* (14 August 1941), 8.

83 “Text of War News Code of Office of Censorship,” *New York Times* (11 December 1943), 8.

84 William H. Stoneman, "Atomic bomb discovery of German's told," *Los Angeles Times* (20 July 1945), 5.

85 George Axelsson, "Nazis talk less of new V weapons," *New York Times* (14 January 1945), B5.

86 E.g., Peter Lyne, "V-2 proves 'bust'?—Nazis turn to atom splitter," *Christian Science Monitor* (7 October 1944), 7; Ralph McGill, "One more word: Now it's atoms?" *Atlanta Constitution* (7 January 1944), 6; Associated Press, "Nazis boast of atomic bomb," *New York Herald Tribune* (28 December 1944), 2B; Philip Heisler, "Same old song sung by Nazis," *(Baltimore) Sun* (16 May 1944), 2.

87 Arthur Compton, "What science requires of the new world," *Science* 99, no. 2559 (14 January 1944), 23–28. Newspaper reporters at the time directly connected these lines to atomic bomb rumors: Associated Press, "Americans are reassured on danger of secret weapon," *Christian Science Monitor* (14 January 1944), 8.

88 MDH, Book 1, Volume 14, "Intelligence and Security," 6.5.

89 The only real exceptions to this were Japanese balloon bombs, released to dumbly follow the wind, that occasionally were spotted near the Hanford site from early 1944 through mid-1945. By total coincidence, one happened to strike power lines that supplied the Hanford reactor cooling facilities with electricity in March 1945. The reactors were briefly shut down as a result. See Schwartz, *Atomic audit*, 271, fn. 1.

90 Groves, *Now it can be told*, ch. 26; Jones, *Manhattan*, 272–274.

91 Vannevar Bush to Franklin D. Roosevelt (16 December 1942), BCF, Roll 1, Target 5, Folder 4, "S-1 Reports to and Conferences with the President (1942–44)."

92 Bush, *Pieces of the action*, 134; Vannevar Bush to James B. Conant (14 February 1944), BCF, Roll 1, Target 5, Folder 4, "S-1 Reports to and Conferences with the President (1942–44)."

93 Riddle, *The Truman committee.*

94 Rudolf Halley to W. S. Carpenter, Jr. (8 June 1943), HBF, Roll 4, Target 8, Folder 62, "Security (Manhattan Project)."

95 Harvey H. Bundy to Henry L. Stimson (11 June 1943), HBF, Roll 4, Target 8, Folder 62, "Security (Manhattan Project)."

96 Henry L. Stimson, diary entry for 17 June 1943, in *The Henry Lewis Stimson Diaries*, microfilm edition retrieved from the Center for Research Libraries, original from Manuscripts and Archives, Yale University Library, New Haven, Connecticut. Further references to Stimson's diary in this chapter come from this source.

97 McCullough, *Truman*, 289.

98 James F. Byrnes to Henry L. Stimson (11 September 1943), HBF, Roll 1, Target 8, Folder 8, "Manhattan (District) Project."

99 Harvey H. Bundy to Henry L. Stimson (27 November 1943), HBF, Roll 4, Target 8, Folder 62, "Security (Manhattan Project)."

100 Julius H. Amberg to Harvey H. Bundy (10 December 1943), HBF, Roll 4,

Target 8, Folder 62, "Security (Manhattan Project)"; Jerry Klutz, "The federal diary," *Washington Post* (10 December 1943), 1B; see also John Lansdale Jr., "Newspaper publicity of HEW," (18 December 1943), Hanford DDRS, D4792277.

101 Robert A. Taft to Henry L. Stimson (4 January 1944), W. B. Persons to Harvey H. Bundy (31 March 1944); Harry S. Truman to Henry L. Stimson (10 March 1944); Robert P. Patterson to Henry L. Stimson (11 March 1944); all in HBF, Roll 4, Target 8, Folder 62, "Security (Manhattan Project)." See also Stimson diary entries for 12 March 1944 and 13 March 1944, which includes Stimson's reply to Truman.

102 Albert J. Engel to Henry L. Stimson (23 February 1945), HBF, Roll 1, Target 2, Folder 2, "Appropriations."

103 Albert J. Engel to Henry L. Stimson (23 February 1945), HBF, Roll 1, Target 2, Folder 2, "Appropriations."

104 James F. Byrnes to Franklin D. Roosevelt (3 March 1945), CTS, Roll 3, Target 5, Folder 20, "Miscellaneous."

105 Leslie R. Groves, "Discussion with Secretary of War on 6 or 7 March 1945," (7 April 1945), CTS, Roll 3, Target 5, Folder 20, "Miscellaneous."

106 This particular quote is used several times in congressional publications, but this particular phrasing comes from Hearings on the First Supplemental Appropriation Bill for 1945 before the Subcommittee of the Committee of Appropriations, House of Representatives, 78th Congress, 2nd Session (November 1944), on 600. The use of full-text searching of congressional records via ProQuest makes it quite easy to see things that would have otherwise been buried in a mountain of paper; for example, a report on minerals from 1943 noted that "the uranium industry was greatly stimulated by a Government program having materials priority over all other military procurement, but most of the facts were buried in War Department secrecy," and noted that "most of the 1943 uranium supply was used by physics laboratories for research on uranium isotopes as a source of energy." C. E. Needham, ed., "Minerals Yearbook, 1943," republished as 79th Congress, 1st Session, House Document No. 35 (1 January 1945), 825–828. This particular "leak" was noted later: Frank R. Kent, Jr., "Atom secret told by U.S. publication," *(Baltimore) Sun* (28 November 1945), 1.

107 Vannevar Bush to James B. Conant (14 February 1944), BCF, Roll 1, Target 5, Folder 4, "S-1 Reports to and Conferences with the President (1942–44)."

108 Robert P. Patterson to Henry L. Stimson (25 February 1944), HBF, Roll 1, Target 2, Folder 2, "Appropriations."

109 Vannevar Bush to Harvey H. Bundy (24 February 1944), CTS, Roll 2, Target 8, Folder 14, "Budget and Fiscal." Over the course of 1944, the Americans became less certain that they were in any real race with Germany. By November 1944, Samuel Goudsmit, scientific head of the Alsos mission in Europe to evaluate

the state of German work, reported that, "Germany had no atom bomb and was not likely to have one in a reasonable time." Goudsmit, *Alsos*, 71.

110 Vannevar Bush, memorandum (10 June 1944), CTS, Roll 2, Target 8, Folder 14, "Budget and Fiscal."

111 Leslie R. Groves to Henry L. Stimson (6 March 1945), CTS, Roll 2, Target 8, Folder 14, "Budget and Fiscal."

112 Groves, *Now it can be told*, 360; Goldberg, "General Groves and the atomic West," 70.

113 See Bok, *Secrets*, ch. 4.

114 McCullough, *Truman*, 289–291.

115 On the diminished accountability in wartime and postwar nuclear programs, see esp. Schwartz, *Atomic audit*, esp. ch. 8.

116 See, e.g., Smith, *A peril and a hope*, 59–61.

117 Henry D. Smyth, "Preliminary Report on Peacetime Plans for the DSM Projects," (8 January 1944), BCF, Roll 11, Target 1, Folder 165, "Reports—Compton No. 2 (1944)."

118 Henry D. Smyth, "Preliminary Report on Peacetime Plans for the DSM Projects," (8 January 1944), BCF, Roll 11, Target 1, Folder 165, "Reports—Compton No. 2 (1944)."

119 Zay Jeffries (Chairman), Enrico Fermi, James Franck, T. R. Hogness, R. S. Mulliken, R. S. Stone, and C. A. Thomas, "Prospectus on Nucleonics," (18 November 1944), BCF, Roll 3, Target 4, Folder 17, "S-1 Technical Reports (1942–44)."

120 Henry D. Smyth, "The Problem of Secrecy and the Future of the DSM Project," (15 March 1945, revised 25 April 1945), CTS, Roll 4, Target 6, Folder 3, "Interim Committee and Scientific Panel."

121 Henry D. Smyth, "The Problem of Secrecy and the Future of the DSM Project," (15 March 1945, revised 25 April 1945), CTS, Roll 4, Target 6, Folder 3, "Interim Committee and Scientific Panel."

122 Wellerstein, "Patenting the bomb," 68–76.

123 William A. Shurcliff, "William A. Shurcliff: A brief autobiography" (unpublished manuscript 15 December 1992), Houghton Library, Harvard University.

124 William A. Shurcliff to Carroll L. Wilson, "Suggestions on Post-War Policies re: S-1," (27 March 1944), BCF, Roll 4, Target 1, Folder 19, "S-1 Interim Committee—Postwar (1944–45)."

125 Vannevar Bush to William A. Shurcliff (17 April 1944), BCF, Roll 4, Target 1, Folder 19, "S-1 Interim Committee—Postwar (1944–45)."

126 Vannevar Bush to James B. Conant (17 April 1944), BCF, Roll 4, Target 1, Folder 19, "S-1 Interim Committee—Postwar (1944–45)."

127 William A. Shurcliff to Richard C. Tolman (8 December 1944), HBF, Roll 6, Target 4, Folder 75, "Interim Committee—Publicity."

128 On tacit knowledge in relation to nuclear weapons, see esp. MacKenzie and

Spinardi, "Tacit knowledge, weapons design, and the uninvention of nuclear weapons."

129 William A. Shurcliff to Richard C. Tolman (8 December 1944), HBF, Roll 6, Target 4, Folder 75, "Interim Committee—Publicity."

130 Vannevar Bush to Harvey H. Bundy (n.d., ca. December 1944), HBF, Roll 6, Target 4, Folder 75, "Interim Committee—Publicity."

131 Richard C. Tolman (chairman), W. K. Lewis, E. W. Mills, Henry D. Smyth, "Report of Committee on Postwar Policy," (28 December 1944), CTS, Roll 4, Target 6, Folder 3, "Interim Committee and Scientific Panel."

132 Richard C. Tolman (chairman), W. K. Lewis, E. W. Mills, Henry D. Smyth, "Report of Committee on Postwar Policy," (28 December 1944), CTS, Roll 4, Target 6, Folder 3, "Interim Committee and Scientific Panel."

133 Edward Teller to Leo Szilard (2 July 1945), JRO, Box 71, "Teller, Edward, 1942–1963." On Szilard's petition efforts, see esp. Lanouette, *Genius in the shadows*, 277–284.

134 Cf. Gieryn, "Boundaries of science."

135 See, for example, Schuessler, "The deception dividend."

136 Truman, *Memoirs*, Vol. 1, 20.

137 To avoid arousing public attention, Groves was ferried through "underground passages" into the White House (to further avoid attracting press attention, Chief of Staff George Marshall skipped the meeting in its entirety).

138 Stimson diary (25 April 1945); Leslie Groves, "Report of Meeting with The President," (25 April 1945), CTS, Roll 3, Target 7, Folder 24, "Memorandums to (Gen.) L. R. Groves Covering Two Meetings with the President (Dec. 30, 1944, and Apr. 25, 1945)."

139 Henry L. Stimson, "Memorandum discussed with the President," (25 April 1945), attached to Stimson diary (25 April 1945).

140 Leslie R. Groves, "Atomic Fission Bombs," (23 April 1945), included as Document 3a in Burr, "The atomic bomb and the end of World War II."

141 Norris, *Racing for the bomb*, 375–376; Leslie R. Groves, "Report of the Meeting with the President," (25 April 1945), CTS, Roll 3, Target 7, Folder 24, "Memorandums to (Gen.) L. R. Groves Covering Two Meetings with the President (Dec. 30, 1944, and Apr. 25, 1945)"; Stimson diary (25 April 1945).

142 Leslie R. Groves to Richard H. Groves (27 June 1958), LRG, Entry 7530B, Box 4, "Groves, Richard"; Norris, *Racing for the bomb*, 250–254.

143 Samuel Goudsmit, head of the Alsos project to assess German advancement, noted in December 1944: "The lack of secrecy in Germany with regard to nuclear physics matters is striking. Letterheads and envelopes bear the title of the Reichmarshall's Deputy for Nuclear Physics and stationary for the KWI for physics at Hechingen carries the complete address. According to the OSS in Switzerland, an envelope of the KWI Physics of Berlin was mailed into Switzerland bearing a Hechingen post mark." Samuel Goudsmit to Richard C. Tolman (8 December 1944), BCF, "S-1 Intelligence (1943–45)."

144 The narrative that the bomb was meant to "shock" the Japanese into surrender is present in some of the pre-surrender discussions about the use of the atomic bomb, but was greatly increased in the post-surrender analysis and justification for the use of the bomb. See Gordin, *Five days in August*, and also Asada, "The shock of the atomic bomb and Japan's decision to surrender."

145 "Report on the Present Status and Future Program on Atomic Fission Bombs," (23 August 1943), appears to be the first of this regularly updated report that included discussion about Soviet attempts to infiltrate at Berkeley. CTS, Roll 5, Target 13, Folder 4, "Miscellaneous File 4."

146 "Atomic bomb held 'best-kept secret,'" *New York Times* (9 August 1945), 8.

147 For the Germans, see esp. Bernstein, *Hitler's uranium club*, 115–140, which includes the German physicists' incredulous reaction to the news of Hiroshima while they were interned by the British. Another indication that the Germans were unaware comes from the one espionage attempt we know of, in which several German spies were landed in the US (and immediately captured), with instructions to learn about US nuclear research. The topics they were instructed to investigate indicate that the Germans assumed the US had a modest research program (like the Germans), not a weapon-production program. See J. Edgar Hoover to Harry L. Hopkins (9 February 1945), CTS, Roll 2, Target 6, Folder 12: "Intelligence and Security." The Japanese case is more murky, but the reaction of their cabinet to the news of Hiroshima does not make it appear that they had any foreknowledge. See Hasegawa, *Racing the enemy*, 183–185.

148 Kojevnikov, *Stalin's great science*, ch. 6; Gordin, *Red cloud at dawn*, ch. 3.

149 This rough estimate comes from assumptions of how many people at the major project laboratories, like Los Alamos, Berkeley, Chicago, and Columbia, were given this information, who would have made up the bulk of those with the knowledge of its purpose.

CHAPTER THREE

1 "Statement by the President Announcing the Use of the A-Bomb at Hiroshima," (6 August 1945), copy online at https://www.presidency.ucsb.edu/documents/statement-the-president-announcing-the-use-the-bomb-hiroshima, accessed 23 February 2018. As this chapter will make clear, attributing this statement to Harry S. Truman is not entirely straightforward.

2 Scholars have been slow to look at the Smyth Report's history and importance. The most important single work on the report, to which I am indebted, is the dissertation by Schwartz, "The making of the history of the atomic bomb." This section has benefited immensely in conversations with Schwartz, as well as her written output.

3 Groves would also lean on on his "Manhattan District History" heavily in the writing of his memoirs: Groves, *Now it can be told*. See also, E. H. Mars-

den, "Manhattan District History Preparation Guide," (1 August 1945), NTA, document NV0727839. A heavily redacted version of the Manhattan District History was released on microfilm in 1977, whereas a much more complete (though still redacted) version was released by the Department of Energy starting in 2013. For the former, see: *Manhattan Project: Official history and documents*; for the latter, see MDH.

4 Vannevar Bush to James B. Conant, "Historian for Manhattan Project," (9 March 1944), BCF, Roll 3, Target 6, Folder 18, "S-1 Military Policy Committee (1942–45)."

5 James B. Conant to Vannevar Bush, "Re: Historian for Manhattan Project," (15 March 1944), BCF, Roll 3, Target 6, Folder 18, "S-1 Military Policy Committee (1942–45)."

6 Ibid.

7 James B. Conant to Vannevar Bush, "Some thoughts concerning the correspondence between the President and the Prime Minister on S-1," (25 March 1943), BCF, Roll 2, Target 5, Folder 10, "S-1 British Relations Prior to the Interim Committee (Fldr.) No. 2."

8 Henry D. Smyth, "Preliminary Report on Peacetime Plans for the DSM Projects," (8 January 1944), BCF, Roll 11, Target 1, Folder 165, "Reports—Compton No. 2 (1944)."

9 Smyth, "The Smyth Report," esp. 176–179. See also MDH, Book 1, Volume 4, Chapter 13, "Preparation and Publication of the Smyth Report," 13.2; and Groves, *Now it can be told*, 348.

10 Schwartz, "The making of the history of the atomic bomb," ch. 2, esp. 64–65; Smyth, "The Smyth Report," 182.

11 J. Robert Oppenheimer to Henry D. Smyth (14 April 1945), NTA, document NV0125249.

12 MDH, Book 1, Volume 4, Chapter 13, "Preparation and Publication of the Smyth Report," 13.8–13.10, and Schwartz, "The making of the history of the atomic bomb," 70.

13 "The above account of the work of the Theoretical Physics, Experimental Nuclear Physics, and Chemistry and Metallurgy Divisions is very incomplete because important aspects of this work cannot be discussed for reasons of security. For the same reasons none of the work of the Ordnance, Explosives, and Bomb Physics Divisions can be discussed at all." Smyth, *Atomic energy for military purposes*, paragraph 12.49.

14 Henry D. Smyth to J. Robert Oppenheimer (6 April 1945), NTA, document NV0125250.

15 Leslie R. Groves to installation heads (13 May 1944), quoted in MDH, Book 1, Volume 4, Chapter 13, "Preparation and Publication of the Smyth Report," 13.3.

16 See Schwartz, "The making of the history of the atomic bomb."

17 These interlinkages are vividly illustrated in Galison, *Image and logic*, ch. 4, esp.

section 4.2. As discussed in Galison's chapter, at the same time he was writing the Smyth Report, Smyth was also making big plans for postwar physics, envisioning investments into the subject (and exclusively physics) that would continue the wartime patronage well into the postwar. See section 4.5.

18 In August 1944, for example, the Experimental Physics and Theoretical Physics divisions at Los Alamos together accounted for 19% of laboratory staff, while the Chemistry Division had 21%, and Engineering had 28%, at a time when the total laboratory staff was on the order of 1,100 people (the remainder either were administrative staff or worked in the shops to fabricate components for other divisions). These numbers become harder to divine from August 1944 onward because of a reorganization at the lab (itself a partial realization of the need for better cross-disciplinary collaboration, centered less around disciplinary distinctions and more around problems; e.g., G Division surrounded practical physics, chemistry, and engineering problems related to actual assembly of the "Gadget"), but the relative size of the Theoretical Physics division only shrank as the work got more concerned with the more practical work of weapon design (the Experimental Physics and Engineering divisions overlapped to such a degree after the reorganization that it is hard to know where one could imagine drawing a line between them). These percentages come from MDH, Book 8, Volume 2, "Los Alamos—Technical," graphs 5 and 6. On the reorganization, see Hoddeson et al., *Critical assembly*, ch. 12.

19 This is, again, an elaboration on the argument presented by Schwartz in her insightful dissertation. See Schwartz, "The making of the history of the atomic bomb."

20 Vannevar Bush and James B. Conant to Henry L. Stimson, "The need for: (1) Release of information to the public . . ." (19 September 1944), CTS, Roll 4, Target 1, Folder 26, "Files Received from Col. Seeman's Section (Foreign Intelligence)," Subfile 26L, "Miscellaneous (Breach of Quebec Agreement and French Scientists)."

21 Groves noted to Smyth, "With respect to II d there is, of course, nothing sacred in the figure five." There were some small changes to the wording of the "rules" made in the summer of 1945 (e.g., from "likely to be truly helpful" to "likely to be of distinct help"), but the essential nature of them is the same. Leslie R. Groves to Henry D. Smyth (21 May 1945), CTS, Roll 2, Target 6, Folder 12, "Intelligence and Security." A somewhat later copy can be found as "Rules Governing the Scientific Release," (31 July 1945), MEDR, Box 64, "Security."

22 Testimony of Leo Szilard, in Hearings on H.R. 4280, "An Act for the Development and Control of Atomic Energy," Committee on Military Affairs, House of Representatives, 79th Congress, 1st Session (18 October 1945), 80. See also Gordin, *Red cloud at dawn*, ch. 4.

23 Henry D. Smyth and Richard C. Tolman to Leslie R. Groves, "Application of Security Rules to Smyth Report" (31 July 1945), RCT, Box 4, "00033–00038."

24 Shurcliff, "William A. Shurcliff," 61–63.

25 Smyth, "The Smyth Report," 185; MDH, Book 1, Volume 4, Chapter 13, "Preparation and Publication of the Smyth Report," 13.8.

26 Harvey H. Bundy to Henry L. Stimson (16 December 1944), CTS, Roll 4, Target 1, Folder 26, "Files Received from Col. Seeman's Section (Foreign Intelligence)," Subfile 26L, "Miscellaneous (Breach of Quebec Agreement and French Scientists)."

27 Blank spot and question mark ("194?") are in original. "Possible Statement by the President," (13 February 1945), HBF, Roll 6, Target 3, Folder 74, "Interim Committee—President Truman's Statement."

28 Harvey H. Bundy to Henry L. Stimson (3 March 1945), CTS, Roll 2, Target 3, Folder 9, "Committees," Subfile 9A, "Interim Committee."

29 Leslie R. Groves to George C. Marshall (26 March 1945), CTS, Roll 1, Target 6, Folder 5, "Events Preceding and Following the Dropping of the First Atomic Bombs at Hiroshima and Nagasaki," Subfile 5B, "Directives, Memorandums, etc., to and from the Chief of Staff, Secretary of War, etc."

30 Weart, *Nuclear fear*, 98–103; Lifton and Mitchell, *Hiroshima in America*, ch. 1; Laurence, *The reminiscences of William L. Laurence*, 283–288.

31 William L. Laurence, "The atom gives up," *Saturday Evening Post* 213, no. 10 (7 September 1940), 12–13, 60–63, on 60.

32 Laurence, *The reminiscences of William L. Laurence*, 283–288.

33 "Notes of an Informal Meeting of the Interim Committee," (14 May 1945), CTS, Roll 4, Target 6, Folder 3, "Interim Committee and Scientific Panel."

34 Memorandum on Trinity test press release (14 May 1945), CTS, Roll 1, Target 5, Folder 4, "TRINITY Test (at Alamogordo, July 16, 1945)." My possible attribution of the memo to Groves is based on its style of writing, but also the topics covered: the memo shared all of Groves' specific pet concerns, like congressional interference, interference with negotiations to secure uranium and thorium supplies globally, the Quebec Agreement, and the notion that the Joint Chiefs would disapprove of them releasing "operational" information without consulting them first, along with the various technical issues. It is not impossible that someone else might have been placed highly enough to worry about all such things simultaneously, but it sounds like Groves.

35 William A. Consodine to George L. Harrison (18 June 1945), HBF, Roll 6, Target 4, Folder 75, "Interim Committee—Publicity."

36 Trinity test press releases (14 May 1945), CTS, Roll 1, Target 5, Folder 4, "TRINITY Test (at Alamogordo, July 16, 1945)."

37 "Notes of an Informal Meeting of the Interim Committee" (18 May 1945), CTS, Roll 4, Target 6, Folder 3, "Interim Committee and Scientific Panel."

38 Copies of the local newspapers can be found in MEDR, Box 66, "Security (317–2)," while accounts of the security men are summarized in several memos dating from late July 1945, in CHNSA, Box 11, "1945–1949," Folder 3.

39 Laurence draft of presidential statement (14 May 1945), CTS, Roll 1, Target 5, Folder 4, "TRINITY Test (at Alamogordo, July 16, 1945)."

40 James B. Conant to Vannevar Bush (18 May 1945), BCF, Roll 4, Target 1, Folder 19, "S-1 Interim Committee—Postwar (1944–45)."

41 Griese, *Arthur W. Page*, esp. ch. 7; see also Forrest C. Pogue, "Interview with Mr. Arthur W. Page," (3 April 1959), George C. Marshall Foundation, online at http://marshallfoundation.org/library/wp-content/uploads/sites/16/2014/06/Page_Final_148N.pdf, accessed 1 April 2018.

42 "1000 to 1" is a reference to the comparison between the explosive power of the atomic bomb as compared to a conventional explosive of the same mass. The "bigger ones" comment likely refers to the fact that pre-Trinity, they believed plutonium bomb designs would be lower in yield to the gun-type bomb, at least initially. The "second story" relates to the pending statement by the secretary of war. The concern with the "international organization" makes it seem as though the Interim Committee were involved. "Objectives," (ca. 29 May 1945), MEDR, Box 31, "Releasing information." Cf. Alperovitz, *The decision to use the bomb and the architecture of an American myth*, 594–595.

43 Maj. John A. Derry and Norman F. Ramsey to Leslie R. Groves, "Summary of Target Committee Meetings on 10 and 11 May 1945," (12 May 1945), CTS, Roll 1, Target 6, Folder 5D, "Selection of Targets."

44 Cf. Alperovitz, *The decision to use the bomb and the architecture of an American myth*, 595–596. It is curious that Nagasaki was the named target at this point, since it was not one of the targets considered by the second Target Committee meeting, and was not added to the target list until July 24, after Kyoto was decisively removed from the list.

45 "Notes of the Interim Committee meeting," (21 June 1945), CTS, Roll 4, Target 6, Folder 3, "Interim Committee and Scientific Panel."

46 Arthur W. Page to George L. Harrison (18 July 1945), HBF, Roll 6, Target 3, Folder 74, "Interim Committee—President Truman's Statement." There is the possibility that Page is also talking about the Potsdam Declaration (not yet issued) and its likely rejection by the Japanese.

47 William L. Laurence to Leslie R. Groves, "Plans for Future Articles on Manhattan District Project," (17 May 1945), CTS, Roll 1, Target 6, Folder 5, "Events Preceding and Following the Dropping of the First Atomic Bombs at Hiroshima and Nagasaki," Subfile 5A, "Publicity."

48 "Releases" (n.d., ca. July 1945), MEDR, Box 31, "Releasing information."

49 William L. Laurence, "Rocket to the Moon (Story No. 3)," (n.d., ca. July 1945), MEDR, Box 31, "Laurence Stories."

50 The edited drafts of Laurence's stories are found in MEDR, Box 31, "Laurence Stories." Laurence would revive "Atomland-on-Mars," and his breathless style, for a later monograph on his experience in the project: Laurence, *Dawn over zero*.

51 E.g., Keever, *News zero*; Goodman and Goodman, *The exception to the rulers*, 295–301.

52 For drafts of the Stimson statement, and British comments on it, see HBF,

Roll 6, Target 2, Folder 73, "Interim Committee—Secretary of War Statement." On the authorship of the statement, see William A. Consodine to Arthur W. Page (19 June 1945), MEDR, Box 31, "Releasing information." For the Quebec Agreement, see "Agreement Relating to Atomic Energy [Quebec Agreement]," (19 August 1943), *Foreign Relations of the United States. Conferences at Washington and Quebec, 1943* (Washington, DC: US Government Printing Office, 1970), 1117–1119.

53 Henry L. Stimson to Harry S. Truman (31 July 1945), and Harry S. Truman to Henry L. Stimson (31 July 1945), both in HBF, Roll 6, Target 3, Folder 74, "Interim Committee—President Truman's Statement." The formal use order was from one general to another, drafted by Groves, but ultimately given formal written approval by Stimson, not Truman. See Norris, *Racing for the bomb*, 412, and George Marshall to Thomas Handy, VICTORY 281, (25 July 1945), HBF, Roll 4, Target 10, Folger 64, "Interim Committee—Potsdam Cables."

54 William A. Consodine to Leslie R. Groves, "MED Public Relations Program," (27 June 1945), MEDR, Box 31, "Releasing information."

55 Consodine to Groves, "MED Public Relations Program."

56 Leslie R. Groves to Kenneth D. Nichols, "Manhattan Engineer District Public Relations Program," (26 July 1945), MEDR, Box 31, "Releasing information."

57 Leslie R. Groves to Robert Oppenheimer (30 July 1945), MEDR, Box 31, "Releasing information."

58 J. Robert Oppenheimer to Los Alamos personnel (6 August 1945), RFB, Box 17, Folder 8.

59 R. G. Arneson, "Notes of the Interim Committee Meeting," (31 May 1945), CTS, Roll 4, Target 6, Folder 3, "Interim Committee and Scientific Panel." On Stimson's complex relationship with the bombings, see Malloy, *Atomic tragedy*, and Wellerstein, "The Kyoto misconception."

60 Discussion of future (thermonuclear) weapons in the range of 10 to 100 megatons was already present in the May 31, 1945, meeting of the Interim Committee, ibid. See also Sherwin, *A world destroyed*, ch. 8.

61 For more on the construction of the "special" nature of the bomb, see Gordin, *Five days in August*.

62 The articles drafted by Laurence had recommendations for how many days after "R-Day" they should be let out. MEDR, Box 31, "Laurence stories."

63 James B. Conant to Vannevar Bush (18 May 1945), BCF, Roll 4, Target 1, Folder 19, "S-1 Interim Committee—Postwar (1944–45)."

64 Lt. Walter A. Parish, Jr. to Leslie R. Groves, "Security at Naval Installations following Publicity Day," (4 August 1945), MEDR, Box 31, Folder "Releasing Information."

65 Thomas Handy to Carl Spaatz (25 July 1945), CTS, Roll 1, Target 6, Folder 5B, "Directives, Memos, Etc. to and from C/S, S/W, etc.)."

66 On the disconnect between the work on Tinian and the head planners, see Gordin, *Five days in August*.

67 Groves, *Now it can be told*, 322. The bombing of Hiroshima took place on 6 August 1945 at 8:15 A.M. local (Japanese) time, which was 5 August at 7:15 P.M. in Washington, DC (Eastern War Time, the equivalent of Eastern Daylight Time). In his memoirs, Groves was unsure why it took over three hours for the message to reach him. Dates used in the text will be according to Eastern War Time.

68 Hersey, *Hiroshima*; Hasegawa, *Racing the enemy*, 184.

69 Leslie R. Groves to George C. Marshall, "Memorandum to the Chief of Staff" (6 August 1945), CTS, Roll 1, Target 6, Folder 5, Subfile 5B, "Directives, Memorandums, etc., to and from the Chief of Staff, Secretary of War, etc."

70 "Statement by the President of the United States" (6 August 1945), in *Manhattan Project: Official history and documents*, Reel 1, Book 1, Volume 4, Chapter 8, "Press Releases, Part 1." Truman recorded a further abbreviated version of the statement for radio and newsreel circulation. His elation at the news is palpable in the footage—it was, as he had announced to the crew of the *Augusta*, "the greatest thing in history." "Log of the President's Trip to the Berlin Conference, 6 July to 7 August 1945," on 49, Harry S. Truman Library, Independence, MO, online at https://www.trumanlibrary.org/calendar/travel_log/documents/index.php?pagenumber=96&documentid=17&documentdate=1945-08-06&studycollectionid=TL&nav=ok, accessed 1 June 2017.

71 "Statement by the President of the United States" (6 August 1945).

72 "Statement of the Secretary of War," (6 August 1945), in *Manhattan Project: Official history and documents*, Reel 1, Book 1, Volume 4, Chapter 8, "Press Releases, Part 1."

73 "Statement by the Honourable C. D. Howe, Canadian Minister of Munitions and Supply," (6 August 1945), in *Manhattan Project: Official history and documents*, Reel 1, Book 1, Volume 4, Chapter 8, "Press Releases, Part 2."

74 Included in "Statements Relating to the Atomic Bomb," (London: His Majesty's Stationary Office, 1945), in *Manhattan Project: Official history and documents*, Reel 1, Book 1, Volume 4, Chapter 8, "Press Releases, Part 2."

75 Secretary of War Statement (draft of 6 July 1945), HBF, Roll 6, Target 2, Folder 73, "Interim Committee—Secretary of War Statement."

76 Kenneth D. Nichols, "Security Message to the Press," (6 August 1945), in *Manhattan Project: Official history and documents*, Reel 1, Book 1, Volume 4, Chapter 8, "Press Releases, Part 1."

77 Harold Johnson, "Death will saturate bomb targets for 70 years, atomic expert says," *Atlanta Constitution* (8 August 1945), 1.

78 "200,000 believed dead in inferno that vaporized city of Hiroshima," *Boston Globe* (9 August 1945), 2. This number, it should be noted, is considerably higher than most Hiroshima fatality estimates, which generally range from

60,000 to 100,000. The inflation comes from of the newspapermen equating a percentage of area destroyed (60%) with percentage of casualties. Nuclear mortality curves are not the same as the damage curves, but it would take considerable research to establish better estimates of the casualties (and such estimates are still controversial today).

79 Paul M. A. Linebarger, "Memorandum for Colonel Buttles: Identification of atomic bomb targets as being military in character," (9 August 1945), Paul M. A. Linebarger Papers Prepared During World War II, Vol. 5, Hoover Institution Archives on War, Peace and Revolution, Stanford, CA. I am grateful to Sean Malloy in providing me with a copy of this document. On the behind-the-scenes considerations of Hiroshima as a "military" target, see Wellerstein, "The Kyoto misconception."

80 Had the bombs been detonated at lower altitudes, this might not have been the case (the Trinity site is still several times more radioactive than background radiation, all these decades later, and would have required substantial remediation after the attacks). On the dosages Japanese victims received from fallout, see Radiation Effects Research Foundation, "RERF's Views on Residual Radiation" (8 December 2012), online at https://www.rerf.or.jp/news/pdf/residualrad_ps_e.pdf, accessed 13 June 2018.

81 The only papers I have found that carried Jacobson's story in various online newspaper databases are the *Atlanta Constitution* and the *Twin Falls (Idaho) Telegram*. By contrast, the denial stories appear to have run in virtually every major and minor newspaper in the databases.

82 "Memorandum to the Press [on radioactivity]," (8 August 1945), in *Manhattan Project: Official history and documents*, Reel 1, Book 1, Volume 4, Chapter 8, "Press Releases, Part 1."

83 Robert S. Stone to Lt. Col. H. L. Friedel (9 August 1945), NTA, document NV0715871.

84 He was correct that there was little long-term contamination at Hiroshima; he was wrong that there was not a significant amount of radioactivity that would prove a health hazard to people who were in the city or who entered into it shortly after detonation. On later surveys of the effects of radioactivity at the bombing sites, see Lindee, *Suffering made real.*

85 Lifton and Mitchell, *Hiroshima in America*, 40–42; "Atom explosion is above ground to lose its deadly rays in the sky," *New York Herald Tribune* (12 August 1945), 1.

86 Malloy, "'A very pleasant way to die.'"

87 "Minutes of Combined Policy Committee Meeting Held at the Pentagon on July 4th, 1945—9:30 A.M.," BCF, Roll 2, Target 6, Folder 11, "S-1 Combined Policy Committee."

88 "Memorandum" (16 July 1945), CTS, Roll 2, Target 6, Folder 12, "Intelligence and Security."

89 Col. W. H. Kyle, "Notes of a Meeting on the Smyth Report in the Office of the

Secretary of War 11:30 A.M. to 1:15 P.M." (2 August 1945), CTS, Roll 2, Target 6, Folder 12, "Intelligence and Security."

90 Col. W. H. Kyle, "Notes of a Meeting on the Smyth Report in the Office of the Secretary of War 11:30 A.M. to 1:15 P.M." See also Hewlett and Anderson, *The new world*, 400–401.

91 James Chadwick to Field Marshal Sir Henry Wilson (4 August 1945), CTS, Roll 2, Target 6, Folder 12, "Intelligence and Security."

92 Bush, *Pieces of the action*, 294; Hewlett and Anderson, *The new world*, 406–407.

93 The copy filed for copyright purposes with the Library of Congress is the only one that bears the red stamp. On the different versions that were published, see Coleman, "The 'Smyth Report.'"

94 Press Release, War Department Bureau of Public Relations (Press Branch) (11 August 1945), CTS, Roll 5, Target 12, Miscellaneous File 3; Schwartz, "The making of the history of the atomic bomb," ch. 3.

95 Smith, "The publishing history of the 'Smyth Report.'"

96 Herbert F. York, oral history interview with author (24 April 2008), OHCNBL.

97 Arnold Kramish to Harold A. Fidler, "Russian Smyth Report" (18 September 1948), RCT, Box 5, Folder 4.

98 Solzhenitsyn, *The Gulag Archipelago*, 598–599.

99 Richard C. Tolman to Leslie R. Groves (n.d., ca. 20 August 1945), MEDR, Box 32, "Censorship"; Cf. illustrations in *Life* 19, no. 8 (20 August 1945), on 87D and 90. The work of John Coster-Mullen in unearthing the probable design of Little Boy makes quite clear how far the divergence of the *Life* depiction was from the realities of the bomb's construction. Cf. Coster-Mullen, *Atom bombs*; Samuels, "Atomic John."

100 For example, a rough discussion of the implosion method was included in a September 1945 issue of *Newsweek*, which curiously cited "authorized publications" to confirm its description. "'Fat Boy' atom bomb hit Nagasaki: 'Big Boy' was ready . . ." *Newsweek* (18 September 1945), 44–45. This example comes from David B. Langmuir, "Control of Information Under Atomic Energy Act of 1946," AEC 111/4 (13 January 1949), OSAEC46, Box 41, "Basic Security Policy—Control of Info—Vol. 1." On the whole, however, it is worth noting that nearly every description of bomb design in the public domain from 1945 to 1951 was of the "gun-type" described in the Smyth Report.

101 Press Release, War Department Bureau of Public Relations (Press Branch) (11 August 1945), CTS, Roll 5, Target 12, Miscellaneous File 3.

102 "Memorandum of telephone conversation between General Groves and Lt. Col. Rea, Oak Ridge Hospital, 9:00 A.M.," (25 August 1945), CTS, Roll 1, Target 6, Folder 5, "Events Preceding and Following the Dropping of the First Atomic Bombs at Hiroshima and Nagasaki," Subfile 5G, "Radiological Effects." See also Malloy, "'A very pleasant way to die,'" and Hacker, *The dragon's tail*, ch. 5.

103 E.g., J. Robert Oppenheimer, "Memorandum for Brigadier General Farrell," (11 May 1945), and "Toxic Effects of the Atomic Bomb," (12 August 1945), both in

CTS, Roll 1, Target 6, Folder 5, "Events Preceding and Following the Dropping of the First Atomic Bombs at Hiroshima and Nagasaki," Subfile 5G, "Radiological Effects."

104 Lindee, *Suffering made real.*

105 On the censorship within Japan, see Braw, *The atomic bomb suppressed.*

106 "Atomic bomb was a pain to censors," *Washington Post* (7 August 1945), 7; "Atom bomb held 'best-kept secret,'" *New York Times* (9 August 1945), 8; "Elaborate security measures protected secret of atomic bomb," (11 August 1945), in *Manhattan Project: Official history and documents*, Reel 1, Book 1, Volume 4, Chapter 8, "Press Releases, Part 1." In a later memo, Groves indicated that the security story was issued "in an effort to stop the release of continued inaccurate information by agencies and persons outside the War Department control which was leading to security difficulties," but did not elaborate. Leslie R. Groves to Clayton Bissell, (5 December 1945), MEDR, Box 32, "Censorship."

107 See, e.g., Joseph Marshall, "How we kept the atomic bomb secret," *Saturday Evening Post* 218, no. 19 (10 November 1945), 14–15, 49.

108 John Weckerling to Leslie R. Groves, "Publication of Information Concerning Intelligence and Counterintelligence Activities of Personnel of the Manhattan Engineer District," (7 November 1945), MEDR, Box 32, "Censorship."

109 Alperovitz, *The decision to use the bomb and the architecture of an American myth*, part 4. On the complexity of the narratives surrounding the bomb, see esp. Boyer, *By the bomb's early light*; Weart, *Nuclear fear*; and Lifton and Mitchell, *Hiroshima in America*. On the film *The Beginning or the End?*, see Mitchell, *The beginning or the end.*

CHAPTER FOUR

1 Harry S. Truman, "Message from the President of the United States Transmitting Request for the Enactment of Legislation to Fix a Policy Covering the Use and Development of the Atomic Bomb," H.Doc. 301, 79th Congress, 1st Session (3 October 1945).

2 On Bohr's role at Los Alamos, see Robert F. Bacher oral history interview with Lillian Hoddeson and Alison Kerr (30 July 1984), 61–62, RFB, Box 48, Folder 5; Hoddeson et al., *Critical assembly*, 317, and MDH, Book 8 (Los Alamos Project), Volume 2 (Technical), pages II-2 to II-3. For an excellent account of Bohr's early thoughts on the postwar nuclear world, Smith, *A peril and a hope*, part 1.

3 Niels Bohr, "Memorandum," (3 July 1944), CTS, Roll 3, Target 5, Folder 20, "Miscellaneous"; see also Smith, *A peril and a hope*, 9–10.

4 Frankfurter had learned of the project in the summer of 1943 when a Metallurgical Laboratory scientist, Irving S. Lowen, had attempted to contact Roosevelt about his feeling that the project was not going well. Vannevar Bush, "Memorandum of Conference," (22 September 1944), BCF, Roll 4, Target 1, Folder 19,

"S-1 Interim Committee—Postwar (1944–45)"; Felix Frankfurter memo (26 May 1945), HBF, Roll 2, Target 2, Folder 19, "Bohr, Dr. Niels"; James B. Conant to Vannevar Bush, "Complaints about the S-1 Project at Chicago reaching the President," (31 July 1943), BCF, Roll 1, Target 5, Folder 4, "S-1 Reports to and Conferences with the President (1942–44)."

5 Formato, "Writing the atom."

6 Vannevar Bush, "Memorandum of Conference," (22 September 1944), BCF, Roll 4, Target 1, Folder 19, "S-1 Interim Committee—Postwar (1944–45)."

7 Niels Bohr, "Addendum to Memorandum of July 3rd, 1944," (24 March 1945), CTS, Roll 3, Target 5, Folder 20, "Miscellaneous."

8 For more on Bohr's attempts to influence high-level policy with his ideas for "openness," see esp. Rhodes, *The making of the atomic bomb*, 527–538.

9 Hewlett and Anderson, *The new world*, 329–330.

10 Vannevar Bush and James B. Conant to Henry L. Stimson (19 September 1944), CTS, Roll 4, Target 1, Folder 26, "Files Received from Col. Seeman's Section (Foreign Intelligence)," Subfile 26L, "Miscellaneous (Breach of Quebec Agreement and French Scientists)."

11 Vannevar Bush and James B. Conant to Henry L. Stimson, "Salient Points Concerning Future International Handling of Subject of Atomic Bombs," and summary sheet (30 September 1944), CTS, Roll 2, Target 4, Folder 10, "International Control of Atomic Energy."

12 Bush and Conant seem to have developed these ideas independent from Bohr. In early October 1944, Bush met with Bohr, and would afterwards write to Conant on how "extraordinarily close" Bohr's position on openness was "to the point of view that you and I arrived at a week or two ago." Vannevar Bush to James B. Conant (11 October 1944), BCF, Roll 5, Target 8, Folder 38, "Bush, V. (1944–45)."

13 James B. Conant and Vannevar Bush to Henry L. Stimson, "Supplementary memorandum giving further details concerning military potentialities of atomic bombs and the need for international exchange of information," (30 September 1944), CTS, Roll 2, Target 4, Folder 10, "International Control of Atomic Energy."

14 Hewlett and Anderson, *The new world*, 344–345.

15 Sherwin, *A world destroyed*, ch. 8.

16 Vannevar Bush to Harry S. Truman, "Scientific Interchange on Atomic Energy," (25 September 1945), BCF, Roll 1, Target 6, Folder 5, "S-1 Intelligence (1943–45)."

17 Irvin Stewart to Vannevar Bush (25 August 1944), with attached "Outline of a bill to be introduced to Congress, 1st Draft," BCF, Roll 8, Target 13, Folder 97, "Legislation S-1 (1944–45)." On wartime use of patents for control by Bush and Conant, see Wellerstein, "Patenting the bomb."

18 James B. Conant to Vannevar Bush (15 September 1944), BCF, Roll 4, Target 1, Folder 19, "S-1 Interim Committee—Postwar (1944–45)"; James B. Conant and

Vannevar Bush to Henry L. Stimson (19 September 1945), CTS, Roll 4, Target 1, Folder 26, "Files Received from Col. Seeman's Section (Foreign Intelligence)," Subfile 26L, "Miscellaneous (Breach of Quebec Agreement and French Scientists)."

19 R. Gordon Arneson, "Notes of the Interim Committee Meeting," (21 June 1945), CTS, Roll 4, Target 6, Folder 3, "Interim Committee and Scientific Panel"; Hewlett and Anderson, *The new world*, 412.

20 "Atomic Research Act of 1945 (Draft)," (18 July 1945), BCF, Roll 4, Target 1, Folder 19, "S-1 Interim Committee—Postwar (1944–45)."

21 Vannevar Bush to George L. Harrison, "Comments on draft of legislation of 18 July 1945," (19 July 1945), BCF, Roll 4, Target 1, Folder 19, "S-1 Interim Committee—Postwar (1944–45)."

22 Vannevar Bush to George L. Harrison, "Comments on draft of legislation of 18 July 1945," (19 July 1945), BCF, Roll 4, Target 1, Folder 19, "S-1 Interim Committee—Postwar (1944–45)."

23 R. Gordon Arneson, "Notes of the Interim Committee Meeting," (19 July 1945), CTS, Roll 4, Target 6, Folder 3, "Interim Committee and Scientific Panel"; see also Hewlett and Anderson, *The new world*, 412–413.

24 Hewlett and Anderson, *The new world*, 414.

25 George S. Allen and George M. Duff, "Memorandum Concerning Third Statutory Draft," (n.d., ca. August 1945), HBF, Roll 6, Target 1, Folder 72, "Interim Committee—Legislation (Draft Bills)."

26 Royall-Marbury bill (Third Draft), (n.d., ca. August 1945), HBF, Roll 6, Target 1, Folder 72, "Interim Committee—Legislation (Draft Bills)."

27 James B. Conant, "Dr. Conant's Notes on the Draft Bill (3rd)," (n.d., ca. August 1945), HBF, Roll 6, Target 1, Folder 72, "Interim Committee—Legislation (Draft Bills)." Emphasis in original.

28 Ibid.

29 George S. Allan and George M. Duff to Kenneth C. Royall, "Proposed Security Provisions of Statute," (6 August 1945), HBF, Roll 6, Target 1, Folder 72, "Interim Committee—Legislation (Draft Bills)." Awkward grammar is in the original.

30 George S. Allan and George M. Duff to Leslie R. Groves, "Security Provisions of Proposed Statute," (28 August 1945), CTS, Roll 2, Target 7, Folder 13, "Legislation (for Atomic Energy)."

31 George S. Allan and George M. Duff to Kenneth C. Royall, "Fifth Draft of Proposed Atomic Energy Act," (29 August 1945), CTS, Roll 2, Target 7, Folder 13, "Legislation (for Atomic Energy)."

32 Oscar M. Ruebhausen to Vannevar Bush (7 August 1945), BCF, Roll 4, Target 1, Folder 19, "S-1 Interim Committee—Postwar (1944–45)." See also Vannevar Bush to George L. Harrison (7 August 1945), BCF, Roll 4, Target 1, Folder 19, "S-1 Interim Committee—Postwar (1944–45)."

33 On the uncontrollability of nuclear imagery, see esp. Weart, *Nuclear fear*.

34 Gordin, *Red cloud at dawn*, ch. 2, esp. 72–77; Ziegler and Jacobson, *Spying without spies*, ch. 2; Helmreich, *Gathering rare ores*.

35 George L. Harrison to Vannevar Bush (with enclosure of final draft), (5 September 1945), BCF, Roll 4, Target 1, Folder 19, "S-1 Interim Committee—Postwar (1944–45)."

36 Hewlett and Anderson, *The new world*, 422–423. On the details of the "Scientists' Movement" against the Royall-Marbury/May-Johnson bills, see esp. Smith, *A peril and a hope*; Wang, *American science in an age of anxiety*; Kaiser, "The atomic secret in Red hands?"; Boyer, *By the bomb's early light*, ch. 4–9, and 13. On the Ad Council campaign ("No secret, no defense, international control"), see Barnhart, "To secure the benefits of science to the general welfare," esp. 139–140.

37 Harry S. Truman, "Message from the President of the United States Transmitting Request for the Enactment of Legislation to Fix a Policy Covering the Use and Development of the Atomic Bomb," H.Doc. 301, 79th Congress, 1st Session (3 October 1945); Hewlett and Anderson, *The new world*, 425–427.

38 Hewlett and Anderson, *The new world*, 428–431.

39 Sections 16–18 in "A bill for the development and control of atomic energy," H.R. 4280 (3 October 1945), 79th Congress, 1st Session. Copies of this (and other contemporaneous atomic energy bills) are available in HBF, Roll 8, Target 10, Folder 96, "Bills (US Senate and House of Representatives)."

40 Hearings on H.R. 4280, "An Act for the Development and Control of Atomic Energy," Committee on Military Affairs, House of Representatives, 79th Congress, 1st Session (9 and 18 October 1945).

41 On the rising power of "the secret" as a trope, see Kaiser, "The atomic secret in Red hands?"

42 Hearings on H.R. 4280, "An Act for the Development and Control of Atomic Energy," Committee on Military Affairs, House of Representatives, 79th Congress, 1st Session (9 and 18 October 1945).

43 Testimony of Leslie R. Groves, Hearings on H.R. 4280, "An Act for the Development and Control of Atomic Energy," Committee on Military Affairs, House of Representatives, 79th Congress, 1st Session (9 and 18 October 1945), 12–19.

44 On the Scientists' Movement and their motivations, see Smith, *A peril and a hope*, and Kevles, *The physicists*, ch. 12.

45 All from testimony in, Hearings on H.R. 4280, "An Act for the Development and Control of Atomic Energy," Committee on Military Affairs, House of Representatives, 79th Congress, 1st Session (9 and 18 October 1945).

46 Hewlett and Anderson, *The new world*, 435–439.

47 Edward U. Condon to J. Robert Oppenheimer (26 April 1943), MEDR, Box 101, "Investigation Files"; Leslie Groves to Harvey Bundy (15 June 1945), HBF, Roll 4, Target 8, Folder 62, "Security (Manhattan Project)." Condon would later suffer a number of lengthy investigations into his loyalty; see Wang, *American science in an age of anxiety*.

48 "Senate Atomic Energy Manual" (15 November 1945), CTS, Roll 2, Target 7, Folder 13, "Legislation (for Atomic Energy)."
49 James R. Newman and Edward U. Condon to Brien McMahon (24 November 1945), SCAE, Box 2, "Groves, Maj. Gen. (Dispute With)."
50 Newman and Anderson, *The new world*, 450.
51 Leslie R. Groves to Robert P. Patterson (23 November 1945), CTS, Roll 2, Target 7, Folder 13, "Legislation (for Atomic Energy)."
52 Robert P. Patterson to James F. Byrnes (23 November 1945), CTS, Roll 2, Target 7, Folder 13, "Legislation (for Atomic Energy)."
53 Edward U. Condon and James R. Newman to Brien McMahon (3 December 1945), SCAE, Box 2, "Groves, Maj. Gen. (Dispute With)."
54 Hewlett and Anderson, *The new world*, 452–453.
55 Hewlett and Anderson, *The new world*, 454–455.
56 Newman and Miller, *The control of atomic energy*, 4.
57 On the evolution of the "basic" and "applied" distinction, see Kline, "Construing 'technology' as 'applied science.'"
58 "A bill for the development and control of atomic energy," S. 1717 (20 December 1945), reprinted in Hewlett and Anderson, *The new world*, appendix 1.
59 Congress is not beholden to the Freedom of Information Act, so there is no way to compel their release. One can try (as I have) to force a Mandatory Declassification Review of closed hearings, but my effort does not appear to have generated any results as of this writing with regard to the Senate Committee on Atomic Energy.
60 Lt. Walter A. Parish, Jr., to Leslie R. Groves (11 February 1946), MEDR, Box 65, "Security of Info. Re Atomic Bomb."
61 Hewlett and Anderson, *The new world*, 501; Kaiser, "The atomic secret in Red hands?," 36; and, for the best evidence regarding Groves' involvement in the leak, see Herken, *The winning weapon*, ch. 5.
62 Leslie R. Groves, "Loss of Security," (19 February 1945), MEDR, Box 65, "Security of Info. Re Atomic Bomb."
63 Groves, "Loss of Security."
64 Harry S. Truman, "Memorandum to All Departments and Agencies of the Federal Government," (21 February 1946), CTS, Roll 2, Target 6, Folder 12, "Intelligence and Security." See also Norris, *Racing for the bomb*, ch. 22.
65 Hewlett and Anderson, *The new world*, 501–502.
66 Hewlett and Anderson, *The new world*, 502–503; testimony of Leslie R. Groves, "Hearings on S.1717, Part 4," Special Committee of Atomic Energy, United States Senate, 79th Congress, 2nd Session (18, 19, and 27 February 1946), 467–496.
67 S.171, Confidential Committee Print No. 4 (27 March 1946), copies of which are loose in SCAE, Box 32. See also the papers discussing the provisions of the bill in SCAE, Box 32, "Analysis of Bill."

68 SCAE, Box 32, "Analysis of Bill."

69 S.171, Confidential Committee Print No. 5 (11 April 1946), a copy of which is loose in SCAE, Box 32. Hewlett and Anderson, *The new world*, 512. See also Smith, *A peril and a hope*, 410–411.

70 The "born secret" interpretation of the "restricted data" clause seems to have come only somewhat after the Atomic Energy Act was passed. See the discussion in *The Government's Classification of Private Ideas* (House Report 96–1540), United States House of Representatives, Committee on Government Operations, 96th Congress, 2nd Session (22 December 1980), which dates all "born secret" discussions to the mid-1950s, around the time in which the AEC attempted to stimulate the private sector. For a divergent view, see Richard G. Hewlett, "The 'born-classified' concept in the U.S. Atomic Energy Commission" (May 1980), a copy of which is included as an appendix in the House Report, on 173–187. Hewlett's analysis suffers from a sort of presentism: though he acknowledges there is no evidence that the McMahon committee or the early Atomic Energy Commission conceived of the clause as a "born secret" clause until the 1950s, he nonetheless asserts that the AEC adopted the idea as a "working assumption." See chapters 6 and 7 of this book for further discussion.

71 Hewlett and Anderson, *The new world*, 514.

72 Specifically, the Espionage Act applies to "document, writing, code book, signal book, sketch, photograph, photographic negative, blue print, plan, map, model, instrument, appliance, or note relating to the national defense." Section 10(b)(2) of the final version of the McMahon Act uses similar language, however, regulating the communication of "any document, writing, sketch, photography, plan, model, instrument, appliance, note or information involving or incorporating restricted data." The only major difference is the inclusion of "information" as a category of regulation.

73 Brien McMahon, Statement to the President on S.1717, *Congressional Record—Senate* (1 June 1946), on 6096.

74 From this point onward in the book, "Restricted Data" shall always refer to this unique legal conception of nuclear secrets, whether capitalized or not. It was almost always referred to in lower-case in this early period; as it became more formalized it became almost exclusively capitalized. Unless I am quoting someone else, I will use the latter convention.

75 "Report to accompany S.1717," 23.

76 Hewlett and Anderson, *The new world*, 515.

77 Testimony of Robert P. Patterson, "Hearings on S.1717," Committee on Military Affairs, House of Representatives, 79th Congress, 2nd Session (11, 12, and 26 June 1946), on 23.

78 Quoted in Hewlett and Anderson, *The new world*, 525.

79 Hewlett and Anderson, *The new world*, 526–530; Atomic Energy Act of 1946,

S. 1717 (Public Law 585), 79th Congress, 2nd Session (1 August 1946). For far more detail on the legislative debate, see Hewlett and Anderson, *The new world*, ch. 14.

80 Byron S. Miller, "A law is passed—The Atomic Energy Act of 1946," *University of Chicago Law Review* 15, no. 4 (Summer 1948), 799–821, on 821.

81 James R. Newman, "Control of information relating to atomic energy," *Yale Law Journal* 56, no. 5 (May 1947), 769–802, on 775, 782.

82 Newman, "Control of information relating to atomic energy," 783.

83 Quoted in Hewlett and Anderson, *The new world*, 433.

84 The linguistic trickiness comes from the fact that "declassification" is a form of classification; instead of classifying the document or information as "secret," for example, it is classified as "unclassified." Within a large classification system, nothing is truly unclassified, as even the designation of "unclassified" needs to be applied. Terms like "downgrading" refer to classifying a given document as a less-stringent form of classification designation (e.g., from "Top Secret" to "Secret"). In this book, "classification" always means "security classification," and can be thought of as synonymous with the structured application of secrecy practices. The best overall discussion of classification principles is Quist, *Security classification of information*, vol. 1.

85 Quist, *Security classification of information*, vol. 1, ch. 1–2.

86 Office of War Information Regulation No. 4 (13 March 1944), quoted in Quist, *Security classification of information*, vol 1, ch. 2.

87 Executive Order 9568, "Providing for the Release of Scientific Information," 10 F.R. 6917 (8 June 1945); Stewart, *Organizing scientific research for war*, 287–289. A further Executive Order issued in August 1945 provided for the release of enemy scientific and technical information, as well. Executive Order 9604, "Providing for the Release of Scientific Information (Extension and Amendment of Executive Order 9568)," 10 F.R. 10960 (25 August 1945).

88 Stewart, *Organizing scientific research for war*, 291–293.

89 See, e.g., Jesse W. Beams to Henry T. Wensel (27 August 1945), Gerhard H. Dieke to Kenneth D. Nichols (22 October 1945), and Francis G. Slack to E. J. Bloch (22 October 1945), all in RCT, Box 5, Folder 13.

90 M. D. Whitaker (Monsanto) to E. J. Murphy, "Request for Clinton Laboratories Products" (14 August 1945), RCT, Box 5, Folder 13; S. E. Quinn (Dow Chemical) to James B. Conant (14 September 1945), RCT, Box 5, Folder 13; C. G. Suits (General Electric) to Kenneth D. Nichols (20 September 1945), RCT, Box 5, Folder 13; Nichols to W. G. Green (Wells Surveys, Inc.) (27 September 1945), RCT, Box 5, Folder 13; James Bailey (Plax Corporation) to Neil J. Carothers (23 October 1945), RCT, Box 5, Folder 13; and P. P. Alexander (Metal Hydrides, Inc.) to Ernest O. Lawrence (17 October 1945), EOL, Box 29, Folder 20, microfilm reel 44, frame 097028.

91 William A. Borden to Leslie R. Groves, "Release of Scientific and Technical Information," (8 October 1945), MEDR, Box 31, "Releasing information";

Leslie R. Groves to William A. Borden (18 October 1945), MEDR, Box 32, "Censorship."

92 J. Robert Oppenheimer to Leslie R. Groves (23 October 1945), MEDR, Box 65, "Declassification."

93 In his later testimony at Robert Oppenheimer's security hearing, Groves would claim that he had the Committee on Declassification "under pretty rigid control" on account that Tolman "was in completely sympathy" with his own views on classification. This is probably an exaggeration, but Groves did align himself with their conclusions. USAEC, *In the matter of J. Robert Oppenheimer*, 175–176.

94 TWX Leslie R. Groves to Richard C. Tolman (2 November 1945), RCT, Box 5, Folder 1.

95 Groves to Tolman (2 November 1945); Compton, *Atomic quest*, 95.

96 Richard C. Tolman to Warren Weaver (20 February 1945), CTS, Roll 2, Target 6, Folder 12, "Intelligence and Security."

97 Arthur H. Compton to Henry A. Wallace (27 September 1945), EOL, Box 28, Folder 23, microfilm reel 42, frame 093402.

98 Eugene P. Wigner to Richard C. Tolman (6 November 1945), RCT, Box 5, Folder 13; Arthur H. Compton to Richard C. Tolman (18 February 1946), RCT, Box 4, Folder "00097."

99 Dictation by Ernest Lawrence (27 November 1945), EOL, Box 29, Folder 4, microfilm reel 43, frame 095985.

100 Lilienthal et al., *A report on the international control of atomic energy*, 60.

101 Committee on Declassification, "Report of Committee on Declassification" (17 November 1945), MEDR, Box 49, "Committee on Declassification." Richard C. Tolman to Robert F. Bacher (17 November 1945), RCT, Box 18, Folder 7. Excerpts from the first "Report of the Committee of Declassification" were not declassified until 1965, at the instigation of Arnold Kramish, who wished to quote it a review of a book, because he felt that "the Tolman Report finally deserves a public airing." Arnold Kramish to Harold A. Fidler (6 April 1965), RCT, Box 5, Folder 14. The full report does not appear to have been declassified until 1990, judging by the declassification stamps on it.

102 USAEC, *In the matter of J. Robert Oppenheimer*, 36.

103 "Report of Committee on Declassification."

104 Richard C. Tolman to Robert F. Bacher (11 December 1945), RFB, Box 18, Folder 7; Tolman to representatives from Westinghouse Electric Corp., Tennessee Eastman Corp., Harshaw Chemical Co., Hooker Electrochemical Co., Union Carbide & Carbon Co., Allis-Chalmers Mfg. Co., Chrysler Corp, S&W Engineering Corp., Crane Co., General Electric Co., Kellex Corp., and DuPont Corp. (12 December 1945), RCT, Box 5, Folder 1; John R. Ruhoff to Tolman (17 December 1945), RCT, Box 5, Folder 10.

105 Richard C. Tolman to Robert F. Bacher (3 January 1946), RFB, Box 18, Folder 17; duplicate letter sent to Ernest Lawrence in RCT, Box 5, Folder 10.

106 "Narrative Report of Meetings in Wilmington and New York" (n.d., ca. 6 January 1946), RCT, Box 4, Folder "00009."

107 Committee on Declassification, "Statement of Recommendations on Release of Atom Bomb Project Information" (4 February 1946), RCT, Box 5, Folder 8. One of the few stories about the letter was William S. Barton, "More atom energy data to be disclosed to public," *Los Angeles Times* (5 February 1946), 2.

108 Richard C. Tolman to Robert F. Bacher (7 January 1946), RFB, Box 18, Folder 7; Committee on Declassification, "Second Report of Committee on Declassification" (20 January 1946, sent 16 February 1946), RFB, Box 18, Folder 7; Committee on Declassification, "Justification for Items Recommended for Declassification in Memorandum of 17 November" (19 February 1946), RFB, Box 18, Folder 8; Committee on Declassification, "Information Previously Known about the Properties of Certain Substances and Further Disclosures which would Result from Declassification, as Proposed in Memorandum of 17 November" (18 February 1946); Committee on Declassification, "Third Report of Committee on Declassification (21 May 1946), RCT, Box 5, Folder 7.

109 "Second Report of the Committee on Declassification"; Alton P. Donnell to Robert F. Bacher, "Transmittal of Declassification Guide for Responsible Reviewers" (3 April 1946), RFB, Box 18, Folder 8.

110 Leslie R. Groves to Robert P. Patterson (25 February 1946), MEDR, Box 65, "Publications."

111 "Draft Minutes of Combined Policy Committee Meeting Held at the State Department on April 15th, 1946," (29 July 1946), HBF, Roll 3, Target 6, Folder 42, "(Minutes of) Meeting, Apr. 15, 1946, Combined Policy Committee."

112 William S. Hutchinson to Robert F. Bacher (4 June 1946), RFB, Box 18, Folder 8.

113 Emphasis in original. John H. Manley to Kenneth D. Nichols (11 July 1946), NTA, document NV0311404. Manley would later attempt to get an account of the work of the Committee on Declassification published in 1949, but it was held up in declassification review. It eventually appeared in print in late 1950. John H. Manley, "Secret science," (14 July 1949, revised 6 January 1950), RFB, Box 18, Folder 17; John H. Manley to Morse Salisbury (6 January 1950), RFB, Box 18, Folder 17; John H. Manley, "Secret science," *Physics Today* (November 1950), 8–16.

114 Warren Johnson, Willard F. Libby, John H. Manley, and R. L. Thornton to Leslie R. Groves (23 August 1946), RCT, Box 5, Folder 8.

115 Frank H. Spedding to John R. Ruhoff (24 November 1946), RCT, Box 4, Folder "00027."

116 Frank H. Spedding to Fletcher Waller (3 November 1947), RCT, Box 5, Folder 14.

117 "Manual for the Declassification of Scientific and Technical Matters" (1 May 1946), NTA, document NV0713951, and AHC, Series 02, Box No. S02B07, "Wartime Files," Folder on "Declassification Committee."

118 Even technical data could, of course, get into complex and thorny problems. In an early AEC evaluation of the first guides created to implement this system,

it was noted, for example, that while much information about fusion reactions could probably be declassified as basic physics, being selective about what to declassify would, by itself, end up highlighting, in relief, the thing they were trying to hide: interest in a thermonuclear bomb. "Analysis of Declassification Guide topics," (20 February 1947), NTA document NV0727261, on 12.

119 Minutes of Meeting no. 18 of the Atomic Energy Commission (2 January 1947), NTA, document NV0068544; Hewlett and Anderson, *The new world*,1.

120 Hewlett and Anderson, *The new world*, 534–545; Bernstein, "The quest for security," esp. 1029–1032. See also, Neuse, *David E. Lilienthal*, 168–174.

121 Lilienthal et al., *A report on the international control of atomic energy*.

122 Lilienthal et al., *A report on the international control of atomic energy*, 53.

123 Lilienthal et al., *A report on the international control of atomic energy*, 57.

124 Confusion is further added in that the "March 16" date usually appears on the published versions of the report, the final version as released was completed on March 17, and it constitutes a different draft. Hewlett and Anderson, *The new world*, 553.

125 David E. Lilienthal, Chester I. Barnard, J. Robert Oppenheimer, Charles A. Thomas, and Harry A. Winne, "A report on the international control of atomic energy," (16 March 1946), CTS, Roll 2, Target 4, File 10 ("International Control of Atomic Energy"), Tab G, 69–70. Groves' original, "Top Secret" copy of the Acheson-Lilienthal Report draft of March 16 is enclosed in this folder. The passages on secrecy quoted here are marked as "out in released report."

126 Lilienthal et al., "A report on the international control of atomic energy," (16 March 1946), 69–74.

127 Lilienthal et al., "A report on the international control of atomic energy," (16 March 1946), 74.

128 Lilienthal et al., *A report on the international control of atomic energy*, 59.

129 Hewlett and Anderson, *The new world*, 551–554.

130 See esp. Helmreich, *Gathering rare ores*.

131 Lilienthal et al., *A report on the international control of atomic energy*, 61.

132 See, for example, US Department of Energy, "Handbook for Notification of Exports to Iraq, Annex 3," SARC-001/98 (April 1998), archived online at: https://web.archive.org/web/20160306230736/http://www.iraqwatch.org/government/us/doe/doe-annex3.htm, which shows how a modern inspection regime conceptualizes bomb programs: almost exclusively as materials and machines, with computer simulations and data being the only "knowledge."

133 See, e.g., Hewlett and Anderson, *The new world*, ch. 15 and 16; Gordin, *Red cloud at dawn*, epilogue.

CHAPTER FIVE

1 David E. Lilienthal's notes on "The Conference on Atomic Energy Control at the University of Chicago, September 20–21, 1945," in *The journals of David E.*

Lilienthal, Volume 2, 642. Further references to diary entries in Lilienthal, *Journals* will refer to this volume.

2 Neuse, *David E. Lilienthal*. Lilienthal's *Journals* are a particularly useful resource in probing his mindset: though edited prior to publication, they still highlight his mindset and even naiveté to an unusual degree of candor.

3 Diary entry (7 August 1945), in Lilienthal, *Journals*, 1.

4 Notes on "The Conference on Atomic Energy Control at the University of Chicago, September 20–21, 1945," in Lilienthal, *Journals*, 637–645, on 642. Emphasis in original.

5 Diary entry (20 January 1946), in Lilienthal, *Journals*, 11–12.

6 Diary entry (28 January 1946), in Lilienthal, *Journals*, 16–17.

7 Diary entry (28 February 1946), in Lillienthal, *Journals*, 25.

8 Hewlett and Anderson, *The new world*, ch. 17; Hewlett and Duncan, *Atomic shield*, ch. 1.

9 "Hearings on the Confirmation of Atomic Energy Commission and General Manager," Senate Section of the Joint Committee on Atomic Energy, 80th Congress, 1st Session (January, February, March 1947), on 32.

10 Diary entry (2 February 1947), in Lilienthal, *Journals*, 138.

11 Diary entry (9 February 1947), in Lilienthal, *Journals*, 140.

12 Atomic Energy Commission, Minutes of Meeting No. 1 (13 November 1946), OSAECM, Box 1; "Classification and Handling of 'Restricted Data,'" (11 November 1946), and Carroll L. Wilson to all AEC personnel, "Security of 'Restricted Data,'" (27 November 1946), both in OSAEC46, Box 41, "Security Regulations, Volume 1."

13 Atomic Energy Commission, Minutes of Meeting No. 14 (23 December 1946), OSAECM, Box 1.

14 Edward R. Trapnell, "Information Program," (7 January 1947), DEL, Box 7, "Information Division—Memoranda—1947."

15 AEC staff paper, "Notes on Information Control in the Atomic Energy Commission" (n.d., but distributed June 1947), OSAEC46, Box 25, "Division of Information Services."

16 Atomic Energy Commission, Minutes of Meeting No. 61 (5 June 1947), OSAEC46, Box 25, "Division of Information Services"; William Waymack to David E. Lilienthal, "Panel on Director of Information," (6 June 1947 [mislabeled as 1946]), DEL, Box 8, "Information Panel."

17 William Waymack to Warren Johnson (19 June 1947), DEL, Box 8, "Information Panel."

18 Milton Eisenhower (chair), George Gallup, Warren Johnson, Eric Hodgins, Raymond P. Brandt, and John Dickey memorandum to AEC (28 June 1947), DEL, Box 8, "Information Panel."

19 John Dickey to David Lilienthal (7 July 1947), DEL, Box 8, "Information Panel."

20 AEC staff paper, "Question of Policy on Organization of Information Services," (n.d., but early 1947), OSAEC46, Box 25, "Division of Information Services."

21 AEC staff paper, "Policy Questions Concerning the Organization of Information Services," (n.d., but post-July 1947), OSAEC46, Box 25, "Division of Information Services."

22 Ibid.

23 Atomic Energy Commission, Minutes of Meeting No. 99 (23 September 1947), OSAECM, Box 25, "Division of Information Services."

24 AEC Press Release, "Morse Salisbury Named Director of Public and Technical Information for U.S. Atomic Energy Commission," (27 September 1947), OSAEC46, Box 25, "Division of Information Services"; G. Lyle Belsley to AEC, "Background Information on Morse Salisbury" (3 September 1947), DEL, Box 8, "Information, Public & Technical, Miscellaneous."

25 Morse Salisbury, "Public and Technical Information Service," (29 September 1947), OSAEC46, Box 25, "Division of Information Services."

26 An additional caveat was that sometimes "security" would be interpreted as "physical security," as it is meant here, and other times as "national security," a much broader category (in which excessive secrecy could actually lessen "security"). On the former, a concise expression of the different roles for the Declassification Branch and the Division of Security is in James B. Fisk, "Classification Procedure," AEC 121 (21 July 1948), OSAEC46, Box 41, "Basic Security Policy, Control of Info, Vol. 1."

27 K. E. Fields to AEC, "Final report of the JCS Evaluation Board for Operation Crossroads" (4 August 1947), DEL, Box 7, "Information Division—Memoranda—1947."

28 United States Atomic Energy Commission, Public and Technical Information Service, Staff Conference Minutes, No. 61 (6 December 1948), DEL, Box 7, "Information Division—Memoranda—1948."

29 William E. Webster (chairman, Military Liaison Committee [MLC]) to David E. Lilienthal (20 July 1949); David E. Lilienthal to William E. Webster (draft) (21 July 1949); David E. Lilienthal to William E. Webster (21 July 1949); all in NTA, document NV0411330.

30 See, e.g., Lilienthal's testimony in Minutes of an Executive Session of the Joint Committee on Atomic Energy (17 March 1949), in which he drew explicit contrast with the Smyth Report.

31 US Atomic Energy Commission, Minutes of Meeting No. 224 (10 December 1948), OSAEC46, Box 41, "Security Violations." Parsons refers to it as his "dragon-slaying" article in a letter to Oppenheimer, where he also complained that it was "tangled in the stratosphere" of bureaucracy. In 1953, he was still trying to get it cleared. William S. Parsons to J. Robert Oppenheimer (27 December 1948) and William S. Parsons to J. Robert Oppenheimer (25 September 1953), both in JRO, Box 56, Folder, "Parsons, William S."

32 Welsome, *The plutonium files*; Moss and Eckhardt, "The human plutonium injection experiments."

33 Carroll L. Wilson to Robert S. Stone (5 June 1947), NTA, document NV0727439.

34 Carroll L. Wilson to Robert S. Stone (5 November 1947), NTA, document NV0137031.

35 Walker, *The road to Yucca Mountain*, ch. 1, quote from 17–18.

36 Stephens, *Nuclear fission and atomic energy*, vii–viii.

37 Testimony of Henry D. Smyth, in Hearings on S. 1297 and Related Bills, "Authorizing a Study of the Possibilities of Better Mobilizing the National Resources of the United States, Part 3," Subcommittee of the Committee on Military Affairs, U.S. Senate, 79th Congress, 1st Session (22–26 October 1945), on 648.

38 Kenneth D. Nichols to Carroll L. Wilson, "Narrative History of Review of Pennsylvania Volume," (22 January 1947), MEDR, Box 65, "Publications."

39 Kenneth D. Nichols to Carroll L. Wilson, "Narrative History of Review of Pennsylvania Volume," (22 January 1947), MEDR, Box 65, "Publications"; Undated notes (c.a. summer 1947) by William Waymack in July 1946 draft copy of "Nuclear Fission and Atomic Energy," RFBAEC, Subject File 1947–1949, Box 3, "Philadelphia Story."

40 Norris Bradbury and John Manley in telegram to Manhattan Engineer District headquarters at Oak Ridge (22 May 1946), NTA, document NV0311374.

41 Morse Salisbury to William E. Stephens (20 February 1948), RFBAEC, Subject File 1947–1949, Box 3, "Philadelphia Story"; Stephens, *Nuclear fission and atomic energy*, ix–x. For more details of the case of *Nuclear fission and atomic energy*, see Wellerstein, "A tale of openness and secrecy."

42 AEC Info Memo 153, "Correspondence with International News Service Regarding Draft Article by Professor J.A. Campbell," (8 February 1949), and AEC Info Memo 153/1, "Correspondence with J. A. Campbell Regarding Draft Article 'The Secrets of the Atomic Bomb,'" (23 February 1949), both in OSAEC46, Box 41, "Security Violations."

43 AEC Info Memo 153/2, "Correspondence with 'Harper's Magazine' Regarding Draft Article by J. A. Campbell," (28 October 1949), OSAEC46, Box 41, "Security Violations." *Harper's* had been in touch in August 1949.

44 Los Alamos scientists had specifically complained about the *Time* article for its double standard. Major General Harold R. Bull's remarks were meant in the context of civil defense; the AEC was as mad at *Time* for writing a story about them as they were at Bull for saying them. Of note is that Bull's remarks were *before* Operation Sandstone (April–May 1948), where the new weapons were actually demonstrated for the first time. Teletype Carroll L. Tyler to AEC (4 March 1948); Morse Salisbury to Carroll L. Tyler, "Remarks by General Bull and Time Magazine Report Thereon," (17 March 1948); Morse Salisbury to David Lilienthal (24 March 1948), Department of the Army, Public Informa-

tion Division, "Address by Major General Harold R. Bull, Deputy Chief, Organization and Training Division, General Staff, U.S. Army, Before the United States Conference of Mayors, Waldorf-Astoria Hotel, New York City," (17 February 1948); "Progress," *Time* (1 March 1948), 68; all in DEL, Box 7, "Information Division—Memoranda—1948."

45 Harry Truman, Executive Order 9835 (21 March 1947), copy available at the American Presidency Project, University of California, Santa Barbara, online at https://www.presidency.ucsb.edu/documents/executive-order-9835-prescribing-procedures-for-the-administration-employees-loyalty, accessed 10 December 2018. See also Wang, *American science in an age of anxiety*, 85–86.

46 Section 10(b)(5)(B), Atomic Energy Act of 1946, S. 1717 (Public Law 585), 79th Congress, 2nd Session (1 August 1946). For an excellent discussion of American counterintelligence targeting of several students of J. Robert Oppenheimer, and the perils of FBI files for historians, see Mullet, "Little man."

47 Hewlett and Duncan, *Atomic shield*, 24–25.

48 Diary entry (7 June 1947), in Lilienthal, *Journals*, 189. It was not lost on Lilienthal that he was forced to review files where the advocacy of international control of atomic energy was considered "derogator information."

49 Wang, *American science in an age of anxiety*, 149; Kaiser, "The atomic secret in Red hands?"

50 Wang, *American science in an age of anxiety*, esp. ch. 5; Kaiser, "The atomic secret in Red hands?"

51 The AEC eventually opted to avoid the classification designation of "Restricted" because of the obvious room for confusion with Restricted Data, and instead opted to use "Official Use Only" as a substitution. US Atomic Energy Commission, Program Council Minutes (23 June 1949), OSAEC46, Box 41, "Basic Security Policy, Control of Info, Vol. 1."

52 Diary entry (30 May 1947), in Lilienthal, *Journals*, 185.

53 Hewlett and Duncan, *Atomic shield*, 129–132, 521–525, 537–539, 585.

54 Carroll L. Wilson, "U.S. Atomic Energy Commission, The Security Program, 1947–49," (29 July 1949), OSAEC46, Box 41, "Basic Security Policy, Control of Info, Vol. 1."

55 Hewlett and Duncan, *Atomic shield*, 324–325, 334.

56 Hewlett and Duncan, *Atomic shield*, 332–334.

57 GAC report quoted in Carroll L. Wilson, "Note by the General Manager: Basic Security Policy," AEC 111 (18 June 1948), OSAEC46, Box 41, "Basic Security Policy, Control of Info, Vol. 1."; Hewlett and Duncan, *Atomic shield*, 340.

58 Glenn T. Seaborg diary entry (6 June 1948), in *Journal of Glenn T. Seaborg, Vol. 2*, 180. A (somewhat mildly redacted) copy of the final report for the Tenth Meeting of the General Advisory Committee is included in this source as well, on 180A–180D.

59 Carroll L. Wilson, "Note by the General Manager: Basic Security Policy," AEC 111 (18 June 1948); and Edward R. Trapnell to Carroll L. Wilson, "Classification

of Administrative Papers" (21 June 1948), both in OSAEC46, Box 41, "Basic Security Policy, Control of Info, Vol. 1." "Official Use Only" was used as an alternative to "Restricted" within the AEC from 1949–1951, to avoid confusion with "Restricted Data." See Quist, *Security classification of information*, vol. 1, 113–114. Today, "Official Use Only" is an even lower level of classification, used only as a form of administrative classification (a government employee can lose their job for disseminating something marked as such, but they cannot go to jail for it, and there is no punishment for the unauthorized possession of it).

60 Diary entries (17–18 June 1948), in Lilienthal, *Journals*, 360–361; Hewlett and Duncan, *Atomic shield*, 340; US Atomic Energy Commission, Minutes of Meeting No. 161 (24 June 1948), OSAEC46, Box 41, "Basic Security Policy, Control of Info, Vol. 1."

61 E.g., T.O. Jones to David B. Langmuir, "Planned Commission Action on Security and Secrecy," (27 July 1948), OSAEC46, Box 41, "Basic Security Policy, Control of Info, Vol. 1."

62 US Atomic Energy Commission, Executive Office, Program Council, "General Discussion on Security Problems in the AEC," AEC 111/1 (21 July 1948), OSAEC46, Box 41, "Basic Security Policy, Control of Info, Vol. 1."

63 Ibid.

64 Ibid.

65 David E. Lilienthal to Owen J. Roberts (and others, with slight differences), (27 July 1948), OSEAC46, Box 41, "Basic Security Policy, Control of Info, Vol. 1."

66 David E. Lilienthal to Robert Oppenheimer (27 July 1948), OSAEC46, Box 41, "Basic Security Policy, Control of Info, Vol. 1."

67 US Atomic Energy Commission, "Minutes of an AEC Meeting to Discuss General Problems of Security and Declassification," Info Memo 48–53/1 (30 July 1948), OSAEC46, Box 41, "Basic Security Policy, Control of Info, Vol. 1."

68 Willard F. Libby to David E. Lilienthal (16 August 1948), DEL, Box 16, "Reorganization of Security Set-up." James B. Conant in particular overtly declined to give any feedback, claiming he was too busy to work on it, and had insufficient knowledge to do a meaningful job. He concluded: "This is no place for an organic chemist!" James B. Conant to David B. Langmuir (n.d., c.a. August 1948), DEL, Box 16, "Reorganization of Security Set-up."

69 While a valid observation, it is of interest to note that the Soviet atomic bomb project was largely explicitly modeled on the Manhattan Project, both through information acquired through illicit sources, but also through the broad programmatic information divulged in the Smyth Report. See Gordin, *Red cloud at dawn*, for considerable discussion of this.

70 Oppenheimer also noted that the Tolman Committee had deliberately stuck to technical discussion because at the time they had felt that administrative matters, like bomb stockpiles, were a matter of policy, not technical declassification. J. Robert Oppenheimer to David E. Lilienthal (16 August 1948), DEL, Box 11, "Correspondence—Robert Oppenheimer."

71 David E. Lilienthal to AEC Commissioners (27 September 1948), DEL, Box 16, "Reorganization of Security Set-up."
72 David B. Langmuir, "Objects and Methods of Secrecy" (15 September 1948), OSAEC46, Box 41, "Basic Security Policy, Control of Info, Vol. 1."
73 Ibid.
74 David B. Langmuir, "Proposal for Study of Secrecy Problems" (15 September 1948), OSAEC46, Box 41, "Basic Security Policy, Control of Info, Vol. 1."
75 David E. Lilienthal to J. Robert Oppenheimer (6 October 1948), DEL, Box 16, "Reorganization of Security Set-up."
76 David B. Langmuir, "A Proposed System for Secrecy," AEC 111/2 (14 October 1948), OSAEC46, Box 41, "Basic Security Policy, Control of Info, Vol. 1."
77 Ibid.
78 US Atomic Energy Commission, Minutes of Meeting No. 207 (15 October 1948), OSAEC46, Box 41, "Basic Security Policy, Control of Info, Vol. 1."
79 See Galison's critique of "atomic" (meant in the sense of "indivisible") secrecy systems in Galison, "Removing knowledge," esp. 241–243. This approach to secrecy seems particularly attractive to physicists. By comparison, the anthropological and sociological analyses of secrecy—even in the early Cold War—had a very different flavor. Cf. Simmel, "The sociology of secrecy and of secret societies"; Weber, "Bureaucracy [1920]," and even Shils, *The torment of secrecy*.
80 AEC Report, "Control of Information under the Atomic Energy Act of 1946," AEC 111/3 (7 January 1949), OSAEC46, Box 41, "Basic Security Policy, Control of Info, Vol. 1."
81 Appendix A, "Draft Statement by the Atomic Energy Commission on a Policy for Control of Information," included in AEC Report, "Control of Information under the Atomic Energy Act of 1946," AEC 111/3 (7 January 1949), OSAEC46, Box 41, "Basic Security Policy, Control of Info, Vol. 1."
82 See esp. Kaiser, "The atomic secret in Red hands?," and Kofksy, *Harry S. Truman and the war scare of 1948*.
83 US Atomic Energy Commission, Minutes of Meeting No. 234 (19 January 1949), OSAECM, Box 1.
84 "Policy on Control of Information," AEC 111/5 (21 March 1949), OSAEC46, Box 41, "Basic Security Policy, Control of Info, Vol. 1."
85 On the failures of international control, see Bernstein, "The quest for security"; Hewlett and Anderson, *The new world*, ch. 16;. Hewlett and Duncan, *Atomic shield*, 264–273.
86 Memo by David E. Lilienthal (24 June 1947), in Lilienthal, *Journals*, 207.
87 And, on top of this, the Korean War began in the summer of 1950—signaling to many in the US policy world a new, expansionist Soviet policy.
88 Hewlett and Duncan, *Atomic shield*, ch. 9–10; for a longer discussion of US-UK interactions in this period, see Paul, *Nuclear rivals*.
89 Hewlett and Duncan, *Atomic shield*, 253, 341–342, 450–451; Wang, *American science in an age of anxiety*, ch. 7.

90 Hearing before the Joint Committee on Atomic Energy, 81th Congress, 1st Session (2 February 1949), 8–9.
91 Hewlett and Duncan, *Atomic shield*, 342–345, 352, 355–358.
92 See esp. Kaiser, "The atomic secret in Red hands?" and Wang, *American science in the age of anxiety*, esp. ch. 2–5.
93 Diary entry (16 March 1949), and (30 March 1949), in Lilienthal, *Journals*, 487 and 496.
94 Diary entry (19 September 1949), in Lilienthal, *Journals*, 566. Cf. Hewlett and Duncan, *Atomic shield*, 360–361.
95 Diary entry (21 September 1949), in Lilienthal, *Journals*, 569; Gordin, *Red cloud at dawn*, 224–225.
96 Gordin, *Red cloud at dawn*, 217–222.
97 Quoted in Gordin, *Red cloud at dawn*, 223–224.
98 Gordin, *Red cloud at dawn*, 223–229.
99 Diary entry (21 September 1949), in Lilienthal, *Journals*, 571; Gordin, *Red cloud at dawn*, 312–313.
100 Gordin, *Red cloud at dawn*, ch. 2, esp. 80.
101 Diary entry (21 September 1949), in Lilienthal, *Journals*, 572.
102 David E. Lilienthal, "Where do we go from here?," speech at the Freedom House, New York (13 October 1949), reprinted in *Bulletin of the Atomic Scientists* 5, no. 11 (November 1949), 294, 308, on 294 and 308. An unexpurgated version is available in JRO, Box 46, Folder, "Lilienthal, David E., Correspondence to Oppenheimer, 1947–1950."
103 William W. Waymack to Robert F. Bacher (13 October 1949), quoted in Gordin, *Red cloud at dawn*, 252.
104 Diary entry (29 September 1949), in Lilienthal, *Journals*, 575.
105 US Atomic Energy Commission, Minutes of Meeting No. 326 (28 October 1949), OSAEC46, Box 41, "Basic Security Policy, Control of Info, Vol. 1."
106 Bacher had retired as of May 10, 1949; Waymack as of December 21, 1948. Smyth began on May 30, 1949; Dean on May 24, 1949.
107 US Atomic Energy Commission, Minutes of Meeting No. 326 (28 October 1949), OSAEC46, Box 41, "Basic Security Policy, Control of Info, Vol. 1."
108 Ibid.
109 Gordon E. Dean, "Memorandum Concerning the Interpretation of Section 10 of the McMahon Act," AEC 236 (21 July 1949), OSAEC46, Box 41, "Basic Security Policy, Control of Info, Vol. 1."
110 Emphasis in original. "Third Declassification Conference, Chalk River, Ontario, September 26–28, 1948, Consideration of Basic Policy, Second Draft," included in James G. Beckerley, "Policy on Control of Information—Reformulation of Secrecy Practices," AEC 111/10 (16 November 1949), OSAEC46, Box 41, "Basic Security Policy, Control of Info, Vol. 1."
111 Carroll L. Wilson, "Review of Current Secrecy Practices" (20 October 1949), and "Third Declassification Conference, Chalk River, Ontario, September

26–28, 1948, Consideration of Basic Policy, Second Draft," both included in James G. Beckerley, "Policy on Control of Information—Reformulation of Secrecy Practices," AEC 111/10 (16 November 1949), OSAEC46, Box 41, "Basic Security Policy, Control of Info, Vol. 1."

112 Minutes of the Joint Military Liaison Committee-Atomic Energy Commission Meeting No. 39 (6 December 1949), OSAEC46, Box 41, "Basic Security Policy, Control of Info, Vol. 1."

113 Roy B. Snapp, "Atomic Energy Commission Policy for Control of Information," AEC 111/11 (5 January 1950), OSAEC46, Box 41, "Basic Security Policy, Control of Info, Vol. 2."

114 Rhodes, *Dark sun*, 381.

115 Tritium (H-3) is bred from lithium-6 in a nuclear reactor. Any neutron that formed tritium is a neutron that might have instead formed plutonium. Because of the mass difference, every gram of tritium produced is 80 grams of plutonium not produced.

116 Rhodes, *Dark sun*; Herken, *Brotherhood of the bomb*; Hansen, *The swords of Armageddon*; Fitzpatrick, "Igniting the light elements"; and Galison and Bernstein, "In any light."

117 For an early, independent derivation, see Thirring, *Die Geschichte der Atombombe*, ch. 42, translated and reprinted as Hans Thirring, "The Super Bomb," *Bulletin of the Atomic Scientists* 6, no. 3 (March 1950), 69–70. For a "leak," by the physicist and Responsible Reviewer Philip Morrison, see "Superbomb seen," *Christian Science Monitor* (28 March 1947), 3.

118 Diary entry (29 October 1949), in Lilienthal, *Journals*, 581.

119 Editorial, "Secrets will out," *Bulletin of the Atomic Scientists* 6, no. 3 (March 1950), 67–68, on 67.

120 Radio Reports, Special for the Atomic Energy Commission, "Senator Johnson Charges War-Time Atomic Security Lax [transcript]," *Radios Reports, Inc.* (1 November 1949), in DEL, Box 16, "Radio—Matters, Re." Also quoted in "The president orders exploration of the Super Bomb," *Bulletin of the Atomic Scientists* 6, no. 3 (March 1950), 66.

121 Alfred Friendly, "New A-bomb has 6 times power of 1st," *Washington Post* (18 November 1949), 1. Hewlett and Duncan, *Atomic shield*, 394. In Friendly's article, Johnson limply insisted that all of his knowledge of the "Super" came from public reports—something that was on the face of it untrue.

122 Diary entry (27 November 1949), in Lilienthal, *Journals*, 601.

123 Hewlett and Duncan, *Atomic shield*, 394.

124 Brien McMahon to Harry S. Truman (21 November 1949), copy reprinted in Executive Hearing, JCAE, "Development of a Super Weapon," (9 January 1950), 34–35.

125 Hewlett and Duncan, *Atomic shield*, 400; Executive Hearing, JCAE (9 January 1950), 6–8; US Atomic Energy Commission, Minutes of Meeting No. 351 (5 January 1950), GDDRS.

126 Senator Millard Tydings, quoted in Executive Hearing, JCAE (9 January 1950), 22–23.
127 David E. Lilienthal, "Memorandum of Meeting with the President," (31 January 1950), copy in Lilienthal, *Journals*, 632.
128 Statement by President Harry S. Truman (31 January 1950), reprinted in Williams and Cantelon, *The American atom*, 131.
129 Galison and Bernstein, "In any light," 306.
130 E.g., Louis N. Ridenour, "The hydrogen bomb," *Scientific American* 182, no. 3 (March 1950), 11–15; see also, York, *The advisors*.
131 Harry S. Truman to David E. Lilienthal (31 January 1950), CHNSA, Box 13, "1950–1954," Folder 1.
132 Special Committee of the National Security Council to the President, "Development of Thermonuclear Weapons" (30 January 1950), CHNSA, Box 13, "1950–1954," Folder 1.
133 On the Berkeley accusations, see esp. Mullet, "Little man"; Herken, *Brotherhood of the bomb*, esp. ch. 10–11; Kaiser, "The atomic secret in Red hands?," 38–39.
134 Williams, *Klaus Fuchs, atom spy*, ch. 1. Fuchs' confessions—one to William Skardon (27 January 1950) and a more technical one to Francis Perrin (30 January 1950), are available as appendices A and B in Williams' text. On Venona (the cryptanalysis project), see Haynes, *Venona* and West, *Venona*. Recently, two excellent books on Fuchs have been written with the advantage of files declassified in the postwar: Close, *Trinity*, and Greenspan, *Atomic spy*.
135 L. Whitson to H. B. Fletcher (22 September 1949); H. B. Fletcher to D. M. Ladd (27 September 1949); J. Edgar Hoover to C. A. Rolander, Jr. (21 October 1949); J. Edgar Hoover to Francis R. Hammack (1 February 1950), all in KFFBI, Volume 1. On the FBI holding back information not to compromise its source, see H. B. Fletcher to D. M. Ladd (3 February 1950), KFFBI, Volume 3. On the AEC November discussion, see US Atomic Energy Commission, Minutes of Meeting No. 327 (2 November 1949), OSAECM, Box 1. In a later review of the Fuchs chronology, the compiler put a note about the November 1949 AEC meeting: "Significance evidently not apparent." "Draft Chronology of the Fuchs Case," (ca. 30 September 1954), in OSAEC51, Box 204, "S & I: Thermonuclear Information." It is a minor contention in some of the literature that Truman's decision on the H-bomb was made with Fuchs' espionage in mind. I've found no primary source evidence of this, and much to suggest that no one involved in the H-bomb decision knew about Fuchs until after January 31.
136 Diary entry (2 February 1950), in Lilienthal, *Journals*, 634.
137 Diary entry (3 February 1950), in Lilienthal, *Journals*, 635.
138 US Atomic Energy Commission, Minutes of Meeting No. 364 (3 February 1950), OSAECM, Box 2; AEC Press Release on "Dr. Karl Fuchs," (3 February 1950), KFFBI, Volume 3; diary entry (3 February 1950), in Lilienthal, *Journals*, 635.
139 Executive Hearing, JCAE, "Concerning Dr. Klaus Fuchs, British Spy," (3 February 1950).

140 Executive Hearing, JCAE (3 February 1950), 403. Diary entry (3 February 1950), in Lilienthal, *Journals*, 635.

141 Executive Hearing, JCAE (3 February 1950), 404.

142 Executive Hearing, JCAE (3 February 1950), 419–420.

143 Executive Hearing, JCAE (3 February 1950), 419. An initiator is a neutron source that sits at the center of the implosion bomb and releases a burst of neutrons upon detonation. Core levitation involves adding an air gap between the core and the pusher or tamper, which allows the latter to accelerate before contact with the core; it was explored during the war but not used until the postwar period. Overall these two innovations increase the efficiency of implosion bombs. See Hansen, *The swords of Armageddon*, vol. 2.

144 On Fuchs' work on the H-bomb, see Williams, *Klaus Fuchs, atom spy*, 153; Goodman, "Who is trying to keep what secret from whom and why?," 140; Goodman, "The grandfather of the hydrogen bomb?"; Bernstein, "John von Neumann and Klaus Fuchs." On his role as a babysitter and editor, see Joseph O. Hirschfelder, "The scientific and technological miracle at Los Alamos," in Badash, Hirschfelder, and Broida, eds., *Reminiscences of Los Alamos*, 67–88, on 86. On his work on gaseous diffusion, see Karl Cohen to Brien McMahon (19 March 1951), RJCAE, Appendix I, Unclassified General Subjects, Box 316, "Foreign Activities—Great Britain—Fuchs."

145 Norris Bradbury, "Los Alamos—The first 25 years," in Badash, Hirschfelder, and Broida, eds., *Reminiscences of Los Alamos*, 161–175, on 167.

146 Executive Hearing, JCAE (3 February 1950), 433.

147 Executive Hearing, JCAE (3 February 1950), 436.

148 Executive Hearing, JCAE, "Study of the British Disclosure," (4 February 1950), 437.

149 Executive Hearing, JCAE (4 February 1950), 441. Fuchs was notoriously shy, bookish, and unassuming; Groves' lack of any personal memory of him is not surprising.

150 Executive Hearing, JCAE (4 February 1950), 454; Groves, *Now it can be told*, 143–144. MED records do back up Groves' insistence that the British had told him that they had already cleared their participants for security. Leslie R. Groves, "Amplifying Notes by General Groves with Respect to Minutes of Meeting Held 8 January 1945," (9 January 1945), CTS, Roll 2, Target 5, Folder 11, "Correspondence with Foreign Nations." Fuchs himself would later offer to the FBI that this would have been the easiest way to detect him—had they known about his political inclinations in Germany, it would have been obvious that he was being "out of character" when he suddenly became apolitical when he went to the UK. Hugh H. Clegg and Robert J. Lamphere, "Report . . . Covering Interviews with Klaus Fuchs in London, England," (4 June 1950), CHNSA, Box 13, "1950–1954," Folder 2.

151 Executive Hearing, JCAE (4 February 1950), 466–468.

152 Executive Hearing, JCAE (4 February 1950), 459.

153 Executive Hearing, JCAE (4 February 1950), 505.

154 Executive Hearing, JCAE (4 February 1950), 519.

155 Executive Hearing, JCAE, "Testimony of Mr. J. Edgar Hoover Re: Klaus Fuchs," (6 February 1950). This particular hearing, in contrast with the others on Fuchs, is available only as a heavily redacted summary, not a transcript.

156 Executive Hearing, JCAE (6 February 1950), 527–528.

157 Executive Hearing, JCAE (6 February 1950), 529.

158 See Joint Committee on Atomic Energy, "Soviet Atomic Espionage," Joint Committee Print, 82nd Congress, 1st Session (April 1951), 33–37, esp. 34.

159 Senator Millard Tydings, quoted in Executive Hearing, JCAE, "Discussion in Re: Supplemental Budget, Fuchs Case, British Mission, Joan Hinton Case, Hydrogen Bomb Development," (10 February 1950).

160 A previous author has also speculated about Soviet penetration within MI-5 as a possible source for their apparent laxity: Williams, *Klaus Fuchs, atom spy*, 141–144. The JCAE members would not receive the contents of Fuchs' confessions for another month—the UK was not fully cooperating, and the FBI was holding onto the information while it attempted to run down Fuchs' accomplices. Executive Hearing, JCAE (3 March 1950), 745.

161 Executive Hearing, JCAE (10 February 1950), 588.

162 Executive Hearing, JCAE (10 March 1950), 776.

163 Among other measures, it is the only instance I have found of JCAE members swearing ("God damn it") on the record. Executive Hearing of JCAE (10 March 1950), 806–807.

164 Executive Hearing, JCAE (3 February 1950), 408.

165 See esp. Holloway, *Stalin and the bomb*; Kojevnikov, *Stalin's great science*, ch. 6; Gordin, *Red cloud at dawn*, ch. 3–4.

166 Max Lerner, "The G-Mind and the H-Bomb," *New York Post* (7 February 1950), copy in KFFBI, Volume 4.

167 Frederic de Hoffmann, W. C. Johnson, J. M. B. Kellogg, W. F. Libby, and R. L. Thornton to Sumner T. Pike, "Evaluation of Fuchs Case by the Committee of Senior Responsible Reviewers," AEC 273/16 Series A (25 April 1950), RJCAE, Declassified Materials from Classified Boxes (Series 2), Box 11.

168 Sumner T. Pike to Brien McMahon (30 June 1950), DELCFCRF, Box 1, "Classified Reading File No. 2."

169 William W. Waymack to Louis N. Ridenour (17 February 1948), WWWAEC, Box 2, "Security." Waymack's letter was in response to Ridenour's complaint against a "flat statement" the former had made expressing confidence in the AEC's assertions that it had good understanding of the biological effects of radiation, probably in regard to nuclear testing. Louis N. Ridenour to William W. Waymack (12 February 1948), WWWAEC, Box 2, "Security."

CHAPTER SIX

1 Strauss, *Men and decisions*, 252.

2 Harry S. Truman to David E. Lilienthal (31 January 1950), CHNSA, Box 13, "1950–1954," Folder 1.

3 US Atomic Energy Commission, Minutes of Meeting No. 363 (2 February 1950), OSAECM, Box 2.

4 Hewlett and Duncan, *Atomic shield*, 374.

5 The earliest public mention of said rumors appears to come from a press conference held at Oak Ridge in late September 1945 with Secretary of War Robert P. Patterson and Leslie Groves. "Question: Is there anything to the rumor that you are making a super bomb that would make the Nagasaki bomb look small? Patterson: I don't know. Groves: I don't think the Nagasaki bomb was made obsolete. That bomb could never be made obsolete. Those we used are pretty super. . . . This thing has just started and no one knows just what will develop." Press Conference, Secretary of War Robert P. Patterson, Clinton Engineering Works (29 September 1945), ARC, Identifier 281581. In 1947, even the JCAE thought such ideas were fantasies. In a discussion about atomic rumors, Brien McMahon noted: "We had a Senator say on the floor the other day that we have got the super-bomb now. He said that we have got a thousand times more powerful bomb than we had at Nagasaki. (Laughter)." Executive Hearing, JCAE, "Proceedings of an Executive Meeting of the Joint Committee on Atomic Energy," (1 July 1947).

6 "Report of the General Advisory Committee to the US Atomic Energy Commission (30 October 1949), reprinted in Williams and Cantelon, *The American atom*, 120–127. See esp. part 3 of the Majority section of the report.

7 US Atomic Energy Commission, Minutes of Meeting No. 351 (5 January 1950), GDDRS; Hewlett and Duncan, *Atomic shield*, 394.

8 US Atomic Energy Commission, Minutes of Meeting No. 367 (8 February 1950), OSAECM, Box 2; Carroll L. Wilson to Principal Washington Staff and Managers of Operations (10 February 1950), CHNSA, Box 13, "1950–1954," Folder 1.

9 "Notes of a Briefing Held at the Los Alamos Scientific Laboratory" (23 February 1950), CHNSA, Box 13, "1950–1954," Folder 1; US Atomic Energy Commission, Minutes of Meeting No. 381 (14 March 1950), OSAECM, Box 2.

10 US Atomic Energy Commission, Minutes of Meeting No. 382 (15 March 1950), OSAECM, Box 2.

11 US Atomic Energy Commission, Minutes of Meeting No. 383 (17 March 1950), OSAECM, Box 2.

12 The three statements were as follows: 1. Bethe's emphasis on the importance of the deuterium-tritium reaction, which had implications for tritium production; 2. The fact that the fusion reactions would definitely require a fission bomb to start them, which was seen as confirming the direction of the US pro-

gram; and 3. That building such a bomb was likely to be accomplished in "years rather than months," which was seen as giving away too much information on the US position in the field. Piel, *Science in the cause of man*, 4–5. A copy of Bethe's original article, with security annotations, is found in OSAEC46, Box 107, "Publication of Information Regarding the Thermonuclear Bomb (Bulky Package/Secret volume)."

13 Gerard Piel, "Reminiscences of Gerard Piel," interview by Frederick Peterson Jessup (1982), Columbia Center for Oral History, Columbia University, 4–207. William R. Conklin, "US censors H-bomb data," *New York Times* (1 April 1950), 1; Piel, *Science in the cause of man*, ch. 1.

14 Piel, *Science in the cause of man*, 5.

15 Piel's address is reprinted in Piel, *Science in the cause of man*, ch. 1.

16 Hans Bethe, oral history interview with Charles Weiner (8 May 1972), OHCNBL.

17 Conklin, "US censors H-bomb data." On Piel and *Scientific American* in the Cold War, see esp. Benintende, "Who was the Scientific American?"

18 John H. Holmes, Patrick M. Malin, and Arthur G. Hays to AEC (18 May 1950), and Carroll L. Wilson to ACLU (draft) (29 May 1950), both in DEL, Box 8, "Information Division—Memoranda—1950."

19 Henry D. Smyth, "Secret Weapons and Free Speech" (21 April 1950), OSAEC46, Box 41, "Basic Security Policy, Control of Info, Vol. 2."

20 "Information Control Policy," AEC 111/12 (15 May 1950), OSAEC46, Box 41, "Basic Security Policy, Control of Info, Vol. 2."

21 Editorial, "It's not what's said, it's who says it," *Bulletin of the Atomic Scientists* 6, no. 5 (May 1950), 130–132.

22 General Advisory Committee of the US Atomic Energy Commission, Minutes of Meeting No. 20 (31 March 1950), CHNSA, Box 13, "1950–1954," Folder 1; James G. Beckerley to Henry D. Smyth, "Guide on H-bomb Classification," (19 April 1950), DEL, Box 8, "Information Division—Memoranda—1950"; US Atomic Energy Commission, "Guide for Project Personnel on Classification of H-bomb Information," (11 May 1950), DEL, Box 10, "MLC Correspondence—1950"; Morse Salisbury to AEC, "Publication of H-bomb information," (24 May 1950), DEL, Box 8, "Information Division—Memoranda—1950"; "Proposed H-bomb Statement," AEC 298/11 (22 May 1950), GDDRS; General Advisory Committee of the US Atomic Energy Commission, Minutes of Meeting No. 21 (1 June 1950), CHNSA, Box 13, "1950–1954," Folder 2; US Department of Defense, "Atomic Energy Commission Proposed H-bomb Statement (draft)," (6 June 1950), GDDRS; Morse Salisbury to James S. Lay (12 June 1950), GDDRS.

23 "Security of Information Regarding Thermonuclear Weapons," (19 June 1950), in National Security Council, "List of Policies Approved by the President in the Atomic Energy Field on the Recommendation of the National Security Council

or the Special Committee of the National Security Council on Atomic Energy," (n.d., last entry 26 June 1951), GDDRS.

24 Joint Committee on Atomic Energy, Committee Print, "The Hydrogen Bomb and International Control: Technical and Background Information" (July 1950).

25 Hewlett and Duncan, *Atomic shield*, covers the Dean years very well, and the flavor of his thinking is reflected in Anders, *Forging the atomic shield*.

26 Rhodes, *Dark sun*, ch. 23.

27 Ford, *Building the H-bomb*, ch. 14.

28 Hans Bethe to Gordon Dean (9 September 1952), NTA, document NV0409418.

29 Ibid.

30 Hewlett and Duncan, *Atomic shield*, 590–591.

31 Declassified testimony of Vannevar Bush *In the Matter of J. Robert Oppenheimer* [Un-redacted version], vol. 17 (23 April 1954), DSAEC, Records Re: Oppenheimer Hearing, Transcripts of Personnel Security Board Hearing, Box 4, on 1976–1977. This discussion was resurrected in an article from the 1990s: Hirsch and Mathews, "Who really gave away the secret?" Whether the Soviets did get information or not from the fallout is disputed, but it has been demonstrated that considerable information about design is in theory derivable from such fallout: De Geer, "The radioactive signature of the hydrogen bomb." See also the Memorandum of a Panel of Consultants on Disarmament ("The Timing of the Thermonuclear Test"), undated (ca. September 1952), in Rose and Petersen, *Foreign relations of the United States, 1952–1954*, Volume 2, Part 2, 994–1008.

32 Joint Task Force 132, "Operation Ivy, Final Report," (9 January 1953), N-1, online at: http://www.dtic.mil/dtic/tr/fulltext/u2/a995443.pdf, accessed 11 December 2018.

33 Gusterson, "Death of the authors of death," 289; Holloway, *Stalin and the bomb*, 312. On the circulation of Mike fallout, see L. Matcha, R. J. List, and L. F. Hubert, "World-wide travel of atomic debris," *Science* 124, no. 3320 (14 September 1956), 474–477.

34 Rhodes, *Dark sun*, ch. 24; seismograph anecdote on 511.

35 Roy B. Snapp, "Note by the Secretary—Letter to J. Edgar Hoover, Operation Ivy, AEC 483/33," (18 November 1952), NTA, document NV0409009.

36 Gordon Dean, "Announcement by the Chairman, United States Atomic Energy Commission," (16 November 1952), in Rose and Petersen, *Foreign relations of the United States, 1952–1954*, Volume 2, Part 2, 1042.

37 Gordon Dean to Dwight D. Eisenhower (7 November 1952), NTA, document NV0409014; Hewlett and Holl, *Atoms for peace and war*, 3–6. Newspapers do not seem to have reported on the artificial "earthquake" of November 1, 1952; however, coincidentally, a massive underwater earthquake did occur on November 4 off the shore of the Kamchatka peninsula. One newspaper did

report on the possibility that it was a hydrogen bomb test: "13-foot Hawaii waves follow Siberia quake," *Daily Boston Globe* (5 November 1952), 1.

38 Draft of the January 1953 State of the Union address (undated, ca. 5 January 1953), Office Files of Henry DeWolf Smyth, Records of the Atomic Energy Commission, Record Group 326, NARACP, Box 1, "Thermonuclear program." Some of this language clearly (particularly the "in a new unit" part) derives from a draft forwarded to the White House by Gordon Dean in late December 1952. Gordon Dean to Henry DeWolf Smyth (29 December 1952), OSAEC51, Box 35, "Legal—State of the Union Message."

39 Henry D. Smyth, "Draft of State of the Union Message," (undated, ca. 5 January 1953), OSAEC51, Box 35, "Legal—State of the Union Message."

40 Thomas Murray, "Memorandum," (7 January 1953), OSAEC51, Box 35, "Legal—State of the Union Message."

41 Harry Truman, "Annual Message to the Congress on the State of the Union," (7 January 1953), copy online at the American Presidency Project, University of California, Santa Barbara: https://www.presidency.ucsb.edu/documents/annual-message-the-congress-the-state-the-union-18, accessed 11 December 2018.

42 For examples of the skimpy coverage, see "First H-blast in 1952 is confirmed by Ike," *Washington Post* (3 February 1954), 6 (three paragraph story); "Hydrogen device test at Eniwetok confirmed," *New York Times* (3 February 1954), 3 (six paragraphs); "Ike identifies US H-blast as first such in history," *Austin Statesman* (3 February 1954), 14 (a longer story, but buried in the back pages).

43 Weisgall, *Operation Crossroads.*

44 On AEC approaches to nuclear testing, see esp. Hacker, *Elements of controversy.*

45 The scientist was Mark M. Mills, who died in 1958. Another scientist in the same helicopter escaped with minor injuries. "Eniwetok copter falls, drowns scientist," *New York Herald Tribune* (8 April 1958), 1.

46 On the Nevada Test Site, see Hacker, *Elements of controversy*, esp. ch. 2 and 7; Hewlett and Duncan, *Atomic shield*, 535.

47 See, e.g., Miller, *Under the cloud.*

48 In 1953, the yield of the total US nuclear arsenal was 73 megatons; by 1960, it would total over 20,400 megatons, equivalent to approximately 1.4 million Hiroshima-sized bombs. US Department of Energy, Office of the Press Secretary, "Declassification of Certain Characteristics of the United State Nuclear Weapon Stockpile," (27 June 1994), online at https://www.osti.gov/includes/opennet/document/press/pc26.html, accessed 11 December 2008.

49 For examples of such speculative articles, see "H-bomb test?," *Christian Science Monitor* (3 March 1954), 10; Jamees Lee, "Super 'H-bomb' may have been detonated by US," *Atlanta Daily World* (2 March 1954), 3; "Atomic 'device' explosion starts new Pacific tests," *Chicago Daily Tribune* (2 March 1954), 8; "Reveal new A-bomb, maybe H-bomb tests," *Newsday* (2 March 1954), 5; Neil Mac-

Neil, "'Atomic device' is detonated by US in new Pacific tests," *Washington Post* (2 March 1954), 2.

50 Rhodes, *Dark sun*, 541–543; Barker, *Bravo for the Marshallese*; and esp. Thomas Kunkle and Byron Ristvet, "Castle Bravo: Fifty Years of Legend and Lore," DTRIAC SR-12-001 (January 2013), online at https://apps.dtic.mil/docs/citations/ADA572278, accessed 11 December 2018.

51 Hines, *Proving Ground*; Lapp, *The voyage of the Lucky Dragon*; Higuchi, *Political fallout*, ch. 2.

52 Statement by Lewis Strauss (31 March 1954), NTA, document NV0049192. Rhodes, *Dark sun*, 542. On the causes of the fallout contamination, see esp. Kunkle and Ristvet, "Castle Bravo: Fifty Years of Legend and Lore."

53 Joseph Rotblat, "The Hydrogen-Uranium Bomb," *Bulletin of the Atomic Scientists* 11, no. 5 (1955), 171–172, 177.

54 On the declassification of the Operation Ivy film, see "Report by the Directors of Classification and Information Service regarding the Film on Operation Ivy," AEC 483/47, (8 December 1953), NTA, document NV0074012. For the review, see Jack Gould, "Government film of H-bomb blast suffers from theatrical tricks," *New York Times* (2 April 1954), 35. The show "Racket Squad" starred Reed Hadley, who was also the narrator of the Operation Ivy documentary. A news report claimed that only black and white footage of the documentary was initially released overseas, because it was determined that color footage might have "too great an emotional impact" abroad. "Fear banned bomb film," *New York Times* (4 April 1954), 86.

55 Hans Bethe to William Borden (10 December 1952), CHNSA, Box 15, "1952."

56 This particular sentence was declassified by the Department of Energy in 1979, as will be discussed in chapter 8. US Department of Energy, Office of Declassification, "Restricted Data Declassification Decisions 1946 to the Present (RDD-7)," (1 January 2001), online at: https://fas.org/sgp/othergov/doe/rdd-7.html, accessed 11 December 2018.

57 On the Soviet discovery of the Teller-Ulam idea—which appears to have been independent of espionage—see Wellerstein and Geist, "The secret of the Soviet H-bomb."

58 Frank Cotter to Corbin Allardice, "Second Fuchs," JCAE document 4888 (7 September 1954), in JCAE, Series 2: General Subject Files, Box 53, "AEC Security." For more details on the case of Theodore Hall, see Albright and Kunstel, *Bombshell*.

59 This account naturally simplifies the Rosenberg casework considerably. For more detail, see Radosh, *The Rosenberg file*, and Weinstein and Vassiliev, *The haunted wood*. On Gold, see Hornblum, *The invisible Harry Gold*. On the fraught FBI-MI5 cooperation on Fuchs, see Goodman, "Who is trying to keep what secret from whom and why?"

60 Roberts, *The brother*.

61 Moynihan, *Secrecy*, 70–71.

62 Moynihan, *Secrecy*, 73–74.

63 For a contemporary analysis of the legal issues, see "Secret documents in criminal prosecutions," *Columbia Law Review* 47, no. 8 (December 1947), 1356–1364. In 1980, the threat of "graymail"—where witnesses would effectively hold the prosecution hostage by threatening to divulge classified information in their testimony—led to a new law (the Classified Information Procedures Act) that made it possible for judges to review classified evidence and testimony *in camera* for criminal cases. As will be discussed in chapter 8, there are separate possible procedures for handling such evidence in non-criminal cases.

64 "Suggestions for Improvement of Atomic Energy Commission Security," (5 February 1950), Klaus Fuchs FBI file, Federal Bureau of Investigation FOIA Vault, Part 109 of 111, online at: https://vault.fbi.gov/rosenberg-case/klaus-fuchs/klaus-fuchs-part-109-of/view, accessed 11 December 2018. There were at least ten such "GIs taking photos" cases in the immediate postwar; some managed to be prosecuted, others did not. For an overview of several of them, see Joint Committee on Atomic Energy, "Soviet Atomic Espionage," 82nd Congress, 1st Session (April 1951), 195. Most of the cases (six out of ten) were alleged to have been taken as souvenirs, and guilty pleas were negotiated.

65 Anders, "The Rosenberg case revisited."

66 "Space ship, proximity fuse details obtained; Greenglass testifies against his sister and others on trial as wartime agents," *Washington Post* (13 Mar 1951), 1; Stenographers Minutes, *USA. vs. Julius Rosenberg, et al.*, vol. 2 (12 March 1951), 719.

67 William R. Conklin, "Atom bomb secret described in court," *New York Times* (13 Mar 1951), 1.

68 "Spy's version of the A-bomb," *Life* (26 March 1951), 51.

69 Peter Kihss, "New bid planned for Sobell trial," *New York Times* (1 Aug 1965), 66.

70 Sidney E. Zion, "'A-bomb' sketch becomes public," *New York Times* (5 August 1966), 8; Sidney E. Zion, "2 scientists denounce evidence against Sobell," *New York Times* (23 Aug 1966).

71 Quoted in, "Sketch of A-bomb taken, US charges," *Washington Post* (8 Mar 1951), 5.

72 *In the Matter of J. Robert Oppenheimer* [Un-redacted version], vol. 4 (15 April 1954), DSAEC, Records Re: Oppenheimer Hearing, Transcripts of Personnel Security Board Hearing, Box 3, on 570.

73 Executive Order 9835, "Prescribing Procedures for the Administration of an Employee Loyalty Program in the Executive Branch of the Government," (21 March 1947).

74 Sub-category (B).9 of AEC Criteria of derogatory information, quoted in Gordon Dean to Sterling Cole (8 June 1953), GD, Box 2, "Classified Reader File, January, 1953, through June, 1953 (Vital)."

75 On the persecution of homosexuals during the Cold War, see Johnson, *The lavender scare.*

76 Executive session of the Joint Committee on Atomic Energy (8 February 1951).

77 Gordon Dean to Sterling Cole (8 June 1953), GD, Box 2, "Classified Reader File, January, 1953, through June, 1953 (Vital)."

78 Executive session, JCAE (9 June 1953). Through an unexpected leap of logic, Morrison's loss of clearance was ultimately put at the doorstep of Klaus Fuchs: Strauss argued that "since a considerable part of the information" that they were worried about Morrison knowing "was presumably now compromised by Dr. Fuchs, this reason for continuing Mr. Morrison's clearance had lost its force." Extract from AEC meeting No. 372 (15 February 1950), OSAEC46, Box 7, "Correspondence re Persons Me thru O."

79 American sources, unsurprisingly, saw Pontecorvo's flight as being evidence of guilt. The opening of the Soviet archives has not indicated this. Pontecorvo's situation in 1950 (and afterwards) was complicated. See Close, *Half-life*. See also Daniels, "Controlling knowledge, controlling people," for discussions of attempts to limit scientific travel.

80 Kaiser, "The atomic secret in Red hands?"

81 Oppenheimer's account is described in an FBI summary of Robert Serber made on 5 June 1947 (NY 116–1533), contained in Serber's FBI file (1121281-000-116-HQ-7681—Section 1 862767, obtained by FOIA). See also Serber and Crease, *Peace and war.*

82 Oppenheimer, as an aside, would credit Charlotte highly in the postwar, noting that in her time at Los Alamos not a single report had gone astray, "where a single serious slip might not only have caused us the profoundest embarrassment but might have jeopardized the successful completion of our job." J. Robert Oppenheimer to Charlotte Serber (2 November 1945), in Smith and Weiner, *Robert Oppenheimer, letters and recollections*, 313–314.

83 *Life* magazine (10 October 1949).

84 Bird and Sherwin, *American Prometheus*; Herken, *Brotherhood of the bomb.*

85 Executive Session, United States House of Representatives, Committee on Un-American Activities (7 June 1949), on 14.

86 Bird and Sherwin, *American Prometheus*; Herken, *Brotherhood of the bomb*; McMillan, *The ruin of J. Robert Oppenheimer and the birth of the modern arms race.*

87 See esp. McMillan, *The ruin of J. Robert Oppenheimer and the birth of the modern arms race.*

88 John Walker, "Policy and Progress in the H-Bomb Program: A Chronology of Leading Events," (1 January 1953), JCAE Document 584, RJCAE.

89 For a later autobiographical account, see Wheeler, *Geons, black holes, and quantum foam*, 284–286. See also Herken, *Brotherhood of the bomb*, 260, 268; Hewlett and Holl, *Atoms for peace and war*, 37–41. For a detailed discussion

of the "Wheeler incident" in the context of Cold War secrecy, see Wellerstein, "John Wheeler's H-bomb blues."

90 John A. Wheeler to Thomas Lovering, "Loss of Classified Document," (3 March 1953), copy obtained from Los Alamos National Laboratory Archives via Freedom of Information Act. A somewhat less redacted version was available via the National Nuclear Security Administration's Freedom of Information Act Virtual Reading Room until 2017, when the latter was taken offline, as document RR00184. An archived copy can be viewed on the Internet Wayback Machine: https://web.archive.org/web/*/https://nnsa.energy.gov/sites/default/files/nnsa/foiareadingroom/RR00184.pdf, accessed 11 December 2018.

91 Ford, *Building the H-bomb*, ch. 1 and 9.

92 Fuchs' knowledge of radiation implosion was very dissimilar to the way it was used in the Teller-Ulam design, however, but this was not comforting to those who feared he had compromised the H-bomb "secret." See Bernstein, "John von Neumann and Klaus Fuchs."

93 Herken, *Brotherhood of the bomb*, 260, 268; Hewlett and Holl, *Atoms for peace and war*, 37–41.

94 Ibid.

95 Herken, *Brotherhood of the bomb*, 267–271.

96 Lilienthal diary entry (5 May 1947), in Lilienthal, *Journals*, 175–176; see also Hewlett and Duncan, *Atomic shield*, 92. The appointment of a permanent Personnel Security Review Board was announced in the AEC's Semiannual Report, no. 6 (July 1949), 12.

97 Carson and Hollinger, *Reappraising Oppenheimer*, esp. the discussions by Gregg Herken, Kai Bird, Martin Serwin, and Barton J. Bernstein in part 2. For far more comprehensive coverage of the Oppenheimer affair, see: Bird and Sherwin, *American Prometheus*; Herken, *Brotherhood of the bomb*; McMillan, *The ruin of J. Robert Oppenheimer and the birth of the modern arms race*; Cassidy, *J. Robert Oppenheimer and the American century*; and Thorpe, *Oppenheimer*, among many other excellent Oppenheimer biographies.

98 There are many accounts of the Chevalier affair; among the most compelling are those in Herken, *Brotherhood of the bomb*, and Bird and Sherwin, *American Prometheus*.

99 USAEC, *In the matter of J. Robert Oppenheimer*, 137.

100 USAEC, *In the matter of J. Robert Oppenheimer*, 153–154.

101 USAEC, *In the matter of J. Robert Oppenheimer*, 710.

102 Hargittai, *Judging Edward Teller*; cf. Goodchild, *Edward Teller*. On Lawrence, see Hiltzik, *Big science*, ch. 19.

103 Quoted in Polenberg, *In the matter of J. Robert Oppenheimer*, 383.

104 McMillan, *The ruin of J. Robert Oppenheimer and the birth of the modern arms race*, 12, 199–200.

105 Quoted in Polenberg, *In the matter of J. Robert Oppenheimer*, 393.

106 Cartoon drawn by Hugh Haynie, *The Greensboro Daily News*, reprinted in "Oppenheimer case—five views," *New York Times* (6 June 1954), E5.

107 USAEC, *In the matter of J. Robert Oppenheimer*, 465.

108 *In the Matter of J. Robert Oppenheimer* [Un-redacted version], vol. 8 (21 April 1954), DSAEC, Records Re: Oppenheimer Hearing, Transcripts of Personnel Security Board Hearing, Box 3, on 1554. I have written about the declassification of these files: Wellerstein, "Oppenheimer, unredacted."

109 On Beckerley's views on secrecy, see "Atomic secrecy deemed overdone," *New York Times* (17 March 1954). On his quitting the AEC see, "Critic of Strauss to quit atom post," *New York Times* (22 May 1954), 6.

110 Sterling Cole to Lewis Strauss (6 May 1954), in OSAEC51, Box 151, "Security—Oppenheimer—Vol. 1."

111 Hewlett and Holl, *Atoms for peace and war*, 105.

112 Ibid.

113 Extracts from AEC meetings No. 1003 (12 June 1954), 1005 (14 June 1954), and 1006 (15 June 1954) are in OSAEC51, "Security—Oppenheimer—Vol. 1."

114 A longhand description of the work by Harry S. Traynor (12 June 1954) can be found in DSAEC, Records Re: Oppenheimer Hearing, Transcripts of Personnel Security Board Hearing, Box 5, "Deletions," and the box in general chronicles much of the work of deletion. AEC meeting 1017 (23 July 1954) also contains substantial discussion by Beckerley of the work he and his staff did, noting that it was anything but "slap-dash." Excerpted copy of AEC meeting 1017 (23 July 1954), in OSAEC51, Box 151, "Security—Oppenheimer—Vol. 2."

115 Extract from AEC meeting No. 1005 (14 June 1954), OSAEC51, "Security—Oppenheimer—Vol. 1."

116 One radio news correspondent willingly violated the embargo, with his publisher backing him up on the grounds that news could not be bottled up; the next day he attacked the AEC for even complaining about the embargo, and implied that Morse Salisbury, the director of information, was potentially associated with Alger Hiss. One cannot generalize from a single case of a New York radioman, but the AEC's relationship with the press was not at this point ideal. Morse Salisbury to Milton Burgh (16 June 1954) and Burgh to Salisbury (16 June 1954), and "Fulton Lewis Comment" (16 June 1954), all in OSAEC51, Box 151, "Security—Oppenheimer—Vol. 1."

117 See esp. Polenberg, "The fortunate fox."

118 Executive session of the JCAE (24 March 1953), RJCAE, Box 14.

119 On radioisotopes in the Cold War, see esp. Creager, *Life atomic*.

120 Executive Hearing, JCAE, "Discussion in Re: Supplemental Budget, Fuchs Case, British Mission, Joan Hinton Case, Hydrogen Bomb Development," (10 February 1950).

121 Hewlett and Holl, *Atoms for peace and war*, ch. 2; Balogh, *Chain reaction*.

122 J. Robert Oppenheimer et al., "Armaments and American Policy," (January

1953), Rose and Petersen, *Foreign relations of the United States, 1952–1954*, Volume 2, Part 2, 1056–1091, on 1078.

123 Chernus, "Operation Candor." See also Hewlett and Holl, *Atoms for peace and war*, 42–60.

124 Hewlett and Holl, *Atoms for peace and war*, 42–60; Bird and Sherwin, *American Prometheus*, ch. 33. On the nature of the 1953 test, see Wellerstein and Geist, "The secret of the Soviet H-bomb."

125 Bird and Sherwin, *American Prometheus*, 470; Freedman, *The evolution of nuclear strategy*, ch. 6.

126 Henry D. Smyth, "Analysis of Secrecy," AEC 111/25 (17 June 1953), OSAEC51, Box 144, "Security and Intelligence 2—Records and Information—Volume 1."

127 Hewlett and Holl, *Atoms for peace and war*, 65–67, 71–72. On Eisenhower's frustrations, see also Wang, *In Sputnik's shadow*, part 2.

128 Jasanoff and Kim, "Containing the atom."

129 Dwight D. Eisenhower, "Atoms for Peace" address to the United Nations General Assembly (8 December 1953), reprinted in Williams and Cantelon, *The American atom*, 104–111.

130 Hewlett and Holl, *Atoms for war and peace*, ch. 7–9. See also Krige, "Atoms for Peace, scientific internationalism, and scientific intelligence."

131 Lewis L. Strauss, "A First Step Toward the Peaceful Use of Atomic Energy," (19 April 1954), reprinted in United Nations, Report of the International Conference on the Peaceful Uses of Atomic Energy, Geneva, Switzerland, August 8–20, 1955, Volume 2, on 421–429, NTA, document NV0707833. Emphasis in the original.

132 Transcript of a Press Conference with I. I. Rabi and George T. Weil (12 May 1955), reprinted in United Nations, Report of the International Conference on the Peaceful Uses of Atomic Energy, Geneva, Switzerland, August 8–20, 1955, Volume 2, on 751–766, NTA, document NV0707833.

133 Transcript of a Press Conference with George T. Weil and John P. McKnight (6 August 1955), reprinted in United Nations, Report of the International Conference on the Peaceful Uses of Atomic Energy, Geneva, Switzerland, August 8–20, 1955, Volume 2, on 800–815, NTA, document NV0707833.

134 The speeches of Bohr and Bhabha are reprinted in United Nations, Report of the International Conference on the Peaceful Uses of Atomic Energy, Geneva, Switzerland, 8–20 August 1955, Volume 1, NTA, document NV0707832. Cf. Krige, "Atoms for Peace, scientific internationalism, and scientific intelligence."

135 Strauss, *Men and decisions*, 252.

136 Hewlett and Holl, *Atoms for peace and war*, ch. 5; "Atomic Energy Act of 1954," Public Law 703, 83rd Congress, 2nd Session Chapter 1073 (30 August 1954).

137 Quist, *Security classification of information*, vol. 2, ch. 4.

138 This change to the Civil Service Commission had been first implemented in 1953, as part of Public Law 164, 83rd Congress, 1st Session, Chapter 283 (31 July 1953).

139 The death penalty was removed as a punishment from the Atomic Energy Act in 1969. Public Law 161, 91st Congress (24 December 1969). It is still a possible sentence under the Espionage Act, however.

140 See esp. Turchetti, "'For slow neutrons, slow pay'"; Boskey, "Inventions and the atom." Between 1947 and 1980, only 38 cases would be considered by the Patent Compensation Board. Statement of Eric Fygi in "The Government's Classification of Private Ideas," Hearings before a Subcommittee of the Committee on Government Operations, House of Representatives, 96th Congress, 2nd Session (28 February, 20 March, and 21 August 1980), 317–321, on 321.

141 For an extensive discussion of the 1954 Act's patent clauses, which are considerably complex, see Boskey, "Patents under the new Atomic Energy Act."

142 Hewlett and Holl, *Atoms for peace and war*, ch. 7–9, 15, 18–19; Balogh, *Chain reaction.*

143 Charles L. Marshall, "Monthly Classification Bulletin, No. 1," (9 November 1956), DOEA (FOIA 2007-000504).

144 Charles L. Marshall, "Monthly Classification Bulletin, No. 4a," (7 March 1957), DOEA (FOIA 2007-000504). Emphasis in original.

145 Charles L. Marshall, "Monthly Classification Bulletin, No. 12," (16 October 1957), DOEA (FOIA 2007-000504).

146 Charles L. Marshall, "Monthly Classification Bulletin, No. 17," (23 March 1958), DOEA (FOIA 2007-000504).

147 Balogh, *Chain reaction*; and on diplomacy, see esp. Krige, *Sharing knowledge, shaping Europe*, ch. 1, 2, and 5.

148 *Semiannual Report of the Atomic Energy Commission* (Washington, DC: United States Government Printing Office, 1950–1959). Patents started to be listed in appendices to the reports with some regularity starting in 1950 but escalating during the Atoms-for-Peace years.

149 Kaiser, "The atomic secret in Red hands?"

150 For two sides of the modern discussion, see Kroenig, *Exporting the bomb* (which generally concludes that Atoms for Peace did not encourage proliferation by itself), and Fuhrmann, *Atomic assistance* (which argues the contrary).

151 See esp. Weart, *Nuclear fear.*

152 Lewis L. Strauss, "The US Atomic Energy Program, 1953–1958," *Bulletin of the Atomic Scientists* 14, no. 7 (September 1958), 256–258.

153 See esp. Richelson, *Spying on the bomb.*

154 Richelson, *Spying on the bomb*, esp. ch. 4–6; Abraham, *The making of the Indian atomic bomb*; Perkovich, *India's nuclear bomb.*

CHAPTER SEVEN

1 Hans Morgenthau, "Introduction," to Nieburg, *Nuclear secrecy and foreign policy*, xiii–xiv.

2 The Soviets did distribute some knowledge and technology, although hesitantly

and often not completely. The case of the People's Republic of China, and the role of Soviet assistance (which was abruptly withdrawn, with severe consequences, after the Sino-Soviet split) is interesting, but beyond the scope of this work. See Lewis and Litai, *China builds the bomb*, 60–72; and Montgomery, "Stop helping me."

3 Krige, *Sharing knowledge, shaping Europe*; Krige, "Atoms for peace, scientific internationalism, and scientific intelligence." A full list of bilateral agreements be found in Appendix 6 of Hewlett and Holl, *Atoms for peace and war*, on 581.

4 Krige, *Sharing knowledge, shaping Europe*, 39.

5 A good primer on centrifuge operation and history is Wood, Glaser, and Kemp, "The gas centrifuge and nuclear weapons proliferation."

6 Kemp, "The end of Manhattan," 272–279; Reed, "Centrifugation during the Manhattan Project."

7 Kemp, "The end of Manhattan," 281–286; Kemp, "Opening a proliferation Pandora's box."

8 Central Intelligence Agency, "The problem of uranium isotope separation by means of ultracentrifuge in the USSR," (8 October 1957), available from the Federation of American Scientists: https://fas.org/irp/cia/product/zippe.pdf, accessed 11 December 2018.

9 Kemp, "The end of Manhattan," 288–289; Wood, "The history of the gas centrifuge and its role in nuclear proliferation."

10 Wood, "The history of the gas centrifuge and its role in nuclear proliferation."

11 Gernot Zippe, "Development and Status of Gas Centrifuge Technology," *Separation phenomena in liquids and gases, seventh workshop proceedings, Moscow, July 24–28, 2000* (Moscow: Moscow Engineering Physics Institute, 2000), 35–53; copy online at "Orphaned (Re)Source: The Gernot Zippe Files," *Atomic Reporters* (December 2013), online at http://www.atomicreporters.com/2013/12/orphaned-resource-the-gernot-zippe-files/, accessed 11 December 2018.

12 Kemp, "Opening a proliferation Pandora's box," 120.

13 On the "Russian centrifuge" vs. "Zippe centrifuge," see Bernstein, *Nuclear weapons*, 261. On the Degussa connection, see Zippe, "Development and Status of Gas Centrifuge Technology," and Kemp, "Opening a proliferation Pandora's box," 110. On Zippe's report, see Kemp, "Opening a proliferation Pandora's box," 110–112, as well as the rest of the article, which traces the distribution of the report globally.

14 Testimony of George Kolstad in Executive Hearing, JCAE (30 August 1960), included as Document 23 in Burr, "The gas centrifuge secret"; Kemp, "The end of Manhattan," 288; Streefland, "Putting the lid on the gas centrifuge."

15 W. B. McCool, "Gas Centrifuge Method of Isotope Separation," AEC 610/15 (9 April 1960), in OPENNET, OSTI-ID 1048782.

16 Klaus Knorr, "Nuclear weapons: 'Haves' and 'have-nots,'" *Foreign Affairs* 36, no. 1 (October 1957), 167–178; Burr, "The 'Labors of Atlas, Sisyphus, or Hercules'?"

17 Testimony of John McCone, Executive Hearing, JCAE (30 August 1960), included as Document 23 in Burr, "The gas centrifuge secret."

18 S. A. Levin, D. E. Hatch, and E. Von Halle, "Production of Enriched Uranium for Nuclear Weapons by Nations X, Y, and Z by Means of the Gas Centrifuge Process," Union Carbide Nuclear Company, Report KOA-662 (26 February 1960), in OPENNET, OSTI-ID 1048388.

19 W. B. McCool, "Gas Centrifuge Method of Isotope Separation," AEC 610/15 (9 April 1960), in OPENNET, OSTI-ID 1048782. Brazil had, in fact, attempted to purchase centrifuges from West Germany in 1954, a sale that the US was able to stop. Despite this intervention, the German firm Sartorious sold three centrifuges to Brazil in 1957. Burr, "The 'Labors of Atlas, Sisyphus, or Hercules'?" For a longer discussion of Brazil's centrifuge work, see Kemp, "Opening a proliferation Pandora's box," 113–115. On Japan, see Kemp, "Opening a proliferation Pandora's box," 122–124.

20 W. B. McCool, "Gas Centrifuge Method of Isotope Separation."

21 McCone preferred the US simply undercutting any European markets by offering discounted uranium to its allies. Memorandum of a Conversation between Secretary of State Christian Herter, John McCone, and Philip Farley (8 June 1960), included as Document 14 in Burr, "The gas centrifuge secret."

22 Arthur Hartman to Howard Meyers (10 March 1960), included as Document 9 in Burr, "The gas centrifuge secret."

23 W. B. McCool, "Gas Centrifuge Method of Isotope Separation."

24 Department of State, Memorandum of a Conversation, "German Classification of Centrifuge Work," (4 October 1960), included as Document 25 in Burr, "The gas centrifuge secret"; A. A. Wells memo to AEC, "Classification Meeting Regarding Centrifuge Technology" (3 March 1962), in OPENNET, OSTI-ID 1046218; see also Appendix A in Howard C. Brown Jr. et al., "A study of gas centrifuge as it relates to the proliferation of nuclear weapons," (4 January 1967), in OPENNET, OSTI-ID 1095798, also available (in a slightly less redacted form) as Document 28 in Burr, "The gas centrifuge secret." For details on the guide's contents, see Streefland, "Putting the lid on the gas centrifuge," 83. Only one item in the guide—the fact that classified work was being done on centrifuges—was considered unclassified, and the guide itself was classified "confidential."

25 It was claimed to be "strictly an error." Anderson then opined: "But it will become Secret from there on out and they will defend it to their dying day." Executive Hearing, JCAE (30 August 1960), included as Document 23 in Burr, "The gas centrifuge secret."

26 All quotes in ibid.

27 See Appendix B in Howard C. Brown Jr. et al., "A study of gas centrifuge as it relates to the proliferation of nuclear weapons."

28 Hewlett, "'Born classified' in the AEC." Hewlett argues that the "born classified" concept was "an underlying factor" (22) in AEC deliberations, and a "working assumption without ever using the term" (21). The examples he gives

to support this are not very compelling (they are not specific to the question of whether the *origin* of the data matters), and even he notes that "all reports, testimony, and hearings on the 1946 and 1954 Acts were silent on this point" (24). It seems more probable that this specific interpretation of the Act had not, in fact, been considered explicitly until the 1960s. For another approach (which comes to similar conclusions as Hewlett, though is similarly unclear about whether the "born classified" idea applied to privately generated data or not in historical discussions), see Cheh, "The *Progressive* case and the Atomic Energy Act," esp. 164–180.

29 See Appendix A in Howard C. Brown Jr. et al., "A study of gas centrifuge as it relates to the proliferation of nuclear weapons"; on Dow Chemical and General Electric, see Memorandum of a Conversation between Secretary of State Christian Herter, John McCone, and Philip Farley (8 June 1960), included as Document 14 in Burr, "The gas centrifuge secret."

30 Hewlett, "'Born classified' in the AEC," 22–24.

31 Green, "The AEC proposals."

32 Green, "AEC information control regulations," 42. Green elaborated: "This seems to be an inherent consequence of abandoning the simplistic and absolute position that all information within the restricted data definition, wherever and however originated, is subject to AEC information controls."

33 Hewlett, "'Born classified' in the AEC," 27.

34 Howard C. Brown Jr. et al., "A study of gas centrifuge as it relates to the proliferation of nuclear weapons."

35 Green, "The AEC proposals," 16; Green, "'Born classified' in the AEC"; Hewlett, "'Born classified' in the AEC," 27.

36 On AEC goals, see Streefland, "Putting the lid on the gas centrifuge," 86; on the use of Restricted Data as a means to restrict the British, see Twigge, "A baffling experience," 155–156, with quote of Tony Benn on 156; on US-UK centrifuge collaboration, see Krige, "Hybrid knowledge," esp. 341–343.

37 Kemp, "Opening a proliferation Pandora's box," 120–121.

38 Streefland, "Putting the lid on the gas centrifuge," 89–90.

39 A. A. Wells memo to AEC, "Centrifuge Classification Meeting" (21 March 1962), in OPENNET, OSTI-ID 1046219; A. A. Wells memo to AEC, "Proposed Four Power Meeting on Handling Centrifuge Information" (14 January 1964), in OPENNET, OSTI-ID 1046221.

40 A. A. Wells memo to AEC, "Proposed Four Power Meeting on Handling Centrifuge Information"; Myron B. Kratzer to AEC, "Classification of gas centrifuge information," (5 March 1964), in OPENNET, OSTI-ID 1046222. On Dutch plans, see S. A. Levin, L. R. Powers, E. Von Halle, "Nth Power Evaluation," Union Carbide Corporation, Report KOA-1237 (4 March 1964), in OPENNET, OSTI-ID 1048389.

41 Krige, "US technological superiority and the special nuclear relationship," 234–235; Schrafstetter and Twigge, "Spinning into Europe," esp. 260.

42 Howard C. Brown Jr. et al., "A study of gas centrifuge as it relates to the proliferation of nuclear weapons."

43 Krige, "The proliferation risks of gas centrifuge enrichment at the dawn of the NPT."

44 W. B. McCool, "Highlights of December 5–6 Discussions with UK on Gas Centrifuge Classification," AEC 610/147 (16 December 1968), in OPENNET, OSTI-ID 1129538; Schrafstetter and Twigge, "Spinning into Europe," 269–270; Krige, "Hybrid knowledge," 343–357.

45 Kemp, "The nonproliferation emperor has no clothes."

46 For a semi-popular treatment of the history of fusion, see Seife, *Sun in a bottle*; for a scholarly treatment of magnetic confinement fusion in particular, see Bromberg, *Fusion*.

47 Bromberg, *Fusion*, chapter 2.

48 For further details on the dual work at Princeton, see esp. Ford, *Building the H-bomb*, ch. 11.

49 Homi J. Bhabha, "The peaceful uses of atomic energy," *Bulletin of the Atomic Scientists* 11, no. 8 (October 1955), 280–284.

50 AEC Press conferences of August 8, 1955, and August 11, 1955, both reprinted in United Nations, Report of the International Conference on the Peaceful Uses of Atomic Energy, Geneva, Switzerland, August 8–20, 1955, Volume 2, on 421–429, NTA, document NV0707833.

51 Henry DeWolf Smyth, Speech before National Industrial Conference Board's 4th Annual Conference on Atomic Energy in Industry, "Controlled Fusion Reactions," (28 October 1955), NTA, document NV0057611.

52 Carl Hinshaw to Lewis Strauss (23 January 1956), and Lewis Strauss to Carl Hinshaw (13 February 1956), both in JBC, Box 1, Folder 6, "Correspondence 1951–1959."

53 Norris Bradbury to Lewis Strauss (20 December 1957), in JBC, Box 1, Folder 6, "Correspondence 1951–1959."

54 US Atomic Energy Commission Press Conference (23 January 1958), in JBC, Box 3, Folder 5, "Reports 1957–1961."

55 On the creation of Livermore, see esp. Gusterson, *Nuclear rites*, 20–25. Nuckolls' biographical information comes largely from Nuckolls, "Contributions to the genesis and progress of ICF."

56 On Project Plowshare, see Kaufman, *Project Plowshare*; see esp. 208–216 for a similar sort of project (Rio Blanco) in which an underground nuclear detonation would be used to generate steam for a power system.

57 Nuckolls, "Contributions to the genesis and progress of ICF," 5–7.

58 Ibid.

59 Kidder, "Laser fusion," esp. 50–53. On the development of the laser (with a brief cameo from laser fusion), see Bromberg, *The laser in America, 1950–1970*.

60 Nuckolls, "Contributions to the genesis and progress of ICF"; Kidder, "Laser fusion." There are forms of inertial confinement fusion that do not use lasers,

and there are forms of laser fusion that do not use inertial confinement (e.g., laser heating). For this section, unless otherwise specified, I will be using inertial confinement fusion and laser fusion synonymously to refer to laser-driven inertial confinement fusion.

61 Nuckolls, "Contributions to the genesis and progress of ICF," 12.

62 Nuckolls, "Contributions to the genesis and progress of ICF," 13.

63 Ray E. Kidder, "Evaluation of Sanders Associates Proposal of 6/12/63 and General Comments (Laser Heating of Thermonuclear Fuel)," UCRL-ID-124776 (12 August 1963), LLNLA.

64 The committee membership was Kidder (Livermore), Francis T. Byrne (Department of Defense, Office of Naval Research), Spitzer (Project Sherwood), and Peter Franken (University of Michigan). Charles L. Marshall to AEC Commissioners, "Classification Policy for Lasers," (16 July 1964), DOEA (FOIA-2008–000766), Record Group 326, Collection 20 Secretariat, Box 1428, Folder 6, "Sec 2–1 Classification Vol. 4." See also letter from Robert E. Hollingsworth to Melvin Price, included in Hearings before the Subcommittee on Research, Development, and Radiation of the Joint Committee on Atomic Energy, 92nd Congress, 1st Session, 10–11 November 1971, Part 2, on 619.

65 For work in other national laser fusion programs at this time, see Velarde and Santamaría, *Inertial confinement nuclear fusion*, which is a remarkable volume of participant accounts from many different countries. There is at least one program conspicuously missing from their volume (the French), owing to its continued classification, or so a former participant told me.

66 Herbert F. York, oral history interview with author (24 April 2008), OHCNBL.

67 KMS Fusion, "Partial History of Laser Fusion as it Pertains to KMSF," (10 November 1971), KMSFD. On Brueckner's background, see also Keith A. Brueckner, oral history interview with Finn Aaserud (2 July 1986), OHCNBL.

68 US Atomic Energy Commission, "AEC Policy and Action Paper on Controlled Thermonuclear Research," TID-23277 (June 1966), copy available online at https://www.osti.gov/servlets/purl/4508284/, accessed 12 December 2018.

69 KMS Fusion, "Partial History of Laser Fusion as it Pertains to KMSF"; Keith A. Brueckner, oral history interview with author (30 May 2008), OHCNBL; Brueckner, "A beginning for ICF by laser."

70 Robert L. Hirsch to Keith A. Brueckner (7 March 1969), KAB, Box 37, Folder 10; Bromberg, *Fusion*, 183–185.

71 On the debates about classified research on campuses, see Leslie, *The Cold War and American science*. On Siegel, see "Keeve M. Siegel," in KMS Industries, *People & Publications* (Ann Arbor, MI: KMS Industries, 1971), KAB, Box 32, Folder 3; KMS Fusion, "Ten Years of Fusion Research at KMS (FE-5036)," (25 April 1984), KMSFD. "Gambler" quote is from Keith A. Brueckner, oral history interview with author (30 May 2008), OHCNBL.

72 "Partial History"; on the specifics of the study, sponsored specifically by the

Defense Atomic Support Agency, see Peter Hammerling, "Early History of the Fusion Project, DASA Contract: 'Laser Heating of Deuterium,'" (22 March 1975), KMSFD; Keith A. Brueckner, oral history interview with author (30 May 2008), OHCNBL.

73 KMS Fusion, "Ten Years"; KMS Fusion, "Partial History." For a description of the patents filed in 1969, see Keith A. Brueckner to William D. Metz (14 March 1975), KAB, Box 38, Folder 9; Keith A. Brueckner, oral history interview with author (30 May 2008), OHCNBL.

74 Keith A. Brueckner to Keeve A. Siegel (7 November 1973), KAB, Box 38, Folder 5. See also Brueckner, "A beginning for ICF by laser," 95; Keith A. Brueckner, oral history interview with author (30 May 2008), OHCNBL; Robert E. Hollingsworth to Keith A. Brueckner (26 November 1969), KAB, Box 37, Folder 10.

75 Charles L. Marshall to AEC Commissioners, "Laser classification panel," (14 May 1971), DOEA (FOIA-2008–000766), Record Group 326, Collection 6 Secretariat, Box 7832, Folder 6, "O&M 7 Laser Classification Panel," and Summary Notes of Commissioners' Meeting with Laser Classification Panel (21 May 1971), DOEA (FOIA-2008–000766), Record Group 326, Collection 1801 GC, Box 94, Folder 7, "O&M 7 Laser Classification Advisory Board." A "pure-fusion" weapon would be a weapon that did not require fission reactions to get fusion energy output. Sometime in the mid-1970s, the AEC seems to have decided that the probability of a pure-fusion weapon was very low, and its role in their discussions decreased. In 1998, the Department of Energy authorized the declassification of the fact that "DOE made a substantial investment in the past to develop a pure fusion weapon," but that "no credible design for a pure fusion weapon resulted from the DOE investment." US Department of Energy, Office of Declassification, "Restricted Data Declassification Decisions, 1946 to the Present," section C.

76 Robert E. Hollingsworth to Keith A. Brueckner (26 November 1969), KAB, Box 37, Folder 10.

77 On late-AEC patent policy, see Walterscheid, "The need for a uniform government patent policy."

78 Charles L. Marshall to AEC Commissioners, "Lubin's Research at the University of Rochester," (23 September 1970), DOEA (FOIA-2008–000766), Record Group 326, Collection 30 Ramey, Box 8043, Folder 23, "Laser Classification Panel." On Lubin's work at Rochester more generally, see McCrory, "Highlights of the history of the University of Rochester."

79 Brueckner, "A beginning for ICF by laser," 95. The details of these applications are not all entirely known, and some may still be classified. A summary of the 1969 patents is in Keith A. Brueckner to William D. Metz, (14 March 1975), KAB, Box 38, Folder 9.

80 Joseph F. Hennessey to Glenn T. Seaborg, et al. (14 August 1970), DOEA (FOIA-2008–000766, final response).

81 Keeve M. Siegel to Glenn T. Seaborg (14 September 1970), Part of SECY-440, DOEA (FOIA-2008–000766, final response).

82 Philip F. Belcher (Assistant Director for Security & Legal Liaison, Los Alamos) to All Consultants and Visiting Staff Members, "KMS Industries, Inc. and Peaceful Applications of Laser-Pellet-Fusion Technology" (18 June 1971), KAB, Box 42, Folder 9. KMS Press Release (4 February 1971), NTA, document NV0913856; AEC Press Release, "AEC Approves KMS Industries Request to Conduct Fusion R&D Program," (4 February 1971), NTA, document NV0147046; "AEC approves restricted KMS Industries controlled-fusion program," *Nucleonics Week* 12, no. 6 (11 February 1971), NTA, document NV0052318.

83 Robert L. Hirsch to Roy W. Gould, "Lasers for CTR as Discussed and Presented at the IV European Conference on Plasma Physics and Controlled Fusion," (29 September 1970), DOEA (FOIA-2008–000766, final response).

84 Frederick Seitz to Glenn T. Seaborg (5 February 1971), DOEA (FOIA-2008–000766), Record Group 326, Collection 6 Secretariat, Box 7836, Folder 6, "O&M 7 Laser Classification Panel."

85 W. B. McCool, "Summary Notes of Commissioners' Meeting With Laser Classification Panel" (28 June 1971), DOEA (FOIA-2008–000766), Record Group 326, Collection 1801 GC, Box 94, Folder 7, "O&M 7 Laser Classification Advisory Board"; Charles L. Marshall to AEC Commissioners, "Laser Classification Panel," (14 May 1971), DOEA (FOIA-2008–000766), Record Group 326, Collection 6 Secretariat, Box 7832, Folder 6, "O&M 7 Laser Classification Panel."

86 Carl Haussmann to Jack W. Rosengren, "Classification of Laser-Induced Thermonuclear Explosives" (15 December 1970), DOEA (FOIA-2008–000766), Record Group 326, Collection 30 Ramey, Box 8043, Folder 23, "Laser Classification Panel."

87 John Nuckolls and Lowell Wood, "Prospects for Unconventional Approaches to Controlled Fusion (UCRL-74488)," (29 December 1971), received from the National Technical Information Service, US Department of Commerce, on 10.

88 Ibid., 3.

89 Keith A. Brueckner, Notes on the Nuckolls-Wood paper (dated 8 February 1972), KAB, Box 37, Folder 12.

90 Nuckolls, "Contributions to the genesis and progress of ICF," 20.

91 John Nuckolls, Lowell Wood, Albert Thiessen, and George Zimmerman, "Laser compression of matter to super-high densities: Thermonuclear (CTR) applications," *Nature* 239 (15 September 1972): 139–142.

92 Keith A. Brueckner, oral history interview with author (30 May 2008), OHCNBL.

93 "Extract of SECY-2469—General Manager's Info. Rpt. #98," (5 May 1972), DOEA (FOIA-2008–000766), Record Group 326, Collection 6 Secretariat, Box 7848, Folder 3, "Security 3–1 Guides."

94 Nuckolls, "Contributions to the genesis and progress of ICF."

95 Charles L. Marshall to D. R. Cotter (15 July 1974), DOEA (FOIA-2008–000766), RG 326, Collection 9 Secretariat, Box 7994, Folder 7, "Security Classification."

96 Roy B. Snapp to Henry J. Gomberg (7 August 1972), Keith A. Brueckner papers, Box 37, Folder 14.

97 Stuart Symington to Dixy Lee Ray (16 March 1973), DLR, Box 29, Folder "KMS Fusion, Inc."; Gordon M. Grant, "Notes of the Meeting Commencing at 2:20 P.M., April 20, 1973, in the Chairman's Office, Concerning KMS' Patents," (23 April 1973), DLR, Box 29, Folder "KMS Fusion, Inc."

98 KMS Press Release (13 May 1974), DLR, Box 29, Folder "KMS Fusion, Inc."

99 See, e.g., "Fusion through use of lasers reported; Firm calls it step toward thermonuclear electricity," *Los Angeles Times* (14 May 1974), A1.

100 "Laser report," *Laser Focus Magazine* 10, no. 6 (June 1974), copy in KAB, Box 38, Folder 7.

101 William D. Metz, "Laser fusion: One milepost passed—millions more to go," *Science* 186, no. 4170 (27 December 1974), 1193–1195.

102 Keith A. Brueckner, oral history interview with author (30 May 2008), OHCNBL.

103 Keith Boyer interview with Robert W. Seidel (5 November 1984), transcript, OHCNBL, 33–34. For a very even-handed assessment of the feasibility and difficulties of laser fusion at the time, see Soloman Buchsbaum, chair, "Final Draft report of the Ad Hoc NSC Panel on Laser Applications to Energy R&D (Laser Fusion and Laser Isotope Separation)," (8 April 1974), DOEA (FOIA-2008–000766).

104 Kidder, "Laser fusion," 49–68; Keith A. Brueckner, oral history interview with author (30 May 2008), OHCNBL. Because Brueckner's initial calculations were all one-dimensional, the immense problem posed by asymmetry would not have been visible. (I thank David Kaiser for this observation.) An AEC evaluation from 1974 further noted that their one-dimensional calculations had major flaws in them as well. John A. Erlewine, AEC Information Report: "AEC Review of the KMSF Experimental Program" (8 August 1974), DOA (FOIA-2008–000766, final response).

105 Lindl, "Development of the indirect-drive approach to inertial confinement fusion and target physics basis for ignition and gain"; Keith A. Brueckner, oral history interview with Author (30 May 2008), OHCNBL.

106 "Laser report," *Laser Focus Magazine*; Ernest Graves to Seymour Schwiller (28 January 1975), DLR, Box 29, Folder "KMS Fusion, Inc."

107 Gene Bylinsky, "KMS Industries bets its life on laser fusion," *Fortune* (December 1974), 148–156.

108 Richard M. Cohen, "The words stuck in his mouth; Stroke kills Hill energy witness," *Washington Post* (15 March 1975), A1. I have been unable to find any other historical examples of Congressional witnesses dying mid-testimony.

109 Specifically, the government desired to have KMS Fusion perform a "redirec-

tion" of its effort toward "automated (high volume) techniques for fabrication of pellets." Martin Stickley (Director, Division of Laser Fusion, ERDA) to Robert C. Seamans, Jr., (Administrator, ERDA), "Interim Extension of Contract with KMS Fusion, Inc." (29 June 1976), DOEA, Collection 1246, Box 3, Folder 15, "Laser Fusion Division."

110 See, e.g., Walker, *The road to Yucca Mountain*, ch. 4; and Balogh, *Chain reaction*, ch. 9.

111 G. S., "Any books on atomic power?" *New York Times Book Review* (18 November 1945), 2.

112 Martha Coble, "Boy's bomb thesis floors most readers," *Washington Post* (22 January 1946), 5. I tracked down "Jimmy" in 2010, who asked me to not use his full name. He noted that the author of the story was a relative, and that while he "rather enjoyed the notoriety at first," he later came to resent the "undertone of hostility and resentment that seemed to come my way" from unsupportive peers. He considers the "Boy genius atomic bomb book" story to be "the monster that dogged me through my adolescence." "Jimmy," email correspondence with the author (3 July 2010).

113 Peter Arno, *New Yorker* (6 April 1963), 42; Richard Lemon, "What happened when Albie Watkins got the bomb," *Saturday Evening Post* 240, no. 13 (1 July 1967), 20.

114 W. J. Frank, ed., "Summary Report of the Nth Country Experiment," UCRL-50249 (March 1967), copy included in Burr, "1960s 'Nth Country Experiment' foreshadows today's concerns over the ease of nuclear proliferation." On the kiloton-range assessment, see Oliver Burkemann, "How two students built an A-bomb," *Guardian* (23 June 2003); see also Stober, "No experience necessary."

115 CIA, "Soviet Capabilities for Clandestine Attack Against the US with Weapons of Mass Destruction and Vulnerability of the US to such Attack (mid-1951 to mid-1952)," National Intelligence Estimate 31 (4 September 1951). Online at http://foia.cia.gov, last accessed 21 June 2010. See also Zenko, "Intelligence estimates of nuclear terrorism."

116 For Taylor's motivations and background, see McPhee, *The curve of binding energy*, and Testimony of Theodore B. Taylor, in "An Act to Combat International Terrorism," Hearings before the Committee on Governmental Affairs, United States Senate, 95th Congress, 2nd Session (23, 25, 27, 30 January 1978; 22 February 1978; 22, 28 March 1978), 254–255.

117 Alan M. Adelson, "Please don't steal the atomic bomb," *Theory into Practice* 8, no. 2 (April 1969) [originally published in the May 1969 issue of *Esquire*], 62–66.

118 McPhee, *The curve of binding energy*. The *New Yorker* articles appeared on 3 December, 10 December, 17 December 1973.

119 Edmund Fuller, "Dangers of a homemade bomb," *Wall Street Journal* (10 May 1974), 10; Sandra Schmidt Oddo, "How not to make an atomic bomb," *New*

York Times Book Review (23 June 1974), 4; Christopher Lehmann-Haupt, "An atom bomb in every home," *New York Times* (9 July 1974), 35; and Richard Rhodes, "A do-it-yourself atom bomb? It's closer than you think," *Chicago Tribune* (30 June 1974), F3.

120 Edward Edelson, "A 'guide' to nuclear blackmail," *Washington Post* (1 June 1974), C4.

121 Willrich and Taylor, *Nuclear theft*, 59. As of 1974, the AEC had projected that there would be 150 civilian nuclear plants in the United States by 1980, and 1,000 by the year 2000. By 2019, there are 98 operating nuclear reactors in the US, and 450 operating globally.

122 Willrich and Taylor, *Nuclear theft*, 21; McPhee, *The curve of binding energy*, 89.

123 Testimony of J. Carson Mark, in "An Act to Combat International Terrorism," Hearings before the Committee on Governmental Affairs, US Senate, 95th Congress, 2nd Session (23, 25, 27, 30 January 1978; 22 February 1978; 22, 28 March 1978): 255–256.

124 On the breakdown of the notion of "containment" in nuclear matters, see Jasanoff and Kim, "Containing the atom."

125 Cf. the way in which American media characterized the news of the Indian blast: "India's nuclear bid—contrast in land of poverty," *Chicago Tribune* (21 May 1974), 6; Editorial, "The sixth atomic power," *Chicago Tribune* (21 May 1974), 10; Editorial, "India joins the club," *Los Angeles Times* (21 May 1974); Editorial, "India's nuclear bomb," *Washington Post* (21 May 1974), A16; Victor K. McElheny, "India test linked to power plants," *New York Times* (23 May 1974), 7; and James Reston, "The nuclear nightmare," *New York Times* (21 May 1974), 33. Taylor himself used the example of the Indian test to push his broader agenda: Lee Dye, "Atoms for Peace may supply atoms for war, experts fear," *Los Angeles Times* (21 May 1974), part 2, page 1.

126 FBI, "Total of Domestic Terrorist Acts, 1971–1977," included as an appendix to "An Act to Combat International Terrorism," Hearings before the Committee on Governmental Affairs, US Senate, 95th Congress, 2nd Session (23, 25, 27, 30 January 1978; 22 February 1978; 22, 28 March 1978), on 664.

127 "Historical Overview of DOE Response to Terrorism," (n.d., ca. May 1978), entered into the public record as part of "An Act to Combat International Terrorism," Hearings before the Committee on Governmental Affairs, US Senate, 95th Congress, 2nd Session (23, 25, 27, 30 January 1978; 22 February 1978; 22, 28 March 1978), on 365.

128 Testimony of Donald Kerr (Acting Assistant Secretary for Defense Program, DOE), in "An Act to Combat International Terrorism," Hearings before the Committee on Governmental Affairs, US Senate, 95th Congress, 2nd Session (23, 25, 27, 30 January 1978; 22 February 1978; 22, 28 March 1978), on 282.

129 For Taylor's views on the efficacy of secrecy in this case, see esp. Willrich and Taylor, *Nuclear theft*, 131. For the AEC view as of 1978, see Testimony of Donald

Kerr, 280–281. The legal issue revolved around whether the Atomic Energy Act allowed for information, once designated by the AEC to be safe, to be designated as secret in a way that was legally binding. The answer: probably not.

130 For likely diplomatic reasons, the Apollo affair has been shrouded with secrecy and the smell of a cover up. Recently declassified documents highlight significant disagreements within the government as to the likelihood of a diversion of material. Roger J. Mattson, "The NUMEC Affair: Did Highly Enriched Uranium from the US Aid Israel's Nuclear Weapons Program?," National Security Archive, Electronic Briefing Book 565 (2 November 2016): https://nsarchive.gwu.edu/briefing-book/nuclear-vault/2016-11-02/numec-affair-did-highly-enriched-uranium-us-aid-israels, accessed 11 December 2018.

131 Ralph E. Lapp, "The ultimate blackmail," *New York Times Magazine* (4 February 1973), 13, 29–34; Mark Andrews, "Orlando's H-bomb threat in 1970 had officials seriously concerned," *Orlando Sentinel* (4 June 1995); Testimony of Delmar L. Crowson (Director, Office of Safeguards and Materials Management, AEC), in "Public Works for Water and Power Development and Atomic Energy Commission Appropriations for Fiscal Year 1972," Hearings Before a Subcommittee of the Committee on Appropriations, US Senate, 92nd Congress, 1st Session (28 April 1971), 188–190. Most of the relevant facts of the case are taken from the Crowson testimony and the Andrews article. See also McPhee, *The curve of binding energy*, 85–86; "Amateur A-bomb?" *Time* (13 May 1974), 87–88; and Walker, "Regulating against nuclear terrorism," esp. 113.

132 John C. Stennis, quoted in Testimony of Delmar L. Crowson.

133 Walker, "Regulating against nuclear terrorism."

134 Walker, *The road to Yucca Mountain*, 83–87, 103–105.

135 John Angier, "The Plutonium Connection," VHS, prod. NOVA/WGBH-TV (New York: Time/Life Films, 1975); see also Jack Hemstock, "Now it's easy to make your own A-bomb," *Chicago Tribune* (9 March 1975), 16.

136 Phillips and Michaelis, *Mushroom*, 85–86.

137 Phillips and Michaelis, *Mushroom*, 86–90; Dyson, *Disturbing the universe*, 163–165.

138 The easy availability of the *Los Alamos Primer* is a recurrent theme in discussing the diffusion of nuclear secrets even at the present time, despite the fact that its treatment of atomic weapons is elementary. Copies of the *Primer* had been circulating privately amongst scientists for years, and were later collected, edited, and published with the help of Richard Rhodes. On the provenance of the *Primer*, as well as the best discussion of its contents, see Serber, *The Los Alamos Primer*. It is of note that as early as 1948, the *Primer* was not considered to contain sensitive information. H. T. Wensel of the AEC remarked in classified congressional testimony to HUAC that the *Primer* could not give away the bomb in any way: "It doesn't contain what we call the real secrets of the process or the know-how of the bomb. It is not that kind of a bomb." Executive Session,

Subcommittee on National Security of the Committee on Un-American Activities, House of Representatives (4 May 1948), 86.

139 The Princeton University library catalog lists the thesis as: John Aristotle Phillips, "The Fundamentals of Atomic Bomb Design: An Assessment of the Problems and Possibilities Confronting a Terrorist Group or Non-Nuclear National Attempting to Design a Crude Pu239 Fission Bomb," (Princeton, N.J.: Department of Physics, 1976). A further note explains: "Restricted. Permission of Physics Department Chair required." I have been allowed to see the paper; all mysticism about secrecy aside, it is good undergraduate work, but still undergraduate work. It does not contain any insights or ideas that a modern researcher would find surprising, and the specific design used for achieving implosion looks unlikely to succeed if implemented literally. On the methods, see Phillips and Michaelis, *Mushroom*, 86–90, 96–97, 101, 111–113. On the grade, see Dyson, *Disturbing the universe*, 164.

140 Phillips and Michaelis, *Mushroom*, 143; Dyson, *Disturbing the universe*, 164.

141 Phillips and Michaelis, *Mushroom*, 151.

142 "Student designs $2,000 atom bomb," *New York Times* (9 October 1976), 16; "College student's project a real bomb," *Los Angeles Times* (10 October 1976), A2; Phillips and Michaelis, *Mushroom*, 151–155.

143 Phillips and Michaelis, *Mushroom*, 184–206 (*Mushroom* is the book in question, co-written with his Princeton roommate); "Student says Pakistanis sought bomb plans," *New York Times* (8 February 1977), 20; "Student's A-bomb plan draws bids," *Los Angeles Times* (8 February 1977), B6; Joseph F. Sullivan, "Nations beat path to door of Princeton senior for his atom bomb design," *New York Times* (10 February 1977), 41; Karen De Witt, "Star student John Phillips; How he started worrying and learned to hate the bomb," *Washington Post* (10 June 1977), B1; and Don Kladstrup, "Headline: Atomic bomb / Princeton student," CBS Evening News (10 February 1977), Record 250349, and James Walker, "Headline: Atomic bomb," ABC Evening News (10 February 1977), Record 47229, both in VTNA. The made-for-television movie was not made, in the end, but the film *Manhattan Project* (1986), in which a high-school student actually builds an atomic bomb for the science fair, was apparently inspired by Phillips' story. James Verini, "Big Brother Inc." *Vanity Fair* (web exclusive) (13 December 2007). Online at http://www.vanityfair.com/politics/features/2007/12/aristotle200712, last accessed 11 June 2010.

144 Dyson, *Disturbing the universe*, 164.

145 The same problem was also evident in Livermore's 1967 "Nth Country Experiment."

146 Testimony of Dmitri A. Rotow, in "An Act to Combat International Terrorism," Hearings before the Committee on Governmental Affairs, United States Senate, 95th Congress, 2nd Session (23, 25, 27, 30 January 1978; 22 February 1978; 22, 28 March 1978), 269, 272–274.

147 Testimony of Theodore B. Taylor, in "An Act to Combat International Terrorism," 254–255. By contrast, J. Carson Mark, testifying at the same session, noted again the extreme difficulty in moving from a diagram and any confidence that it would work.

148 Testimony of Donald Kerr, in "An Act to Combat International Terrorism," 305.

149 Testimony of Donald Kerr, in "An Act to Combat International Terrorism," 280–281.

150 Balogh, *Chain reaction*, ch. 9; Walker, *The road to Yucca Mountain*, ch. 4.

151 Mills, *The seventh power*, 43. A number of other pot-boilers about nuclear terrorism soon followed: see James Coates and Eleanor Randolph, "How to make A-bomb: Go to library," *Chicago Tribune* (25 July 1979), 1.

152 "The Majority Report Handywoman's Guide: How to build your own atomic bomb and strike a balance of power with the patriarchy," *Majority Report* (8–21 July, 1978). For some background, see Joseph B. Treaster, "US journals have printed atom bomb directions," *New York Times* (11 March 1979), 21. For additional discussion of the trope, see Larabee, *The wrong hands*.

153 E.g., Adelson, "Please don't steal the atomic bomb."

154 This debate continued on into the post–Cold War era as well, centered instead around the possibility of inventing or maintaining nuclear weapons without recourse to nuclear testing. See MacKenzie and Spinardi, "Tacit knowledge, weapons design, and the uninvention of nuclear weapons," 44–99; and Gusterson, *People of the bomb*, ch. 9.

155 "A. Q. Khan Chronology," *Carnegie Endowment for International Peace Issue Brief* 7, no. 8 (7 September 2005); Richelson, *Spying on the bomb*, ch. 8, 11, 14.

CHAPTER EIGHT

1 Edward Teller, "How many secrecies?," *New York Times* (1 December 1971), 47.

2 See, e.g., Schoenfeld, *Necessary secrets*, ch. 8; Bok, *Secrets*, ch. 14 and and 16. For earlier critiques of secrecy, see Shils, *The torment of secrecy*, and Rourke, *Secrecy and publicity*, neither of which, despite their titles, are quite yet examples of "anti-secrecy" to my mind.

3 See testimony of J. Dexter Peach, in "Erroneous Declassification of Nuclear Weapons Information, Part 2," Hearing before the Subcommittee on Energy, Nuclear Proliferation and Federal Services of the Committee on Governmental Affairs, United States Senate, 96th Congress, 1st Session, Part 1 (2 October 1979), 3–14; and prepared statement of Duane Sewell, in "Erroneous Declassification of Nuclear Weapons Information, Part 2," 21.

4 AEC Press Release, "AEC in Midst of Declassification Drive"; AEC Press Release, "AEC Declassification Drive Moves to Chicago," (7 September 1973), NTA, document NV0148223; prepared statement of Duane Sewell, in "Erroneous Declassification of Nuclear Weapons Information, Part 2," 21.

5 Executive Order 11652, "Classification and Declassification of National Security Information and Material," 37 F.R. 5209 (8 March 1972); Quist, *Security Classification of Information*, vol. 1, 53–54; Theoharis, "The Freedom of Information Act versus the FBI."

6 *New York Times Co. v. United States*, 403 US 713 (1971); Halperin and Hoffman, *Top secret*, ch. 2, and appendix B; Schoenfeld, *Necessary secrets*, ch. 9–10, and Ellsberg, *Secrets*.

7 On the nuclear material, see Ellsberg, *The doomsday machine*, 5–17. Ellsberg would become a prominent anti-nuclear activist, but it was not until 2017 that he set down his full story in an easy-to-access form.

8 See, e.g., Bernstein and Woodward, *All the President's men*; Kutler, *Watergate*.

9 For a critical view of this overall trend, see, e.g. Schoenfeld, *Necessary secrets*, ch. 10.

10 In the case of *Mink v. Environmental Protection Agency*, 410 U.S. 73 (1973), the court denied a group of congressional representatives attempting to obtain information about the 1971 nuclear test "Cannikin," detonated in Amchitka island, Alaska. The majority ruling found that the executive branch could withhold secret information from FOIA requests simply by filing an affidavit stating that it was, indeed, secret. For background from a major participant in the case, see Mink, "The Cannikin Papers."

11 Lilienthal, *Change, hope, and the bomb*, 117.

12 "A touch of the sun," reprinted in Eliot, *The H bomb*, 48, art by R. M. Chapin, Jr.; "A possible destroyer of the world illustrated diagrammatically," *Illustrated London News* (4 February 1950), art by G. H. Davis.

13 See, e.g., Wellerstein and Geist, "The secret of the Soviet H-bomb."

14 "Why the H-bomb is now called the 3-F," *Life* magazine (5 December 1955), 54–55, credited to Adolph E. Brotman.

15 The idea that graphical tropes can correspond with epistemological claims is explored in the context of patent drawings in Rankin, "The 'person skilled in the art' is really quite conventional."

16 On the Union of Concerned Scientists, see esp. Downey, "Reproducing cultural identity in negotiating nuclear power."

17 Much of this account of Morland's career and motivations comes from Morland, *The secret that exploded*, as well as conversations and e-mails I have had with Morland over the years. See also Morland, "The Holocaust bomb."

18 Morland, *The secret that exploded*, 39.

19 York, *The advisors*, 8.

20 Morland, *The secret that exploded*, 40 and 51.

21 Morland, *The secret that exploded*, 54–55.

22 Morland, *The secret that exploded*, 54–55. For a comparison of the *Encyclopedia Americana* (1974) and *Merit Students Encyclopedia* (1964) drawings, see De Volpi et al., *Born secret*, figs. 2 and 3.

23 Morland, "The Holocaust bomb."

24 Howard Morland, in "Transcript of *Weapons of mass destruction*," 1366–1378.
25 Morland, *The secret that exploded*, 20. Emphasis in original.
26 Howard Morland, e-mail correspondence with author (1 January 2005).
27 Morland, *The secret that exploded*, 60.
28 Morland, *The secret that exploded*, 107.
29 Morland, *The secret that exploded*, 256–257.
30 Morland, *The secret that exploded*, 8.
31 Morland, *The secret that exploded*, 133, 138, 120, 132, respectively.
32 Day, *Crossing the line*, 101.
33 Day, *Crossing the line*, 103. Cf. Morland, *The secret that exploded*, 130–131, 141.
34 Morland, *The secret that exploded*, 142.
35 Morland, in "Transcript of *Weapons of mass destruction*," 1370.
36 Day, *Crossing the line*, 104.
37 Day, *Crossing the line*, 105–107.
38 Day, *Crossing the line*, 106; Morland, *The secret that exploded*, 142.
39 Day, *Crossing the line*, 108.
40 Morland, *The secret that exploded*, 140.
41 Day, *Crossing the line*, 108.
42 Day, *Crossing the line*, 111.
43 Day, *Crossing the line*, 113–116.
44 James Schlesinger, in "Transcript of *Weapons of mass destruction*," on 1345–1346.
45 Day, *Crossing the line*, 116.
46 Hewlett, "'Born classified' in the AEC," 22–24.
47 Schlesinger, in "Transcript of *Weapons of mass destruction*," 1350–1351.
48 Ibid., 1346–1347.
49 "Plaintiff's Statement of Points and Authorities in Support of Application for a Temporary Restraining Order and Motion for a Preliminary Injunction," *US v. The Progressive, Inc.* (3 March 1979), HMNSA, Box 6. The most detailed, document-by-document discussion of both sides of the *Progressive* case is found in De Volpi et al., *Born secret*, chapters 4, 9, and 10.
50 Bell and Ostrow, *Taking care of the law*, 132; Day, *Crossing the line*, 120.
51 Thomas S. Martin, interview with author (10 September 2010).
52 Concurring (majority) opinion of Thurgood Marshall, *New York Times Co. v. United States*, 403 US 713 (1971), footnote 3.
53 Quoted in De Volpi et al., *Born secret*, 61–62.
54 See, e.g., De Volpi et al., *Born secret*, 189–208.
55 De Volpi et al., *Born secret*, ch. 4.
56 Extended excerpts can be found in De Volpi et al., *Born secret*, ch. 4, esp. note IV-5. Copies of the original affidavits are in HMNSA, Box 6.
57 Affidavit of Jack W. Rosengren (13 March 1979), in HMNSA, Box 6.
58 Supplemental for Affidavit for John A. Griffin (13 March 1979), in HMNSA, Box 6.

59 De Volpi et al., *Born secret*, 67.

60 Norman Dorsen, in "Transcript of *Weapons of mass destruction*," on 1352–1354.

61 Day, *Crossing the line*, 108; Morland, *The secret that exploded*, 166–167; De Volpi et al., *Born secret*, 65–66.

62 Quoted in De Volpi et al., *Born secret*, 69.

63 Affidavit of Hans A. Bethe (22 March 1979), HMNSA, Box 6. Morland preferred the government's description of his work to that of his defense lawyers, who "practically accused me of plagiarizing children's encyclopedias." Morland, "The Holocaust bomb."

64 Quoted in De Volpi et al., *Born secret*, 72.

65 See Pozen, "The mosaic theory, national security, and the Freedom of Information Act."

66 Testimony of James C. Goodale, in Hearings on H.R. 12004 before a Subcommittee of the Committee on Government Operations, House of Representatives, 93rd Congress, 2nd Session (11 and 25 July, and 1 August 1974): 396–413, quote from 409–410.

67 Editorial, "John Mitchell's dream case," *Washington Post* (11 March 1979), C6.

68 Editorial, "Public bombs, and minds born secret," *New York Times* (25 March 1979), E18.

69 Douglas E. Kneeland, "Furor over H-Bomb article astonishes magazine editor," *New York Times* (18 March 1979), 24.

70 Morland, *The secret that exploded*, 79 and 162; Day, *Crossing the line*, 139.

71 Morland, in "Transcript of *Weapons of mass destruction*," 1373.

72 Ray E. Kidder, oral history interview with author (29 April 2008), OHCNBL. See also, Kidder, "Weapons of mass destruction, national security, and a free press." An excellent example of Kidder's mastery of the public domain literature is in Kidder's correspondence with Hans Bethe from 1979, which was later released and declassified. See, "The 1979 Bethe-Kidder Correspondence," online at http://fas.org/sgp/eprint/bethe-kidder.html, accessed 29 May 2009.

73 Quoted in De Volpi et al., *Born secret*, 243.

74 See Arthur D. Thomas to John A. Griffin (2 November 1977), as reproduced in Kidder, "Weapons of mass destruction, national security, and a free press," exhibit 4, on 1399.

75 Frank Tuerkheimer, in "Transcript of *Weapons of mass destruction*," 1364–1365.

76 Thomas S. Martin, interview with author (10 September 2010).

77 Morland, *The secret that exploded*, 183, 197, 198, 213.

78 Morland, *The secret that exploded*, 196.

79 Jeremy J. Stone, in "Prior restraint: How much to print about the hydrogen bomb," *Problems of Journalism* (1979), 36–40, on 37.

80 Morland, *The secret that exploded*, 206–207; clarification of the ACLU/Rotow connection by email from Howard Morland to author (2 July 2010).

81 Testimony of Robert Thorn, in "Erroneous Declassification of Nuclear Weapons Information, Part 1," Hearing before the Subcommittee on Energy, Nuclear

Proliferation and Federal Services of the Committee on Governmental Affairs, United States Senate, 96th Congress, 1st Session (23 May 1979), on 34.

82 Testimony of Dmitri A. Rotow, in "Erroneous Declassification of Nuclear Weapons Information, Part 1," Hearing before the Subcommittee on Energy, Nuclear Proliferation and Federal Services of the Committee on Governmental Affairs, United States Senate, 96th Congress, 1st Session (23 May 1979), 9.

83 Details on UCRL-4725's contents can be found in Hansen, *The swords of Armageddon*, volume 4.

84 Testimony of Theodore B. Taylor, in "Erroneous Declassification of Nuclear Weapons Information, Part 1," 16–21.

85 US Atomic Energy Commission, "Supplement to Indexes of Limited-Distribution Reports, Changes in classification, Vol. 29, No. 6," (31 December 1973), and testimony of John A. Griffin, both in "Erroneous Declassification of Nuclear Weapons Information, Part 1," Hearing before the Subcommittee on Energy, Nuclear Proliferation and Federal Services of the Committee on Governmental Affairs, United States Senate, 96th Congress, 1st Session (23 May 1979), 29–30, and 38. See also, "Staff Report on the Erroneous Declassification of Nuclear Weapons Information," (September 1979), included in "Erroneous Declassification of Nuclear Weapons Information, Part 2," Hearing before the Subcommittee on Energy, Nuclear Proliferation and Federal Services of the Committee on Governmental Affairs, United States Senate, 96th Congress, 1st Session (2 October 1979), 45–84.

86 E.g., "Hydrogen bomb instructions," *NBC Evening News* (17 May 1979), VTNA, Record 504254; Deirde Carmody, "US library is shut to review secrets," *New York Times* (12 May 1979), 8; "Priceless H-bomb secrets declassified in error, U. S. officials concede," *Los Angeles Times* (24 May 1979), B7; Morton Mintz, "H-bomb report was public 4 years," *Washington Post* (18 May 1979), A1.

87 John Glenn, in "Erroneous Declassification of Nuclear Weapons Information, Part 1," 27, 37.

88 Ronald J. Ostrow and Jim Mann, "Most Justice Dept. lawyers cite weak case, drop opposition to H-bomb article," *Los Angeles Times* (8 June 1979), B14; Morland, *The secret that exploded*, 227. Cf. Tuerkheimer, in "Transcript of *Weapons of mass destruction*," 1364–1365.

89 Quoted in Morland, *The secret that exploded*, 225.

90 See, e.g., Hansen's early work: Chuck Hansen, "US nuclear bombs," *Replica in Scale* 3, no. 3 (January 1976).

91 Quoted in Tom Abate, "The H-bomb 'secret,'" *Daily Californian (UC Berkeley)* (14 September 1979), 1. See also De Volpi et al., *Born secret*, 173–174.

92 Abate, "The H-bomb 'secret,'" 2; Morland, *The secret that exploded*, 222.

93 Morland, *The secret that exploded*, 223.

94 Morland, *The secret that exploded*, 226; De Volpi, et al., *Born secret*, 183.

95 De Volpi et al., *Born secret*, 184.

96 Livermore physicist Hugh DeWitt speculated that part of the government's motivation to declaring the Hansen letter equivalent to the Morland article was an act of "disinformation" meant to undo their earlier "validation" of the Morland article, but to me this suggests more cleverness than is probable. Morland, *The secret that exploded*, 227. A DOJ attorney told me that they were informed by the DOE that the matter had been mooted and that the DOJ's withdrawal of the case did not involve any "strategic" maneuvering on their part. Thomas S. Martin, interview with author (10 September 2010).

97 Philip Taubman, "US drops efforts to bar publication of H-bomb articles," *New York Times* (18 September 1979), 1; De Volpi et al., *Born secret*, 212–213.

98 *Progressive* (November 1979), cover.

99 Morland, *The secret that exploded*, 197.

100 Winterberg would later publish his own book of how H-bombs work under the "Fusion Energy Foundation Frontiers of Science Series": Winterberg, *The physical principles of thermonuclear explosive devices*. For more information about the LaRouche fusion activities, Winterberg, and Parpart, see King, *Lyndon LaRouche and the new American fascism*, 69–70, 79–80, 162, and 171.

101 Uwe Parpart, "Implications of Rudakov disclosures: The Soviet Union is on the verge of a strategic weapons breakthrough," *New Solidarity* (15 October 1976), 1, 5.

102 Morland, *The secret that exploded*, 198.

103 Morland, *The secret that exploded*, 198.

104 Howard Morland, "Errata," *Progressive* (November 1979), 35.

105 Howard Morland, "The H-bomb secret: To know how is to ask why," *Progressive* (November 1979): 3–12.

106 Morland, *The secret that exploded*, 228.

107 E.g., Cheh, "The *Progressive* case and the Atomic Energy Act"; Sobota, "The unexploded bomb"; Tribe and Remes, "Some reflections on the *Progressive* case."

108 It is worth noting that there were significant differences between Ellsberg and Morland. Ellsberg was an insider with legitimate access to secrets who had become a whistle-blower, whereas Morland was an outsider determined to re-derive secrets through clever sleuthing. These two positions are significantly different with respect to the secrecy regime, and put the two in very different legal situations (criminal vs. civil cases).

109 Schlesinger, in "Transcript of *Weapons of mass destruction*," 1347.

110 Thomas S. Martin, interview with author (10 September 2010).

111 US Department of Energy, Office of Declassification, "Restricted Data Declassification Decisions, 1946 to the Present," (1 January 2001), section V.C.e, copy available online at http://www.fas.org/sgp/othergov/doe/rdd-7.html, accessed 14 May 2010.

112 For a fairly comprehensive state of the field as of the mid-1990s, see Lindl,

"Development of the indirect-drive approach to inertial confinement fusion and target physics basis for ignition and gain." See also US Department of Energy, Office of Declassification, "Restricted Data Declassification Decisions, 1946 to the Present," section 9.

113 E.g., Shils, *The torment of secrecy*.

114 See, e.g. William Proxmire, "Let's stop silly secrecy in government," *This Week Magazine* (14 May 1961), 6–7.

115 E.g., Edwards, *The closed world*, ch. 9; Weart, *Nuclear fear*, ch. 19.

116 Broad, *Teller's war*.

117 Norris and Kristensen, "Counting nuclear weapons in the public interest," 85.

118 William M. Arkin, interview with author (13 September 2018); Howard Kurtz, "Explosive analyst," *Washington Post* (24 May 2002), C1.

119 Fred Kaplan, "US secrets sometimes open," *Boston Globe* (8 June 1986), 16.

120 Peter Pringle, "Anti-nuclear US admiral gets cold feet," *Observer* (19 April 1981), 10.

121 William M. Arkin, interview with author (13 September 2018).

122 Cochran, Arkin, and Hoenig, *Nuclear Weapons Databook*, Vol. 1. The final in the series was published in 1994: Norris, Burrows, and Fieldhouse, *Nuclear Weapons Databook*, Vol. 5. The Stockholm International Peace Research Institute (SIPRI) published an annual yearbook on "World Armaments and Disarmament" that also trafficked in similar sorts of statistics since 1969, though in their case they do not seem to have made stockpile estimates until the 1980s, when they received assistance in this from Cochran, Norris, Arkin, et al.

123 McGeorge Bundy, "Deception, self-deception and nuclear arms," *New York Times Book Review* (11 March 1984), 3.

124 Arkin, *Nuclear battlefields*; William M. Arkin, interview with author (13 September 2018); for examples of "localized" stories, see, e.g., James Gerstenzang, "California has most nuclear warfare sites, book says," *Los Angeles Times* (14 June 1985), A8; Joseph F. Sullivan, "Navy depot in Jersey: Concern and complacency" (18 June 1985), B2. The *Tonight Show* monologue was for the 14 June 1985 episode.

125 William M. Arkin, interview with author (13 September 2018); Robert S. Norris, correspondence with author (28 September 2018).

126 William M. Arkin, interview with author (13 September 2018); Michael L. Ross, "Disarmament at sea," *Foreign Policy*, no. 77 (Winter 1989–1990), 94–112; "US reassures Iceland on contingency plans for placing weapons," *Globe and Mail* (15 March 1985), 8.

127 Quote from Hansen, *The swords of Armageddon*, i–iii; Hansen, *US nuclear weapons*.

128 Winterberg, *The physical principles of thermonuclear explosive devices*.

129 Hansen, *US nuclear weapons*, 22 and 121–122.

130 Interview with Mike Wagnon (28 October 2004). The reference photographs

and Hansen's sketches are contained in Box 10, "Artwork," CHNSA. On the visual language of engineering, see Henderson, "Flexible sketches and inflexible data bases."

131 For reception, see, e.g., William J. Broad, "From Cold War to nuclear nostalgia," *New York Times* (12 December 1989), C1. For the quote on motivation, see Erwin Knoll, "Questioner," *Progressive* (June 1992), 4.

132 These details all come from the rather small FBI file that was created for Hansen. Nearly every page within it pertains to the 1993 investigation. Charles "Chuck" Hansen FBI file, FOIAP Request No. 1142316–000 (7 October 2010), Federal Bureau of Investigation.

133 E.g., "Secrecy: The hard way," *Los Angeles Times* (4 February 1982), D10; William Safire, "An obsession with secrecy," *Chicago Tribune* (30 November 1983), 19; Robert G. Kaiser, "Leaks are the target, but our values are the victim," *Washington Post* (18 December 1983), C1; Thomas Collins, "The secrecy syndrome: The Reagan team closes the doors," *Newsday* (24 June 1984), 1. For a book-length treatment, see Curry, *Freedom at risk.*

134 For a particularly florid "connecting the dots" of Reagan's secrecy policies see Floyd Abrams, "The new effort to control information," *New York Times Magazine* (25 September 1983), 22–28, 72–73. Another similar op-ed is Anthony Lewis, "When the president restricts freedom of the press," *Boston Globe* (18 November 1983), 17.

135 Jonathan King, "Energy Department's secrecy epidemic," *Hartford Courant* (28 June 1983), B9; David Burnham, "Plan on restricting nuclear data arousing wide array of protests," *New York Times* (16 August 1983), A1; Matt Yancey, "Nuclear secrecy proposal draws broad opposition," *Boston Globe* (17 August 1983), 3; "Security weaknesses at the nuclear weapons laboratories," Committee on Governmental Affairs, US Senate, 100th Congress, 2nd Session (11 October 1988), esp. 10–19, and 135–146.

136 "2 clash in debate on secrecy rules," *New York Times* (16 December 1983), A27.

137 Quote from "Silly secrecy," *Los Angeles Times* (13 December 1988), D6.

138 John Hanrahan, "Scholars find Reagan's administration the most secretive since WWII," *Los Angeles Times* (30 August 1987), 4.

139 Nancy Blodgett, "Is it espionage? Photo leak to media at issue," *ABA Journal (American Bar Association)* 71, no. 5 (May 1985), 18; Colman McCarthy, "Samuel Morison: A leaker, not a thief," *Washington Post* (17 November 1985), H9; George Lardner Jr., "Morison given 2 years for leaking spy photos," *Washington Post* (5 December 1985), A18.

140 Jack Nelson, "Reagan criticizes disaster secrecy," *Los Angeles Times* (4 May 1986), A1; John O. Pastore and Peter Zheutlin, "'Consent' on weapons can't be based on lies," *Los Angeles Times* (30 October 1988), E5. On Soviet secrecy at Chernobyl, see esp. Plokhy, *Chernobyl.*

141 The most prominent example of this "re-branding" is found in Moynihan,

Secrecy, picking up on Max Weber's definition of secrecy as a form of "regulation," the latter of which Moynihan injected with more 1990s Republican sensibility than Weber's nineteenth-century German context really warranted.

142 Edward Teller, "State secrecy doesn't help national security," *Wall Street Journal* (18 June 1986), 26.

143 Herbert F. York, oral history interview with author (24 April 2008), OHCNBL.

CHAPTER NINE

1 Daniel P. Moynihan, Chairman, "Report on the Commission on Protecting and Reducing Government Secrecy," S. Doc. 105–2, 103rd Congress, (Washington, DC: Government Printing Office, 1997), xxi.

2 US Department of Energy, *Classification policy study* (4 July 1992), online at http://www.fas.org/sgp/othergov/doe/, accessed 10 December 2018. See also Quist, *Security classification of information*, vol. 1, 120.

3 E.g., A.M. Rosenthal "The secrecy game," *New York Times* (13 November 1992), A29; Mary Cheh, "Half a billion for locks," *Washington Post* (29 December 1992), A15; Stanley Kutler, "Mr. Clinton, strike a blow at secrecy," *Wall Street Journal* (14 January 1993), A17; Anthony Lewis, "Someone should tell the CIA and FBI that the Cold War's over," *Seattle Post-Intelligencer* (15 December 1992), A21.

4 Quist, *Security classification of information*, vol. 1, 61.

5 There had been one white, female Chairwoman of the Atomic Energy Commission: Dixy Lee Ray, who served from 1974 to 1975.

6 US Department of Energy, Press conference transcript, "Openness," (7 December 1993), DOEA, Collection 1703, Box 4, Folder 40, "Press Conference 12/7/93."

7 US Department of Energy, Press conference transcript, "Openness."

8 On the experiments and their publication, see: Welsome, *The plutonium files*; Moss and Eckhardt, "The human plutonium injection experiments." On the human radiation experiments in general, see Advisory Committee on Human Radiation Experiments, *Final Report* (October 1995), online at https://bioethicsarchive.georgetown.edu/achre/, accessed 10 December 2018. On the cover-up, see ch. 5 of this book.

9 For an account of O'Leary's first exposure to the human radiations stories, see, Gup, *Nation of secrets*, 114–118.

10 US Department of Energy, Press conference transcript, "Openness."

11 DOE Office of the Secretary, "Openness: Beyond declassification (draft)," (18 May 1994), DOEA, Collection 1769, Box 1, Folder 39, "Declassification Editorials"; "Openness and secrecy: A symbolism on establishing accountability in the nuclear age, National Press Club," (18 May 1994), DOEA, Collection 1704, Box 2, Folder 28, "Openness and Secrecy, 5/18/94."

12 E.g., William J. Broad, "US begins effort to recast the law on atomic secrets," *New York Times* (8 January 1994), 1; Gary Lee, "Letting the nation in on decades of secrets," *Washington Post* (31 March 1994), A29; Matthew Wald, "Millions of secrets burden energy agency," *New York Times* (7 February 1996), A15.

13 Burr, S. Blanton, and Schwartz, "The costs and consequences of nuclear secrecy," on 471–472.

14 US Department of Energy, Press conference transcript, "Release of previously classified information regarding the US nuclear arms program," (27 June 1994), DOEA, Collection 1769, Box 1, Folder 39, "Declassification Editorials."

15 Quist, *Security classification of information*, vol. 1, 128. Like many such rebranding attempts, it was relatively short-lived: by 2004, it was once again the Office of Classification.

16 Quist, *Security classification of information*, vol. 1, 122–125.

17 William J. Broad, "US bomb labs, waging peace, shift their targets," *New York Times* (6 December 1992), E6; John Burgess, "Bombs into bulldozers; US defense labs, in search of new mission, turn to industry," *Washington Post* (23 August 1992), H1; "Taking on the future; Harold Agnew and Los Alamos scientists discuss the potential of the laboratory," *Los Alamos Science* 21 (1993), 4–30; Gusterson, *Nuclear rites*, ch. 9.

18 US Department of Energy, "Declassification stakeholder meeting (transcript)," (28 January 1994), DOEA, Collection 1702, Box 3, Folder 34, "Jan. 28, 1994, Declass. Stakeholder."

19 A. Bryan Siebert, "The DOE Openness Initiative, The Classification System, and Challenges and Issues," part of a meeting of the Department of Energy Openness Advisory Panel (24 June 1996), online at https://www.osti.gov/opennet/forms?formurl=document/oap1.html#ZZ7, accessed 20 October 2019.

20 Moynihan, "Report of the Commission on Protecting and Reducing Government Secrecy," quote on xxxvi (emphasis in original). Moynihan's association with secrecy as a form of regulation was deliberately indebted to the sociologist of bureaucracy Max Weber. Weber's writings on secrecy should be read and used with some caution. Weber's subject was the formation of the nineteenth-century German bureaucracy, not the twentieth-century military-industrial complex. While there are some similarities (secrecy as a form and practice of power), there are many differences (the German bureaucratic example lacks the existential urgency, and military enforcement, of the nuclear situation). Weber, "Bureaucracy [1920]."

21 Stober and Hoffman, *A convenient spy*, 167–168.

22 Report of the Select Committee on US National Security and Military/Commercial Concerns with the People's Republic of China (Report 105–851), House of Representatives, 105th Congress (3 January 1999; declassified version released 25 May 1999).

23 Ibid.

24 George Lardner Jr., "DOE puts declassification into reverse," *Washington Post* (19 May 2001).

25 See esp. Stober and Hoffman, *A convenient spy*; William J. Broad, "Spies vs. sweat: The debate over China's nuclear advance," *New York Times* (7 September 1999), A1; William J. Broad, "Are there any secrets left to steal?" *New York Times* (3 September 2000), WK1. See also Masco, "Lie detectors."

26 Stober and Hoffman, *A convenient spy*, 308–309.

27 See esp. Galison, "Secrecy in three acts."

28 Pollack, "The secret treachery of A. Q. Khan"; Kroenig, *Exporting the bomb*, ch. 4.

29 William J. Broad, "U.S. web archive is said to reveal a nuclear primer," *New York Times* (3 November 2006), A1.

30 Comprehensive Report of the Special Advisor to the Director of Central Intelligence on Iraq's Weapons of Mass Destruction (30 September 2004), Volume 2, Chapter 4, on 42; available online at https://www.cia.gov/library/reports/general-reports-1/iraq_wmd_2004/chap4.html, accessed 20 October 2019. One item of note that emerged in the report on the Iraqi nuclear effort: they had, it reported, used classification extensively on even open-source information not because it was sensitive, but "to create an aura of importance." Ibid., 55.

31 Wellerstein, "Building an H-bomb in plain sight."

32 See, e.g., William J. Broad and Elaine Sciolino, "Iran's secrecy widens gap in nuclear intelligence," *New York Times* (19 May 2006), A1.

33 Loveday Morris and Karen DeYoung, "Israel says it holds a trove of documents from Iran's secret nuclear weapons archive," *Washington Post* (30 April 2018).

34 Israel's "secret" was released in 1986 by a dissident, Mordechai Vanunu, whose publication of extensive photographic documentation of the secret Israeli nuclear facility at Dimona made it clear, beyond a doubt, that Israel had weaponized the atom. Vanunu was severely punished for this, having been kidnapped abroad and smuggled back to Israel for trial. See Cohen, *The worst-kept secret*, esp. ch. 4.

35 Executive Office of the President, "Transparency and Open Government," (27 January 2009), *Federal Register*, FR-Doc. E9–1777, online at https://www.federalregister.gov/documents/2009/01/26/E9–1777/transparency-and-open-government, accessed 10 December 2018.

36 "Increasing Transparency in the US Nuclear Weapons Stockpile," (3 May 2010), US Department of Defense. Previously online at https://www.defense.gov/Portals/1/features/defenseReviews/NPR/10–05–03_Fact_Sheet_US_Nuclear_Transparency__FINAL_w_Date.pdf, since removed. Available on the Internet Archive at https://web.archive.org/web/20170131184122/https://www.defense.gov/Portals/1/features/defenseReviews/NPR/10–05–03_Fact_Sheet_US_

Nuclear_Transparency__FINAL_w_Date.pdf, accessed 10 December 2018. In 2019, when the Trump administration halted this practice, it was greeted as an attack on transparency, as one would expect: Hans M. Kristensen, "Pentagon slams door on nuclear weapons stockpile transparency," *Federation of American Scientists Blogs* (17 April 2019), online at https://fas.org/blogs/security/2019/04/stockpilenumbersecret/, accessed 17 April 2019.

37 See, e.g., Albert R. Hunt, "Under Obama, a chill on press freedom," *New York Times* (8 June 2014).

CONCLUSION

1 Walter G. Andrews, "Atomic Energy Legislation" (draft speech, undated, ca. March 1946), in HBF, Roll 8, Target 7, File 93, "Congressman (Walter G.) Andrews."

2 Samuels, "Atomic John."

3 My sentence is a light paraphrase of a variously attributed quote ("It is easier to imagine an end to the world than an end of capitalism"), but there are indeed teams of scientists working on clever technical ways to verify warhead dismantlement that do not involve learning warhead design secrets, for just this reason: they think disarmament is more likely and achievable than the elimination of secrecy. See Wellerstein, "The virtues of nuclear ignorance."

4 On the latter, see esp. MacKenzie and Spinardi, "Tacit knowledge, weapons design, and the uninvention of nuclear weapons."

5 What factors matter for successful nuclear proliferation is an area of contentious debate amongst scholars of international relations. See, for example, Fuhrmann, *Atomic assistance*; Hymans, *Achieving nuclear ambitions*; Kemp, "The nonproliferation emperor has no clothes," and Kroenig, *Exporting the bomb*, as representative of this literature. In some of these models, the acquisition or transfer of "sensitive information" (variously defined) has a strong role in predicting the success of a proliferation attempt; in others, it is a minor factor compared to organizational and political factors.

6 Executive Hearing, JCAE, "Minutes of an Executive Meeting of the Joint Committee on Atomic Energy," (6 April 1949).

7 See, for example, the interviewees in Galison and Moss, *Secrecy*.

8 "Classification of information; limitations," *Code of Federal Regulations*, title 28 § 17.22 (2019).

9 See esp. Vogel, *Phantom menace or looming danger?*

10 An exception to this *may* be the Nazi program, which never got off the ground in part because of certain technical misunderstandings, but even that feels like a stretch. For a good overview of German misunderstandings, see Popp, "Misinterpreted documents and ignored physical facts."

11 On cyberthreats, see, e.g., Sanger, *The perfect weapon*.

12 See, e.g., Schneier, *Secrets and lies*, esp. ch. 25.

13 See, e.g., Denis Grady and William J. Broad, "Seeing terror risk, US asks journals to cut flu study facts," *New York Times* (20 December 2011), A1.

14 Except, perhaps, for the Manhattan Project itself, which was driven initially by fear of a non-existent Nazi atomic bomb.

15 See, e.g., Hymans, *Achieving nuclear ambitions*.

16 See, e.g., Shils, *The torment of secrecy*, Weart, *Nuclear fear*, and Wills, *Bomb power*.

17 This is a generalized perception, both from the archives of each, but also from discussions I have had with people who interacted with each agency.

18 It is difficult to know the precise "time lag" between "open" and "secret" research, because of course having an entire part of a field be closed to many experts and funding sources necessarily affects it. But in general, my informal observation of the history of nuclear research is that at most the "secret" communities might be around a decade more advanced than the "open" ones, and in some cases they may not diverge significantly at all. This is not proposed as an iron-clad rule, and "progress" is a tricky concept for historians of science and technology in any case.

19 This is not to say, as my colleagues who study these countries (esp. the Soviet Union) would be quick to remind me, that the internal history of secrecy in these countries would be entirely as straightforward as it looks from the outside, or that it would not be interesting in its own ways, different from the US case.

BIBLIOGRAPHY

ARCHIVAL SOURCES AND ABBREVIATIONS

AHC	Arthur Holly Compton papers, University Archives, Washington University in St. Louis, MO.
ARC	National Archives and Records Administration, Archival Research Catalog, online at http://catalog.archives.gov.
BCF	*Bush-Conant File Relating the Development of the Atomic Bomb, 1940–1945*, Records of the Office of Scientific Research and Development, Record Group 227, microfilm publication M1392, Washington, DC: National Archives and Records Administration, n.d. (ca. 1990).
CHNSA	Chuck Hansen papers, Record 408, National Security Archive, Gelman Library, George Washington University, Washington, DC.
CTS	*Correspondence ("Top Secret") of the Manhattan Engineer District, 1942–1946*, microfilm publication M1109, Washington, DC: National Archives and Records Administration, 1980.
DEL	Office files of David E. Lilienthal, Records of the Office of the Chairman, Records of the Atomic Energy Commission, Record Group 326, NARACP.
DELCFCRF	Chairman's (Formerly Classified) Reading File, December 1946–July 1950, Office files of David E. Lilienthal, Records of the Office of the Chairman, Records of the Atomic Energy Commission, Record Group 326, NARACP.
DLR	Dixy Lee Ray papers, Hoover Institution Archives, Palo Alto, CA.
DOEA	Department of Energy Archives, Germantown, MD (accessed with the aid of Terrence Fehner).
DSAEC	Division of Security, Records of the Atomic Energy Commission, Record Group 326, NARACP.
EOL	Ernest O. Lawrence papers, Bancroft Library, University of California, Berkeley.
GD	Office files of Gordon Dean, Records of the Commissioners, Records of the Atomic Energy Commission, Record Group 326, NARACP.

GDDRS U.S. Declassified Documents Reference Online (formerly Declassified Documents Reference System), Primary Source Media, Gale Group, online at https://www.gale.com/c/us-declassified-documents-online.

HBF *Harrison-Bundy Files Relating to the Development of the Atomic Bomb, 1942–1946*, microfilm publication M1108, Washington, DC: National Archives and Records Administration, 1980.

HMNSA Howard Morland papers, Record 48, National Security Archive, Gelman Library, George Washington University, Washington, DC.

JCAE Joint Committee on Atomic Energy, Executive (Unpublished) Hearings. Unless otherwise indicated, these formerly classified transcripts have been accessed through the ProQuest Congressional Legislative & Executive Publications database.

JBC Joan Bromberg collection, "Materials Collected for *Fusion: science, politics, and the invention of a new energy source*, 1922–1982," AR 297 z, Niels Bohr Library and Archives, American Institute of Physics, College Park, MD.

JRO J. Robert Oppenheimer papers, MSS35188, Library of Congress, Washington, DC.

KAB Keith A. Brueckner papers, MSS 0094, Mandeville Special Collections Library, University of California, San Diego.

KFFBI Klaus Fuchs FBI file, 65–58805, Federal Bureau of Investigation, Washington, DC.

KMSFD KMS Fusion documents, Documents Relating to the KMS Fusion Project, Inc., IH301, Niels Bohr Library and Archives, American Institute of Physics, College Park, MD.

LLNLA Lawrence Livermore National Laboratory Archives, Livermore, California. (Some documents available through OPENNET.)

LRG Leslie R. Groves papers, Record Group 200, NARACP.

LSP Leo Szilard papers, MSS 32, Special Collections & Archives, University of California, San Diego Library. Citations refer to folders in the digitized version of the collection, available online at: https://library.ucsd.edu/dc/collection/bb0752385q.

MEDR Manhattan Engineer District records, Records of the Army Corps of Engineers, Record Group 77, NARACP.

MDH *Manhattan District History*, ca. 1947–1948. This document has been partially declassified over the years, first as part of the *Manhattan Project: Official history and documents* microfilm collection in 1977 (see the entry under books), but in 2013 the Department of Energy began to release the remaining books (often heavily redacted), available online at https://www.osti.gov/opennet/manhattan_district. Contextual information (and a mirror of the files) can be found in Wellerstein, "General Groves' secret history."

NARACP	National Archives and Records Administration, Archives II, College Park, MD.
NARADC	National Archives and Records Administration, Archives I, Washington, DC.
NTA	Nuclear Testing Archive, Nevada Site Office of the U.S. Department of Energy/National Nuclear Security Administration, Las Vegas, NV. Documents are indexed on OPENNET, sometimes available online, sometimes a request is necessary.
OHCNBL	Oral History Collection, Niels Bohr Library and Archives, American Institute of Physics, College Park, MD.
OPENNET	U.S. Department of Energy, Office of Environment, Health, Safety and Security, OpenNet System. Online at https://www.osti.gov/opennet/.
OSAEC46	Office of the Secretary, General Correspondence, 1946–1951, Records of the Atomic Energy Commission, Record Group 326, NARACP.
OSAEC51	Office of the Secretary, General Correspondence, 1951–1958, Records of the Atomic Energy Commission, Record Group 326, NARACP.
OSAECM	Office of the Secretary, Minutes of the Meetings of the Atomic Energy Commission, Records of the Atomic Energy Commission, Record Group 326, NARACP.
RCT	Richard C. Tolman papers, 10105-MS, Caltech Institute Archives, Pasadena, CA.
RFB	Robert F. Bacher papers, Caltech Institute Archives, Pasadena, CA.
RFBAEC	Office files of Robert F. Bacher, Records of the Commissioners, Records of the Atomic Energy Commission, Record Group 326, NARACP.
RJCAE	Records of the Joint Committee on Atomic Energy, Record Group 128, NARADC.
SCAE	Records of the Special Committee on Atomic Energy, 79th Congress, Record Group 46, NARADC.
WWWAEC	Office files of William W. Waymack, Records of the Commissioners, Records of the Atomic Energy Commission, Record Group 326, NARACP.
VTNA	Vanderbilt Television News Archive, Vanderbilt University, online at http://tvnews.vanderbilt.edu.

Note that in some instances, specific documents have been retrieved from the aforementioned archives only by Freedom of Information Act (FOIA) requests by the author. These are indicated in the text with their FOIA identification number alongside the archival listing. When possible I have tried to refer to locations of documents that do not require FOIA requests.

Over time, some of the digital resources have, predictably, gone dead. In some cases, this was by deliberate action or neglect by the US government. In practically all cases,

I have maintained a backup copy of any documents cited here, and in some cases, have worked to make copies of entire databases. Should a scholar or other interested party find themselves unable to access the cited information, they should seek me out.

Congressional testimony, including formerly classified sessions of the Joint Committee on Atomic Energy, was obtained through the ProQuest Congressional Legislative & Executive Publications database if no other archival source is given (some Executive Session transcripts are contained only in the NARA Congressional records, and indicated as such). Newspaper accounts from the *Boston Globe*, *Chicago Tribune*, *Los Angeles Times*, *New York Times*, *Wall Street Journal*, and *Washington Post* were obtained through the ProQuest Historical Newspapers database. Some newspaper accounts (smaller newspapers, generally) were obtained through NewspaperArchive.com.

I have generally put only scholarly articles into the bibliography. Articles used primarily as primary sources alone have been cited fully in the footnote text.

In some books that rely on previously classified sources, historians have indicated the original classification markings on whatever documents they used. I have not done so here, both because I am not entirely sure what the value would be (in the few cases where the specific classification is important, I have noted it in the text), and because it is often difficult to know what the "original" classifications were, in any case. By the time I have seen a document, it has likely passed through the hands of several reviewers, at several different times in the past, each potentially changing classification categories (mostly an act of downgrading, but occasionally other activities, like "transclassification," have taken place, like retroactively adding the "Restricted Data" designation to documents produced prior to its legislative creation). In theory, any given document will tell you, in its array of stamps and signatures and other bureaucratic graffiti that adorns it, the history of its own declassification. In practice, this is very inconsistent.

ARTICLES

Anders, Roger M. "The Rosenberg case revisited: The Greenglass testimony and the protection of atomic secrets." *American Historical Review* 83, no. 2 (April 1978): 388–400.

Asada, Sado. "The shock of the atomic bomb and Japan's decision to surrender: A reconsideration." *Pacific Historical Review* 67, no. 4 (November 1998): 477–512.

Badash, Lawrence, Elizabeth Hodes, and Adolph Tiddens. "Nuclear fission: Reaction to the discovery of 1939." *Proceedings of the American Philosophical Society* 130, no. 2 (1986): 196–231.

Berger, Albert. "The *Astounding* investigation: The Manhattan Project's confrontation with science fiction." *Analog Science Fiction/Science Fact* 104, no. 9 (September 1984): 125–137.

Bernstein, Barton J. "The quest for security: American foreign policy and international control of atomic energy, 1942–1946." *Journal of American History* 60, no. 4 (March 1974): 1003–1044.

———. "Reconsidering the 'Atomic General': Leslie R. Groves." *Journal of Military History* 67, no. 3 (July 2003): 883–920.
———. "Roosevelt, Truman, and the atomic bomb, 1941–1945: A reinterpretation." *Political Science Quarterly* 90, no. 1 (Spring 1975): 23–69.
Bernstein, Jeremy. "John von Neumann and Klaus Fuchs: An unlikely collaboration." *Physics in Perspective* 12, no. 1 (March 2010): 36–50.
Boskey, Bennett. "Inventions and the atom." *Columbia Law Review* 50, no. 4 (April 1950): 433–477.
———. "Patents under the new Atomic Energy Act." *Journal of the Patent Office Society* 36, no. 12 (December 1954): 867–881.
Brueckner, Keith A. "A beginning for ICF by laser." In *Inertial confinement nuclear fusion: A historical approach by its pioneers*, edited by Guillermo Velarde and Natividad Carpintero Santamaría, 93–101. London: Foxwell & Davies, 2007.
Burr, William. "1960s 'Nth Country Experiment' Foreshadows Today's Concerns Over the Ease of Nuclear Proliferation," (1 July 2003). National Security Archive, George Washington University, online at https://nsarchive2.gwu.edu/news/20030701/.
———. "The atomic bomb and the end of World War II." *National Security Archive Electronic Briefing Book* no. 162 (5 August 2005, updated 4 August 2015), National Security Archive, George Washington University, online at https://nsarchive2.gwu.edu//NSAEBB/NSAEBB162/.
———. "The gas centrifuge secret: Origins of a U.S. policy of nuclear denial, 1954–1960," *National Security Archive Electronic Briefing Book* no. 518 (29 June 2015), National Security Archive, George Washington University, online at: https://nsarchive2.gwu.edu/nukevault/ebb518-the-gas-centrifuge-secret-origins-of-US-policy-of-nuclear-denial-1954-1960/.
———. "The 'labors of Atlas, Sisyphus, or Hercules'? US gas-centrifuge policy and diplomacy, 1954–60." *International History Review* 37, no. 3 (2015): 431–457.
Burr, William, Thomas S. Blanton, and Stephen I. Schwartz. "The costs and consequences of nuclear secrecy." In *Atomic audit: The costs and consequences of U.S. nuclear weapons since 1940*, edited by Stephen I. Schwartz, 433–483. Washington, DC: Brookings Institution Press, 1998.
Burton, Shirley J., Susan H. Karren, and Joseph D. Suster. "Following the paper trail west: Using archival sources for nuclear history." *Pacific Northwest Quarterly* 87, no. 1 (January 1994): 35–38.
Cheh, Mary M. "The *Progressive* case and the Atomic Energy Act: Waking to the dangers of government information controls." *George Washington Law Review* 48, no. 2 (January 1980): 163–210.
Chernus, Ira. "Operation Candor: Fear, faith, and flexibility." *Diplomatic History* 29, no. 5 (November 2005): 779–809.
Coleman, Earle E. "The 'Smyth Report': A descriptive check list." *Princeton University Library Chronicle* 37, no. 3 (1976): 219–230.
Craig, Campbell. "The atom bomb as policy maker: FDR and the road not taken." In

The Age of Hiroshima, edited by Michael D. Gordin and G. John Ikenberry, 19–33. Princeton, NJ: Princeton University Press, 2020.

Daniels, Mario. "Controlling knowledge, controlling people: Travel restrictions of U.S. scientists and national security." *Diplomatic History* 43, no. 1 (January 2019): 57–82.

De Geer, Lars-Erik. "The radioactive signature of the hydrogen bomb." *Science and Global Security* 2, no. 4 (1991): 351–363.

Dennis, Michael Aaron. "Secrecy and science revisited: From politics to historical practice and back." In *Secrecy and knowledge production*, edited by Judith Reppy, 1–16. Ithaca, NY: Cornell University Peace Studies: Occasional Paper 23, 1999.

Downey, Gary L. "Reproducing cultural identity in negotiating nuclear power: The Union of Concerned Scientists and emergency core cooling." *Social Studies of Science* 18, no. 2 (May 1988): 231–264.

Fox, John F., Jr. "Unique unto itself: The records of the Federal Bureau of Investigation 1908 to 1945." *Journal of Government Information* 30 (2004): 470–481.

Galison, Peter. "Removing knowledge." *Critical Inquiry* 31 (Autumn 2004): 229–243.

———. "Secrecy in three acts." *Social Research* 77, no. 3 (Fall 2010): 941–974.

Galison, Peter, and Barton Bernstein. "In any light: Scientists and the decision to build the Superbomb, 1942–1954." *Historical Studies in the Physical and Biological Sciences* 19, no. 1 (1988): 267–347.

Galison, Peter, and Robb Moss, directors. *Secrecy*. Cambridge, MA: Redacted Pictures, 2008. DVD.

Gieryn, Thomas F. "Boundaries of science." In *Handbook of science and technology studies*, edited by Sheila Jasanoff, 393–444. Thousand Oaks, CA: Sage Publications, 1995.

Goldberg, Stanley. "General Groves and the atomic West: The making and meaning of Hanford." In *The atomic West*, edited by Bruce Hevly and John Findlay, 39–89. Seattle: University of Washington Press, 1998.

———. "Inventing a climate of opinion: Vannevar Bush and the decision to build the bomb." *Isis* 83, no. 3 (September 1992): 429–452.

Goodman, Michael S. "The grandfather of the hydrogen bomb? Anglo-American intelligence and Klaus Fuchs." *Historical Studies in the Physical and Biological Sciences* 34, no. 1 (September 2004): 1–22.

———. "Who is trying to keep what secret from whom and why? MI5-FBI relations and the Fuchs Case." *Journal of Cold War Studies* 7, no. 3 (2005): 124–146.

Green, Harold P. "AEC information control regulations." *Bulletin of the Atomic Scientists* 24, no. 5 (May 1968): 41–43.

———. "'Born classified' in the AEC: A legal perspective." *Bulletin of the Atomic Scientists* 37, no. 10 (December 1981): 28–30.

———. "The AEC proposals—A threat to scientific freedom." *Bulletin of the Atomic Scientists* 23, no. 8 (October 1967): 15–17.

Gusterson, Hugh. "Death of the authors of death: Prestige and creativity among nuclear weapons scientists." In *Scientific authorship: Credit and intellectual prop-*

erty in science, edited by Mario Biagioli and Peter Galison, 281–307. New York: Routledge, 2003.

Hacker, Barton C. "Writing the history of a controversial program: Radiation safety, the AEC, and nuclear weapons testing." *Public Historian* 14, no. 1 (Winter 1992): 31–53.

Henderson, Kathryn. "Flexible sketches and inflexible data bases: Visual communication, conscription devices, and boundary objects in design engineering." *Science, Technology, and Human Values* 16, no. 4 (Autumn 1991): 448–473.

Herken, Gregg. "The University of California, the federal weapons labs, and the founding of the atomic West." In *The atomic West*, edited by Bruce Hevly and John Findlay, 119–135. Seattle: University of Washington Press, 1998.

Hewlett, Richard G. "'Born classified' in the AEC: A historian's view." *Bulletin of the Atomic Scientists* 37, no. 10 (December 1981): 20–27.

Hewlett, Richard G., and Jo Anne McCormick Quatannens. "Richard G. Hewlett: Federal historian." *Public Historian* 19, no. 1 (Winter 1997): 53–83.

Hirsch, Daniel, and William G. Mathews. "Who really gave away the secret?" *Bulletin of the Atomic Scientists* (January 1990): 23–30.

Hollinger, David A. "The defense of democracy and Robert K. Merton's formulation of the scientific ethos." In *Science, Jews, and secular culture: Studies in mid-twentieth-century American intellectual history*, 80–96. Princeton, NJ: Princeton University Press, 1996.

Hull, David. "Openness and secrecy in science: Their origins and limitations." *Science, Technology, and Human Values* 10, no. 2 (Spring 1985): 4–13.

Iliffe, Rob. "In the warehouse: Privacy, property, and priority in the early Royal Society." *History of Science* 30 (1992): 29–68.

Jasanoff, Sheila, and Sang-Hyun Kim. "Containing the atom: Sociotechnical imaginaries and nuclear power in the United States and South Korea." *Minerva* 47 (2009): 119–146.

Jenkins, John. "Atomic energy is 'moonshine': What did Rutherford really mean?" *Physics in Perspective* 13 (2011): 128–145.

Kaiser, David. "The atomic secret in Red hands? American suspicions of theoretical physicists during the early Cold War." *Representations* 90 (2005): 28–60.

Kemp, R. Scott. "The end of Manhattan: How the gas centrifuge changed the quest for nuclear weapons." *Technology and Culture* 53, no. 2 (April 2012): 272–305.

———. "The nonproliferation emperor has no clothes: The gas centrifuge, supply-side controls, and the future of nuclear proliferation." *International Security* 38, no. 4 (Spring 2014): 39–78.

———."Opening a proliferation Pandora's box: The spread of the Soviet-type gas centrifuge." *Nonproliferation Review* 24, no. 1–2 (2017): 101–127.

Kevles, Daniel J. "The National Science Foundation and the debate over postwar research policy, 1942–1945: A political reinterpretation of *Science–The Endless Frontier*." *Isis* 68, no. 1 (March 1977): 4–26.

Kidder, Ray E. "Laser fusion: The first ten years 1962–1972." In *Inertial confinement nuclear fusion: A historical approach by its pioneers*, edited by Guillermo Velarde and Natividad Carpintero Santamaría, 49–68. London: Foxwell & Davies, 2007.

———. "Weapons of mass destruction, national security, and a free press." *Cardozo Law Review* 26, no. 4 (2005): 1389–1399.

Kline, Ronald. "Construing 'technology' as 'applied science': Public rhetoric of scientists and engineers in the United States, 1880–1945." *Isis* 86, no. 2 (June 1995): 194–221.

Krige, John. "Atoms for Peace, scientific internationalism, and scientific intelligence." *Osiris* 21 (2006): 161–181.

———. "Hybrid knowledge: The transnational co-production of the gas centrifuge for uranium enrichment in the 1960s." *British Journal for the History of Science* 45, no. 3 (September 2012): 337–357.

———. "The proliferation risks of gas centrifuge enrichment at the dawn of the NPT." *Nonproliferation Review* 19, no. 2 (2012): 219–227.

———. "US technological superiority and the special nuclear relationship: Contrasting British and US policies for controlling the proliferation of gas-centrifuge enrichment." *International History Review* 36, no. 2 (2014): 230–251.

Lee, Sabing H. "Protecting the private inventor under peacetime provisions of the Invention Secrecy Act." *Berkeley Technology Law Journal* 12, no. 2 (1997): 345–411.

Lindl, John. "Development of the indirect-drive approach to inertial confinement fusion and target physics basis for ignition and gain." *Physics of Plasmas* 2, no. 11 (November 1995): 3933–4024.

MacKenzie, Donald, and Graham Spinardi. "Tacit knowledge, weapons design, and the uninvention of nuclear weapons." *American Journal of Sociology* 101, no. 1 (July 1995): 44–99.

Malloy, Sean. "'A very pleasant way to die': Radiation effects and the decision to use the atomic bomb against Japan." *Diplomatic History* 36, no. 3 (June 2012): 515–545.

Masco, Joseph. "Lie detectors: On secrets and hypersecurity at Los Alamos." *Public Culture* 14, no. 3 (2002): 441–467.

McCrory, Robert L. "Highlights of the history of the University of Rochester." In *Inertial confinement nuclear fusion: A historical approach by its pioneers*, edited by Guillermo Velarde and Natividad Carpintero Santamaría, 127–162. London: Foxwell & Davies, 2007.

Merton, Robert K. "The normative structure of science [1942]." In *The sociology of science: Theoretical and empirical investigations*, ch. 13. Chicago: University of Chicago Press, 1973.

Mink, Patsy T. "The Cannikin Papers: A case study in freedom of information." In *Secrecy and foreign policy*, edited by Thomas M. Franck and Edward Weisband, 114–131. New York: Oxford University Press, 1974.

Montgomery, Alexander H. "Stop helping me: When nuclear assistance impedes nuclear programs." In *The nuclear renaissance and international security*, edited by Adam N. Stulberg and Matthew Fuhrmann, ch. 7. Stanford, CA: Stanford University Press, 2013.

Morland, Howard. "The Holocaust bomb: A question of time." (15 November 1999, revised 5 February 2003), http://www.fas.org/sgp/eprint/morland.html, accessed 23 July 2010.

Moss, William, and Roger Eckhardt. "The human plutonium injection experiments." *Los Alamos Science* no. 23 (1995): 177–233.

Norris, Robert S., and Hans Kristensen. "Counting nuclear weapons in the public interest." *Bulletin of the Atomic Scientists* 71, no. 1 (2015): 85–90.

Nuckolls, John H. "Contributions to the genesis and progress of ICF." In *Inertial confinement nuclear fusion: A historical approach by its pioneers*, edited by Guillermo Velarde and Natividad Carpintero Santamaría, 1–48. London: Foxwell & Davies, 2007.

Owens, Larry. "The counterproductive management of science in the Second World War: Vannevar Bush and the Office of Scientific Research and Development." *Business History Review* 68 (1994): 515–576.

Pearson, J. Michael. "On the belated discovery of fission." *Physics Today* 68, no. 6 (June 2015): 40–45.

Polenberg, Richard. "The fortunate fox." In *Reappraising Oppenheimer: Centennial studies and reflections*, edited by Cathryn Carson and David A. Hollinger, 267–272. Berkeley: Office for History of Science and Technology, University of California, Berkeley, 2005.

Pollack, Joshua. "The secret treachery of A. Q. Khan." *Playboy* (January/February 2012). Online at: http://carnegieendowment.org/files/the_secret%20treachery%20of%20aq%20khan.pdf.

Popp, Manfred. "Misinterpreted documents and ignored physical facts: The history of 'Hitler's atomic bomb' needs to be corrected." *Ber. Wissenschaftsgesch* 39 (2016): 265–282.

Pozen, David E. "The mosaic theory, national security, and the Freedom of Information Act." *Yale Law Journal* 115, no. 3 (December 2005): 628–679.

Rankin, William. "The 'person skilled in the art' is really quite conventional: U.S. patent drawings and the persona of the inventor, 1870–2005." In *Making and unmaking intellectual property: Creative production in legal and cultural perspective*, edited by Mario Biagioli, Peter Jaszi, and Martha Woodmansee, 55–75. Chicago: University of Chicago Press, 2011.

Reed, Bruce Cameron. "Centrifugation during the Manhattan Project." *Physics in Perspective* 11 (2009): 426–441.

Relyea, Harold C. "The evolution of government information security—Classification policy: A brief overview (1775–1973)." Included as Appendix 3 in "Security Classification Reform (H.R. 12004)," Hearings before a House Subcommittee of the Committee Government Operations, 93rd Congress, 2nd Session (11 July 1974, 25 July 1974, and 1 August 1974): 505–597.

———. "Freedom of information, privacy, and official secrecy: The evolution of federal government information policy concepts." *Social Indicators Research* 7, no. 1/4 (January 1980): 137–156.

Roland, Alex. "Science and war." *Osiris* 1 (1985): 247–272.

Samuels, David. "Atomic John." *New Yorker* (15 December 2008): 50–63.

Schrafstetter, Susanna, and Stephen Twigge. "Spinning into Europe: Britain, West Germany and the Netherlands: Uranium enrichment and the development of the gas centrifuge 1964–1970." *Contemporary European History* 11, no. 2 (May 2002): 253–272.

Schuessler, John M. "The deception dividend: FDR's undeclared war." *International Security* 34, no. 4 (2010): 133–165.

Segrè, Emilio G. "The discovery of fission." *Physics Today* 42, no. 7 (July 1989): 38–43.

Seidel, Robert W. "Secret scientific communities: Classification and scientific communication in the DOE and DoD." In *Proceedings of the 1998 Conference on the History and Heritage of Science Information Systems*, edited by Mary Ellen Bowden et al., 46–60. Medford, NJ: Information Today, 1999.

Simmel, Georg. "The sociology of secrecy and of secret societies." *American Journal of Sociology* 11, no. 4 (January 1906): 441–498.

Smith, Datus C. "The publishing history of the 'Smyth Report.'" *Princeton University Library Chronicle* 37, no. 3 (Spring 1976): 190–203.

Smyth, Henry D. "The Smyth Report." *Princeton University Library Chronicle* 37, no. 3 (Spring 1976): 173–189.

Sobota, Lenore. "The unexploded bomb: The *Progressive* and prior restraint." *Southern Illinois University Law Review* 5 (1980): 199–233.

Stober, Dan. "No experience necessary." *Bulletin of the Atomic Scientists* (March/April 2003): 12.

Streefland, Abel. "Putting the lid on the gas centrifuge: Classification of the Dutch ultracentrifuge project, 1960–1961." In *Cold War science and the transatlantic circulation of knowledge*, edited by Jeroen van Dongen, 77–100. Boston: Brill, 2015.

Theoharis, Athan G. "The Freedom of Information Act versus the FBI." In *A culture of secrecy: The government versus the people's right to know*, edited by Athan G. Theoharis, ch. 1. Lawrence: University of Kansas Press, 1998.

"Transcript of *Weapons of Mass Destruction, National Security, and a Free Press: Seminal Issues as Viewed through the Lens of The Progressive Case*," *Cardozo Law Review* 26, no. 4 (2005): 1337–1388.

Tribe, Laurence H., and David H. Remes. "Some reflections on the *Progressive* case: Publish *and* perish?" *Bulletin of the Atomic Scientists* 36, no. 3 (March 1980): 20–24.

Turchetti, Simone. "'For slow neutrons, slow pay': Enrico Fermi's patent and the U.S. atomic energy program, 1938–1953." *Isis* 97, no. 1 (2006): 1–27.

———. "The invisible businessman: Nuclear physics, patenting practices, and trading activities in the 1930s." *Historical Studies in the Physical and Biological Sciences* 37, no. 1 (2006): 153–172.

Twigge, Stephen. "A baffling experience: Technology transfer, Anglo-American nuclear relations, and the development of the gas centrifuge 1964–70." *History and Technology* 19, no. 2 (2003): 151–163.

Vermeir, Koen, and Daniel Margocsy. "States of secrecy: An introduction." *British Journal for the History of Science* 45, no. 2 (June 2012): 153–164.

Walker, J. Samuel. "Regulating against nuclear terrorism: The domestic safeguards issue, 1970–1979." *Technology and Culture* 42, no. 1 (January 2001): 107–132.

Walterscheid, Edward C. "The need for a uniform government patent policy: The D.O.E. example." *Harvard Journal of Law and Technology* 3 (1990): 103–166.

Washburn, Patrick S. "The Office of Censorship's attempt to control press coverage of the atomic bomb during World War II." *Journalism Monographs* 120 (1990): 1–43.

Weart, Spencer. "Scientists with a secret." *Physics Today* 28, no. 2 (February 1976): 23–30.

Weber, Max. "Bureaucracy [1920]." In *Economy and society: An outline of interpretive sociology*, edited by Guenther Roth and Claus Wittich, 956–1005. Berkeley: University of California Press, 1978.

Wellerstein, Alex. "Building an H-Bomb in plain sight." *Atlantic* (5 September 2017). Online at: https://www.theatlantic.com/science/archive/2017/09/the-credible-visible-h-bomb/538863/.

———. "General Groves' secret history." *Restricted data: The nuclear secrecy blog* (5 September 2014). Online at: http://blog.nuclearsecrecy.com/2014/09/05/general-groves-secret-history/.

———. "John Wheeler's H-bomb blues." *Physics Today* 72, no. 4 (2019): 42–51.

———. "The Kyoto misconception: What Truman knew, and didn't know, about Hiroshima." In *The Age of Hiroshima*, edited by Michael D. Gordin and G. John Ikenberry, 34–55. Princeton, NJ: Princeton University Press, 2020.

———. "Oppenheimer, unredacted: Part I—Finding the lost transcripts." *Restricted data: The nuclear secrecy blog* (9 January 2015). Online at: http://blog.nuclearsecrecy.com/2015/01/09/oppenheimer-unredacted-part-i/.

———. "Patenting the bomb: Nuclear weapons, intellectual property, and technological control." *Isis* 99, no. 1 (March 2008): 57–87.

———."A tale of openness and secrecy: The Philadelphia story." *Physics Today* 65, no. 5 (2012): 47–53.

———. "The virtues of nuclear ignorance." *New Yorker* (20 September 2016). Online at: http://www.newyorker.com/tech/elements/the-virtues-of-nuclear-ignorance.

Wellerstein, Alex, and Edward Geist. "The secret of the Soviet H-bomb." *Physics Today* 70, no. 4 (2017): 40–47.

Westwick, Peter. "In the beginning: The origin of nuclear secrecy." *Bulletin of the Atomic Scientists* 56 (November/December 2000): 43–49.

———. "Secret science: A classified community in the national laboratories." *Minerva* 38, no. 4 (2000): 363–391.

Wood, Houston G. "The history of the gas centrifuge and its role in nuclear proliferation." Presentation given at the Woodrow Wilson Center, Washington, DC, 20 January 2010. Slides online at: https://www.wilsoncenter.org/sites/default/files/2010_01_houston_wood_history_gas_centrifuge_role_nuclear_proliferation.pdf, accessed 11 December 2018.

Wood, Houston G., Alexander Glaser, and R. Scott Kemp. "The gas centrifuge and nuclear weapons proliferation." *Physics Today* 61, no. 9 (September 2008): 40–45.

Zenko, Micah. "Intelligence estimates of nuclear terrorism." *Annals of the American Academy of Political and Social Science* 607 (September 2006): 87–102.

BOOKS AND MONOGRAPHS

Abraham, Itty. *The making of the Indian atomic bomb: Science, secrecy and the post-colonial state*. London: Zed Books, 1998.

Albright, Joseph, and Marcia Kunstel. *Bombshell: The secret story of America's unknown atomic spy conspiracy*. New York: Times Books, 1997.

Alperovitz, Gar. *The decision to use the bomb and the architecture of an American myth*. New York: Knopf, 1995.

Anders, Roger M., ed. *Forging the atomic shield: Excerpts from the office diary of Gordon E. Dean*. Chapel Hill: University of North Carolina Press, 1987.

Arkin, William. *Nuclear battlefields: Global links in the arms race*. Cambridge, MA: Ballinger Press, 1985.

Badash, Lawrence, Joseph O. Hirschfelder, and Herbert P. Broida, eds. *Reminiscences of Los Alamos, 1943–1945*. Boston: Reidel, 1980.

Balogh, Brian. *Chain reaction: Expert debate and public participation in American commercial nuclear power, 1945–1975*. New York: Cambridge University Press, 1991.

Barker, Holly M. *Bravo for the Marshallese: Regaining control in a post-nuclear, post-colonial world*. Belmont, CA: Thomson, 2004.

Barnhart, Megan Kathleen. "'To secure the benefits of science to the general welfare': The scientists' movement and the American public during the Cold War, 1945–1960." PhD diss., University of California, Los Angeles, 2007.

Bell, Griffin B., with Ronald J. Ostrow. *Taking care of the law*. New York: Morrow, 1986.

Benintende, Emma. "Who was the Scientific American? Science, identity, and politics through the lens of a Cold War periodical." Senior thesis, Harvard University, Department of the History of Science, 2011.

Bernal, J. D. *The social function of science*. London: Routledge, 1939.

Bernstein, Carl, and Bob Woodward. *All the President's men*. New York: Simon and Schuster, 1974.

Bernstein, Jeremy. *Hitler's uranium club: The secret recordings at Farm Hall*. New York: Springer Verlag, 2001.

———. *Nuclear weapons: What you need to know*. New York: Cambridge University Press, 2008.

Biagioli, Mario. *Galileo's instruments of credit: Telescopes, images, secrecy*. Chicago: University of Chicago Press, 2006.

Biagioli, Mario, and Peter Galison, eds. *Scientific authorship: Credit and intellectual property in science*. New York: Routledge, 2003.

Bird, Kai, and Martin Sherwin. *American Prometheus: The triumph and the tragedy of J. Robert Oppenheimer*. New York: Knopf, 2005.

Bok, Sissela. *Secrets: On the ethics of concealment and revelation*. New York: Pantheon Books, 1982.
Boyer, Paul. *By the bomb's early light: American thought and culture at the dawn of the Atomic Age*. New York: Pantheon, 1985.
Braw, Monica. *The atomic bomb suppressed: American censorship in occupied Japan*. New York: M. E. Sharpe, 1991.
Broad, William J. *Teller's war: The top-secret story behind the Star Wars deception*. New York: Simon and Schuster, 1992.
Bromberg, Joan L. *Fusion: Science, politics, and the invention of a new energy source*. Cambridge, MA: MIT Press, 1982.
———. *The laser in America, 1950–1970*. Cambridge, MA: MIT Press, 1991.
Brown, Kate. *Plutopia: Nuclear families, atomic cities, and the great Soviet and American plutonium disasters*. New York: Oxford University Press, 2013.
Browne, Janet. *Charles Darwin*. Vol. 2, *The power of place*. Princeton, NJ: Princeton University Press, 2002.
Bush, Vannevar. *Pieces of the action*. New York: Morrow, 1970.
Campos, Luis A. *Radium and the secret of life*. Chicago: University of Chicago Press, 2015.
Carson, Cathryn, and David A. Hollinger, eds. *Reappraising Oppenheimer: Centennial studies and reflections*. Berkeley: Office for History of Science and Technology, University of California, Berkeley, 2005.
Cassidy, David C. *J. Robert Oppenheimer and the American century*. New York: Pi Press, 2005.
Certeau, Michel de. *The practice of everyday life*. Berkeley: University of California Press, 1984.
Close, Frank. *Half-life: The divided life of Bruno Pontecorvo, physicist or spy*. New York: Basic Books, 2015.
———. *Trinity: The treachery and pursuit of the most dangerous spy in history*. London: Allen Lane, 2020.
Cochran, Thomas B., William M. Arkin, and Milton M. Hoenig. *Nuclear weapons databook*. Vol. 1, *US nuclear forces and capabilities*. Cambridge, MA: Ballinger, 1984.
Cohen, Avner. *The worst-kept secret: Israel's bargain with the bomb*. New York: Columbia University Press, 2012.
Compton, Arthur H. *Atomic quest: A personal narrative*. New York: Oxford University Press, 1956.
Coster-Mullen, John. *Atom bombs: The Top Secret inside story of Little Boy and Fat Man*. Self-published MS. 2002, rev. 2007 and 2013.
Creager, Angela. *Life atomic: A history of radioisotopes in science and medicine*. Chicago: University of Chicago Press, 2013.
Cronin, Brian. *Was Superman a spy?* New York: Plume, 2009.
Curry, Richard O., ed. *Freedom at risk: Secrecy, censorship, and repression in the 1980s*. Philadephia, PA: Temple University Press, 1988.

Day, Samuel H., Jr. *Crossing the line: From editor to activist to inmate—a writer's journey*. Baltimore: Fortkamp Publishing Co., 1991.

De Volpi, Alexander, Gerald E. Marsh, Theodore A. Postol, and George S. Stanford. *Born secret: The H-bomb, the* Progressive *case and national security*. New York: Pergamon Press, 1981.

Dupree, A. Hunter. *Science in the federal government: A history of policies and activities to 1940*. Cambridge, MA: Harvard University Press, 1957.

Dyson, Freeman. *Disturbing the universe*. New York: Harper and Row, 1981.

Eamon, William. *Science and the secrets of nature: Books of secrets in medieval and early modern culture*. Princeton, NJ: Princeton University Press, 1994.

Edwards, Paul N. *The closed world: Computers and the politics of discourse in Cold War America*. Cambridge, MA: MIT Press, 1996.

Eliot, George Fielding. *The H bomb*. New York: Didier, 1950.

Ellsberg, Daniel. *The doomsday machine: Confessions of a nuclear war planner*. New York: Bloomsbury, 2017.

———. *Secrets: A memoir of Vietnam and the Pentagon Papers*. New York: Viking, 2002.

Ezrahi, Yaron. *The descent of Icarus: Science and the transformation of contemporary democracy*. Cambridge, MA: Harvard University Press, 1990.

Feynman, Richard P. *"Surely you're joking, Mr. Feynman!" Adventures of a curious character*. New York: W. W. Norton, 1985.

Fitzpatrick, Anne. "Igniting the light elements: The Los Alamos thermonuclear weapon project, 1942–1952." PhD diss., George Washington University, 1999.

Formato, Megan. "Writing the atom: Niels and Margrethe Bohr and the construction of quantum theory." PhD diss., Harvard University, 2016.

Foucault, Michel. *Discipline and punish: The birth of the prison*. New York: Vintage Books, 1979.

Ford, Kenneth W. *Building the H-bomb: A personal history*. Hackensack, NJ: World Scientific, 2015.

Freedman, Lawrence. *The evolution of nuclear strategy*. London: MacMillan Press, 1982.

Fuhrmann, Matthew. *Atomic assistance: How "atoms for peace" programs cause nuclear insecurity*. Ithaca, NY: Cornell University Press, 2012.

Galison, Peter. *Image and logic: A material culture of microphysics*. Chicago: University of Chicago Press, 1997.

Goodchild, Peter. *Edward Teller: The real Dr. Strangelove*. Cambridge, MA: Harvard University Press, 2004.

Goodman, Amy, and David Goodman. *The exception to the rulers: Exposing oily politicians, war profiteers, and the media that love them*. New York: Hyperion, 2004.

Gordin, Michael. *Five days in August: How World War II became a nuclear war*. Princeton, NJ: Princeton University Press, 2007.

———. *Red cloud at dawn: Truman, Stalin, and the end of the atomic monopoly*. New York: Farrar, Straus and Giroux, 2009.

Goudsmit, Samuel. *Alsos*.Woodbury, NY: AIP Press, 1996 [1947].

Gowing, Margaret. *Britain and atomic energy, 1939–1945*. New York: St. Martin's, 1964.

Greenspan, Nancy T. *Atomic spy: The dark lives of Klaus Fuchs*. New York: Viking, 2020.

Griese, Noel L. *Arthur W. Page: Publisher, public relations pioneer, patriot*. Atlanta: Anvil Publishers, 2001.

Groueff, Stéphane. *Manhattan Project: The untold story of the making of the atomic bomb*. Boston: Little, Brown, 1967.

Groves, Leslie R. *Now it can be told: The story of the Manhattan Project*. New York: Harper, 1962.

Grunden, Walter E. *Secret weapons and World War II: Japan in the shadow of big science*. Lawrence: University Press of Kansas, 2005.

Gup, Ted. *Nation of secrets: The threat to democracy and the American way of life*. New York: Doubleday, 2007.

Gusterson, Hugh. *Nuclear rites: A weapons laboratory at the end of the Cold War*. Berkeley: University of California Press, 1996.

———. *People of the bomb: Portraits of America's nuclear complex*. Minneapolis: University of Minnesota Press, 2004.

Hacker, Barton. *The dragon's tail: Radiation safety in the Manhattan Project, 1942–1946*. Berkeley: University of California Press, 1987.

———. *Elements of controversy: The Atomic Energy Commission and radiation safety in nuclear weapons testing, 1947–1974*. Berkeley: University of California Press, 1994.

Hales, Peter B. *Atomic spaces: Living on the Manhattan Project*. Urbana: University of Illinois Press, 1997.

Halperin, Morton H., and Daniel N. Hoffman. *Top secret: National security and the right to know*. Washington, DC: National Republic Books, 1977.

Hansen, Chuck. *The swords of Armageddon: U.S. nuclear weapons development since 1945*. CD-ROM. Sunnyvale, CA: Chukelea Publications, 1995, updated (version 2) 2007.

———. *U.S. nuclear weapons: The secret history*. Arlington, TX: Aerofax, 1988.

Hargittai, István. *Judging Edward Teller: A closer look at one of the most influential scientists of the twentieth century*. Amherst, NY: Prometheus Books, 2010.

Hasegawa, Tsuyoshi. *Racing the enemy: Stalin, Truman, and the surrender of Japan*. Cambridge, MA: Harvard University Press, 2005.

Haynes, John Earl. *Venona: Decoding Soviet espionage in America*. New Haven, CT: Yale University Press, 1999.

Haynes, John Earl, Harvey Klehr, and Alexander Vassiliev. *Spies: The rise and fall of the KGB in America*. New Haven, CT: Yale University Press, 2009.

Hecht, David K. *Storytelling and science: Rewriting Oppenheimer in the nuclear age*. Boston: University of Massachusetts Press, 2015.

Heilbron, J. L., and Robert W. Seidel. *Lawrence and his laboratory: A history of the Lawrence Berkeley Laboratory*. Berkeley: University of California Press, 1989.

Helmreich, Jonathan E. *Gathering rare ores: The diplomacy of uranium acquisition, 1943–1954*. Princeton, NJ: Princeton University Press, 1986.

Herken, Gregg. *Brotherhood of the bomb: The tangled lives and loyalties of Robert Oppenheimer, Ernest Lawrence, and Edward Teller*. New York: Henry Holt, 2002.

———. *The winning weapon: The atomic bomb in the Cold War, 1945–1950*. New York: Knopf, 1980.

Hersey, John. *Hiroshima*. New York: Knopf, 1946.

Hershberg, James. *James B. Conant: Harvard to Hiroshima and the making of the nuclear age*. New York: Knopf, 1993.

Hewlett, Richard G., and Oscar E. Anderson Jr. *The new world, 1939–1946*. University Park: Pennsylvania State University Press, 1962.

Hewlett, Richard G., and Francis Duncan. *Atomic shield, 1947–1952*. Washington, DC: US Atomic Energy Commission, 1969.

Hewlett, Richard G., and Jack M. Holl. *Atoms for peace and war, 1953–1961: Eisenhower and the Atomic Energy Commission*. Berkeley: University of California Press, 1989.

Higuchi, Toshihiro. *Political fallout: Nuclear weapons testing and the making of a global environmental crisis*. Stanford, CA: Stanford University Press, 2020.

Hiltzik, Michael A. *Big science: Ernest Lawrence and the invention that launched the military-industrial complex*. New York: Simon and Schuster, 2015.

Hines, Neal O. *Proving ground: An account of the radiobiological studies in the Pacific, 1946–1961*. Seattle: University of Washington Press, 1962.

Hoddeson, Lillian, Paul W. Henriksen, Roger A. Meade, and Catherine Westfall. *Critical assembly: A technical history of Los Alamos during the Oppenheimer years*. New York: Cambridge University Press, 1993.

Holloway, David. *Stalin and the bomb: The Soviet Union and atomic energy, 1939–1956*. New Haven, CT: Yale University Press, 1994.

Hornblum, Allen M. *The invisible Harry Gold: The man who gave the Soviets the atomic bomb*. New Haven, CT: Yale University Press, 2010.

Howes, Ruth and Caroline L. Herzenberg. *Their day in the sun: Women of the Manhattan Project*. Philadelphia, PA: Temple University Press, 1999.

Hunner, Jon. *Inventing Los Alamos: The growth of an atomic community*. Norman: University of Oklahoma Press, 2004.

Hymans, Jacques. *Achieving nuclear ambitions: Scientists, politicians, and proliferation*. New York: Cambridge University Press, 2012.

Isaacson, Walter. *Einstein: His life and universe*. New York: Simon and Schuster, 2007.

Johnson, David K. *The lavender scare: The Cold War persecution of gays and lesbians in the federal government*. Chicago: University of Chicago Press, 2004.

Jones, Vincent C. *Manhattan: The Army and the atomic bomb*. Washington, DC: US Government Printing Office, 1985.

Kaufman, Scott. *Project Plowshare: The peaceful use of nuclear explosives in Cold War America*. Ithaca, NY: Cornell University Press, 2012.

Keever, Beverly Deepe. *News zero: The* New York Times *and the bomb*. Monroe, ME: Common Courage Press, 2004.

Kevles, Daniel. *The physicists: The history of a scientific community in modern America*. Cambridge, MA: Harvard University Press, 1987.

Kiernan, Denise. *The girls of Atomic City: The untold story of the women who helped win World War II*. New York: Simon and Schuster, 2013.

King, Dennis. *Lyndon LaRouche and the new American fascism*. New York: Doubleday, 1989.

Kofsky, Frank. *Harry S. Truman and the war scare of 1948: A successful campaign to deceive the nation*. New York: St. Martin's Press, 1995.

Kojevnikov, Alexei. *Stalin's great science: The times and adventures of Soviet physicists*. London: Imperial College Press, 2004.

Kragh, Helge. *Quantum generations: A history of physics in the twentieth century*. Princeton, NJ: Princeton University Press, 1999.

Krige, John. *Sharing knowledge, shaping Europe: US technological collaboration and nonproliferation*. Cambridge, MA: MIT Press, 2016.

Kroenig, Matthew. *Exporting the bomb: Technology transfer and the spread of nuclear weapons*. Ithaca, NY: Cornell University Press, 2010.

Kutler, Stanley I., ed. *Watergate: A brief history with documents*. Malden, MA: Wiley-Blackwell, 2010.

Lanouette, William. *Genius in the shadows: A biography of Leo Szilard, the man behind the bomb*. New York: Skyhorse Publishing, 2013.

Lapp, Ralph E. *The voyage of the Lucky Dragon*. New York: Harper, 1958.

Larabee, Ann. *The wrong hands: Popular weapons manuals and their historic challenges to a democratic society*. New York: Oxford University Press, 2015.

Laurence, William L. *Dawn over zero: The story of the atomic bomb*. New York: Knopf, 1947.

———. *The reminiscences of William L. Laurence*. Microform. Glen Rock, NJ: Microfilming Corporation of America, 1972.

Lenoir, Timothy. *Instituting science: The cultural production of scientific disciplines*. Stanford, CA: Stanford University Press, 1997.

Leslie, Stuart W. *The Cold War and American science: The military-industrial complex at MIT and Stanford*. New York: Columbia University Press, 1993.

Lewis, John Wilson, and Xue Litai. *China builds the bomb*. Stanford, CA: Stanford University Press, 1988.

Lifton, Robert Jay, and Greg Mitchell. *Hiroshima in America: Fifty years of denial*. New York: Putnam's Sons, 1995.

Lilienthal, David E. *Change, hope, and the bomb*. Princeton, NJ: Princeton University Press, 1963.

———. *The journals of David E. Lilienthal*. Vol. 2, *The atomic energy years, 1945–1950*. New York: Harper & Row, 1964.

Lilienthal, David E., Chester I. Barnard, J. Robert Oppenheimer, Charles A. Thomas, and Harry A. Winne. *A report on the international control of atomic energy*. Washington, DC: U.S. Government Printing Office, 1946.

Lindee, M. Susan. *Suffering made real: American science and the survivors at Hiroshima*. Chicago: University of Chicago Press, 1994.

Long, Pamela. *Openness, secrecy, authorship: Technical arts and the culture of knowledge from Antiquity to the Renaissance*. Baltimore: Johns Hopkins University Press, 2001.

Malloy, Sean. *Atomic tragedy: Henry L. Stimson and the decision to use the bomb against Japan*. Ithaca, NY: Cornell University Press, 2008.

Manhattan Project: Official history and documents. Microform. Washington, DC: University Publications of America, 1977.

McCullough, David. *Truman*. New York: Simon and Schuster, 1992.

McMillan, Priscilla J. *The ruin of J. Robert Oppenheimer and the birth of the modern arms race*. New York: Viking, 2005.

McPhee, John. *The curve of binding energy*. New York: Ballantine Books, 1974.

Miller, Richard L. *Under the cloud: The decades of nuclear testing*. New York: Free Press, 1986.

Mills, James. *The seventh power*. New York: E. P. Dutton, 1976.

Mitchell, Greg. *The beginning or the end: How Hollywood—and America—learned to stop worrying and love the bomb*. New York: The New Press, 2020.

Morland, Howard. *The secret that exploded*. New York: Random House, 1981.

Moynihan, Daniel P. *Secrecy: The American experience*. New Haven, CT: Yale University Press, 1998.

Mullet, Shawn. "Little man: Four junior physicists and the Red scare experience." PhD diss., Harvard University, 2008.

Neuse, Steven M. *David E. Lilienthal: The journey of an American liberal*. Knoxville: University of Tennessee Press, 1996.

Newman, James R., and Byron S. Miller. *The control of atomic energy: A study of its social, economic, and political implications*. New York: McGraw Hill, 1948.

Newman, William R. *Atoms and alchemy: Chymistry and the experimental origins of the Scientific Revolution*. Chicago: University of Chicago Press, 2006.

Nieburg, Harold. *Nuclear secrecy and foreign policy*. Washington, DC: Public Affairs Press, 1964.

Norris, Robert S. *Racing for the bomb: General Leslie R. Groves, the Manhattan Project's indispensable man*. South Royalton, VT: Steerforth Press, 2002.

Norris, Robert S., A. S. Burrows, and R. W. Fieldhouse. *Nuclear weapons databook*. Vol. 5, *British, French, and Chinese nuclear weapons*. Boulder, CO: Westfield Press, 1994.

Paul, Septimus H. *Nuclear rivals: Anglo-American atomic relations, 1941–1952*. Columbus: Ohio State University Press, 2000.

Perkovich, George. *India's nuclear bomb: The impact on global proliferation*. Berkeley: University of California Press, 1999.

Phillips, John A., and David Michaelis. *Mushroom: The story of the A-bomb kid.* New York: William Morrow, 1978.

Piel, Gerard. *Science in the cause of man.* 2nd ed. New York: Knopf, 1962.

Plokhy, Serhii. *Chernobyl: The history of nuclear catastrophe.* New York: Basic Books, 2018.

Polenberg, Richard, ed. *In the matter of J. Robert Oppenheimer: The security clearance hearing.* Ithaca, NY: Cornell University Press, 2002.

Priest, Dana, and William Arkin. *Top secret America: The rise of the new American security state.* New York: Little, Brown, 2011.

Quist, Arvin S. *Security classification of information*, Vol. 1, *Introduction, history, and adverse impacts.* Oak Ridge, TN: Oak Ridge Classification Associates, 2002. Online at: https://fas.org/sgp/library/quist/.

———. *Security classification of information*, Vol. 2, *Principles for classification of information.* Oak Ridge, TN: Oak Ridge K-25 Site, 1993. Online at: https://fas.org/sgp/library/quist2/.

Radosh, Ronald. *The Rosenberg file.* 2nd ed. New Haven, CT: Yale University Press, 1997.

Rhodes, Richard. *Dark sun: The making of the hydrogen bomb.* New York: Simon and Schuster, 1995.

———. *The making of the atomic bomb.* New York: Simon and Schuster, 1986.

Richelson, Jeffrey T. *Spying on the bomb: American nuclear intelligence from Nazi Germany to Iran and North Korea.* New York: W. W. Norton, 2006.

Riddle, Donald H. *The Truman committee: A study in congressional responsibility.* New Brunswick, NJ: Rutgers University Press, 1964.

Roberts, Sam. *The brother: The untold story of atomic spy David Greenglass and how he sent his sister, Ethel Rosenberg, to the electric chair.* New York: Random House, 2001.

Rose, Lisle A., and Neal H. Petersen, eds. *Foreign relations of the United States, 1952–1954, National security affairs*, Vol. 2, Part 2. Washington, DC: United States Government Printing Office, 1984.

Rourke, Francis E. *Secrecy and publicity: Dilemmas of democracy.* Baltimore: Johns Hopkins Press, 1961.

Sanger, David E. *The perfect weapon: War, sabotage, and fear in the Cyber Age.* New York: Crown Publishers, 2018.

Schneier, Bruce. *Secrets and lies: Digital security in a networked world.* New York: John Wiley, 2000.

Schoenfeld, Gabriel. *Necessary secrets: National security, the media, and the rule of law.* New York: W. W. Norton, 2010.

Schwartz, Rebecca Press. "The making of the history of the atomic bomb: Henry DeWolf Smyth and the historiography of the Manhattan Project." PhD diss., Princeton University, 2008.

Schwartz, Stephen, ed. *Atomic audit: The costs and consequences of U.S. nuclear weapons since 1940.* Washington, DC: Brookings Institution Press, 1998.

Seaborg, Glenn T. *Journal of Glenn T. Seaborg, Vol. 2, 1946–1958*. Berkeley: Lawrence Berkeley Laboratory, University of California, 1990.

Seife, Charles. *Sun in a bottle: The strange history of fusion and the science of wishful thinking*. New York: Viking, 2008.

Serber, Robert. *The Los Alamos Primer: The first lectures on how to build an atomic bomb*. Berkeley: University of California Press, 1992.

Serber, Robert, and Robert P. Crease. *Peace and war: Reminsciences of a life on the frontiers of science*. New York: Columbia University Press, 1998.

Shapin, Stephen. *Never pure: Historical studies of science as if it was produced by people with bodies, situated in time, space, culture, and society, and struggling for credibility and authority*. Baltimore: Johns Hopkins University Press, 2010.

Shattuck, Roger. *Forbidden knowledge: From Prometheus to pornography*. New York: St. Martin's Press, 1996.

Sherwin, Martin J. *A world destroyed: Hiroshima and the origins of the arms race*. New York: Vintage, 1987 [1975].

Shils, Edward. *The torment of secrecy: The background and consequences of American security policies*. Glencoe, IL: Free Press, 1956.

Shurcliff, William A. *William A. Shurcliff: A brief autobiography*. Unpublished MS. Cambridge, MA: 15 December 1992. Copy in Houghton Library, Harvard University.

Siegel, Barry. *Claim of privilege: A mysterious plane crash, a landmark Supreme Court case, and the rise of state secrets*. New York: HarperCollins, 2008.

Sime, Ruth Lewin. *Lise Meitner: A life in physics*. Berkeley: University of California Press, 1997.

Smith, Alice Kimball. *A peril and a hope: The scientists' movement in America, 1945–47*. Chicago: University of Chicago Press, 1965.

Smith, Alice Kimball, and Charles Weiner, eds. *Robert Oppenheimer, letters and recollections*. Stanford, CA: Stanford University Press, 1995.

Smith, P. D. *Doomsday men: The real Dr Strangelove and the dream of the superweapon*. London: Allen Lane, 2007.

Smyth, Henry D. *Atomic energy for military purposes: The official report on the development of the atomic bomb under the auspices of the United States government, 1940–1945*. Princeton, NJ: Princeton University Press, 1945.

Solzhenitsyn, Aleksandr I. *The Gulag Archipelago 1918–1956: An experiment in literary investigation*, Vol. 1–2. Translated by Thomas P. Whitney. New York: Harper and Row, 1974.

Stephens, William E., ed. *Nuclear fission and atomic energy*. Lancaster, PA: Science Press, 1948.

Stewart, Irvin. *Organizing scientific research for war: The administrative history of the Office of Scientific Research and Development*. Boston: Little, Brown, 1948.

Stober, Dan, and Ian Hoffman. *A convenient spy: Wen Ho Lee and the politics of nuclear espionage*. New York: Simon and Schuster, 2001.

Strauss, Lewis L. *Men and decisions*. Garden City, NY: Doubleday, 1962.

Sweeney, Michael. *Secrets of victory: The Office of Censorship and the American press and radio in World War II*. Chapel Hill: University of North Carolina Press, 2001.
Thirring, Hans. *Die Geschichte der Atombombe*. Vienna: Neues Österreich Zeitungs- und Verlagsgesellschaft, 1946.
Thorpe, Charles. *Oppenheimer: The tragic intellect*. Chicago: University of Chicago Press, 2006.
Truman, Harry S. *Memoirs*, Vol. 1, *Year of decisions*. Garden City, NY: Signet, 1955.
Ulam, Stanislaw. *Adventures of a mathematician*. New York: Scribner, 1976.
U.S. Atomic Energy Commission. *In the matter of J. Robert Oppenheimer: Transcript of hearing before Personnel Security Board and texts of principal documents and letters*. Washington, DC: US Government Printing Office, 1954. Reprinted by MIT Press, 1971.
Vogel, Kathleen M. *Phantom menace or looming danger? A new framework for assessing bioweapons threats*. Baltimore: Johns Hopkins University Press, 2013.
Walker, J. Samuel. *The road to Yucca Mountain: The development of radioactive waste policy in the United States*. Berkeley: University of California Press, 2009.
Walker, Mark. *German national socialism and the quest for nuclear power, 1939–1949*. New York: Cambridge University Press, 1989.
———. *Nazi science: Myth, truth, and the German atomic bomb*. New York: Plenum Press, 1995.
Walterscheid, Edward C. *To promote the progress of useful arts: American patent law and administration, 1787–1836*. Littleton, CO: Fred B. Rothman, 1998.
Wang, Jessica. *American science in an age of anxiety: Scientists, anticommunism, and the Cold War*. Chapel Hill: University of North Carolina Press, 1999.
Wang, Zuoyue. *In Sputnik's shadow: The President's Science Advisory Committee and Cold War America*. New Brunswick, NJ: Rutgers University Press, 2008.
Weart, Spencer. *Nuclear fear: A history of images*. Cambridge, MA: Harvard University Press, 1988.
———. *Scientists in power*. Cambridge, MA: Harvard University Press, 1979.
Weart, Spencer, and Gertrude Weiss Szilard, eds. *Leo Szilard: His version of the facts: Selected recollections and correspondence*. Cambridge, MA: MIT Press, 1978.
Weinstein, Allen, and Alexander Vassiliev. *The haunted wood: Soviet espionage in America*. New York: Random House, 1999.
Weisgall, Jonathan M. *Operation Crossroads: The atomic tests at Bikini Atoll*. Annapolis, MD: Naval Institute Press, 1994.
Welsome, Eileen. *The plutonium files: America's secret medical experiments in the Cold War*. New York: Dial Press, 1999.
West, Nigel. *Venona: The greatest secret of the Cold War*. London: HarperCollins, 1999.
Westfall, Richard S. *Never at rest: A biography of Isaac Newton*. Cambridge: Cambridge University Press, 1983.
Wheeler, John A., with Kenneth Ford. *Geons, black holes, and quantum foam: A life in physics*. New York: W. W. Norton, 1998.

Williams, Robert C. *Klaus Fuchs, atom spy*. Cambridge, MA: Harvard University Press, 1987.

Williams, Robert C., and Philip L. Cantelon, eds. *The American atom: A documentary history of nuclear policies from the discovery of fission to the present, 1939–1984*. Philadelphia: University of Pennsylvania Press, 1984.

Willrich, Mason, and Theodore B. Taylor. *Nuclear theft: Risks and safeguards*. Cambridge, MA: Ballinger, 1974.

Wills, Gary. *Bomb power: The modern presidency and the national security state*. New York: Penguin, 2010.

Winterberg, Friedwardt. *The physical principles of thermonuclear explosive devices*. New York: Fusion Energy Foundation, 1981.

York, Herbert. *The advisors: Oppenheimer, Teller, and the superbomb*. Stanford, CA: Stanford University Press, 1976.

Zaloga, Steven J. *Target America: The Soviet Union and the strategic arms race, 1945–1964*. Novato, CA: Presidio, 1993.

Ziegler, Charles A., and David Jacobson. *Spying without spies: Origins of America's secret nuclear surveillance system*. Westport, CT: Praeger, 1995.

INDEX

Page numbers in italics refer to figures or tables.

"absolute secrecy," 177, 282; demanded by Franklin D. Roosevelt, 43; difficulties of achieving, 51, 93; implemented, 45; and the "problem of secrecy," 82, 86; and "Publicity," 97–8
Acheson, Dean, 145, 171, 181, 212, 272
Acheson-Lilienthal Report, 171, 174–75, 181, 208, 335
"active measures," 61–62
Ad Council, 145
Advisory Committee on Uranium. *See* Uranium Committee
AEC. *See* US Atomic Energy Commission
Aerofax, Inc., 376
Albuquerque Tribune (newspaper), 387
alchemy, 19
Alsos mission, 62, 438n109, 440n143
Alvarez, Luis, 218
American Association for the Advancement of Science, 76, 314
American Bar Association, 379
American Civil Liberties Union, 238, 353–54, 360, 366–67, 403
American Legion, 71
American Physical Society, 99
American Society of Newspaper Editors, 237
Amin, Idi, fears of, 352
Anderson, Clinton P., 295
Andrews, Walter, 397
anti-secrecy: defined, 335–37; discourse of, 335–37, 398, 401–2, 405, 415; and international control, 173–75; in the 1970s, 7, 349; in the 1980s, 380–82; politics of, 381–83, 389, 395; and the *Progressive* case, 349, 356, 367–71; as "secret seeking" motivation, 377
anti-war activism, 7, 341–42
Apollo affair, 325, 492n130
Argentina, 281; fears of, 295, 299; and nuclear fusion, 301
Argonne National Laboratory, 153, 210, 348, 354, *367*
Arkin, William M., 372–75
Army Corps of Engineers, 47
artificial intelligence, 399, 407
Associated Press, 67
Astounding Science Fiction (magazine), 74
AT&T, 107, 109
Atomic Energy Act of 1946, 176, 397, 404; calls for revision, 215–16, 225; compared to Atomic Energy Act of 1954, 276–79; early draft legislation, 140–46; FBI objections to, 195; as interpreted by Atomic Energy Commission, 184, 186, 189, 191, 215, 240; modified by Congress, 157–58; and Restricted Data, 296, 414; revisions (*see* Atomic Energy Act of 1954); and security clearances, 192
Atomic Energy Act of 1954, 404; calls for revision, 385, 387; and capital punishment, 481n139; compared to Atomic Energy Act of 1946, 276–79; and declassification, 279,

Atomic Energy Act of 1954 (*continued*) 296; and exchange with the United Kingdom, 278; impetus for, 276, 296; invoked as threat, 310, 312; legal test during the *Progressive* case, 349–51, 362–63, 366; and nuclear power, 296; and patenting, 279; and Restricted Data, 296, 366

Atomic Energy Commission, 7–8, 234; abolished, 319, 339; adopts Tolman Committee guidelines, 171, 179; Advisory Committee on Biology and Medicine, 188; ambiguous mandate of, 156, 177; assumes control of US nuclear complex, 179; and Atomic Energy Act of 1954, 277–79; attempts at secrecy reform, 197–207, 200, 214–15, 229–30; attempt to coordinate classification policy with West Germany and the Netherlands, 294–95; calls for abolishment, 339; and censorship, 189–91; and classification, 184, 188; Classification Branch, 184, 236–37, 265, 316; conflicts with military, 192; Controlled Thermonuclear Research, 308–9, 313; created, 158; culture of secrecy, 412; and declassification, 169, 179, 184–87, 188, 277, 279–82, 336; Declassification Branch, 186, 191, 198; Declassification Drive, 336, *337*, 361; Director of Information, 184; Division of Research, 185, 314; Division of Security, 185; early concepts of, 133, 140–41, 146; and espionage, 221–28; first commissioners, 182; and gas centrifuge research, 291, 293, 295–99; General Advisory Committee (*see* General Advisory Committee); and H-bomb gag order, 221, 235–36, 239; and homosexuality, 255; and human radiation experiments, 188; and the hydrogen bomb, 220–23, 226, 235; and "information control," 183–86, 189, 206; Intelligence Branch, 222; internal criticisms of secrecy, 196–97; as "island of socialism," 151, 176; and Klaus Fuchs, 222–23; and KMS Industries, 310–19; and laser fusion, 308, 310–18; under David Lilienthal, 182–83, 196; under the May-Johnson bill, 146; and the McMahon Act, 151, 154–56; *Monthly Classification Bulletins*, 279–81; "no comment" policy, 191, 242, 341, 348, 366, 376 (*see also* Department of Energy: "no comment" policy); and nuclear fusion, 301–4, 308, 309; and nuclear proliferation, 300; and nuclear terrorism, 322, 324–26; and nuclear testing, 246; Office of Information Control, 185; Office of Technical and Public Information, 186–87; and Operation Candor, 273; and the Oppenheimer affair, 264, 267–69; Patent Compensation Board, 278; and patents, 278–79; Personnel Security Board (*see* Personnel Security Board); Public Information Branch, 186; relations with Joint Committee on Atomic Energy, 196, 209, 211; relations with private industry, 310–11; and Restricted Data, 183, 186, 189; and the Rosenberg trial, 252; and secrecy under Lewis Strauss, 282; and security clearances, 194–97, 210, 255–56; statement on secrecy, 200–207, 217, 238; Technical Information Branch, 186

atomic kitsch, 246

"Atoms for Peace," 288, 333; conferences, 274–75, 302–3; and declassification, 169, 274, 319; and nuclear terrorism, 322, 324, 330; origins, 273–74; and secrecy, 273, 275

Austria, scientists recruited by Soviet Union, 289, 300

Bacher, Robert F.: as AEC commissioner, 182, 188, 207, 214; and Tolman Committee, 161, 168

background checks, 32–33. *See also* security clearances

Baker, Charles P., 40

balance; and secrecy, 210, 403

Banks, Charles H., 193

Baruch, Bernard, 175

Baruch Plan, 174–75, 208

Beams, Jesse W., 289–91, 293

Beckerley, James G., 236–37, 265, 267

Beginning or the End?, The (Mitchell), 450n109

Belgium, fears of, 299

Bell, Griffin B., 350, 359, 367

Bequerel, Henri, 16

Beria, Lavrenty, 227, 290

Berlin Airlift, 207

Bernal, J. D., views on secrecy in science, 19–20
Bernstein, Carl, 370
"best-kept secret of the war," 51, 93, 130, 224, 227, 450n106
Bethe, Hans: and the hydrogen bomb, 241, 249; and the *Progressive* case, 354–55; and *Scientific American* article, 236–39, 341, 471n12
Bhabha, Homi J., 275, 284, 302
Bhatnagar, Shanti S., 436n78
bilateral agreements, 288
black budget, 37, 77
Blackett, P. M. S., 24–25
Bohr, Niels, 16, 427n60; and "Atoms for Peace," 275; and international control, 136–39, 451n12; at Los Alamos, 58; and self-censorship, 24; and scientific openness, 137; views on secrecy, 89; and John Wheeler's theory of fission, 27
Borden, William, 259–61
"born secret" concept. *See under* Restricted Data
Boston Globe (newspaper), 122
Bowen, Harold, 28
Bradbury, Norris, 224, 303
Brandt, Raymond, 184
Brazil: fears of, 299; and gas centrifuge research, 293
Breit, Gregory, 35, 66; complaints about secrecy, 45–46; and publication censorship, 28–29
Briggs, Lyman J., 27, 34–35, 37, 45
Brueckner, Keith A., 308–18, 489n104; and patents, 310–11, 315, 317
Bull, Harold R., 462n44
Bulletin of the Atomic Scientists (journal), 218, 239, 374
Bundy, Harvey H., 105–6
Bundy, McGeorge, 374
Bureau of Budget, 159
Bush, George H. W., 375, 385
Bush, George W., 394
Bush, Vannevar, 411; and the Acheson-Lilienthal Report, 174; and censorship, 67–68; and Congress, 44, 77, 80–82, 92, 141; and declassification, 159; and hydrogen bomb testing, 241–42; and international control, 117–18, 120, 137–40, 451n12; and patenting, 42; and postwar legislation, 140–42, 144–45; and the presidential statement about the atomic bomb, 109; and press censorship, 72–73, 107; and "problem of secrecy," 86–89; and "Publicity," 106; as scientist-administrator in World War II, 6, 34–48, 52; secrecy practices, 39–42, 55, 66–68, 77; and the Smyth Report, 98–100, 102, 124, 126, 182; views on postwar secrecy, 86–89
Byrnes, James F., 78–80, 150; and the Acheson-Lilienthal Report, 174; and the Smyth Report, 126

Caldicott, Helen, 379
California Institute of Technology, 34, 53
Canada: and declassification, 199, 216; and international control, 146; and laser fusion, 358; and the Manhattan Project, 61; press release after Hiroshima, 119–20; spy ring, 152, 221
capital punishment, 1, 7, 157, 277, 481n139
Carnegie Institution, 25, 34
Carson, Johnny, 374
Carter, James (Jimmy) E., 326, 386; and the *Progressive* case, 350, 367
Castle "Bravo," 129, 243, 246–48, 267, 282, 340
Center for Defense Information, 372
Central Intelligence Agency, 325, 386, 395; culture of secrecy, 412
centrifugal enrichment. *See* gas centrifuge
Chadwick, James, and the Smyth Report, 124–26
chain reaction, discovery of, 17
Chernobyl accident, 380–81
Chevalier, Haakon, 262
Chevalier affair, 263
Chicago Sun (newspaper), 70
Chicago Tribune (newspaper), 321
children: fears of, 320, 325–27; and nuclear knowledge, 320
China, People's Republic of, 373, 374, 404; and espionage, 390; fears of, 282, 390, 395, 399; and proliferation, 284; and Soviet nuclear assistance, 481n2

Churchill, Winston: and international control, 137; press release after Hiroshima, 120–21
civil defense, 462n44
Civil Service Commission, 192, 277
classification categories: "Confidential," 33, 39, 49, 64, 85, 194, 205, 429n82; "Official Use Only," 196, 463n51, 463n59; in the OSRD, 41, 429n82; "Restricted," 33, 64, 85, 194, 429n82, 463n51, 463n59; Restricted Data (*see* Restricted Data); "Secret," 33, 39, 49, 64, 85, 194, 196, 205, 280, 295, 429n82; "Secret-Limited," 64, 104; "Top Secret," 1, 33, 64, 85, 93, 150, 153–54, 162, 194, 221, 259, 265, 270; in World War II, 64
"classification pendulum," 336, 400, 404
Classified Information Procedures Act, 476n63
classified journals, 29
Cleveland Press (newspaper), 75
climate geoengineering, 399
Clinton, William J., 386, 390
Cockcroft, John, 427n60
code-names, 4, 39, 52, 136, 280
Cohen, Avner, 393
Cold War: impact on secrecy reform, 207, 230; secrecy regime (*see* secrecy regime: Cold War)
Cole, Sterling, 267
college students: drawing nuclear weapons, 327–31, 342, 362, 371; fears of, 327
Columbia University, 21–25, 40–41
Combined Development Trust, 144
Combined Policy Committee, 168
command and control, 283
Committee on Declassification. *See* Tolman Committee
Committee on Postwar Policy, 87–89, 102, 161
Committee on Uranium. *See* Uranium Committee
Communism, fears of, 41, 61–63, 192, 194, 255, 261–62, 412–13
Communist Party of the United States, 256–57, 262
compartmentalization: criticisms of, 60, 95; and Leslie Groves, 55, 61, 92; at Los Alamos, 57; before the Manhattan Project, 45; in the Manhattan Project, 4, 55, 61, 83, 92–93; and safety, 57; and the Smyth Report, 100, 127
Compton, Arthur H.: and acceleration of fission research, 36–38; clearance problems, 41; lax attitude toward secrecy, 45–46; and the Metallurgical Laboratory, 36–38, 40, 74, 84, 100; speech alluding to weapons, 76; testimony on domestic control, 148; and the Tolman Committee, 161–62
Compton, Karl T., 34–35
Conant, James B.: and code names, 39; concerns about Arthur Compton, 45–46; experience in World War I, 46, 430n99; and the hydrogen bomb, 218; and international control, 117–18, 120, 138–40, 451n12; and postwar legislation, 140–45, 147; and the presidential statement about the atomic bomb, 109; and "Publicity," 106; as scientist-administrator during World War II, 6, 34–42, 47, 87, 411; secrecy practices, 39–40, 42, 54–55, 68, 107; and the Smyth Report, 98–100, 102, 125–26, 182; views on postwar secrecy, 89
Condon, Edward U., 40, 58, 60; and Leslie Groves, 149; and Special Senate Committee on Atomic Energy, 149–50
"Confidential." *See under* classification categories
Congress, 77–81, 98, 142; and the Atomic Energy Commission, 182; and Chinese espionage, 390; and domestic control, 135, 146–52, 154, 156–58; fears of, 410; and the Freedom of Information Act, 454n59; and the hydrogen bomb, 220, 245; and laser fusion, 316; and the Manhattan Project, 44, 51, 80–82, 108; as motivation for secrecy, 44, 77, 81, 92; and nuclear secrecy, 157, 339; and nuclear terrorism, 330; passes the McMahon Act, 157; and secrecy, 158
Congressional Joint Committee on Atomic Energy. *See* Joint Committee on Atomic Energy
Connally, Tom, 210, 271
Consodine, William A., *113*, 115–16
Constitution, 398. *See also* First Amendment
Controlled Thermonuclear Research. *See*

Controlled Thermonuclear Research; US Atomic Energy Commission
Cornell University, 40, 256
Coster-Mullen, John, 398, 449n99
court system, and secrecy, 251–52, 254, 357, 476n63
Cox, Christopher, 390
Cox Committee, 390
critical mass, 161
cryptography, 408–9
Cuba; fears of, 295
"cult of secrecy," 379
culture of secrecy: at Central Intelligence Agency, 412; at US Atomic Energy Commission, 412; at US Department of Defense, 412. *See also* secrecy practices: secrecy culture
Curie, Marie, 16
Curie, Pierre, 16
Curve of Binding Energy, The (McPhee), 321–22, 326–27, 330
cyber-warfare, 408
Czechoslovakia, fears of, 299

Daily Californian (newspaper), 363
Darrow, Karl K., 99
Darwin, Charles, 20
Day, Samuel H., Jr., 347–49, 357, 366–67
DC Armory, 33
DC Comics, 74
D-Day invasion, 64
Dean, Gordon, 184; as AEC chairman, 231, 240–41, 243, 251, 255–56, 269–70; as AEC commissioner, 214–17; and the Atomic Energy Act of 1946, 215–16, 240; and the Rosenberg trial, 251
declassification, 10, 414; and the AEC, 184–87, 279–81, 336; in the Army and Navy, 160; and the Atomic Energy Act of 1954, 276–77; and "Atoms for Peace," 274; in the Cold War, 233; compared with classification, 456n84; creation of concept, 158–59; guides, 166–68, 187, 293, 412, 458n118; of the implosion design, 251; and laser fusion, 314–16, 358, 369, 387; and nuclear fusion, 303–4; and the nuclear industry, 287–88; and nuclear power, 279; and nuclear proliferation, 282; and nuclear terrorism, 322, 324, 330; after the Pentagon Papers, 336; in the post-Cold War, 386–89; rationales for, 163–64; of the Teller-Ulam design, 368–69
Defense Nuclear Agency, 377–78
Defense Secrets Act, 30–31
Degussa, 290–91
democracy, and secrecy, 405
Denmark, fears of, 299
Department of Defense, 216, 239, 308–10, 314, 394; and the AEC, 217; and the Atomic Energy Act of 1954, 278; culture of secrecy, 412; and declassification, 277; and the Rosenberg trial, 252
Department of Energy, 267, 329–30, 339–40, 373, 379, 394; "no comment" policy, 348 (*see also* Atomic Energy Commission: "no comment" policy); Office of Classification, 388; Office of Declassification, 388–89; and the *Progressive* case, 348–59, 361, 364, 366–68, 499n96; review of Restricted Data guidelines, 385–86; as successor to AEC, 319
Department of Justice, 338, 349; and the *Progressive* case, 351–52, 355–56, 359, 366–67, 499n96
Department of State, 239, 295, 373, 394
Department of War, 43, 72, 78–79, 81, 121–22, 125–26, 130, 141
De Volpi, Alexander, 354, *367*
DeWitt, Hugh, 499n96
Dickey, John, 184
domestic control, 138, 140–45
downgrading. *See* declassification
dual-use technology, 7, 293, 307
DuPont, 71, 78, 328
Dyson, Freeman, 328–29

East Germany, 290, 400; fears of, 299; gas centrifuge research, 295
Egypt, fears of, 295
Einstein, Albert, letter to Franklin D. Roosevelt, 27
Eisenhower, Dwight D., 378; and "Atoms for Peace," 169, 273–74; and declassification, 271; and the hydrogen bomb, 243, 245,

Eisenhower, Dwight D. (*continued*) 247, 260; and nuclear power, 276; and Operation Candor, 272–73; and J. Robert Oppenheimer, 261
Eisenhower, Milton, 184
Eisenhower Panel, 184–85
Ellsberg, Daniel, 338–39, 366, 370, 379, 424n11, 499n108
Elston, Charles, 224
encyclopedias, and nuclear weapons design information, 343–45, 354, 357, 369
Energy Research and Development Administration, 319, 339
Engel, Albert J., 79–80
Enlightenment, the, 3–4
enriched uranium, 277. *See also* uranium enrichment
environmentalism, 313
espionage, 93
espionage, 93; in Canada, 152; fears of, 42; German, 73; industrial, 395; and the People's Republic of China, 390; by the Soviet Union (*see* Soviet Union: espionage)
Espionage Act of 1917, 151, 404; and capital punishment, 481n139; categories of secrets, 31–32, 455n72; enacted by Congress, 31–33; use as threat, 58, 82, 117, 123, 131, 144
Esquire (magazine), 321–22
Euratom, 288, 294, 298–99
European Atomic Energy Committee. *See* Euratom
evidence, secret. *See* secret evidence
Executive Orders: number 8381 ("Defining Certain Vital Military and Naval Installations and Equipment," 1940), 33; number 9568 ("Providing for the Release of Scientific Information," 1945), 159–60; number 11652 ("Classification and Declassification of National Security Information and Material," 1972), 336; as part of American secrecy regime, 32–33, 386, 404
explicit knowledge, 88
export control, 379, 390, 405

fallout. *See* nuclear fallout
fears, 397, 401; of Argentina, 295, 299; of Belgium, 299; of Brazil, 299; of children, 320, 325–27; of college students, 327; of Communism, 41, 61–63, 192, 194, 255, 261–62, 412–13; of Congress, 410; and control, 398; of Cuba, 295; of Czechoslovakia, 299; of Denmark, 299; of East Germany, 299; of Egypt, 295; of espionage, 42; of France, 91, 255; of Idi Amin, 352; of India, 390, 491n125; of Iran, 395; of Israel, 295; of Italy, 299; of Japan, 76, 90–91, 295; of laser fusion, 308; of "loose nukes," 392; of Nazi Germany (*see* Nazi Germany: fears of); of North Korea, 319, 395; of Norway, 299; of nuclear blackmail, 322; of nuclear black market, 323; of nuclear proliferation (*see* nuclear proliferation: fears of); of nuclear smuggling, 320–21; of nuclear terrorism (*see* nuclear terrorism: fears of); of Pakistan, 390; of the People's Republic of China, 282, 390, 395, 399; of Portugal, 299; of pure-fusion weapons, 310; of the Russian Federation, 391, 399; of sabotage, 62, 77, 91; of scientists, 229, 237–39, 256–57; of the Soviet Union (*see* Soviet Union: fears of); of Spain, 299; of the United Kingdom, 91, 93; of West Germany, 292
Federal Bureau of Investigation, 328, 377–78, 386, 395, 403; and anti-Communism, 255; and the Atomic Energy Act of 1954, 277; and German espionage, 73; and the Manhattan Project, 73–74, 123; and the McMahon Act, 157; and oaths, 40; objections to the Atomic Energy Act of 1946, 195, 277; and J. Robert Oppenheimer, 258, 267–68; and security clearances, 32–33, 192–94, 210, 258, 277; and Soviet espionage, 93, 225, 250–51; and terrorism, 324–26
Federation of American Scientists, 360, 375
Fermi, Enrico, 16, 196, 281, 400; and self-censorship, 21–24; and Leo Szilard, 21–24
Fidler, Harold, 198
First Amendment, 154, 338–39, 350, 353–56, 366, 398
fissile material safeguards, 322–27, 330–35, 368, 391; and international control, 172, 175; vs. secrecy, 176–77, 331, 333, 335, 340, 402, 405. *See also* materiality

Fleröv, Georgii, 67
Food and Drug Administration, 142
foreign intelligence, 199–200
France, 294, 299, 373; fears of, 91, 225, 432n28; first atomic bomb, 292; interest in fission, 25; and laser fusion, 308, 313, 333; and nuclear proliferation, 281, 284; participation in Manhattan Project, 61, 432n28
Franck, James, 180
Frankfurter, Felix, 137, 450n4
Frazier, Thomas A., 71
Freedom of Information Act, 9–11, 336–37, 374–79, 401; and Congress, 454
freedom of speech, 3–4, 143. *See also* First Amendment
Frisch, Otto, 16
Fuchs, Klaus, 256–57, 408, 469n150; and the hydrogen bomb, 227, 468n135; impact on secrecy reform, 227–28; and radiation implosion, 260, 478n92; revelation as a spy, 222–28; as spy at Los Alamos, 65, 249–50
Fusion Energy Foundation, 364, 499n100

Galileo, Galilei, 20
Gallup, George, 184
gas centrifuge, 289–95, 298–300, 407; as challenge to nuclear monopoly, 289; during Manhattan Project, 38, 165, 289; and the Nuclear Non-Proliferation Treaty, 299; and nuclear proliferation, 299, 332, 392; and private industry, 297; and Restricted Data, 296
gaseous diffusion, 38, 289, 291
gas warfare, 30, 46
General Advisory Committee, 261, 314; criticisms of AEC secrecy, 196–98, 208; and hydrogen bomb, 218, 236, 239, 258
General Motors, 307, 316
Germany. *See* East Germany; Nazi Germany; West Germany
Glasstone, Samuel, 344
Glenn, John, 329–30, 361
Gold, Harry, 250, 257
Goodale, James, 356
Gorbachev, Mikhail, 381, 383
Goudsmit, Samuel, 438n109, 440n143
"graymail," 476n63
Greenglass, David, 65, 250, 252–54, 257; drawing of "the very atomic bomb itself," 252, 253
Greenglass, Ruth, 250
Greenhouse "George," 249
Greenpeace, 375, 388
Gregg, Alan, 198
Grenada, war in, 379
Griffin, John A., 350, 353
Groves, Leslie R., 87, 95, 251, 257; and the Acheson-Lilienthal Report, 174; and Canadian spy ring, 152, 414; and censorship, 66–69, 72, 75; on compartmentalization, 55, 60–61, 92; and Congress, 80–82, 92, 98; and declassification, 158, 160, 167–68, 170; deflection of Congressional investigations into Manhattan Project, 80; and Edward Condon, 149; heads Manhattan Project, 52; and Hiroshima attack, 118; and the hydrogen bomb, 471n5; initiates postwar declassification, 160–61; justifications for secrecy, 77, 91–92, 404; and David Lilienthal, 181, 195; and postwar legislation, 115, 143–44, 147, 149; postwar testimony, 65, 147, 153, 224–25; practices of secrecy, 53–55, 58, 60–64, 68–69, 71, 73, 75, 80, 83, 91–92; and "Publicity," 107, 111–12, 115–23, 128–29; responding to revelation of Klaus Fuchs as a spy, 63; on the Rosenbergs, 255, 267; on scientists, 1, 11, 53, 55; and "the secret," 147; and security clearances, 193; and the Smyth Report, 98, 101–4, 124–26, 128–29, 131, 182; and Soviet espionage, 63, 65–66, 224–25; and the Special Senate Committee on Atomic Energy, 149–53; and Leo Szilard, 61; and the Tolman Committee, 167, 457n93; and Harry Truman, 90; and uranium control, 144–45
gun-type design, 128, 254, 445n42, 449n100. *See also* nuclear weapon design

Hadley, Reed, 475n54
Hahn, Otto, 16, 19
Haig, Alexander, 373
Halban, Hans van, 432n28
Hall, Theodore, 65, 250

Hanford, 203; attacked by Japanese, 437n89; and Klaus Fuchs, 227; investigated by Congress, 79, 81; labor issues, 56; land seizures, 68; during Manhattan Project, 54–56, 61, 71, 83, 101; press censorship, 68; in "Publicity," 112, *113*, 115; rumors about, 71, 78–79; secrecy practices, 54–55, 61; and the Smyth Report, 127–28
Hansen, Chuck, 362–63, 375–78, *377*, 417
Harper's Magazine, 191
Harrison, George L., 126
Harvard University, 34, 107, 329
Haussmann, Carl, 314
H-bomb. *See* hydrogen bomb
heavy water, 72–73
Heisenberg, Werner, 16
Hersey, John, 130
Hewlett, Richard G., 8
Hickenlooper, Bourke, 210
Hinshaw, Carl, 303
Hiroshima: bombing of, 1, 95, 118; news coverage of bombing, 112, 447n78; press reactions to, 121–23; radioactivity, 122–23, 129, 164, *165*; statements about bombing (*see* "Publicity": presidential press release about Hiroshima); strike order, 118, 446n53
Hiroshima (Hersey), 130
Hiss, Alger, 479n116
Hitler, Adolf, 23
Hodgins, Eric, 184
homosexuality, and security clearances, 255–56
Hoover, Herbert, 182
Hoover, J. Edgar, 225, 260–61
House Un-American Activities Committee, 208, 210–11, 221, 224–25; and the Joint Committee on Atomic Energy, 258; and J. Robert Oppenheimer, 258
Howe, Clarence D., 120
HUAC. *See* House Un-American Activities Committee
Hughes Aircraft, 307
human radiation experiments, 188–89, 387
hydrogen bomb, 320; and classification, 161, *166*, 471n5, 471n12; as crash program ordered, 220–22; debate leaks, 219; debate over crash program, 7, 180, 209, 217–21, 234–35, 304; design of (*see* Teller-Ulam design); and espionage, 226, 390; fears of news leaking about, 241–42; "gag order," 221, 234–39, 243–45; impact on postwar planning, 118, 138; impact on secrecy discourse, 230, 233; impact on secrecy reform, 221, 228; invention of, 240–41; JCAE history of, 259–61; layer-cake design, 340; and North Korea, 392; and nuclear terrorism, 325; and nuclear testing, 241–42, 246–48; origins, 217–18, 446n60; rumors about, 236, 471n5; secrecy of design, 10; and Soviet espionage, 226; Soviet test of, 272; speculation on design, 340–58, 362–65, 376; "ultimate secret," 234–35, 249, 333, 413

Iceland, 375
Illustrated London News (magazine), 340–41
implosion design, 161, 223, 253–54, 393, 449n100, 469n143; declassification of, 251–52. *See also* nuclear weapon design
independent reinvention, 86–87, 307, 310
India: delegation of scientists visiting the United States during World War II, 74–75, 436n78; fears of, 390, 491n125; and nuclear proliferation, 284, 324; nuclear testing, 284, 332
industrial secrecy, 216
inertial confinement fusion, 306–8, 310–18, 332–33, 369; vs. laser fusion, 485n60; origins of, 304–5; and the *Progressive* case, 358
"information control," 179, 184–86, 206, 239, 393, 399; in AEC, 187; challenges of, 189; contrasted with Cold War approach to secrecy, 235; idea of, 185
informed consent, 188, 387
insider threats, 323
Interim Committee, 106, 109, 111, 139, 141–43; and "Publicity" 106–7
International Atomic Energy Agency, 275, 288, 299, 309
international control, 91, 110, 136–40, 145–46, 194, 208, 243, 335, 401, 411, 451n12; Acheson-Lilienthal Report, 171–75; and

secrecy, 139, 171–72; and the Tolman Committee, 162
International Declassification Conferences, 199, 216
International News Service, 191
International Quantum Electronics Conference, 315
Iran: fears of, 395; and gas centrifuges, 332, 392; and nuclear proliferation, 393; and Stuxnet, 408
Iran-Contra affair, 379
Iraq, 400; war in, 392
Israel, 374; and the Apollo affair, 325, 492n130; fears of, 295; and Iran, 393; and laser fusion, 308; and nuclear proliferation, 284; and nuclear weapons, 393–94, 504n34
Italy, fears of, 299
Ivy "Mike," 241–46, 248, 260; leaks to press, 242

Jacobson, Harold, 122–23, 130
Jane's Defence Weekly (magazine), 380
Jane's Information Group, 371
Japan, 375; and Castle "Bravo," 247–48; fears of, 76, 90–91, 295; and gas centrifuge research, 293; interest in fission, 25; knowledge of Manhattan Project, 64, 76, 436n78; and laser fusion, 308, 358; as "nuclear victim," 129, 247–48; occupation of, 129, 247
JCAE. *See* Joint Committee on Atomic Energy
Jeffries, Zay, 84
Jeffries Committee, 84
Jewett, Frank B., 34, 44
"Jimmy" (boy genius), 320, 490n112
"Joe-1," 180, 208–9; announcement of, 213; debates about announcing, 212–13; detection of, 212; and the hydrogen bomb, 217–19, 221; impact on declassification philosophy, 216; impact on secrecy reform, 215, 217, 221, 228
Johnson, Edwin C., 146, 219–20
Johnson, Louis, 212
Johnson, Lyndon B., 338
Johnson, Warren, 184
Joint Committee on Atomic Energy: and the Atomic Energy Commission, 158, 183, 187, 194, 196, 209–11, 215, 271; and declassification, 303; and espionage, 221, 252; and gas centrifuges, 295; and the House Un-American Activities Committee, 225, 258; and the hydrogen bomb, 219–20, 226, 239; and laser fusion, 317; and the Rosenberg trial, 252; and security clearances, 255, 267; and Soviet espionage, 223–27; staff's lost H-bomb history, 259–61, 270
Joliot-Curie, Frédéric, 16, 42, 48, 225, 432n28; and self-censorship, 22–23, 25, 27, 426n33; and Leo Szilard, 22–25
Joliot-Curie, Irène, 16, 22
Jones, Thomas O., 193
journalists: interest in nuclear fission, 66–67; and Los Alamos, 75; and the Manhattan Project, 115, 122; and the Rosenberg trial, 252, 254; speculation about atomic bombs, 38, 66–67

Kaiser, David I., 256–57
Kaufman, Irving, 252, 254
Kennedy, John F., 374
Khan, Abdul Q., 332, 392
Kidder, Ray, 307–8; and the *Progressive* case, 358, 366–67
Kistemaker, Jaap, 290
KMS Fusion. *See* KMS Industries
KMS Industries, 309; and the Atomic Energy Commission, 310–19, 489n109
Knoll, Erwin, 348, 357, 366–67
"know-how," 147, 211, 222, 409. *See also* tacit knowledge
knowledge and power, 400–401
Knoxville, Tennessee, 52
Kokura, 118
Korean War, 465n87
Koval, George, 65
Kyl-Lott Amendment, 390–91

land seizures, 68
Langmuir, David B., 200–208; secrecy diagrams, 201–5, *201*, *202*, *204*, *205*
Lansdale, John, Jr., 436n75
Lapp, Ralph, 343–44
"laptop of death," 393
LaRouche, Lyndon, 364, 371, 376

laser, invention of, 305, 307. *See also* laser fusion
laser fusion, 305–7, 309–18, 332–33; and classification, 308; classification guidelines for, 313–14; and declassification, 314–15, 369, 387; fears of, 308; independent reinvention of, 307; vs. inertial confinement fusion, 485n60; and patents, 310; and the *Progressive* case, 358, 366–68, 383
Laurence, William L., 38, 106–7, *113*, 118–21, 184, 235, 428n66; controversy over Pulitzer Prize, 114; disinformation about the Trinity test, 108; draft press release on the use of the atomic bomb, 109; stories written for Manhattan Project, 111–12
lavender scare, 255. *See also* homosexuality
Lawrence, Ernest O., 26, 36–38, 40, 46, 51, 70–71, 281, 434n57; and the hydrogen bomb, 218, 258; and the Tolman Committee, 161–62
Lawrence Livermore National Laboratory, 264, 304, 314, 342, 352, 358, 370, 389; and laser fusion, 305, 309, 311, 313, 315, 317–18; and nuclear proliferation, 320
leaks, 71, 74–75, 378–80, 391, 394; of hydrogen bomb, 219–20, 242; of Manhattan Project, 75, 78, 130, 152
Lee, Wen Ho, 391
legislation, 138, 140–2, 145–46; and early American secrecy, 30–31; postwar, 143–44
Lehrer, Tom, 292, 320
Libby, Willard F., 199
Libya, and gas centrifuges, 332, 392
Life (magazine), 128, 254, 340–41, 343
Lilienthal, David E.: and the Acheson-Lilienthal Report, 171–73; as AEC chairman, 179–83, 207, 209, 231, 240, 251, 258–59; and AEC scandals, 210; and the Atomic Energy Act of 1946, 215; attempts at secrecy reform, 196–201, 203, 207–10, 215–17, 230, 400; becomes chairman of AEC, 182; calls for abolition of AEC, 339; and declassification, 187–88; early ideas about nuclear secrecy, 180–81; goals for AEC, 179; and the hydrogen bomb, 218–21, 235; "incredible mismanagement" hearings, 211; and "information control," 171–73, 181–84, 186–87, 194; and "Joe-1," 211–14; and the Joint Committee on Atomic Energy, 187, 209, 211, 221, 224; and Leslie Groves, 181; and Lewis Strauss, 271; and the Manhattan Project, 182; and the "problem of secrecy," 180; and Restricted Data, 195; and security clearances, 193–95, 261; and Soviet espionage, 222–24, 227; and Tennessee Valley Authority, 180; retirement, 221
Linschitz, Henry, 254
Livermore. *See* Lawrence Livermore National Laboratory
"loose nukes," fears of, 392
Los Alamos. *See* Los Alamos National Laboratory
Los Alamos National Laboratory: badge system, 57–58; and Niels Bohr, 136; compartmentalization, 57–58, 60, 83; and Congress, 81; creation of, 53–55; division of labor, 443n18; espionage at during Manhattan Project, 65, 222 (*see also* Soviet Union: espionage); and laser fusion, 305, 309, 313–14; and Lawrence Livermore National Laboratory, 304; library, 257, 360, 477n82; lost hard drives, 391; and J. Robert Oppenheimer, 62, 117; physical security, 57; in post-Cold War, 389–90; press leaks about, 70; and "Publicity," 111, 117; secrecy practices, 54–58; site chosen for laboratory, 53; and the Smyth Report, 100–101, 127, 442n13; and the University of California, 70, 434n57; wartime colloquia series, 58
Los Alamos Primer (Serber), 328, 492n138
Los Alamos Ranch School, 53
Los Angeles Times (newspaper), 76, 328, 361
Lowen, Irving S., 450n4
loyalty investigations, 192, 255, 262–63. *See also* security clearances
Loyalty-Security program, 192, 255, 262–63
Lubin, Moshe J., 311, 313

MacArthur, Douglas, 129
magnetic confinement fusion, 301–2, 308; declassified, 303–4
Maiman, Theodore, 305

Majority Report (newspaper), 331
Manhattan Engineer District, 52, 70–71, 140–42, 153, 189–90. *See also* Manhattan Project
Manhattan Project, 4, 6, 11, 44, 51–95, 410; attempts to audit, 78–81; authorization of, 47; as "best-kept secret of the war" (*see* "best-kept secret of the war"); Congress and, 44; Counter-Intelligence Corps, 62; creation of Declassification Office, 167; and declassification, 159, 160; Declassification Organization, 169, *170*; and democracy, 82; espionage, 62–66, 180, 208–9, 221–28 (*see also* Soviet Union: espionage); fears of sabotage, 62, 77, 91; gas centrifuge research during, 289; and human radiation experiments, 188; Indian scientists asking about, 74–75, 436n78; leaks, 75, 78, 130, 152; and nuclear testing, 245; number of Soviet spies inside, 65, 433n41; number of workers, 56; origin of name, 52; oversight of, 44; in postwar, 152; postwar plans for secrecy, 135–36; press releases (*see under* "Publicity"); "Publicity" policy (*see* "Publicity"); Public Relations Organization, *113*, 114–17, 130, 160; rumors about, 71–72, 74, 81; secrecy goals of, 63; secrecy practices, 42, 63–64, 73, 83; secret sites of, 53–54; security investigations, 62, 66, 71, 74, 77; security organization of, 62; Smyth Report (*see* Smyth Report)
Manley, John H., 168, 458n113
Manning, Chelsea, 391
Marbury, William L., 141
Mark, J. Carson, 323–24
Marshall, Charles L., 337
Marshall, George C., 37, 81, 106, 119
Marshall, James C., 52
Marshall, Thurgood, 351
Marshall Islands, 241–42, 245–48, 361, 388
Martin, Thomas S., 350–51, 359, 367
Massachusetts Institute of Technology, 34, 327, 347
materiality, 176–77, 333, 335, 401, 410
Matthias, Franklin T., 68
MAUD Committee, 36; origin of name, 427n60
May, Andrew Jackson, 146
May-Johnson bill, 146, 148–51, 154, 156–58
McCarthyism, 63, 233, 250, 255–56, 279, 335, 370, 412–13
McCone, John, as AEC chairman, 292, 295
McCormack, James, 211
McCoy Air Force Base, 326
McMahon, Brien, 149–51, 167, 182, 184, 214, 229, 414; and the hydrogen bomb, 218–20; and Restricted Data concept, 155; and Soviet espionage, 223–27
McMahon Act, 150–57, 182, 192; comparison to May-Johnson bill, 151, 154; and secrecy, 151–55
McMahon Committee. *See* Special Senate Committee on Atomic Energy
McPhee, John, 321–23, 326
Meitner, Lise, 16, 19, 427n60
Merton, Robert K., views on secrecy in science, 20
Mertonian norms, 20
Metallurgical Laboratory, 100, 123, 162, 188; peacetime planning at, 83–86, 89
MGM Studios, 130
Military Affairs Committee (House of Representatives): attempt to audit Manhattan Project, 79; and domestic control, 147
Military Liaison Committee (of the AEC), 187, 216
Miller, Byron, 157
Mills, Mark M., 474n45
misinformation. *See under* secrecy practices
Montreal laboratory, 61
Moody Bible Institute, 74
Morgenthau, Hans, 287
Morland, Howard, 341–69, 383, 497n63, 499n108; drawings of the Teller-Ulam design, *346*, *365*; and other "secret seekers," 371, 373–74, 376. *See also* US v. the *Progressive*, Inc.
Morrison, Philip, 254; clearance issues, 256, 477n78
mosaic theory, 355
Moynihan, Daniel P., 385, 389, 502n141, 503n20
Moynihan Report, 385, 389
Munich Olympics, 324

Murphy, Charles, 244
Murray, Thomas E., 244, 268

Nagasaki, 110, 118, 445n44; bombing of, 4, 11, 112, 121, 126, 129, 135, 160
nanotechnology, 407
Nashville Banner (newspaper), 71
National Academy of Sciences, 28, 31, 34, 36, 44
National Advisory Committee for Aeronautics, 34
National Association of Science Writers, 38, 76
National Bureau of Standards, 27, 149
National Defense Research Committee (NDRC), 34–35, 37, 49
National Nuclear Security Agency, 394
National Research Committee, 209–10
National Research Council, 28, 35, 66
National Research Foundation, 141
National Security Agency, 394–95, 403, 408
National Security and Resources Center, 360. *See also* Los Alamos National Laboratory, library at
National Security Council, 220–21, 236, 239
National Security Decision Directive, 84, 378
national security state, 412–13
National Technical Information Service, 328
NATO, 278
Natural Resources Defense Council, 360, 371, 374–75
Nature (journal), 24–25, 315, 407; and self-censorship, 24
Navy Bureau of Ordnance, 182
Nazi Germany: atomic program of, 72, 76, 91, 438n109, 440n143; attempts at espionage, 73; fears of, 1, 6, 15–23, 27, 35, 43–44, 47–48, 51, 63–64, 76, 81, 90–91, 397; invoked as symbolic of secrecy, 20, 162; knowledge of Manhattan Project, 64, 76, 436n81, 441n147; propaganda, 76; scientists recruited by Soviet Union, 289
NDRC. *See* National Defense Research Committee (NDRC)
Near v. Minnesota, 50
“need to know.” *See* compartmentalization
Neptunium, 28
Netherlands: fears of, 299; and gas centrifuge research, 290–91, 293–95, 297–300, 332
Nevada Test Site, 246
New Deal, 180
Newman, James R., 149–51, 154, 157
New Solidarity (newspaper), 364
Newton, Isaac, 20
New Yorker (magazine), 130, 320–22
New York Herald Tribune (newspaper), 38, 73
New York Post (newspaper), 228
New York Times (newspaper): and hydrogen bomb, 235, 238; and Manhattan Project, 38, *70*, 72–73, 76, 107, 111, 121; and Pentagon Papers, 338, 351; and the *Progressive* case, 356–57, *367*; and Rosenberg trial, 254
New Zealand, 375
Nichols, Kenneth D., 122, 190
Niigata, 118
Nixon, Richard M., 324, 335–39, 367
Nobel Prize in Physics, 22
Norsk Hydro, 72
Northern States Power Company, 386
North Korea: fears of, 319, 395; and gas centrifuges, 332, 392; and thermonuclear weapons, 392
Norway, fears of, 299
NOVA, 327
NRDC. *See* Natural Resources Defense Council
“Nth Country Experiment,” 320
Nuckolls, John, 304–7, 314–16, 318
nuclear blackmail, 325–26; fears of, 322
nuclear black market, 332, 392; fears of, 323
nuclear fallout, 247–48; and secrecy, 473n31; and the Trinity test, 108
nuclear fission: discovery of, 15–16, 19, 26; peaceful uses of, 300–301; and secondary neutrons, 19, 21, 23–26
Nuclear Fission and Atomic Energy (Stephens), 189–91
“Nuclear Free Seas” campaign, 375
nuclear fusion, 217–18, 240, 300–301; and classification, 303–4, 458n118; inertial confinement (*see* inertial confinement fusion); laser (*see* laser fusion); magnetic confinement (*see* magnetic confinement fusion); peaceful applications of, 300–304, 318, 333

nuclear industry: and the Atomic Energy Act of 1954, 276, 278–79; and declassification, 287–88; and nuclear terrorism, 321–23; push for in 1950s, 271; and secrecy, 287–88, 294, 297, 312
nuclear monopoly, American: considerations in postwar, 157, 172–74; contemplated in World War II, 91, 118, 144–45, 147; challenges by gas centrifuge, 289, 294; de facto in Cold War, 287–89; end of, 212–12, 216, 229, 248 (*see also* "Joe-1")
Nuclear Non-Proliferation Treaty, 176, 392–93; and gas centrifuges, 299
nuclear power, 271, 276–79, 287–88, 294, 413; fear that secrecy would inhibit, 162; projected capabilities in the 1970s, 491n121
nuclear proliferation, 288, 320, 368, 398, 411; fears of, 281–84, 288, 292–93, 300, 319, 352, 355; and gas centrifuge, 292–95, 299–300, 332; and India, 324; and Iran, 393; and the Khan network, 332, 392; role of secrecy in, 401, 404–5, 408, 505n5; and the United Kingdom, 292
nuclear reactors: and declassification, *165*, 169, 187, 274–76, 278–81; fear that secrecy would inhibit, 162; and nuclear power, 271, 294, 371; patent applications for, 42; and Smyth Report, 128; in World War II, 24, 37, 54, 72–73, 83
Nuclear Regulatory Commission, 189, 339
nuclear reprocessing, 326
nuclear secrecy: conceptual overview, 5–7, 10, 399–400; and fear (*see* fears); lessons from history, 400–401, 405–7, 409–12, 414–15; and the national security state, 412–13; origins of, 15, 400; research and methodological issues, 8–11, 405–7
nuclear smuggling, fears of, 320–21
nuclear terrorism, 319–23, 326–27, 341, 401–2, 405; fears of, 321–27, 331, 333, 352, 379, 390–91, 399; Orlando hoax, 325–26
nuclear testing, 212, 241, 245–46, 306, 371, 386–89, 495n10; and India, 284; and laser fusion, 333; moratorium on, 306; and secrecy, 241–48, 473n31. *See also* nuclear fallout
nuclear waste, 189
nuclear weapon deployments, 372–73, 375
nuclear weapon design: and Cold War secrecy regime, 281–82, 297; computer codes, 390; core levitation, 223, 252, 469n143; and espionage, 223, 251–54, 408; gun-type design (*see* gun-type design); implosion design (*see* implosion design); and laser fusion, 306–7; public speculation about, 326–29, 331, 340–41, 343–45, 347–55, 362–65, 376–77; speculation in press, 128, 191–92, 248; Teller-Ulam design (*see* Teller-Ulam design)
Nuclear Weapons Databook (Cochran, Arkin, and Hoenig), 373–74
nuclear weapons designers, 7, 305–6, 318, 321–23, 331

Oak Ridge: compartmentalization at, 57; and declassification, 387; espionage at, 65; labor issues, 56–57, 210; land seizures, 68; and leaks, 71–72; during the Manhattan Project, 54–57, 81, 101; morale problems, 56–57; and "Publicity," 115, 122; rumors, 71–72, 79; secrecy practice, 54–57; and the Smyth Report, 127
Obama, Barack, 394
Office of Censorship, 67–70, 92, 130
Office of Scientific Research and Development: and atomic bomb development, 36–37, 41, 44, 47, 52, 66; and classification categories, 41, 429n82; Committee on Publications, 159–60; and Congress, 44; created, 36; and declassification, 159–60; Liaison office, 86; as model for Atomic Energy Commission, 141–42; secrecy practices, 41, 49, 62–63
Office of War Information, 64, 122; role in downgrading classification, 159
Office of War Mobilization, 78
Office of War Mobilization and Reconversion, 149
Official Secrets Act, 30
"Official Use Only." *See under* classification categories
Ohio State University, 40
O'Leary, Hazel, 386, 388–90
O'Neill, John, 38

openness: and anti-secrecy, 335–36, 371; criticisms of, 210; and cryptography, 408–9; as Enlightenment ideal, 3–4, 395, 398, 414; and international control, 137, 139; limits of, 190; and Niels Bohr, 137, 139; performative, 392; politics of, 389, 394; practices of, 131. *See also* Openness Initiative
Openness Initiative, 386, 388–90
open source intelligence, 371–75
Operation Candor, 271–73, 335
Operation Castle. *See* Castle "Bravo"
Operation Crossroads, 186–87, 245–46
Operation Dominic, 306
Operation Ivy. *See* Ivy "Mike"
Operation Redwing, 361
Operation Sandstone, 462n44
Oppenheimer, Frank, 258
Oppenheimer, J. Robert: and the Acheson-Lilienthal Report, 171–73; and anti-secrecy, 175–77; and Atomic Energy Commission secrecy reform, 199–200, 206; attempts at secrecy reform, 203, 206; becomes Scientific Director of Manhattan Project, 53; brought into fission work, 46; and the Chevalier affair, 262; clearance issues, 62; concerns about secrecy, 4, 60–61; and declassification, 160–63, 170–71, 175, 177, 403, 414; on domestic control, 148; as head of General Advisory Committee, 196; and House Un-American Activities Committee, 258; and the hydrogen bomb, 218, 241, 263, 347; and international control, 118, 136–37, 171–73, 175–77, 181, 333, 401, 410–11; and "Joe-1," 213; and leak about, 75; and David Lilienthal, 181, 196, 198; and Operation Candor, 271–72; postwar activities, 158; and postwar anti-secrecy efforts, 136, 199, 200; and "Publicity," 116–17, 123; on scientific secrecy, 148; secrecy practices, 58, 64; security clearance issues, 258–61; security hearing, 261–70, 282; and the Smyth Report, 100–101; on Soviet espionage, 224; students of, 63, 210, 256–57; and Jean Tatlock, 61; and the Tolman Committee, 161, 163, 170–71
Oppenheimer, Katherine, 258
Orlando, Florida, nuclear terrorism hoax, 325–26
OSRD. *See* Office of Scientific Research and Development
overclassification, 49, 207, 336, 404
oversight, of Manhattan Project, 44

Pacific Proving Grounds, 241–42, 245–46, 306
Page, Arthur W., 107, 109–11, 114, 119–20
Pakistan: fears of, 390; and gas centrifuges, 332, 392; nuclear program, 392; and John Aristotle Phillips, 328
Parpart, Uwe, 364
Parsons, William S., 188
patents, 18, 21, 32, 425n19; and the AEC, 281, 310; and the Atomic Energy Act of 1946, 278; and the Atomic Energy Act of 1954, 278–79; and censorship, 86; and declassification, 281; and gas centrifuges, 298–99; and laser fusion, 310–11, 315, 317; and nuclear power, 281; and postwar legislation, 140; and secrecy, 18, 21, 32, 42, 281, 310–11
PATRIOT Act, 394
Patterson, Robert P., 145, 150, 157, 167–68, 471n5; on Restricted Data, 156
Pearl Harbor attack, 32, 38, 67
Pegram, George, and self-censorship, 24
Pell, Wilbur, Jr., 362
Penney, William, 58, 292
Pentagon Papers, 335–39, 350–56, 368, 370
Percy, Charles, 362
Perón, Juan, 301
Personnel Security Board, 195, 198, 261–65, 268–69
Personnel Security Questionnaire, 193
Phillips, John A., 327–31, 493n139, 493n143
Physical Review (journal), and self-censorship, 23–28
Piel, Gerard, 237–38
Pike, Sumner T., 182, 207, 215, 229, 237
plutonium: and classification, *165*, *204*, 387; discovery of, 24, 28, 196; and espionage, 227; and fissile material safeguards, 277, 322–28; in human radiation experiments, 188, 387; and implosion, 254; and international control, 172; in Manhattan Project, 36–37, 42, 54, 72, 101, 445n42; and nuclear

proliferation, 293, 392; and the Smyth Report, 104, 127; and the Teller-Ulam design, 364, *365*; and the Tolman Report, *165*; and tritium, 218, 467n115
Poland, gas centrifuge research, 295
poliovirus, 407, 409
Pontecorvo, Bruno, 256, 477n79
Portugal, fears of, 299
Postol, Theodore, 348, 354, 357, 363, *367*
Potsdam conference, 119, 125, 139
practices of secrecy. *See* secrecy practices
press censorship. *See under* secrecy practices
press leaks. *See* leaks
Press-Connection (newspaper), 363
Price, Byron, 67, 130
Princeton University, 17, 40, 99–100, 259, 327–28, 330
Princeton University Press, 127–28
prior restraint, 349–50, 364. *See also* secrecy practices: press censorship
private industry, 310–11
"problem of secrecy," 2, 7, 82–89, 102, 176, 180, 233, 282, 399
Proceedings of the Royal Society of London (journal), 24
programmatic secrecy, 44, 200, 237, 471n12
Progressive (magazine), 347–58, 360, 363–69. *See also* US v. the *Progressive*, Inc.
Project Matterhorn, 259, 301–2
Project Plowshare, 304
Project Y. *See* Los Alamos National Laboratory
"Publicity": and the "best-kept secret of the war," 130; contradictory goals of, 131; contrasted with postwar declassification, 158–59; and Interim Committee, 138; news articles, 111–14, 121; newspaper articles, 106–7; policy conceived, 7, 97–98, 112; presidential press release about Hiroshima, 1–3, 105–6; press releases, 105–6, 108–11, 114–15, 121; "Publicity Day," 118; Public Relations Organization, *113*, 114–17, 130, 160; put into motion after Hiroshima bombing, 118–21; and radioactivity, 122–23, 128–30; Smyth Report (*see* Smyth Report)
Pulitzer Prize, 112, 114
pure-fusion weapons, 487n75; fears of, 310

Q clearance, 193, 255–56. *See also* security clearances
Quebec Agreement, 114, 120, 124, 168

Rabi, Isidor I., 265–67, 274
radiation implosion: and classification, 249; declassified, 368–69; invention, 240, 259–60; and Klaus Fuchs, 260, 478n92; and laser fusion, 304–7, 317; and the *Progressive* case, 347, 358, 364–65. *See also* Teller-Ulam design; laser fusion
Radiation Laboratory (Berkeley), 26, 36, 38, 40, 70–71
radioactive waste, 187
radioactivity, 123, 189; at Hiroshima and Nagasaki, 122–23, 129–30, 164–65, 448n80, 448n84; in the Marshall Islands, 247–48; and nuclear testing, 247–48
radioisotope production, 270
Ramey, James, 299
Ramsey, Norman, 58
RAND Corporation, 338
Rankin, John E., 157
Raper, John, 75
Rathjens, George W., 347–48
Ray, Dixy Lee, 502n5
Ray, Maud, 427n60. *See also* MAUD Committee: origin of name
Rea, Charles E., 129
Reagan, Ronald, 370–74, 378–81
reclassification, 64, 159, 325, 327, 391. *See also* declassification
redaction, 6, 9–10, 268, 402
Rehnquist, William H., 336
Reichstag fire, 17
Report of Committee on Declassification. *See* Tolman Report
reprocessing, nuclear. *See* nuclear reprocessing
Research Corporation, 21
Responsible Reviewers, 166–68, 186, 190, 198–99, 228, 256
"Restricted." *See under* classification categories
Restricted Data: under Atomic Energy Act of 1954, 276–79; and the Atomic Energy Commission, 187, 189, 210; in Hans Bethe's

Restricted Data (*continued*)
Scientific American article, 237–38; "born secret" interpretation, 154–55, 189, 296, 300, 362, 366, 388, 455n70, 483n28; compared with other classification categories, 194; constitutional challenge, 362; creation of legal concept in McMahon Act, 7, 154–58; and declassification, 176; and the First Amendment, 354, 362; and gas centrifuges, 296–97, 300; "inadvertently released," 391 (*see also* UCRL-4725); and laser fusion, 312, 314; and nuclear fusion, 301; and private industry, 296–97, 312; and the *Progressive* case, 348–56, 363, 366; vs. "Restricted," 463n51, 463n59; reviewed in post-Cold War, 385–86; and security clearances, 195; as understood by Brien McMahon, 155
Richard, George J., 81
Richter, Ronald, 301
Ridenour, Louis N., 236–37, 470n169
"RIPPLE" nuclear weapon design, 306–7
Roberts, Owen J., 195, 198
Rockefeller Foundation, 162
Rocky Flats, 388
Röntgen, Wilhelm, 16
Roosevelt, Eleanor, 72
Roosevelt, Franklin D.: and "absolute secrecy," 43–44, 53, 82, 90, 120, 150; approves creating Manhattan Project, 47, 52; and black budget, 37, 77; and Niels Bohr, 137–38; and Congress, 82, 92; creates the National Defense Research Committee, 34; creates the Office of Scientific Research and Development, 36; creates Uranium Committee, 27; death of, 89; Executive Orders of, 33–34; and fission research, 27, 37–38, 42–45, 47, 52; and international control, 91, 137–38; letter from Einstein and Szilard, 27; rumors about, 71–72; secrecy practices, 37, 43–44, 67, 77, 82, 92, 430n92; and Soviet Union, 65, 91; and Harry S. Truman, 89–90; and wartime censorship, 67
Rosenberg, Ethel, 250, 255
Rosenberg, Julius, 250–52, 255, 257
Rosenberg trial, 250–55, 269–70
Rosengren, Jack W., 353
Rotblat, Joseph, 248
Rotow, Dmitri A., 329–31; and the *Progressive* case, 360–61
Royall, Kenneth C., 141–42
Royall-Marbury bill, 141–46
Russian Federation, 400; fears of, 391, 399
Rutherford, Ernest, 16; and Leo Szilard, 18; and "moonshine," 17

safeguards. *See* fissile material safeguards
Saha, Meghnad, 74, 436n78
Salisbury, Morse, 186, 189, 198, 479n116
Sanders Associates, 307
Sandia National Laboratories, 314
Saturday Evening Post (newspaper), 38, 107, 320
Scheffel, Rudolf, 290
Schlesinger, James, 349, 353, 359, 367
Schneir, Miriam, 254
Schneir, Walter, 254
Schwartz, Alvin, 436n75
Schwartz, Rebecca Press, 101–2
Science (journal), 76
science and secrecy, 20, 395, 398, 405, 410, 425n17; J. D. Bernal's views on, 19–20; Robert K. Merton's views on, 20
Science Service (journal), 25
Science: The Endless Frontier (Bush), 141
Scientific American (journal), 236–39, 341
scientists, fears of, 229, 237–39, 256–57
Scientists' Movement, 145, 153, 176, 181–82, 320; and the McMahon Act, 156; mobilization against the May-Johnson bill, 148
Scottsboro boys, 193–94
Seaborg, Glenn, 196; as AEC chairman, 312
Seborer, Oscar, 433n40
secondary neutrons. *See under* nuclear fission
"Second Fuchs." *See* Hall, Theodore
secrecy: American context of, 3, 5, 414–15; and the American press, 335, 337; anthropology of, 464n79; attitudes of scientists towards, 15, 47–48; and balance, 210, 403; conceptual overview, 5–6, 406; and democracy, 3, 82, 403; discourses of, 398; and Enlightenment ideals, 3–4, 413–14; etymology of, 5; measuring harm of, 199, 506n18; morale problems from, 56–57; politics of, 381–83,

389, 395; practices (*see* secrecy practices); and privacy, 395; "problem of" (*see* "problem of secrecy"); as regulation, 382, 389, 404, 501n141, 503n20; and science (*see* science and secrecy); vs. security, 184, 186, 461n26; sociology of, 82, 464n79; and "stickiness," 402; value of, 407–8

secrecy practices: "active measures," 61–62; background checks, 40; centralization of work, 40; code names, 49; compartmentalization, 41–42, 49; contracts, 42, 58–59; counter-intelligence, 62; cover sheets, 93, *94*; document control, 49, 93–94; document tracking, 93–94; homicide allegations, 61–62; misinformation, 39, 49, 73, 107–9; oaths and pledges, 39–40, 49, 58, *59*, 63; physical security, 49, 57; polygraph use, 391; press censorship, 67–70, 72–74, 76, 106, 349, 367; prior restraint (*see* prior restraint); publication censorship, 49 (*see also* prior restraint); secrecy culture, 41, 62; security clearances (*see* security clearance); self-censorship (*see* self-censorship); site-isolation, 49, 54; stamps, 9, 15, 40, 49–50, 279, 405

secrecy reform: difficulties in achieving, 399–404, 406–7; post-Cold War, 386–89, 391; post-Nixonian, 336; in Soviet Union, 383. *See also* Atomic Energy Commission: attempts at secrecy reform

secrecy regimes, 406, 410; "absolute secrecy" (*see* "absolute secrecy"); and anti-secrecy, 338; challenges to Cold War, 287–88, 293, 300, 318–19, 331, 333, 338, 367, 370, 383; Cold War, 7, 176, 231, 233–35, 282–83, 287, 300, 318–19, 324, 331, 333, 338, 370, 383; definition of, 6, 423n6; early self-censorship, 23; eliminating, 400; post-Cold War, 385; "Publicity" (*see* "Publicity")

secrecy stamps. *See* secrecy practices: document control; secrecy practices: stamps

"Secret." *See under* classification categories

"secret, the": totemic approach to nuclear secrecy, 147, 157, 228, 251, 253–54, 273, 342, 345, 395, 412–13

secret evidence, 251–52, 254, 357. *See also* court system, and secrecy

"Secret-Limited." *See under* classification categories

secret seekers, 337, 359–60, 362, 364, 371–75, 376–78

Section S-1, 39–41, 45, 66, 69

security, vs. secrecy, 184, 186

security clearances, 195, 210, 378, 403, 413; under the Atomic Energy Act of 1946, 192; under Atomic Energy Commission, 193; and homosexuality, 255–56; under Manhattan Project, 193; and the Office of Scientific Research and Development, 429n82; P clearance, 193; Q clearance, 193, 255–56; S clearance, 193; and sexuality, 255–56, 263, 403

"security theater," 405

Sedition Act of 1918, 32

self-censorship, 6, 15, 19, 26–27, 48; and biology, 407; and the hydrogen bomb, 218

Selective Service, 71

September 11 attacks, 391

Senate Committee to Investigate the National Defense Program. *See* Truman Committee

Serber, Charlotte, 257, 477n82

Serber, Robert, 257

sexuality, and security clearances, 255, 263, 403

Shurcliff, William A., 86–89, 104

Siebert, A. Bryan, 389

Siegel, Keeve M., 309–13, 316–17, 489n108

"silly secrecy," 370, 402

Site W. *See* Hanford

Site X. *See* Oak Ridge

Site Y. *See* Los Alamos National Laboratory

Smyth, Henry D.: as AEC commissioner, 214, 223–24, 237–39, 244, 268–69, 273, 303; analysis of secrecy, 84–86, 89, 238–39; on compartmentalization, 60; and hydrogen bomb, 244; and nuclear fusion, 303; and Operation Candor, 273; and J. Robert Oppenheimer, 265; and postwar planning, 83, 442n17; and the Smyth Report, 99–105, 124, 131; and Soviet espionage, 223–24; testimony in postwar, 190

Smyth Report: circulation, 127; comparison to later approaches to declassification, 158, 177; debates about releasing, 123–26; de-

Smyth Report (*continued*)
cision to release, 125–26; goals of, 98–99, 101, 116, 131; as illustrative of "Publicity" campaign, 98, 105, 131; origins of, 98–102; postwar criticism, 103–4, 148, 167, 182; and postwar researchers, use by, 189–90, 320; as programmatic guide, 131; role in "Publicity" campaign, 116, 119; and radioactivity, 129; rules for release, *94*, 102–4, *103*, 124, 147, 161, 443n21; and Senate Special Committee on Atomic Energy, use by, 150; in Soviet Union, 127–28; on title, 126, *127*
Snowden, Edward, 391, 394
Sobell, Morton, 254
sociology of science, 19–20, 401, 425n17
Solzhenitsyn, Aleksandr, 128
S-1 Committee. *See* Section S-1
Souers, Sidney, 235
sources and methods, 212, 251
Soviet Union: deduction of Manhattan Project, 66–67; detonation of first atomic bomb (*see* "Joe-1"); dissolution of, 381, 383; espionage, 7, 63–66, 90, 92, 139, 152, 221–28, 250–51, 261–62, 408, 413; fears of, 63, 90–91, 110, 125, 136, 208, 214, 220, 241, 243, 319, 390, 411–12; gas centrifuge research, 289–91, 300; interest in fission, 25; "Joe-1" (*see* "Joe-1"); and international control, 175; and laser fusion, 308, 313; nuclear assistance to the People's Republic of China, 481n2; nuclear capabilities, 283, 373–74; nuclear program, 67, 227, 283, 288, 404, 464n69; nuclear testing, 245, 306; recruitment of German and Austrian scientists, 289, 300; role in AEC discussions about secrecy policy, 197, 199–201, *201*, *202*, 216; and the Smyth Report, 127–28; test of hydrogen bomb, 272; use of espionage data, 227–28
Spain: and laser fusion, 308; fears of, 299
Special Engineer Detachment, 250, 433n41
"special" nature: of atomic bomb, 39, 43, 46, 77, 117–18, 121, 131, 339, 428n72; of nuclear secrecy, 35, 366, 388
Special Senate Committee on Atomic Energy, 149–55, 167
Spedding, Frank, and the Tolman Committee, 161, 168–69
Spitzer, Lyman, Jr., 301–2
Sproul, Robert, 69–70
Sputnik, 303
Stalin, Joseph, 67, 139; is told about bomb from Truman, 139
stamps. *See under* secrecy practices
State Secrets Privilege, 426n43
Steenbeck, Max, 289–90
Stellerator, 302
Stern, Der (magazine), 372
Stewart, Irvin, 140–41
Stimson, Henry L.: attitude towards atomic bomb, 117; and Congress, 78–82; and domestic control, 140–41; early involvement with Manhattan Project, 37–38; and international control, 138–40; press releases after Hiroshima, 105, 109, 114, 120–21; and "Publicity," 106, 111; and the Smyth Report, 102, 125–26; tells Truman about Manhattan Project, 90
stockpile secrecy, 150, 161, 191–92, 200, 272–73, 394, 504n36
Stone, Jeremy, 360
Stone, Robert S., 123, 188–89
Strassmann, Fritz, 16
Strategic Defense Initiative, 370
Strauss, Lewis, L.: as AEC chairman, 231, 240, 247, 258–59, 270–71, 282, 301–2; as AEC commissioner, 188, 198, 207, 209–10, 221, 223, 402; and "Atoms for Peace," 273–74; and Castle "Bravo," 247–48; attitudes towards secrecy, 233, 275; and declassification, 271, 282; and the hydrogen bomb, 217–18, 236, 248; and nuclear fusion, 302–3; and Operation Candor, 272; and J. Robert Oppenheimer, 258–59, 261, 265, 268–69; and security clearances, 194; and Soviet espionage, 226
Strong, George V., 73
Stuxnet, 408
submarines, 30, 32
super bomb. *See* hydrogen bomb
Superman, 74
Supreme Court of the United States, 338–39, 351

Svenska Dagbladet (newspaper), 72
Sweden, 375
Swedish Ministry of Defense, 327
synthetic biology, 399, 407, 409
Szilard, Leo: criticism of the Smyth Report, 103–4, 148, 182; discovery of chain reaction, 17–19; and Albert Einstein's letter to Franklin Roosevelt, 27; failure of self-censorship, 25–26; and Enrico Fermi, 21–22, 281, 400; and Leslie Groves, 61, 83, 149–50; influence of H. G. Wells on, 18; and Frédéric Joliot-Curie, 22–23, 25; learns of nuclear fission, 17; and nuclear fission, 19; and patenting, 18, 21; and publication censorship, 28–29; and Ernest Rutherford, 18; and secrecy, 15, 19, 50; and self-censorship, 15, 19, 21–24, 26–28, 48, 407; testimony on domestic control, 148; and the Uranium Committee, 28

tacit knowledge, 88, 102, 147, 211, 409, 494n154
Taft, Robert, 79
Target Committee, 110
Tatlock, Jean, 61–62, 258, 263
Taylor, Theodore B., 321–31, 340, 361, 374
Teller, Edward, 399, 413; and the hydrogen bomb, 217–18, 240, 242, 258; and Lawrence Livermore National Laboratory, 304; and Howard Morland, 343–45, 352, 355, 357–58; and J. Robert Oppenheimer, 259, 263–64; protégés, 304, 314; and secrecy, 23, 89, 335, 382–83, 388, 413; and self-censorship, 23; and the Strategic Defense Initiative, 370; and Leo Szilard, 23
Teller-Ulam design, 259–60; declassified, 368–69; drawings of, *346*, *365*, *369*; and espionage, 390, 475n57; and laser fusion, 304–7, *306*, 310–11, 315, 318; and Howard Morland, 345–46; origin of, 240; and the *Progressive* case, 352–58, 360–61; tested (*see* Ivy "Mike"); as "ultimate secret" 249, 281, 333, 340, 342, 351, 369, 413
Tennessee Valley Authority, 171, 180
terrorism, 324, 391; nuclear (*see* nuclear terrorism)
thermonuclear weapons. *See* hydrogen bomb
thorium, control of, 144
Time (magazine), 66, 128, 192, 242, 340–41, 462n44
Tolman, Richard C., 34, 87–89, 104, 128; and the Smyth Report, 102–3, 124; and the Tolman Committee, 161–62, 167, 179
Tolman Committee, 164–71, 199, 206, 234; and the Acheson-Lilienthal Report, 174; formation of, 161–62; guidelines adopted by Atomic Energy Commission, 179; and international control, 162; named Committee on Declassification, 162
Tolman Report, 163–67, *164*, *165*, *166*, 199; implementation of, 168; philosophy of, 169–70
Tonight Show, 374
Top Policy Group, 37, 47, 90
"Top Secret." *See under* classification categories
trade secrecy, 316, 318
transparency, 335–36
Trenton Times (newspaper), 328
Trinity test, 116, 118, 245; false press releases, 107–9, 444n34; newspaper stories about, 121, 128; radioactivity, 123
tritium, 218, 235, 304, 307, 467n115, 471n12
Truman, Harry S.: announcement of the bombing of Hiroshima, 1–3, 97, 109–11, 114, 119–21; becomes president, 89; and declassification, 91, 159, 168; and domestic control, 135, 150, 153, 157; and the hydrogen bomb, 209, 218–22, 234–36, 243–45, 468n135; and international control, 139–40, 146; and "Joe-1," 212–14; and leaks, 219–20; learns about Manhattan Project, 90; and David Lilienthal, 181; Loyalty-Security program, 192, 255; and the McMahon Act, 157; and Franklin Roosevelt, 90; as Senator, 78–79, 82; and the Smyth Report, 125–26; tells Joseph Stalin about the atomic bomb, 139; and use of atomic bomb, 115
Truman Committee, 78–79
Trump, Donald, 504n36
Tuerkheimer, Frank, 359
Turner, Louis, 28
Tydings, Millard, 223

UCRL-4725, 360–61
Ulam, Stanislaw, 240, 345, 352, 355
"ultimate secret." *See under* Teller-Ulam design
UN Atomic Energy Commission. *See under* United Nations
Unclassified Controlled Nuclear Information, 379
Underhill, Robert, 434n57
Union Carbide Nuclear Company, 292
Union of Concerned Scientists, 342
United Kingdom, 51, 373; and the Atomic Energy Act of 1954, 278; collaboration with United States, 61, 90, 110, 114, 209, 214–15, 297–99; communication on early fission work, 39; concerns about Smyth Report, 124–26; and declassification, 168, 199, 216; develops own atomic bomb, 288, 292; and gas centrifuges, 292, 297–99, 332; Leslie Groves' fears of, 91, 93; interest in fission, 25; and international control, 146; MAUD Committee (*see* MAUD Committee); MAUD Report, 37; and nuclear fusion, 303; and nuclear proliferation, 292; press censorship, 72; press release after Hiroshima, 119–20; and "Publicity," 114; and Restricted Data, 298–99; and Soviet espionage, 222–26, 250; and Leo Szilard's patent secrecy, 18
United Nations: Atomic Energy Commission of, 175; and "Atoms for Peace," 273; and international control, 171–75
United States: history of secrecy in, 29–30, 32; relationship between science, technology, and governance in, 33–34
United States Navy, 41
United States v. Reynolds, 426n43
University of California, Berkeley, 53, 65, 69–71, 363, 434n57; Soviet espionage attempts at, 63, 91
University of California, San Diego, 308
University of Chicago, 36–38, 60–61, 74, 180; Metallurgical Laboratory, 40, 45–46, 83
University of Michigan, 307, 309
University of Pennsylvania, 189–91
University of Rochester, 311
University of Virginia, 289, 291
University of Wisconsin, 28
uranium, control of, 144–45, 172, 174–75, 209
Uranium Committee, 27–29, 34–37, 48, 73
uranium enrichment, 281, 297; and declassification, *165*, 281; gas centrifuge (*see* gas centrifuge); gaseous diffusion (*see* gaseous diffusion); and international control, 172; Manhattan Project work, 32, 36–38, 44, 52, 54; safeguards, 299
Urenco, 299, 332
Urey, Harold C., 38, 73; and Tolman Committee, 161–62
US Air Force, 258, 264, 308
US Army Corps of Engineers, 6, 42–43, 49, 51–52
US Atomic Energy Commission. *See* Atomic Energy Commission
US Civil War; secrecy during, 30
US Department of Defense. *See* Department of Defense
US Department of Energy. *See* Department of Energy
US Department of Justice. *See* Department of Justice
US Department of State. *See* Department of State
US Department of War. *See* Department of War
US Energy Research and Development Administration, 319, 339
US Office of Censorship, 67–70, 92, 130
US Office of War Information, 64, 122; role in downgrading classification, 159
US Selective Service, 71
US-UK-Canada Declassification Conferences, 199, 216
US v. the *Progressive*, Inc., 352–69, 375, 383, 390

Vanunu, Mordechai, 504n34
Venona project, 222, 250–51
Vietnam War, 335, 338, 341–42, 352

Wagnon, Mike, 376–77, *377*
Walker, J. Samuel, 189, 326
Wallace, Henry, 37, 90
Wall Street Journal, 321

War Department. *See* Department of War
Warren, Robert W., 351, 355
Washington Post (newspaper), 219, 347, 356; Hiroshima coverage, 121
Watergate scandal, 335–36, 339, 368, 370
Waymack, William W., 182, 184, 190, 214, 229–30, 470n169
Weaver, Warren, 162
Weber, Max, 501n141, 503n20
Weil, George T., 275
Weinberger, Casper, 373
Weisskopf, Victor, 24–25
Wells, H. G., 18
West Germany, 372; and gas centrifuges, 290–95, 297–300, 332; and laser fusion, 308, 358; fears of, 292
Wheeler, John A.: and hydrogen bomb, 259–60, 270, 302; and Niels Bohr's theory of fission, 27
WikiLeaks, 394
Willrich, Mason, 322
Wilson, Carroll, 183, 188–89, 223
Wilson, Woodrow, 31
Winterberg, Friedwardt, 364, 376, 499n100
Wood, Lowell, 314–16
Woodward, Bob, 370
World Trade Center, 323
World War I, 34, 46; secrecy and science in, 20; secrecy during, 30–31
World War II; funding of science and technology during, 34; origins of, 25; secrecy during, 32. *See also* Manhattan Project
"worst-kept secret," 393

York, Herbert, 342, 382; and the Smyth Report, 127
yttrium, 280
Yugoslavia, 212

Zippe, Gernot, 289–91, 294, 300
Zippe centrifuge, 290–91, 294
Zuckert, Eugene, 268